Outstanding Contributions to Logic

Volume 3

For further volumes:
http://www.springer.com/series/10033

Sven Ove Hansson
Editor

David Makinson on Classical Methods for Non-Classical Problems

Springer

Editor
Sven Ove Hansson
Division of Philosophy
Royal Institute of Technology (KTH)
Stockholm
Sweden

ISSN 2211-2758 ISSN 2211-2766 (electronic)
ISBN 978-94-007-7758-3 ISBN 978-94-007-7759-0 (eBook)
DOI 10.1007/978-94-007-7759-0
Springer Dordrecht Heidelberg New York London

Library of Congress Control Number: 2013956333

Printed on acid-free paper

Springer is part of Springer Science+Business Media (www.springer.com)

Preface

I had my first contact with David Makinson in 1987 when I was a Ph.D. student. Peter Gärdenfors advised me to send my manuscript on belief change to David, since he was such a good commentator. So I did. Three weeks later when returning home from a short vacation I found three handwritten letters from David, containing detailed comments on different parts of the manuscript. Those comments helped me tremendously in improving what became my first publication in that area (appearing in *Theoria* 1989). Like so many others, I have continued over the years to benefit from David's unusual competence as a logician. I have also been much influenced by his continued efforts to make full use of the resources of classical logic. There has been much discussion about the "unreasonable effectiveness of mathematics" in the empirical sciences. Perhaps, we should also start discussing the unreasonable effectiveness of classical logic in the formal sciences. Several remarkable examples of that effectiveness can be found in David's oeuvre.

So it should be no surprise that I was delighted when (before becoming the series editor) I was offered the task of editing this volume. Editing the book has continued to be a pleasure, not least due to the enthusiasm and cooperative spirit of all the authors. All contributions have been thoroughly peer-reviewed (including David's, at his own insistence). Reviewing is proverbially a thankless task, but on behalf of the authors and myself I would like to thank the reviewers who have done so much for the quality of this book: Alex Bochman, Guido Boella, Richard Booth, Luc Bovens, Richard Bradley, John Cantwell, Samir Chopra, Igor Douven, Didier Dubois, J. Michael Dunn, Eduardo Fermé, André Fuhrmann, Lou Goble, Guido Governatori, Davide Grossi, Lloyd Humberstone, Hannes Leitgeb, Joao Marcos, Tommie Meyer, Jeff Paris, Ramon Pino Perez, Tor Sandqvist, Marek Sergot, John Slaney, Kees van Deemter, Frederik van der Putte, Paul Weirich, Emil Weydert, and Gregory Wheeler.

Very special thanks go to David Makinson for being just as helpful to the authors and editor of this volume as he was to me 26 years ago.

Stockholm, May 24, 2013 — Sven Ove Hansson

Contents

Part III Uncertain Reasoning

Part IV Normative Systems

Part V Classical Resources

Part VI Responses

Contributors

Dov Gabbay is an Augustus de Morgan (emeritus) professor of logic at the King's College London, an Invited Professor at the University of Luxembourg, and a professor at Bar-Illan University in Isreal. He is editor-in-chief of several international journals, and several Handbooks, including the *Handbook of Philosophical Logic*. His research interests include: logic and computation; proof theory and goal-directed theorem proving; nonclassical logics and nonmonotonic reasoning; labeled deductive systems; fibring logics; and reactive systems. His most recent books include, both co-authored with Karl Schlechta: *Conditionals and Modularity in General Logics* (Springer 2011); *Logical Tools for Handling Change in Agent-based Systems* (Springer 2009).

Peter Gärdenfors is a professor of Cognitive Science at Lund University. His previous research focussed on decision theory, belief revision, and nonmonotonic reasoning. His main current research interests are concept formation, cognitive semantics, and the evolution of thinking. His main books are *Knowledge in Flux: Modeling the Dynamics of Epistemic States*, (MIT Press 1988), *Conceptual Spaces* (MIT Press 2000), *How Homo Became Sapiens: On the Evolution of Thinking* (Oxford University Press 2003) and the forthcoming *Geometry of Meaning: Semantics Based on Conceptual Spaces* (MIT Press 2013).

Jörg Hansen studied philosophy and law at the universities of Bielefeld, Hamburg and Leipzig, Germany. He obtained his Ph.D. from the University of Leipzig, Institute of Philosophy, with a thesis on *Imperatives and Deontic Logic*. He works as a lawyer in Eisenach, Germany, and in 2005 was appointed the director of the Bach House, Eisenach, the museum at Johann Sebastian Bach's birthplace. His logical research focusses on deontic logic.

Sven Ove Hansson is a professor in philosophy and head of the Division of Philosophy, Royal Institute of Technology, Stockholm. He is editor-in-chief of *Theoria* and of the book series *Outstanding Contributions to Logic*. He is also member of the editorial boards of several journals, including the *Journal of Philosophical Logic*, *Studia Logica*, and *Synthese*. His logical research has its focus on belief revision, preference logic, and deontic logic. His other philosophical research includes contributions to decision theory, the philosophy of risk and moral, and political philosophy. His books include *A Textbook of Belief*

Dynamics. Theory Change and Database Updating (Kluwer 1999) and *The Structures of Values and Norms* (CUP 2001).

James Hawthorne is a philosophy professor at the University of Oklahoma. His research primarily involves the development and explication of the probabilistic logics, especially Bayesian confirmation theory, logics of belief and comparative confidence, and logics of defeasible support.

John Horty received his BA in Classics and Philosophy from Oberlin College and his Ph.D. in Philosophy from the University of Pittsburgh. He is currently a Professor in the Philosophy Department and the Institute for Advanced Computer Studies at the University of Maryland, as well as an Affiliate Professor in the Computer Science Department. His primary interests are in philosophical logic, artificial intelligence, and cognitive science. He has secondary interests in the philosophy of language, practical reasoning, ethics, and the philosophy of law. Horty is the author of three books as well as papers on a variety of topics in logic, philosophy, and computer science. His work has been supported by fellowships from the National Endowment for Humanities and grants from the National Science Foundation. He has held visiting fellowships at the Netherlands Institute for Advanced Studies, and at the Center for Advanced Studies in Behavioral Sciences at Stanford University.

Hykel Hosni did his Ph.D. at the Uncertain Reasoning Group of the School of Mathematics, University of Manchester. He then moved to the Scuola Normale Superiore in Pisa, where he has been Research and Teaching Fellow since 2008. During 2013–2015, he will be Marie Curie Fellow at the CPNSS of the London School of Economics. His project "Rethinking Uncertainty" develops the choice-problem approach to uncertain reasoning fleshed out in his contribution to this volume. He translated for Springer Bruno de Finetti's Philosophical Lectures on Probability in 2008. Since 2011, Hykel has been secretary of the Italian Association for Logic and Applications.

David Makinson is Guest Professor in the Department of Philosophy, Logic and Scientific Method of the London School of Economics (LSE). Before that, in reverse chronological order, he worked at King's College (London), UNESCO (Paris), and the American University of Beirut, following education in Oxford, UK after Sydney, Australia. He is the author of *Sets Logic and Maths for Computing* (second edition 2012), *Bridges from Classical to Nonmonotonic Logic* (2005), *Topics in Modern Logic* (1973), as well as many papers in professional journals. His research has covered many areas in logic, including the logic of uncertain inference, belief change, and normative systems. The present volume contains a more extended professional biography.

Edwin Mares is a professor of philosophy and a founder and member of the Centre for Logic, Language, and Computation at Victoria University of Wellington. He is the author of *Relevant Logic: A Philosophical Interpretation* (CUP 2004), *Realism and Anti-Realism* (with Stuart Brock) (Acumen 2007),

A Priori (Acumen 2010), and more than fifty articles. Mares works mostly on nonclassical logics and their philosophy, concentrating largely on relevant logics.

Xavier Parent is a research assistant in computer science at the University of Luxembourg. His research has its focus on deontic and nonmonotonic logics. He is a co-editor of the *Handbook of Deontic Logic and Normative Systems* (volume 1 forthcoming in 2013; volume 2 forthcoming in 2014/2015). He obtained his Ph.D. degree in philosophy from the University of Aix-Marseille I, France. His doctoral dissertation, *Nonmonotonic Logics and Modes of Argumentation*, was conferred the 2002 best thesis award in the Humanities and Social Sciences. In 2011, he also received the best paper award at the DEON conference for his paper 'Moral Particularism and Deontic Logic'.

Rohit Parikh is Distinguished Professor in Computer Science, Mathematics and Philosophy at the CUNY graduate Center and Brooklyn College. Parikh was born in India but all his degrees are from Harvard where he studied logic with W. V. Quine, Burton Dreben, and Hartley Rogers (who was at MIT). There was also much early influence from Kreisel. Early areas of research include formal languages (with Chomsky), recursion theory, proof theory, and nonstandard analysis. More recent areas include logic of programs, logic of games, epistemic logic, game theory, and social software (a term which Parikh has coined although the field itself is ancient). He has been editor of the *Journal of Philosophical Logic* and the *International Journal for the foundations of Theoretical Computer Science*.

Pavlos Peppas is an associate professor at the Department of Business Administration, University of Patras. He has been a program committee member in a number of conferences including the *International Conference on Principles of Knowledge Representation and Reasoning (KR),* the *International Joint Conference on Artificial Intelligence (IJCAI)*, and the *European Conference on Artificial Intelligence (ECAI)*, and has served as guest editor for *Studia Logica* and the *Journal of Logic and Computation*. His research interests lie primarily in the areas of *Belief Revision*, *Reasoning about Action*, *Nonmonotonic Reasoning*, and *Modal Logic*. He has published several research articles in conference proceedings (*IJCAI, ECAI, KR*, etc.), and scientific journals such as *Artificial Intelligence Journal*, *Journal of Logic and Computation*, and *Journal of Philosophical Logic*.

Hans Rott is a professor of theoretical philosophy at the Department of Philosophy of Regensburg University. As a logician, he is best known for his studies in belief revision and for his use of choice mechanisms operating on beliefs as a means to unify theoretical and practical reasoning. Much of this work is reported in his book *Change, Choice and Inference: A Study of Belief Revision and Nonmonotonic Reasoning* (Oxford University Press 2001). His other topics of research include analyticity, the relationship between knowledge and belief, the application of philosophical meaning theory in literature studies, and intercultural understanding.

Karl Schlechta is a professor of computer science at Aix-Marseille University in France, and member of the Laboratoire d'Informatique Fondamentale de Marseille. He works on nonmonotonic logics, theory revision, and related subjects. His main interest is in the semantical side of these logics, in particular in preferential structures, and accompanying representation theorems. His books include *Coherent systems* (Elsevier 2004), *Logical tools for handling change in agent-based systems* (Springer 2009), and *Conditionals and modularity in general logics* (Springer 2011), the latter two co-authored with Dov Gabbay.

Wolfgang Spohn studied philosophy, logic and philosophy of science, and mathematics at the University of Munich. Since 1996, after professorships in Regensburg and Bielefeld, he holds a chair of philosophy and philosophy of science at the University of Konstanz. His areas of specializations are formal epistemology, philosophy of science (induction, causation, explanation), philosophy of mind and language, philosophical logic, ontology and metaphysics, philosophy of mathematics, and the theory of practical rationality (decision and game theory), on which he has written many papers. His book *The Laws of Belief. Ranking Theory and Its Philosophical Applications* (OUP 2012) has been distinguished with the Lakatos Award. He has been editor-in-chief of *Erkenntnis* for 13 years and has been serving in many Editorial Boards and philosophical societies. Since 2002, he is a fellow of the *Leopoldina, German National Academy of Sciences.*

Audun Stolpe is a scientist at the Norwegian Defence Research Establishment and holds a Ph.D. in philosophy from the University of Bergen Norway. His research is interdisciplinary and contributes to philosophical logic as well as computer science. His philosophical research focuses mainly on the theory of input/output logic and its relationship to belief revision and propositional relevance, but includes also contributions to the theory of quantified modal logic. His research in computer science focuses on applications of knowledge representation and classical AI in computer networks, particularly on the problem of giving a logical characterisation of query federation on the web.

Leendert van der Torre is a professor in computer science at the University of Luxembourg. He is a co-editor of the *Handbook of Deontic Logic and Normative Systems* (volume 1 forthcoming in 2013; volume 2 forthcoming in 2014/2015), and a Co-editor of the "deontic logic" corner of the *Journal of Logic and Computation.* His research interests include deontic logic, logic in security, compliance, agreement technologies, and most recently cognitive robotics. He has co-authored over 150 papers indexed by DBLP, and received best paper awards at PRIMA07 and KES-AMSTA07.

Part I
Introductory

Preview

Sven Ove Hansson

Abstract This is a brief summary of the chapters in a multi-author book devoted to David Makinson's contributions to logic. The major themes are belief revision, uncertain reasoning, normative systems, and the resources of classical logic.

Keywords David Makinson · Belief change · Uncertain reasoning · Probability · Deontic logic · Normative systems · Input/output logic · Introduction rules · Elimination rules · Relevance logic

Classical logic, suitably elaborated and applied, can be used to investigate many of the topics that non-classical logics have been introduced to deal with. That is a unifying idea in many of David Makinson's contributions to logic, and it is also the underlying theme of this book. In "David Makinson and the Extension of Classical Logic" *Peter Gärdenfors* and *Sven Ove Hansson* relate it to more general developments in logic. The chapter focuses on the "inferential-preferential method", i.e., the combined use of classical logic and mechanisms of preference and choice. This combination has turned out to be surprisingly powerful and versatile in a wide range of applications. The chapter gives examples from Makinson's work in non-monotonic and defeasible reasoning and belief revision. Other chapters provide additional examples.

"A Tale of Five Cities" is a short autobiography by *David Makinson* in which we learn how he discovered modern logic, follow his trails across continents, and are told about many of his intellectual encounters and inspirations. The chapter also contains an unusually explicit statement of his views on the (limited but important) role of logic in philosophy.

After this follow four sections, each devoted to a particular area of logic: belief change, uncertain reasoning, normative systems, and the resources of classical logic.

S. O. Hansson (✉)
Royal Institute of Technology (KTH) Div. Philosophy, Stockholm, Sweden
e-mail: sven-ove.hansson@abe.kth.se

S. O. Hansson (ed.), *David Makinson on Classical Methods for Non-Classical Problems*, Outstanding Contributions to Logic 3, DOI: 10.1007/978-94-007-7759-0_1,

1 Belief Change

In the logic of belief change, David Makinson is best known as the M of AGM, i.e., one of the three authors of the 1985 paper that set what is still the gold standard in belief revision theory, the theory to which all other theories in that area are compared. (The A is Carlos Alchourrón and the G is Peter Gärdenfors.) In "Safe Contraction Revisited" *Hans Rott* and *Sven Ove Hansson* take a close look at a significantly different approach to belief contraction, namely safe contraction which was introduced in a paper by Alchourrón and Makinson, also in 1985. Both approaches are examples of the inferential-preferential method. In AGM contraction a selection mechanism is applied to find the most choice-worthy among the maximal subsets of the original set of beliefs that do not contain the sentence to be removed. In safe contraction, a choice mechanism is instead used to find the most discardable elements of the minimal subsets of the original set of beliefs that imply the sentence to be removed. The chapter provides some new results on safe contraction and its (close but intricate) connections with the AGM model. In conclusion a list of unsolved problems is offered, and it is suggested that safe contraction deserves much more attention than it has received among students of belief change.

The AGM model provides a mechanism for revising the original set of beliefs, but it does not say much on how the revised set should in its turn be changed in response to new information. If we begin with a belief set K and receive an input sentence p, then the AGM model tells us how to determine the outcome $K * p$ of revising K in order to make room for p. However, it does not tell us how to co-ordinate that first revision with subsequent revisions of the output $K * p$ in order to accommodate a further input q. One of the most discussed issues in belief revision is how to account for such, so-called iterated revision. In "A Panorama of Iterated Revision" *Pavlos Peppas* provides a systematic overview and analysis of the major proposals for this in the literature. He shows in particular how all these proposals operate by enriching either the representation of the original state of belief so that it consists of more than just a set of beliefs, or the representation of the input so that in addition to a sentence it contains information indicating the degree of firmness with which the new belief should be held. Both these types of enrichment come in several variants. Peppas concludes that there is still controversy on the relative merits and demerits of these proposals, and that there are, as he says, "no signs of convergence towards a 'standard model' (like AGM is for one-step revision)".

One of the belief revision models discussed by Peppas is more thoroughly introduced in "AGM, Ranking Theory, and the ManyWays to Cope with Examples" by *Wolfgang Spohn*, namely ranking theory, in which the strength of beliefs is represented by cardinal numbers assigned to the propositions representing beliefs. These numbers have much the same role in ranking theory as the preference relations and choice functions used in AGM theory. Spohn outlines the basic principles of ranking theory and shows how the standard AGM operations are related to ranking-theoretical conditionalization. After that he turns to an issue that is central in both approaches, namely whether the usual operations of revision (incorporation of a new belief) and contraction (removal of a belief) have desirable properties. This has mostly been discussed with reference to the standard postulates of AGM revision and contraction. Spohn examines what are arguably the five most contested of these postulates. The resources of ranking theory turn out to be quite useful in elucidating them and analyzing the counterexamples that have been proposed. The chapter defends the standard postulates, but also argues that there is need for closer analysis of the relationship between the pragmatics and semantics of utterances of the types referred to in belief revision.

In AGM theory, there is only one inconsistent belief set, namely the whole language. Once an agent has accepted two mutually inconsistent sentences, for instance both p and $\neg p$, she is committed to believing any sentence q. In "Liars, Lotteries, and Prefaces: Two Paraconsistent Accounts of Belief Change" *Edwin Mares* introduces a paraconsistent approach to belief change that is based on relevance logic. In this approach, an agent has both a belief set consisting of the sentences she believes in and a reject set consisting of the sentences she rejects. Reject sets are duals of theories, which means that they are closed under disjunction and that if a reject set contains a certain sentence then it also contains all sentences that entail it. Both the belief set and the reject set can be subjected to the standard belief revision operations (i.e., contraction, revision, and expansion by a sentence). In this general framework Mares develops two systems that he calls the strong and the weak systems of paraconsistent belief change. In the strong system, an agent who accepts an inconsistent set of beliefs is committed to believing all conjunctions of elements of that set. In the weak system the agent can accept inconsistent beliefs without committing herself to believe in their conjunction. The weak system provides means to resolve (or at least evade) both Kyburg's lottery paradox and Makinson's preface paradox, both of which assume that the set of beliefs is closed under conjunction.

In "Epistemic Reasoning in Life and Literature" *Rohit Parikh* takes a quite different approach to the discussion on the realism of models in formal epistemology. He invites us to have a close look at examples from literature. Using a series taken from as diverse sources as Shakespeare, Sir Arthur Conan Doyle, and George Bernard Shaw, he points out features of actual epistemic reasoning that playwrights and novelists have sometimes treated more deftly than logicians. In conclusion he proposes some areas of improvement for formal epistemology in order to "catch up with these brilliant literary minds". We need better representations of reasoning that is based on non-sentential inputs such as perceptual impressions. We also need formal accounts of how one can infer someone's beliefs from how they act, and more generally of the role of default assumptions in reasoning about other people's beliefs.

2 Uncertain Reasoning

As Makinson tells us in his "Tale of five cities", after publication of the AGM paper in 1985 he turned to the logic of qualitative uncertain inference, aka non-monotonic logic. This is another important application area for the inferential-preferential method, and his contributions have laid bare the surprisingly strong connections between belief revision and non-monotonic logic. In later years he has turned to the connections between belief revision and quantitative uncertain inference, in particular conditional probabilities—thus once again taking up a line of research that combines classical logic with formal tools from outside of logic.

This book contains three chapters that develop themes from Makinson's work on uncertain reasoning. In "New Horn Rules for Probabilistic Consequence: Is O+ Enough?" *James Hawthorne* returns to a topic that he and Makinson investigated in a 2007 paper, namely probabilistic consequence relations. By this is meant a consequence relation $\mid\sim$ such that it holds for all propositions a and x that $a \mid\sim x$ if and only if $p(x \mid a) > t$, for some fixed numerical limit t. (In the limiting case when $p(a) = 0$, $a \mid\sim x$ is defined to hold for all x.) Hawthorne and Makinson found that the set of all such probabilistic consequence relations cannot be completely characterized by any set of finite-premised Horn rules (i.e., rules of the form "If $a_1 \mid\sim x_1, \dots a_n \mid\sim x_n$ then $b \mid\sim y$"). In 2009 Jeff Paris and Richard Simmonds presented an algorithm that produces all finite-premised Horn rules that are sound for all probabilistic consequence relations (i.e., for all values of t) but as they say in their paper, the series of rules thus generated is "somewhat incomprehensible". In this chapter Hawthorne proves the soundness of another, more intuitive series of rules and raises the question whether it is also sufficient for derivation of every finite-premised Horn rule that is sound for all probabilistic consequence relations.

In "Non-monotonic Logic: Preferential Versus Algebraic Semantics" *Karl Schlechta* provides both an overview and a philosophical background to the various ways in which preference structures can be used in non-monotonic logic and related areas. A preference structure is a pair $\langle U, \prec \rangle$ consisting of a set U and a binary relation $\prec$ on U. A common device in such a structure is a choice function μ based on $\prec$. One possible reading puts $\mu(X)$ to be the set of $\prec$-minimal elements of X. Schlechta also emphasizes the usefulness of an operator (or quantifier) ∇ expressing "normally" or "in almost all cases". According to Schlechta, ∇ suits well to express the notion of a default; a system of defaults should reflect how things mostly are. The chapter can be read as a *plaidoyer* for an approach to these issues that is somewhat different from Makinson's. Schlechta prefers to work on the semantical side, and gives precedence to semantical arguments from which he derives conclusions for syntactical principles. In this chapter he warns that logic should speak about something, and that logic without semantics is "an exercise in thin air". Makinson has often begun on the syntactical side, i.e., with rules and axioms he considers to be interesting or desirable, taking them as a starting-point in the search for a suitable semantic account. As I am sure they both agree, these are complementary rather than conflicting approaches.

In "Towards a Bayesian Theory of Second-Order Uncertainty: Lessons from Non-standard Logics" *Hykel Hosni* compares the role that classical logic has in logic with that of Bayesian theory in the study of uncertain reasoning. There are many types of reasoning that cannot be expressed in classical logic. However, this should not lead us to give up classical logic altogether. In his view—a view much in line with the general theme of this book—the most fruitful approach to these limitations has been to extend the resources of classical logic so that it can express a larger variety of forms of rational reasoning. Bayesian theory faces similar problems. In particular, its insistence on exact probabilities seems ill-fitted to deal with the many cases in real life when we do not have enough information to assign precise and non-arbitrary probabilities to that which we are uncertain about. But, says Hosni, this is no reason to give up Bayesian theory. Doing so "would be like saying that logic is irrelevant to rational reasoning because classical logic is inadequate for air-traffic control". More specifically, he argues for extending Bayesian theory with imprecise probabilities. This can be done in several ways. Hosni warns against the idea that it should be done with second-order probabilities. He is more favourable towards the use of convex sets of probability measures, a solution that was already proposed by Savage in 1972 in the second edition of his textbook.

3 Normative Systems

The logic of norms is a topic to which Makinson has repeatedly returned over the years. He has approached the subject in different ways, some of which are discussed in the four chapters that now follow. In "Abstract Interfaces of Input/Output Logic" and "Intuitionistic Basis for Input/Output Logic" the focus is on input/output logic which he developed with Leendert van der Torre in the first years of the present millennium. Input/output logic is a system that takes propositions as inputs and delivers propositions as outputs. The framework can potentially be used for many applications, but it was developed with a focus on normative systems in which the inputs are conditions and the outputs are (conditional) norms.

In "Abstract Interfaces of Input/Output Logic" *Audun Stolpe* applies language splitting to input/output logic. Language splitting is a method devised by Rohit Parikh to divide a set of sentences into disjoint parts that refer to different topics. Stolpe applies splitting to two operations in input/output logic, the simple-minded and the basic operation. The simple-minded operation has a very weak logic. The basic operation is characterizable as having in addition the rule that if the input a yields the output c, and the input b also yields the output c, then the input $a \vee b$ yields the output c. Stolpe proves that every set of norms has a unique finest splitting with respect to simple-minded output. The corresponding question for basic output remains to answer. He also explores some connections between input/output logic and the AGM account of belief change. In the final part of his essay he brings these logical investigations to bear on an issue in conceptual analysis, namely the meaning of positive permission. By a positive permission is meant a permission that has some force in itself and is not just a report of the absence of the corresponding prohibition. Positive permissions have two major functions in a norm system. They can give rise to exemptions from prohibitions (i.e., override these prohibitions), and they can prevent prohibitions that may be added in the future from having effect. The latter function is illustrated by permissions included in a country's constitution. Stolpe investigates these two functions in the framework of input/output logic, making use of some of the results on splitting obtained earlier in the chapter. In addition to new formal results the chapter contains a series of open questions on language splitting and the logic of normative systems.

Input/output logic was constructed and presented as a mechanism that makes use of classical logic. But does it need all the resources of classical logic? In "Intuitionistic Basis for Input/Output Logic" *Xavier Parent*, *Dov Gabbay*, and *Leendert van der Torre* provide important parts of the answer to that question. They investigate whether the axiomatic characterizations available for the four major operations in input/output logic hold also for intuitionistic logic. It turns out that for three of the operations, the same axiomatic characterizations can be used as for classical logic. (Two of the three are the operations referred to in Stolpe's chapter.) For the fourth operation (the basic reusable one), the existing proof of completeness does not carry over to the intuitionistic context, with little prospect for repair. The axiomatic characterization of that operation is still an open issue. In conclusion the authors note that they did not use the whole of intuitionistic logic in their proofs of the three axiomatic characterizations. These results seem to hold also for Ingebrigt Johansson's minimal logic that differs from intuitionistic logic in not satisfying *ex falso quodlibet*.

In "Reasoning About Permission and Obligation" *Jörg Hansen* takes us back to the issue of positive permission. Exactly what courses of action are open to a person when the normative system regulating her behaviour contains both obligations and (positive) permissions? To get a grip on this problem Hansen makes use of von Wright's notion of norm consistency. According to von Wright, a set containing both obligations and permissions is consistent if and only if the contents of each of the permissions can be realized while the contents of all the obligations are also realized. Since we cannot be required to perform the impossible, an inconsistent set of norms will in some way have to be cut down to a consistent subset. If it has only one maximal consistent subset, then that is the obvious choice. But what should we do if there are several? Choose only one of them, or settle for their intersection? And what should the relative priorities of obligations and permissions be in such an exercise; should we for instance follow Alchourrón's and Bulygin's proposal that permissions always prevail over obligations? After carefully investigating what can be done with purely logical resources, Hansen settles for a model in which logic is complemented by a selection mechanism that allows a permission to have either higher, lower, or the same priority as an obligation. Real-life examples are easily found that corroborate the reasonableness of allowing for all three possibilities. In conclusion, Hansen applies the insights gained in this analysis to two classical problems in the logic of permission: free choice permission and David Lewis's so-called problem of permission. The former is concerned with exactly what it means for instance that I am permitted to take an apple or an orange. The latter asks what the effects of a positive permission are on an obligation that it overrides. Suppose I am under an obligation to spend the Sunday on my workplace and answer all calls on my mobile work phone. If my boss allows me to spend the Sunday at home, do I still have to answer those phone calls? Hansen suggests that the problem cannot be solved by logic alone. This has an important practical implication: Speakers wishing to moderate a previous command are well advised to replace that command by a new one, rather than just granting an additional permission. In the final section, Hansen takes another look at von Wright's definition of norm consistency and suggests that perhaps it should be amended to take into account permissions not just individually but collectively, much like mandatory norms.

There are many connections between the logic of belief change and that of norms. Makinson's own work in belief change had its origins in his work with Carlos Alchourrón on changes in legal systems. In "Norm Change in the Common Law", *John Horty* recalls that their work referred to changes in written legal codes, such as those set down by legislatures. That is the dominant form of legal change in civil law countries. In this chapter Horty develops an account of norm change modelled after legal change in a common law system. In such a system, legal change takes place incrementally through court decisions that are constrained by previous decisions in other courts. The assumption is that a court's decision has to be consistent with the rules set out in earlier court decisions. However, it is allowed to make add new distinctions and therefore make a different decision based on factors not present in the previous decision. Horty develops two formal models of this process. The first model is based on refinement of (the set of factors taken into account in) the set of previous cases on which a decision is based. In the second model the focus is on a preference ordering on reasons. The court is allowed to supplement, but not to revise the preference ordering on reasons that can be inferred from previous cases. Interestingly, the two accounts are equivalent. Horty recognizes that these accounts are both idealized, since in practice, the body of previous cases cannot be expected to be fully consistent. But such inconsistencies are local and do not propagate to cases not involving the particular reasons treated differently by previous courts. Therefore a court can make a consistent decision even if the case base is not consistent; the important requirement is that no new inconsistencies should be added to the case base.

4 Classical Resources

The fourth section features two new texts by *David Makinson* in which he explores the resources of classical logic. In "Intelim Rules for Classical Connectives" he investigates introduction and elimination rules (in short: intelim rules) for truth-functional connectives. These rules have an important role in the philosophical discussion on the meaning of the connectives. Two major questions concerning set/formula intelim rules are investigated in the chapter. The first is which truth-functional connectives satisfy at least one (non-degenerate) introduction or elimination rule. The answer is that all truth-functional connectives except verum satisfy at least one non-degenerate elimination rule, and all truth-functional connectives except falsum satisfy at least one non-degenerate introduction rule. The second major question is which truth-functional connectives are uniquely determined by the introduction and elimination rules that they satisfy. This question can be further specified in at least four different ways, depending on how we interprete "uniquely determined". For two of these interpretations a rather complex pattern emerges and, although it answers many questions this chapter also leaves some quite interesting issues open for future investigations.

In "Relevance Logic as a Conservative Extension of Classical Logic" Makinson investigates how a suitable consequence relation for Anderson's and Belnap's system R of relevance logic could be constructed. Unfortunately the most obvious adequacy criteria for such a relation are in conflict, so some kind of compromise seems to be necessary. Makinson discusses several alternative constructions and develops two of them in full detail. Perhaps the more promising of these is based on an extension of the labelling system for natural deduction used by Anderson and Belnap in order to accommodate modus ponens. The consequence relation is defined such that the sentence β is a consequence of the set A of sentences if and only if β is obtainable in a derivation all of whose undischarged premises are in A (irrespective of what labels are attached to β). This consequence relation is a conservative extension of classical consequence, and the consequences that it produces from the empty set are exactly the theorems of R. This shows, says Makinson, that "if one really wants to be a relevantist, then one may opt for the logic R whilst keeping classical logic intact".

Finally, in "Reflections on the Contributions" David Makinson reflects on the contributions of the other authors in this volume.

David Makinson and the Extension of Classical Logic

Sven Ove Hansson and Peter Gärdenfors

Abstract There are two major ways to deal with the limitations of classical logic. It can be replaced by systems representing alternative accounts of the laws of thought (non-classical logic), or it can be supplemented with non-inferential mechanisms. David Makinson has a leading role as proponent of the latter approach in the form of the *inferential-preferential* method in which classical logic is combined with representations of preference or choice. This has turned out to be a highly efficient and versatile method. Its applications in non-monotonic logic and belief revision are used as examples.

Keywords David Makinson · Classical logic · Inferential-preferential method · Nonmonotonic logic · Belief revision · Input/output logic

If we wish to understand the significance of a major researcher's contributions to science we need to see them in a larger context. Logic took big steps forward in the twentieth century, and David Makinson has been a major contributor to its progress. In order to appreciate his role in the development of the discipline, let us begin with a bird's-eye view of how logic has evolved.

1 A Tradition from Antiquity

The logic that we inherited from antiquity had its focus on single steps of reasoning. Its major achievement was the Aristotelian theory of syllogisms, a fairly precise account of steps of thought that take us with certainty from two premises

S. O. Hansson (✉)
Royal Institute of Technology, Stockholm, Sweden
e-mail: soh@kth.se

P. Gärdenfors
Lund University Cognitive Science, Lund, Sweden
e-mail: Peter.Gardenfors@lucs.lu.se

S. O. Hansson (ed.), *David Makinson on Classical Methods for Non-Classical Problems*, Outstanding Contributions to Logic 3, DOI: 10.1007/978-94-007-7759-0_2,

to a conclusion. The argumentation steps represented in the standard syllogisms are those in which the notions "some", "all", and "no" have a pivotal role. These types of argument steps are relatively rare in actual human reasoning, but they are particularly well suited for precise modelling. It was an admirable intellectual achievement to identify them as a starting-point and a model for a rigorous account of reasoning. But in the two millennia that followed, progress towards more generality was remarkably slow. When Kant in 1787 described logic as closed and perfected ("geschlossen und vollendet", AA 3:7), it might have been more adequate to describe it as stuck in a position from which it was difficult to move forward.

As a theory of human rational thought and ratiocination, Aristotle's theory of syllogisms has three major limitations. First, it only treats small steps of argumentation one at a time, rather than complete inferences or lines of argumentation. Therefore it cannot answer questions such as whether a certain conclusion follows from a given set of premises (Hodges 2009). Secondly, it only covers arguments expressible in terms of a small selection of terms, namely "all", "some", and "no". Although Aristotle provided (in *Topics*) logical principles for other terms such as betterness, these attempts did not reach the high level of precision achieved in the syllogisms, and they were not integrated with (or easily integrable into) the theory of syllogisms. Thirdly, the theory of syllogisms only covers (deductive) inferences that can be made with full certainty, and it does not have much to say about inductive and other fallible inferences.

The second of these limitations was attacked in several ways through the centuries, most importantly through the parallel development of propositional logic. The logic of propositions reached a high level of sophistication and formal rigour in George Boole's *The Laws of Thought* (1854)—but it was still unconnected to syllogistic logic and also still subject to the other two limitations.

The first limitation, that to single steps of argumentation, was overcome with the quantifiers that Frege introduced in his *Begriffsschrift* (1879). With these new tools it was possible to develop more comprehensive systems that allowed a general investigation of what can and cannot be inferred from a set of premises. These developments opened the way for mathematically more advanced investigations of logic. Consequently, in the twentieth century logic has attracted some of the most brilliant mathematical minds, but unfortunately also repelled some able but less mathematically minded philosophers.

Although Frege's system was revolutionary in terms of its expressive power, it was still traditional in the sense of retaining the major laws of thought that had been almost universally accepted since antiquity: bivalence, the laws of identity, non-contradiction and the excluded middle, and the elimination of double negations. As could be expected, however, the introduction of more comprehensive logical systems opened up the possibility of constructing precisely defined alternative logics representing divergent specifications of the laws of thought. This gave rise to an abundance of non-classical logics including intuitionistic, many-valued, paraconsistent, and quantum logics.[1]

The new logic developed by Frege and his immediate successors was still subject to the other two limitations, namely the restriction to a small selection of pivotal terms (the logical constants) and the exclusive focus on inferences that can be made with full certainty. In the last century these limitations have to a considerable extent been overcome through a large number of extensions of the logical apparatus that cover new subject areas: inductive logic, semantically based modal logic, the logic of conditionals, epistemic and doxastic logic, deontic logic, preference logic, action logic, belief revision, ... It is in this development that David Makinson has been one of the most influential contributors.

2 The Inferential-Preferential Method

Many logicians have combined the two developments just mentioned, namely the extension of logic to new subject-matter and the construction of non-classical alternatives to classical logic. It has often been believed that in order to extend logic to subject-matter such as inductive and non-monotonic reasoning, classical logic has to be replaced by something else. But there is also another approach to the extension of logic to new areas—an approach that David Makinson has (gently) advocated more consistently and efficiently than anyone else: Instead of giving up classical logic, we can combine it with non-inferential mechanisms, in particular mechanisms of preference and choice. The combination of inferential and preferential mechanisms has turned out to be surprisingly powerful and versatile in a wide range of applications.

This method was used by Lewis (1973) in his pioneering work on the logic of counterfactual conditionals. In collaboration with others Makinson introduced it into belief revision theory. In a joint paper he and Carlos Alchourrón used it to define safe contraction (Alchourrón and Makinson 1985). The central problem in belief contraction is to find out which sentences we should remove from a logically closed set K in order to obtain another logically set $K \div p$ such that $p \notin K \div p \subseteq K$. In safe contraction a sentence q is discarded in this contraction if and only if it satisfies both an inferential and a preferential criterion. The inferential criterion is that q is an element of some subset X of K that implies p but has no proper subset that implies p. The preferential criterion is that q has a bottom position among the elements of X according to a hierarchy $<$ that regulates the contraction (i.e. there is no $r \in X$ with $r < q$). In partial meet contraction, the recipe for belief contraction that Makinson developed together with Alchourrón and Gärdenfors (Alchourrón et al. 1985), the

[1] On the connections between intuitionistic and classical logic, see Gödel (1986, esp. pp. 286–295 and 300–303) and Humberstone and Makinson (2011). On the connections between relevance logic and classical logic, see Friedman and Meyer (1992) and Makinson's chapter on relevance logic in this book, "Relevance Logic as a Conservative Extension of Classical Logic".

selection mechanism is instead applied to maximal subsets of K not implying p, and the contraction outcome is obtained as the meet of the selected such subsets.

Makinson's book *Bridges from Classical to Nonmonotonic Logic* (2005) is highly recommended not only as an unsurpassed presentation of non-monotonic logic but also as an introduction to the inferential-preferential methodology. In this book he constructs non-monotonic logics using what he calls three "natural bridges between classical consequence and the principal kinds of nonmonotonic logic to be found in the literature" (p. ix). One of these bridges is the use of additional background assumptions in the form of a set of propositions that are used along with the current premises in the derivation of consequences. A problem with this method is that even if the set of premises is consistent, it may yield an inconsistent set of outcomes. This can be solved by using only a selection of the background assumptions (presumably selected because they are more highly preferred than the others). In the following section we will present this methodology in greater detail, as applied to non-monotonic reasoning.

The second method is to restrict the set of Boolean valuations on the language that are used to semantically determine the relation of consequence. In other words, some of the possible assignments of truth values to atomic sentences are not used in evaluating logical consequence. This, of course, again requires some mechanism of choice. This is the basic framework that was used for instance in the well-known system presented by Kraus et al. (1990).

The third method is similar to the first in that it also adds background assumptions to be used together with the present premises to derive consequences. However, in this case the background assumptions are rules rather than propositions. By a rule is meant a pair $\langle x, y\rangle$ of propositions. A set A of sentences is closed under the rule $\langle x, y\rangle$ if and only if: if $x \in A$ then $y \in A$. The set of non-monotonic consequences of A can then be characterized by being closed both under classical consequence and under a set of rules. Just as in the first method, in order to avoid the derivation of inconsistences from a consistent set of premises a mechanism of choice is needed, in this case one that makes a choice among the rules. This is the idea behind the default rule systems developed by Reiter (1980), Brewka (1991), and others.

The reader is referred to Makinson's book for a systematic discussion of the intuitive and formal connections between insertions of preferential principles at different places in the inferential system. Other such connections have been obtained in belief revision theory, see for instance Gärdenfors and Makinson (1988), Fuhrmann (1997a, b), Grove (1988), Rott (2001), and Hansson (2008, 2013). The inferential-preferential method has also been used to show how consistent normative recommendations can be obtained from a system of norms containing potentially conflicting elements (Hansson and Makinson 1997).

3 Non-monotonic Reasoning as Classical Logic with Defeasible Premises

The first of the three above-mentioned uses of the inferential-preferential method is arguably the simplest one from a conceptual point of view. Its simplicity makes it particularly well suited for elucidating the philosophical background of that method. Let us therefore have a closer look at this particular bridge from classical to non-monotonic logic.

In classical logic, the focus is on a syntactic relation $P \vdash q$ between a set P of premises and a conclusion q. The validity of an argument that is representable by the relation $\vdash$ is assumed to depend only on the logical structure of q and the sentences in P, and consequently to be independent of their meaning, their truth, and the context. However, actual human reasoning is not based exclusively on premises that are taken as given. It is also influenced by the stronger or weaker expectations that we have about facts of the world. Such expectations can be expressed by sentences that are regarded as plausible enough to be used as a basis for inference so long as they do not give rise to inconsistency. Gärdenfors and Makinson (1994) took this extension of classical logic as a basis for analysing non-monotonic reasoning. The key idea for using expectations can be put informally as follows:

> P non-monotonically entails q (denoted $P \mathrel{|\!\sim} q$) if and only if q follows logically from P together with as many as possible of the set E of our expectations as are compatible with P.

Note that the expectations are represented in the same way as the premises, namely as a set of beliefs. Contrary to many other accounts of non-monotonic reasoning, this approach does not need a special formal representation for default beliefs.

The expectations in E are not premises that have to be accepted. Instead they are defeasible in the sense that if the premises P are in conflict with some of these expectations, then we do not use them when determining whether q follows from P. In order to make this inference recipe precise, the meaning of "as many as possible" must be explicated.

Expectations are suppressed in classical logic, but in practical reasoning they are not. Everyday arguments are full of hidden premises that need to be made explicit in order to make the argument logically valid. In each separate case, it may be possible to add the hidden assumptions to make the derivation performable with classical methods. Expectations are normally shared among speakers, and unless they are contradicted, they serve as a common background for arguments.

To give a concrete example of how this mechanism leads to non-monotonic inferences, suppose that among the expectations are "Swedes are protestants", "Swedes with Italian parents are catholics" and "Protestants are not catholics". Now if "Anton is a Swede" is given as a premise, then one can conclude "Anton is a protestant" using the first expectation. If the premise "Anton has Italian parents" is added, then one can also conclude "Anton is a catholic". This conclusion is inconsistent with the first, given the expectations, so in order to maintain consistency the expectation "Swedes are protestants" is given up and one concludes only "Anton is a catholic".

The reasoning pattern is non-monotonic since a previous conclusion is lost when more premises are added.

In order to make precise "as many as possible of the expectations", Gärdenfors and Makinson (1994) studied two selection mechanisms. The first is based on selection functions of precisely the same type as those that are used in models of belief revision. The second is based on an ordering of "defeasibility" of the expectations that is similar to the entrenchment orderings for belief revision used by Gärdenfors and Makinson (1988). They also formulated a set of postulates for non-monotonic reasoning that is essentially equivalent to the postulates used by Kraus et al. (1990). Gärdenfors and Makinson (1994) could then prove completeness results: (1) The selection mechanism based on selection functions is characterized by a basic set of postulates only involving one expectation. (2) The selection mechanism based on defeasibility orderings is characterized by the full set of postulates also involving several expectations. They also showed that the selection mechanisms based on orderings form a subclass of the preferential models studied by Shoham (1988), Kraus et al. (1990).

On the basis of the similarity between those preferential models based on defeasibility/entrenchment ordering that have been developed for belief revision and those that have been developed for non-monotonic reasoning, the two authors established a method for translating postulates for belief revision into postulates for non-monotonic reasoning, and vice versa (Makinson and Gärdenfors 1990; Gärdenfors and Makinson 1994). The key idea for the translation from belief revision to non-monotonic logic is that a statement of the form $q \in K * p$ is seen as a non-monotonic inference from p to q given the set K of sentences that is then interpreted as the set of background expectations ($K = E$). In this way the statement $q \in K * p$ in belief revision is translated into the statement $p \mid\sim_K q$ in non-monotonic logic (or into $p \mid\sim_K q$, if one wants to emphasize the role of the background beliefs). Conversely, a statement of the form $p \mid\sim q$ for non-monotonic logic is translated into a statement of the form $q \in K * p$ for belief revision, where K is introduced as a fixed belief set. In our view, this example shows that the "bridges" that Makinson has been instrumental in building with the inferential-preferential method can be used not only to connect classical and non-classical logic with each other but also to connect different areas of logical analysis with each other.

4 Logic as an Auxiliary Tool

In the examples we have described above, logical inference is the driving force, but it is supplemented with some non-logical mechanism such as a choice process in order to obtain the desired properties. However, it is not self-evident that the logical component should have the dominant role in such a combination. The opposite relation is also possible, as exemplified by the input/output systems introduced by Makinson and van der Torre (2000).

An input/output system is a mechanism that takes propositions as inputs and delivers propositions as outputs. In most interpretations of this versatile framework the inputs are conditions, whereas the outputs may for instance be norms, preferences, goals, intentions or ideals. The central component of an input/output system is some relation that assigns output propositions to input propositions. In principle, logical inference need not be involved. However, logic can be helpful in at least three ways. First, it can prepare the inputs, for instance by seeing to it that an input consisting of two inputs x and y are treated in the same way as one consisting of the single input $x \& y$. Secondly, it can unpack the outputs or specify their interpretation, for instance by closing them under logical consequence. Thirdly, it can co-ordinate inputs and outputs, for instance by making sure that outputs are reusable as inputs.

Makinson and van der Torre state quite explicitly that in their systems, logic has only an auxiliary role. In a model like this, logic is not the central transformation mechanism. Instead it should be "seen in another role, as 'secretarial assistant' to some other, perhaps non-logical, transformation engine" (p. 384). Furthermore:

> The process as a whole is one of 'logically assisted transformation', and is an inference only when the central transformation is so. This is the general perspective underlying the present paper. It is one of 'logic at work' rather than 'logic in isolation'; we are not studying some kind of non-classical logic, but a way of using the classical one (2000, p. 384).

Logicians tend to think of logic as the leading principle of (rational) human thought. Practical experience as well as results from cognitive science give us reasons to assign a much less dominant role to logic (see e.g. Gärdenfors 1994; Benthem 2008). The concept of "logically assisted" mechanisms is, in our view, well worth investigating as a means to obtain more realistic models of the role of logic in mental processes.

5 Conclusion

As we mentioned above, two major types of solutions have been developed to overcome the limitations of classical logic: replacing it with something else and combining it with non-inferential mechanisms. David Makinson has been a most influential proponent of the second method. Developing the inferential-preferential model, he has shown how it can be applied in a powerful way to expand classical reasoning to a wide range of applications.

References

Alchourrón, C., Gärdenfors, P., & Makinson, D. (1985). On the logic of theory change: Partial meet contraction and revision functions. *Journal of Symbolic Logic*, *50*, 510–530.

Alchourrón, C., & Makinson, D. (1985). On the logic of theory change: Safe contraction. *Studia Logica*, *44*, 405–422.

Boole, G. (1854). *An investigation of the laws of thought, on which are found the mathematical theories of logic and probability*. London: Walton and Maberly.

Brewka, G. (1991). Cumulative default logic: In defense of nonmonotonic inference rules. *Artificial Intelligence, 50*, 183–205.

Frege, G. (1879). *Begriffsschrift, eine der aritmetischen nachgebildete Formelsprache des reinen Denkens*. Halle.

Friedman, H., & Meyer, R. K. (1992). Whither relevant arithmetic? *Journal of Symbolic Logic, 57*, 824–831

Fuhrmann, A. (1997a). Solid belief. *Theoria, 63*, 90–104.

Fuhrmann, A. (1997b). *An essay on contraction*. Stanford: CSLI.

Gärdenfors, P. (1994). The role of expectations in reasoning. In M. Masuch & L. Polos (Eds.), *Knowledge representation and reasoning under uncertainty* (pp. 1–16). Berlin: Springer-Verlag.

Gärdenfors, P., & Makinson, D. (1988). Revisions of knowledge systems using epistemic entrenchment. In M. Y. Vardi (Ed.), *Proceedings of the Second Conference on Theoretical Aspects of Reasoning about Knowledge* (pp. 83–95). Los Altos: Morgan Kaufmann.

Gärdenfors, P., & Makinson, D. (1994). Nonmonotonic inference based on expectations. *Artificial Intelligence, 65*, 197–245.

Gödel, K. (1986). Collected works. In S. Feferman et al. (Eds.), I: Publications 1929–1936. Oxford: Oxford University Press.

Grove, A. (1988). Two modellings for theory change. *Journal of Philosophical Logic, 17*, 157–170.

Hansson, S. O. (2008). Specified meet contraction. *Erkenntnis, 69*, 31–54.

Hansson, S. O. (2013). Blockage contraction. *Journal of Philosophical Logic, 42*(2), 415–442.

Hansson, S. O., & David M. (1997). Applying Normative Rules with Restraint, pp. 313–332. In M. L. Dalla Chiara et al. (Eds.), *The Tenth International Congress of Logic, Methodology and Philosophy of Science, Florence*, August 1995. Vol. 1, Logic and Scientific Methods. Boston, Mass.: Kluwer.

Hodges, W. (2009). Traditional logic, modern logic and natural language. *Journal of Philosophical Logic, 38*, 589–606.

Humberstone, L., & Makinson, D. (2011). Intuitionistic logic and elementary rules. *Mind, 120*(1035–1051), 2011.

Kraus, S., Lehmann, D., & Magidor, M. (1990). Nonmonotonic reasoning, preferential models and cumulative logics. *Artificial Intelligence, 44*, 167–207.

Lewis, D. (1973). *Counterfactuals*. Oxford: Blackwell.

Makinson, D. (2005). *Bridges from classical to nonmonotonic logic*. London: King's College.

Makinson, D., & Gärdenfors, P. (1990). Relations between the logic of theory change and nonmonotonic logic. In A. Fuhrmann & M. Morreau (Eds.), The logic of theory change (pp. 185–205). Berlin: Springer-Verlag.

Makinson, D., & van der Torre, L. (2000). Input/output logics. *Journal of Philosophical Logic, 29*, 383–408.

Reiter, R. (1980). A logic for default reasoning. *Artificial Intelligence, 13*(81–132), 1980.

Rott, H. (2001). *Change, choice and inference: A study of belief revision and nonmonotonic reasoning*. Oxford: Clarendon.

Shoham, Y. (1988). *Reasoning about change*. Cambridge: Cambridge University Press.

Van Benthem, J. (2008). Logic and reasoning: do the facts matter. *Studia Logica, 88*, 67–84.

A Tale of Five Cities

David Makinson

Abstract The author looks back at his work in logic under the headings: early days in Sydney, graduate studies in Oxford, adventures in Beirut, midnight oil in Paris, and last lap in London.

Keywords Logic · Life

1 Early Days in Sydney

Try as I may, I just can't remember when logic first began to interest me. Little in high school seems to have prepared me for it, apart from a diet of good old-fashioned plane geometry with attention to the proofs of its theorems. But the idea of standing back to study the inference tools themselves never occurred to anyone in the classroom.

School exposure to mathematics was rather limited. At a certain stage I made the decision of taking an option of 'General Mathematics' rather than the double dose that was called Maths I and II. Actually, I had not been doing too badly and in retrospect there seem to have been two reasons for the unfortunate choice. One was an immense but distant admiration for my father, who was a physicist; being reluctant to compete with him in any way, I gave his territory a wide berth. The other was a succession of exciting teachers of history and literature who fired up my adolescent curiosity in a way that their colleagues in the sciences could not do.

I still remember the names of those inspiring teachers, presumably now deceased. They included Lynn Bottomley for history, who patiently put up with, and gently probed, the Marxist slant that I was giving to just about everything; Claude Doherty for drills in English grammar and style, for which I am eternally grateful; and Neil Hope, alias Soap, for literature. I recall how, in the very first day of class, Soap gave

D. Makinson (✉)
LSE, London, UK
e-mail: david.makinson@gmail.com

S. O. Hansson (ed.), *David Makinson on Classical Methods for Non-Classical Problems*, Outstanding Contributions to Logic 3, DOI: 10.1007/978-94-007-7759-0_3,

out a list of a hundred or so great works with a single instruction: read whatever you like from this list or beyond it, as much as you can, and give me a report every week.

Later I discovered that Soap also happened to be a figure in a down-town group calling themselves 'libertarians'—in the old-fashioned sense of being fundamentally sceptical of all forms of political and social authority, rather than in the now dominant sense of advocating unregulated markets. Also known locally as 'the push', it was the movement with which, as an undergraduate a few years later, I most empathized. Not long afterwards, Soap left the shores of Australia for what he hoped would be a less philistine society in Italy, only to die there in a motor-scooter accident.

Perhaps one thing did make me susceptible to logic. My parents were both intellectual Marxists—a very rare species in 1950s Australian society, which was alarmed to the point of panic by the real and imaginary dangers of the Cold War. The country's tiny Communist party, of which my parents were both members, was an easy target for local versions of American spy-hunting. While I was still very young, a member of the Federal Parliament and acolyte of Senator Joe McCarthy in the USA, denounced my father as a dangerous Russian spy. Parliamentary immunity shielded the politician from legal challenge, and although my father did not actually lose his academic post, his career was blighted thereafter.

Anyway, to get back to the subject, my avid schoolboy reading of Marxist fare led me to works such as Engels' *Anti-Dühring* and *Dialectics of Nature*, where one is asked to accept the presence and dynamic role of contradictions in nature. That puzzled me, understanding a contradiction as an assertion that cannot possibly be true, and my natural reaction was to conclude that the term was being used in a figurative manner, referring to unstable situations with causal currents running in opposite directions. But no, the prophets of the end of history insisted that when they said 'contradiction' they a meant real, unmitigated contradiction—leaving me quite nonplussed and with an ear ready for logical clarification. Strange are the starting points of long journeys.

Entering Sydney University in 1958, I enrolled in a programme whose first year was made up of courses in Economics, Psychology, History and Philosophy, with 'logic' making up a considerable part of the fare of the last of these. But I put the word in inverted commas, for the syllabus was nothing like what one would expect today. No truth-functional connectives or first-order quantification; just unremitting Aristotelian syllogism and immediate inference, accompanied by defences against their nineteenth-century critics such as John Stuart Mill. It was delivered in tediously dictated lectures that had initially been prepared by the ageing Scottish–Australian metaphysician John Anderson and were relayed to us by his disciples.

This, of course, was already a hundred years after Boole's *Laws of Thought* of 1854 and a quarter of a century since publication of the first German edition of Hilbert and Ackermann's *Mathematical Logic*, not to speak of the English translation of its second edition in 1950. Even in courses for Philosophy students, a new world had begun to appear with textbooks such as Quine's *Elementary Logic* of 1941 and his *Methods of Logic* of 1950, followed by Fitch's *Symbolic Logic* of 1952. But none of that had filtered through to the course being taught in the Philosophy Department of Sydney University in 1958.

However, one should not be too scornful of such resistance; something rather analogous may also be found today. Throughout the second half of the twentieth century, and even into the present one, there has been an unfortunate tendency for introductory logic courses in departments of Philosophy to consist of just that—logic. But there are other formal instruments that students of Philosophy, and their counterparts in any other discipline requiring abstract thought, should be able to use—notably the basic tools of sets, relations, functions, recursion/induction, trees and finite probability. Even within logic itself, there is more to offer the neophyte than skill in constructing symbolic natural deductions that follow one or another system of rules, a drill that still dominates many courses.

Moreover, experience in the classroom has convinced me that in practice, such formal tools should be made available *before* getting systematically into logic. Despite its simplicity, if not because of it, logic can be quite elusive as it vacillates between the crashingly obvious and the baffling. For most students, it acquires intuitive meaning only as independently motivated examples of its use are reviewed. Moreover, it turns out that a clear understanding of the basic concepts of logic requires some familiarity with all but the last of the mathematical notions mentioned above; it is difficult for an instructor to say anything penetrating about the subject without having recourse to them. So, what the apprentice student of Philosophy needs is not a course about logic alone, but one on formal methods of conceptual reasoning.

A very early text that took some steps in this direction was Patrick Suppes' *Introduction to Logic*, back in 1957. But it still *began* with logic, leaving naïve set theory, relations and functions to its often neglected second part; nor did it give any attention to structural recursion and induction. As far as I am aware, the first textbook written explicitly for Philosophy students with a broad range of formal methods culminating, rather than beginning, with logic was John Pollock's *Technical Methods in Philosophy* of 1990. But it appeared before its time, does not seem to have been widely used, and is long out of print. Even there, structural recursion and induction are covered in the shortest chapter—the only one without sections—of an already brief text.

The situation is now beginning to change, although the old paradigm has not been dislodged from the practice of most Philosophy departments. My own *Sets, Logic and Maths for Computing*, whose first edition of 2008 was written before I stumbled across Pollock's text, was intended as a further step in this direction. Although it appears in a series aimed at students of Computing, as is reflected in the title, it was written also to serve students of Philosophy and any other discipline that demands abstract thought. They all need to know something about sets, relations, functions, structural recursion as a form of definition and induction as a method of proof piggy-backing on it, as well as directed trees and finite probability.

But let's return to the narrative and that first year of dictated notes in 1958. Quite ignorant of what was missing from them, I was nevertheless fascinated with what I was getting, and as the four years of undergraduate study went by I slowly came to discover more. In part, this was thanks to instructors who began to break out of the mould: Tom Rose, who gave seminars on Boole and de Morgan; John Mackie, who sought to use modern logic in the formulation of traditional philosophical problems; and Richard Routley, freshly imported from the USA with the gospel of relevance logic in his hand, later to be celebrated for the Meyer-Routley three-place relational models for such systems. But much of my journey of logical discovery was self-taught, which usually meant badly taught. Thanks to cheap paperbacks from the Dover publishing company, then a godsend for impecunious students in outlandish places, I wrestled rather unsuccessfully with Carnap's ponderous *Introduction to Symbolic Logic with Applications* and Paul Rosenbloom's sibylline *Elements of Mathematical Logic*.

Somewhat more success came with Kleene's monumental *Introduction to Metamathematics*. Having graduated from Sydney University at the end of 1961 and awaiting departure to Oxford on a Commonwealth Scholarship the following September, I realized that my future classmates in the UK might be better prepared than myself, and so began the lonely labour of working systematically through Kleene. I certainly did not finish that massive and rigorous volume in nine months of untutored reading, but did manage to get some understanding of parts of it which served me well when I landed in England, never to return to Australia for more than short visits.

2 Graduate Studies in Oxford

Already determined to specialize in logic, my first concern was to wriggle out of the designated path towards a B. Phil. This degree had been designed to produce all-round teachers of Philosophy who could be exported to other universities in the United Kingdom and indeed the far corners of the British Commonwealth, to carry the glad tidings of philosophy according to the Oxford paradigm. My goal was to transfer to a doctoral program, which was still a rather unusual thing to do. In the meantime, Freddy Ayer was my supervisor. He had little interest in logic, notwithstanding his title of Wykeham Professor of the same; and I did not have enough sparkle or repartee to make him look twice in my direction. There was a brief flicker of interest when I produced a short piece (which became my first publication) showing the continued collapse of successive refinements of an attempt, back in his *Language, Truth and Logic* of 1936, to define a notion of verifiability in terms of a purely deductive relationship to 'observation statements'. But perhaps he was tired of the question and already inclining to the view that he expressed in a later interview (see Bryan Magee *Talking Philosophy: Dialogues with Fifteen Leading Philosophers*, new edition 2001, p. 107). Speaking in the 1970s of both his own book and the logical positivist movement in general, he confided: "Well, I suppose the most important of the defects was that nearly all of it was false". I, on the other hand, was still taking it seriously.

Thankfully, after a year I was allowed to switch to a doctorate, and serious work began under Michael Dummett. I have an enormous debt to him for forcing me onwards to the point of exhaustion with reading, thinking and at least the illusion of understanding a broad sweep of logic. I still remember arriving for a session after a week in which I had been asked to read Smullyan's *Formal Systems* and grasp its approach to the concept of decidability. "Well, how did it go? Any difficulties?" Too embarrassed to admit to sweat and tears, my stammer let him continue with "Good, let's go on to the next topic..."—and another week of plunging into the unknown. It took me to the edge of nervous breakdown, but I did survive, and learned more about logic than I had from anyone else.

Yet I could never enter Dummett's world, with its idea that mathematical statements may be regarded as true or false only in so far as we can construct intuitively convincing proofs or counter-examples for them. And without that perspective there seemed little reason for contending that logical connectives should be understood by means of rules of proof rather than of truth. Perhaps most of all, I was uncomfortable with the suggestion that independently of any epistemological arguments, intuitionistic logic finds some purely formal justification in the fact that it can neatly be characterized using introduction and elimination rules for connectives, compared with less elegant characterizations in that style for classical logic. However, I was not able to articulate the discomfort into something more substantial, and the issue went into the bottom drawer of my mind. It was dug out again nearly fifty years later in a joint paper with Lloyd Humberstone, 'Intuitionistic logic and elementary rules', which finally came to grips with unfinished business.

Behind such academic differences there were also deep contrasts of world outlook. They were never discussed, as they did not form part of our mutual task, but they were certainly present and made it difficult to come together on a personal level. On the one hand, the tutor was a believing catholic and combined a fierce defence of human rights with what seemed to be a deep cultural conservatism; on the other hand, his student was a committed atheist from the 'colonies', leaning to the left but with a scepticism about all exercise of political power that derived from his Sydney libertarian days.

Returning to academic matters, Dummett cultivated to perfection a style of writing that was popular in British philosophy at the time, but which I found exasperating. The discussion would be conducted without an explicitly stated goal but only a spiral of approximations, objections, and modifications from which, after many iterations, a picture would emerge implicitly.

There were also methodological features that are still common today. One is a tendency to blur the edges between formal points and their philosophical interpretations, creating an entangled skein in which the former are expressed less succinctly and rigorously than they might be while the latter seek to introduce themselves with the authority of the formal material. Another is a tendency to see philosophical perspectives as matters on which one is either right or wrong, and to take the task of the philosopher as one of finding arguments that establish or refute them.

To my mind, these are both recipes for disaster. To be sure, in logic formal results gain interest in so far as they suggest or illustrate broad philosophical perspectives, which may conversely point us in directions that help us articulate formal observations. But it is essential to keep the two levels clearly apart, never confusing one with the other. Whereas we can establish formal results by rigorous proofs using mathematical tools, the general perspectives that hover around them need to be assessed in a quite different way. It is not a matter of being correct, seldom even of being incorrect, but rather one of the advantages and disadvantages, costs and benefits, insights and shortcomings, of different ways of conceptualizing a situation. One result of this is that competing conceptual organizations will sometimes be in something like a tie; another is that their relative merits may change as technical results reveal unanticipated benefits for one of them. Choosing a philosophical perspective in logic is more like devising a good diagram, selecting an elegant notation, or finding the 'best' level of generality at which to work, than like stating a doctrine.

This is not to ignore the importance of argument in developing a general viewpoint, nor to suggest that 'anything goes'. It is to be modest about the character of that argumentation and what it can hope to achieve. Forget about 'proofs' in philosophy, and regard knock-down refutations as secondary phenomena that can often be sidestepped by a more careful formulation of the impugned position. Take seriously the role of argument in philosophical logic (indeed, in philosophy in general), but avoid the perennial temptation to over-estimate its capacities.

To resume the narrative: E. J. Lemmon's lectures on modal logic, given shortly before he left in 1963 for the United States and early death, played an important part in my studies. While Dummett was on leave for a while, he was replaced as interim adviser by William Kneale, co-author of *The Development of Logic*, providing a breathing space of avuncular chats after the intense working sessions that had been driving me to the edge. And in the final period, when the material for the dissertation was beginning to come together but Dummett was travelling again, Arthur Prior, then in Manchester, kindly accepted that I mail him chapter drafts each month before joining him for a long weekend in his family home or country retreat to go over what I had written. Fond memories, of wonderful people.

While in Oxford I had the privilege of meeting other personalities who also exercised an influence in one way or another. The young Saul Kripke was participating in Dummett's seminars, and it was from his work in modal logic, particularly his 1963 *Zeitschrift* paper explained in one of the seminar sessions, that I was led to the central ingredient of my own doctoral thesis. Kripke had established completeness theorems for modal propositional logics in terms of his relational models, with proofs that elaborated classical semantic decomposition-trees (or Beth tableaux as they were then called). My idea, which came in the middle of a sleepless night in an uncomfortable bed visiting friends in London, was to do it by developing the maxi-set constructions that had originally been introduced by Lindenbaum and Tarski for classical propositional logic and had already been extended by Henkin to cover classical first-order logic.

As I later learned, several others were having the same idea. Indeed, Arnould Bayart had already anticipated some of it as early as 1959 in a rather impenetrable study of first and second order S5 (where, however, no accessibility relation is needed in the semantics). Dana Scott presented it orally in a summer school in Leicester only a month or so after I completed the doctorate; and David Kaplan reported on Scott's work the following year in a review in the *Journal of Symbolic Logic*.

Of course, each of the two methods has its advantages and disadvantages: the decomposition-tree method is constructive and computationally suggestive, while the maxi-set one is elegant, transparent and very succinct.

While in Oxford, quite by happenstance, I was one day struck by the disclaimer in the preface of a textbook that I was studying. The author thanked colleagues for their assistance, and absolved them from responsibility for the errors that must surely remain. This reminded me of what I had felt myself while on a vacation job a couple of years earlier, proof-reading the Sydney telephone book. There I had seen that despite the team's best efforts, too many typos were getting through. Reflecting on the epistemic oddity of the situation led to a second paper, 'The paradox of the preface'. It passed into the shadows for a couple of decades, to emerge again towards the end of the century alongside Kyburg's closely related lottery paradox in the newly labelled area of formal epistemology. The paper drew attention to a problem rather than provide a solution, thus leaving more unfinished business.

Suddenly, in June 1965 it was all over, with the D. Phil in my hand. What to do? There was no hesitation about that question: I thought of nothing else than working in university philosophy departments. But where? That was much more difficult to answer.

A first impulse was to return to Sydney, but that was vetoed by my recently acquired wife from Buenos Aires, who regarded it as the end of the world. My own letters of enquiry to Argentina remained unanswered. I had been asked discretely whether I would be interested in a job in Oxford itself, but despite the evident attractions I was impatient with a culture where so much appeared to be an exercise in appearing clever rather than clarifying and, if possible, settling matters. So, perhaps arrogantly, I declined.

In those days of economic growth and educational expansion following the Second World War academic jobs hung like apples on trees, and I was offered positions in Sussex and Leeds. But English academic salaries were dismal and the climate depressing for much of the year. I felt that it was time to do something more exciting. However, the United States was right in the middle of the Vietnam War, to which I was very much opposed, so that was out; and Canada sounded far too cold.

By chance, while visiting Prior in Manchester, I met someone who had spent a few years at the American University of Beirut, briefly AUB, in Lebanon. That sounded exciting enough, though I admit that I had to look up an atlas to learn the location. So I wrote a letter of inquiry and, amazingly for even those times, received in reply a thick envelope containing a three-year contract with double the salary that I would have received in England, plus a place to rent in a brand new faculty apartment on campus. How could I refuse?

3 Adventures in Beirut

Three years went by and became fifteen. They might perhaps have stretched on to now had not the Lebanese civil way intervened and forced my departure. But this part of the story should be told from its beginning.

Unknown to me, I was flying into a hornet's nest. Not long before, one of the staff, Roland Puccetti, had angered the governing board with his rather sceptical views on personal identity, the mind-body problem and philosophy of religion. The institution had, after all, been founded a century earlier as the Syrian Protestant College and although the name had long changed something of the earlier mind-set still prevailed among board members. Puccetti's contract was not renewed, and the Head of Department, a conservative metaphysician steeped in Aquinas and Heidegger, began searching for a solid believer as replacement. But some colleagues took advantage of his momentary absence from the country and the coincidental arrival of a letter of enquiry from the UK, to sneak in a substitute from the neutral territory of logic.

So one may imagine the atmosphere in which I was received. Well before my departure from England, the departmental Head, who also happened to be a former foreign minister of the host country and so had some strings to pull, had set the security services to investigating my background and that of my wife in an attempt to bar entry into the country. He managed to delay the issue of our visas, but it turned out that his influence was rather less than he had imagined and entry papers came through just in time for me to take up the job in September 1965. However, the episode did leave me on the security watch list, not for anything that I had done, nor even for my parents' political affiliations (which apparently had not been noticed), but rather because of my wife's family origins. So, every year the renewal of my work visa would take a month longer than all the others, until eventually the friendly Lebanese fixer handling such matters for the university got tired of the hassle and managed to get the file closed.

Ironically, it was not long before Aquinas was on the receiving end. A course on the history of western thought included in its mimeographed readings some pages from the writings of the *Doctor Angelicus*, and within those pages was a paragraph with some animadversions on the Prophet. These days, one can immediately guess the reaction, but in the Lebanon of 1966 it came as rather a surprise. Denunciations were made in the mosques, and the police came on campus searching for Thomas Aquinas. They did not find him, and they did not want to touch his local representative the former foreign minister; so a junior instructor who happened to be among those teaching the multi-section course was summarily expelled from the country.

One might wonder how it was possible to work in such a climate. I maintained a low profile, kept my personal views to myself, focussed on my interests in logic, and all went well. To be honest, the fifteen years in Lebanon were among the most interesting and exciting of my life, getting acquainted with local life, exploring the country and surrounding region, wandering in a camping car each summer from Beirut through Turkey, the Balkans and onwards to France, as well as enjoying a relaxed Mediterranean outdoor life-style, swimming much of the year and in winter learning to ski just an hour's drive away.

More often than not, philosophy majors at the AUB were there as a result of some personal crisis, whether in religion, culture, politics, family—or all or the above. In general, logic was the last thing that they wanted to know about. But eager students flocked from elsewhere. For the university saw itself as a liberal arts college, and those heading for the sciences or the professions—physics, chemistry, engineering, medicine, agriculture etc.—were obliged to take a certain number of humanities options. The problem for them was that while generally bright, they were often weak in English. To take an extreme example, a student may have been speaking Armenian at home, gone to a school where the language of instruction was Arabic and the main foreign language French, and boned up on English just enough to meet the entry requirements. So when I organized my introductory logic course as a purely problem-solving effort with no essays, they were delighted. A few were even able to wangle permission to carry over into a second, more advanced course; but all ultimately vanished into their careers when the party was over.

As far as my work in logic was concerned, I was certainly isolated. Those were the days before internet, and the only means of correspondence abroad was by letter. Of my colleagues at the AUB, those with whom I had most in common professionally were in the Mathematics Department: Fawzi Yaqub, an algebraist who worked a little in Boolean algebras; Azmi Hanna who was fascinated by category theory; and Ahmed Chalabi, also in abstract algebra and who was much later to play a shadowy role in the second American invasion of Iraq.

But I did manage to keep my work going. At first, this was by following up the investigations in modal logic begun in the doctoral thesis: constructing a simple and quite well motivated example of a modal propositional logic lacking the finite model property (the first of its kind), examining the structure of the upper end of the lattice of all modal logics, counting prime filters in infinite Boolean algebras and distributive lattices, and so on. But as time went by, I felt that fresh pastures were needed. I was wondering where to find them when, by chance, I met someone who would point the way: Carlos Alchourrón.

In Buenos Aires, where I was on vacation visiting my wife's family, I was invited to give a talk on deontic logics to the SADAF (Sociedad Argentina de Análisis Filosófíco), then leading a rather precarious existence quite separate from the universities, which were under the thumb of the military regime and its ideology. From past experience of giving talks in other places I had assumed that nobody in the audience would know much about the topic and so prepared an elementary and rather superficial talk. To my surprise and embarrassment, an individual in the audience began asking the most acute and well-informed questions. As he did so, it became clear that he had thought about the subject more deeply than myself. That was Carlos, and we became firm friends.

In our discussions he raised a question that had long been bothering him. Suppose that we have a code of norms, and decide to derogate one of its consequences. If we are to be coherent, we need also to annul, or at least qualify, the norm that gives rise to it. So far so good; but what if the item being rejected was obtained by combining two or more of the norms that figure explicitly in the code. Which are affected by the derogation? My initial response was to shrug shoulders, saying that there is no formal answer; rather it is a pragmatic matter to be answered from case to case. But slowly I began to realize that nevertheless there are interesting logical questions to pursue. In particular, we may ask whether there are any regularity conditions that one might require—for example comparing the contraction of different items from a single code or the sequential contraction of one item after another—as well as whether one can give some kind of formal rationale for such regularities.

Thus began our work together on the logic of what we then called theory change, now usually known as belief change, in which we were joined by Peter Gärdenfors coming into the same territory from a different direction, eventually leading to the joint paper setting out the AGM account in terms of partial meet contraction and revision. The story of this collaboration has recently been told by Peter in his 'Notes on the history of ideas behind AGM' (*J. Phil. Logic* 2011, 40: 115–120), and I have also recalled various aspects of it in scattered places, notably an obituary for Carlos, who died of cancer of the pancreas in 1996, in the *Nordic Journal of Philosophical Logic*, a paper 'Ways of doing logic: what was different about AGM 1985?', and a foreword to the reprinting by College Publications of Peter's *Knowledge in Flux*. So I will not say more about it here.

Early in the 1970s life at the American University of Beirut gradually became more difficult. The first sign was a prolonged student strike, ostensibly about fees but in reality an exercise in student mobilization on the part of political groups whose militias would soon be playing their parts in the Lebanese civil war. All teaching stopped for a semester, and I took refuge in the basement of the library, shielded from interminable speeches and martial music blaring over loud-speakers, to write my first book *Topics in Modern Logic* based on notes for cancelled lectures.

The war itself began in April 1975, and on 30th September of the same year these memories nearly came to a sticky end: my plane returning to Beirut from vacation in France blew up in mid-air just off the coast of Lebanon, with no survivors. As Wikipedia comments, "no official statement was ever made on the crash and its cause has never been publicly revealed". Malfeasance is presumed. But I was not aboard, having the night before reluctantly succumbed to persuasion from my wife and some friends to wait until the situation on the ground in Beirut became clearer. Once again I thank all concerned for their persistence, although they were in one respect mistaken—the scythe cut in the air, not on the ground. From that day on, I felt that I was in a second life, whose flavour has been distinctly different from the first one.

As the war gathered momentum, people managed their anxiety with dubious cognitive stratagems, whether by declaring that things were not really as bad as the overseas press made out, or by opining that the situation must soon get better because it couldn't get any worse. One thing I learned from the experience is that it always can get worse. Although the University limped on, life became more confined and dangerous. In 1980 I eventually left on leave without pay for Paris and UNESCO.

4 Midnight Oil in Paris

I could write at length about UNESCO, but little of it would be complimentary and this is not the place to do it. Suffice it to say that in its offices I found a day-job that kept body and soul together so that I could continue with serious work in logic in the evenings, lunch-times, and weekends. As one might imagine, this was rather stressful and after two years, during a relative calm following the Israeli invasion of Lebanon, I took another leave without pay and returned to the American University of Beirut for the academic year 1982–1983. It was in that year, while alone in Beirut with the family in Paris, that the joint work with Carlos and Peter came together into a manuscript, submitted to the *Journal of Symbolic Logic* in October 1983 where it appeared in 1985. But the situation in the country was deteriorating again by spring 1983 and I had to leave once more, this time with a definitive resignation. Several friends and colleagues who did not leave ended up kidnapped or assassinated, usually both, over the next two years.

It was in UNESCO that I came to appreciate Seneca's dictum 'It is not that we have a short time to live, but that we waste a lot of it', although I presume that he had other forms of waste in mind. To preserve sanity in the face of the mind-numbing pointlessness that was typical of UNESCO activities, I found it essential to partition the mental disk. One part was for office labour, to be carried out as efficiently as possible but without investing any emotional capital, which would only be lost unless one deceived oneself about the real value of what one was doing. Above all, under no circumstances should I allow any of my own investigations in logic to filter across the partition membrane, as they would be subject to what I liked to call the 'reverse Midas' effect: everything that the organization touched would turn not into gold, but something else less pleasant.

With time, I have come to see that I may have been rather unfair in this view. Not that the endless paper-churning, speechifying, and jet-conferencing were any more productive than they had seemed at close quarters. But they were merely an example, perhaps rather extreme, of what happens in any large bureaucracy with a vague mandate, buffeted by political winds and used as a ladder for personal ambitions.

While there I learned the delicate art of being able to interrupt an intellectual task abruptly without undue frustration and take up the thread again hours later—a skill that I am now losing from lack of practice. After ten years, however, I managed to switch from the surrealistic world of the Philosophy Division to the job of editing a semi-popular periodical of 'social sciences for the world citizen', where the company was more agreeable and the work no longer absurd; where, moreover, I could learn some of the skills of editing and become acquainted with the surprising diversity of methodologies and professional standards in the disciplines that make up the social sciences. The downside of that change was that the new work actually became rather interesting, and I found myself inadvertently giving it some of the emotional commitment that might have gone elsewhere.

After the publication of the AGM paper on belief change in 1985, my thoughts began turning to the adjacent area of so-called nonmonotonic reasoning—more descriptively, the logic of qualitative uncertain inference. I could sense that there were close similarities between the two, but postponed any attempt to articulate a formal connection while trying to understand nonmonotonic reasoning in its own terms with its particular motivations, insights, and jargon. In this task I was immensely encouraged by a generous and risky gesture of Dov Gabbay, who engaged me to write an overview of the subject for a projected volume of his *Handbook of Logic in Artificial Intelligence and Logic Programming*.

I say 'risky', because there were plenty of people who knew a lot more about the subject than I did, and I still wince when I remember the naiveté of my first drafts and their ignorance of much of the literature. Fortunately, Dov-the-magician could obtain enough grant money to organize a series of meetings between contributing authors. The criticisms that my miserable beginnings received there, as well as the personal contact with others active in the area, helped get a proper show on the road. Fortunately too, some of the authors ended up dragging their feet for submission, providing me with the opportunity to revise and update the text for several years. All this led to my first really long paper 'General patterns in nonmonotonic reasoning', which appeared in the *Handbook* in 1994.

While that was marinating, and once I felt that I understood what was going on in non-monotonic inference—or at least the kind based on Shoham-style preferential models—I turned attention to clarifying the formal relationship between it and the logic of belief change on which I had previously been working. The connection with belief revision turned out to be surprisingly direct and simple, and was written up in a joint paper with Gärdenfors, 'Relations between the logic of theory change and nonmonotonic logic' which appeared very rapidly in 1991.

Although living in Paris, or rather in its leafy outskirts at Ville d'Avray, the constraints of my desk job at UNESCO meant that I had very little contact with French academia. My main interlocutor in Paris was Michael Freund, a mathematician who had been convinced by his brother-in-law Daniel Lehmann that nonmonotonic reasoning was something interesting to get into. Daniel himself had emigrated from France to Israel years before, but often returned to his family in Paris. The three of us collaborated a good deal through the 1990s, although this is not very visible from the sole publication (or more accurately semi-publication) in which all three names figure. Our exchanges often took the form of 'OK, you think that this might work and I suspect it won't, so you search for a proof while I look for a counterexample and let's see what happens'. I won a couple of bottles of wine for counterexamples.

My last years in Paris were marked by a renewed interest in deontic logic or, as I prefer to call it, the logic of norms and normative systems. The driving problem was one that had been at the back of my mind for a very long time. If we accept the idea that statements of obligation and permission—indeed norms in general—lack truth-values, why are we analyzing them using a Kripke-style semantics in which deontic formulae are attributed truth-values at points (or states or worlds if one prefers those terms) linked by an accessibility relation? Is not there a coherent way of treating the semantics of normative systems that does not accord truth-values to statements of obligation?

Technically, the difficult part of the task lies with conditional norms, and after wresting with it alone in a 1999 paper entitled 'On a fundamental problem of deontic logic', I joined forces with Leendert (Leon) van der Torre, then in Amsterdam, to construct the abstract mechanisms that we dubbed 'input/output logics' published in a series of papers from 2000 to 2003. These have not received as much attention as I think they deserve. The basic constructions fulfill their purpose in a simple and elegant manner, I believe, although the elaborations needed to handle contrary-to-duty obligation and above all several different varieties of conditional permission are less straightforward and much more debateable.

At that time, personnel of the United Nations system were subject to compulsory retirement at the age of sixty. For me, this was something of a release from servitude and an occasion to look for a way back into academia. Fortune smiled, again in the person of Dov Gabbay who had just moved from Imperial to King's College London and, with Tom Maibaum, was building an exciting logic group in its Computer Science Department. Once again he made a generous and risky offer, and in January 2002, on the very day that euros made their appearance on the streets of France, I took the train to what is still the land of the pound sterling.

5 Last Lap in London

At last I had an opportunity to talk in the classroom about the material on non-monotonic reasoning on which I had been working for nearly a decade. This I did in graduate seminars, writing up the product, exercises and all, as a textbook *Bridges from Classical to Nonmonotonic Logic*. Thanks to Jane Spurr's magic LaTeX manipulations, it came out as a cheap print-on demand book in a series, College Publications, managed with great success by Dov and Jane. To be accurate, the series was then called 'King's College Publications': jealousies from the central administration led to demands to hand over or quit, which Dov adroitly side-stepped by rebadging.

Seen from a distance, that was just an incident in a cold war. The administration had decided on budgetary grounds that the logic group so recently built up should be dismantled and replaced by more down-to-earth software engineering in what was to become a Department of Informatics. Without ever declaring its goals, it chipped away at courses and staff. Given the reduction in logic courses, I was asked to switch to an 'introduction to finite mathematics' unit for more than a hundred first-year students. Actually, it turned out to be a very positive experience, and brought me to think more clearly about what should go into introductory courses, whether for computer science students or others. This led to the textbook mentioned earlier, *Sets Logic and Maths for Computing*, which my then colleague Ian Mackie commissioned for a Springer series that he was setting up. The book itself came out in 2008—initially with the index to somebody else's text on economics! The publisher agreed to pulp the printing, but by then hundreds of copies were on the shelves of bookshops over the world. If you were landed with one of those, do not bin it; after all, postage stamps with lesser imperfections have sometimes become valuable! A considerably revised second edition, thankfully with the correct index, appeared in 2012.

In 2006, shortly before my sixty-fifth birthday, the five-year contract with King's reached its end. For the administration, there was no question of any kind of extension for one logician too many, so I found myself retiring for the second time. Logicians are rather like bedouin, pitching their camps in the gardens of departments of Computer Science, Mathematics, and Philosophy, where they may be welcome guests or undesirable aliens, and it was clear what category I belonged to. But this second retirement turned out to be an extraordinary opportunity. Thanks to a kind offer by Colin Howson, I walked across the Strand to the London School of Economics, taking up residence as a kind of permanently-temporary visiting professor, unsalaried and without any formal obligations, but participating in the life of the Department as well as doing some bought-in logic teaching. I have to say that this has been the most satisfying and enjoyable working environment I have ever experienced—and not just for the absence of administrative duties! Most of all, for the intellectual environment in LSE's quaintly named Department of Philosophy, Logic and Scientific Method, with congenial colleagues of distinction. All this, plus access to a good library and to other logic activities around London. What more could one want?

At LSE I found the courage, at last, to enter the non-man's land on the border between qualitative uncertain reasoning (which I had been studying for about fifteen years) and probabilistically based inference (which I had not). What exactly are the differences, and how well do the two perspectives mesh with each other? Such questions had been raised in one chapter of *Bridges from Classical to Nonmonotonic Logic* but with only some first steps towards clarification. Work in this area brought me into collaboration with Jim Hawthorne of Oklahoma, leading to our 2007 joint paper 'The quantitative/qualitative watershed for rules of uncertain inference'. That was followed by my own more recent 'Conditional probability in the light of qualitative belief change' and the long-delayed 'Logical questions behind the lottery and preface paradoxes: lossy rules for uncertain inference'. The last of those three papers takes up questions raised by the paradox of the preface, articulated so many years before when a graduate student in Oxford. In particular, it considers inference rules under which some probability may be given up, but with an upper bound on the amount that may be lost in each application. Such 'lossy' rules are studied further in a paper with Jim, scheduled to appear in 2015.

Indeed, one could say that a fair part of my most recent work takes up shelved puzzles from yesteryear; thus does the past grow more present as the future contracts. As mentioned earlier, 'Intuitionistic logic and elementary rules' examines attempts to justify intuitionistic logic through its generation by introduction and elimination rules for its connectives. In turn, that prompted the formal work in a chapter 'Intelim rules for classical connectives' for the present volume, and a close study of Sandqvist's inferential semantics for classical logic scheduled to appear about the same time. Another recent paper looks at the concept of propositional relevance through letter-sharing; in a rudimentary form, this concept is familiar from the early days of relevance logic, but it takes on quite a new dimension when it is understood modulo a background belief set. 'Friendliness and sympathy in logic' also took its origin in undergraduate curiosity when studying Boole, this time about his notion of the elimination of a term in an equation of the algebra of logic. While the concept of 'friendliness' (with a perfectly good formal definition) does not seem destined to play an important role in logical theory, it is agreeably elegant in behaviour and I would recommend it for recreation. Finally, another paper contributed to the present volume attempts to clarify another puzzle that already bothered me back in student days. It seeks to answer the question: how far does a commitment to relevance logic really oblige one to give up the full force of classical consequence? Readers should be warned, however, that it does nothing to untangle or justify relevance logic itself—goals that I suspect to be unattainable.

Part II
Logic of Belief Change

Safe Contraction Revisited

Hans Rott and Sven Ove Hansson

Abstract Modern belief revision theory is based to a large extent on partial meet contraction that was introduced in the seminal article by Carlos Alchourrón, Peter Gärdenfors, and David Makinson that appeared in 1985. In the same year, Alchourrón and Makinson published a significantly different approach to the same problem, called safe contraction. Since then, safe contraction has received much less attention than partial meet contraction. The present paper summarizes the current state of knowledge on safe contraction, provides some new results and offers a list of unsolved problems that are in need of investigation.

Keywords Safe contraction · Hierarchy · AGM · Epistemic entrenchment · Kernel contraction · Supplementary postulates · Iterated safe contraction · Multiple safe contraction · Non-prioritized safe contraction · Safe revision

1 Introduction

Modern belief revision theory is based to a large extent on the seminal article by Carlos Alchourrón, Peter Gärdenfors, and David Makinson in the *Journal of Symbolic Logic* in 1985. Alchourrón and Makinson had previously co-operated in studies of changes in legal codes. Gärdenfors was investigating the relationship between belief changes and conditional sentences. Combining forces, they developed a theory of rational belief change whose central construction is partial meet contraction. In the same year, Alchourrón and Makinson published a significantly different approach

H. Rott (✉)
University of Regensburg, Regensburg, Germany
e-mail: hans.rott@ur.de

S. O. Hansson
Royal Institute of Technology (KTH), Stockholm, Sweden
e-mail: soh@kth.se

S. O. Hansson (ed.), *David Makinson on Classical Methods for Non-Classical Problems*, Outstanding Contributions to Logic 3, DOI: 10.1007/978-94-007-7759-0_4,

to the same problem, called safe contraction. Even though Makinson has always had "a particular affection for it", safe contraction has received much less attention than partial meet contraction.[1] The purpose of the present paper is to summarize the current state of knowledge on safe contraction, provide some new results and offer a list of unsolved problems that are in need of investigation.

The two approaches have the same purpose and are based on the same general assumptions. In both cases the agent's present set of beliefs is represented by a set of sentences. In the AGM approach this set is assumed to be logically closed. In the paper introducing safe contraction that assumption was not made, but most of the formal results did in fact refer to the changes of logically closed sets. In what follows we will focus on logically closed belief-representing sets that will as usual be called *belief sets*. The letter K is reserved to denote a belief set. In contraction the agent receives an input consisting in a sentence ϕ to be contracted, and then performs an operation whose goal is to find a plausible outcome $K \dot{-} \phi$ that is a logically closed subset of the prior belief set K not implying ϕ. In order to achieve this we need to remove not only ϕ itself but also additional sentences that would in combination with other sentences imply ϕ. This involves a selection among the elements of K.

The two operations perform this selection in quite different ways. Partial meet contraction is based on selection of sentences to be retained. This is done with a selection function γ. However, γ does not perform a selection directly among sentences. Instead it is applied to the set of maximal consistent subsets of K not implying ϕ. These are called the *ϕ-remainders* of K. Let $K \perp \phi$ be the set of ϕ-remainders of K.[2] Then $\gamma(K \perp \phi)$ is a non-empty subset of $K \perp \phi$, unless ϕ is a tautology so that $K \perp \phi$ is empty, in which case $\gamma(K \perp \phi)$ is put equal to K. Intuitively, that a remainder (i.e., an element of $K \perp \phi$) is in $\gamma(K \perp \phi)$ means that it belongs to those elements of $K \perp \phi$ that are most "plausible" or "secure" or "valuable".

Definition 1. *The partial meet contraction associated with γ is defined to be equal to the intersection of the selected remainders, i.e.,*

$$K \dot{-} \phi = \bigcap \gamma(K \perp \phi)$$

A partial meet contraction is relational *if and only if it can be constructed from a selection function γ that is based on a relation $\sqsubseteq$ on $\bigcup\{K \perp \psi : \psi \in K\}$, such that $\gamma(K \perp \phi) = \{X \in K \perp \phi : Y \sqsubseteq X \text{ for all } Y \in K \perp \phi\}$ whenever $K \perp \phi$ is non-empty. If $\sqsubseteq$ is transitive, then the operation is a* transitively relational *partial meet contraction.*

Safe contraction is based on a selection of sentences to be removed, rather than retained. Therefore, the focus is instead on the set of minimal subsets of K that

---

[1] The quote is taken from Carnota and Rodríguez (2011, p. 8). According to Google Scholar, in December 2012 the AGM paper on partial meet contraction had been quoted thirteen times more often than Alchourrón and Makinson's paper on safe contraction.

[2] Thus $X \in K \perp \phi$ if and only if (1) $X \subseteq K$, (2) $X \nvdash \phi$, and (3) $X' \vdash \phi$ whenever $X \subset X' \subseteq K$. It is easy to check that a ϕ-remainder of K is logically closed if K is.

imply ϕ. These sets have been called the *ϕ-kernels* of K. We use the notation $K \perp\!\!\!\perp \phi$ to denote the set of ϕ-kernels of K.[3]

Obviously, in order for a contraction outcome $K \dot{-} \phi$ not to imply ϕ, at least one element from each of the ϕ-kernels has to be removed. This is done by means of a binary relation $\prec$ on K. Intuitively, $\rho \prec \psi$ means that ρ is more "exposed" or "vulnerable" than ψ, or alternatively that it is less "secure or reliable or plausible" (Alchourrón and Makinson 1985). From each ϕ-kernel, the most vulnerable elements are selected for removal. More precisely, if there is no ρ such that $\rho \prec \psi$ then ψ is selected for removal. A sentence $\psi \in K$ is *$\prec$-safe with respect to ϕ* if and only if it is not selected for removal in any of the ϕ-kernels.[4] Let K/ϕ be the set of sentences in K that are safe with respect to ϕ. Clearly, if ϕ is a tautology then the set of ϕ-kernels is $\{\emptyset\}$ and thus K/ϕ is equal to K.

Definition 2. *The safe contraction associated with $\prec$ is defined to be equal to the set of consequences of the safe elements*[5]*:*

$$K \dot{-} \phi = \mathrm{Cn}(K/\phi)$$

In summary, partial meet and safe contraction as introduced by Alchourrón, Gärdenfors and Makinson (1985) and Alchourrón and Makinson (1985), respectively, were introduced for precisely the same purpose. They have some things in common, but they also differ in several respects:

1. Partial meet contraction proceeds by selecting what to retain, safe contraction by selecting what to remove.
2. In partial meet contraction, the selection mechanism selects among (logically closed) sets of sentences. In safe contraction it selects among single sentences.
3. In partial meet contraction, the selection mechanism is a choice function (that may or may not be relational, i.e., based on a relation $\sqsubseteq$ between sentences). In safe contraction the selection mechanism is always based on a relation $\prec$ between sentences.
4. In relational partial meet contraction the relations are non-strict, i.e., of the "less-than-or-equal-to" type, and they are used for optimization (something is chosen if it is at least as good as any other option). The relations used in safe contraction

[3] Thus $X \in K \perp\!\!\!\perp \phi$ if and only if (1) $X \subseteq K$, (2) $X \vdash \phi$, and (3) $X' \nvdash \phi$ whenever $X' \subset X$. Kernels, and in particular consistent kernels, have played an important role in a certain brand of argumentation theory, see the definition of an "argument" and of a "contour" in Besnard and Hunter (2008, pp. 39, 176). Besnard and Hunter's theory, however, does not use any orderings of certainty, reliability or priority over the sentences of the language.

[4] Thus the sentences not included in any ϕ-kernel are safe with respect to ϕ. They correspond to the full meet contraction by ϕ.

[5] Here we tacitly assume that logically equivalent sentences are interchangeable. See Sect. 2 for a more precise definition. For sets of sentences H that are not logically closed, Alchourrón and Makinson define safe contraction as $H \dot{-} \phi = H \cap \mathrm{Cn}(H/\phi)$.

are of the strict, i.e., of the "less-than" type, and they are used for minimization (something is chosen if it is better than no other option).[6]

5. In partial meet contraction, the selection mechanism is specific to a particular belief set K, and the same selection mechanism cannot be applied to two different belief sets. In safe contraction, one and the same selection mechanism can be applied to different belief sets.

These properties are not necessarily tied to each other. In what follows we will argue that the basic idea of safe contraction can be combined with the choice function approach (second difference), that partial meet contraction may (and perhaps should) use relations of the "less-than" type (fourth difference), and that, with a minor adjustment, one and the same partial meet contraction can be applied to different belief sets (fifth difference).

It is the purpose of the present contribution to provide an overview and some new insights on safe contraction, and in particular to show how its basic principles can be further developed in different directions. In Sect. 2 we introduce safe contraction in more detail. The properties of hierarchies and the effects of these properties are investigated in Sect. 3 that also provides some new results on the properties of some types of safe contraction and their relations with major classes of partial meet contraction. Section 4 investigates the relationships between safe contraction and entrenchment-based contraction as introduced by Gärdenfors and Makinson (1988). In Sect. 5 the focus is on iterated safe contraction and in Sect. 6 on kernel contraction, the choice-functional generalization of safe contraction. In Sect. 7 we consider more briefly three further developments that have as yet not been much investigated: multiple safe contraction, non-prioritized safe contraction, and safe revision. In the concluding Sect. 8 a list of unsolved problems for safe contraction is offered. All formal proofs are deferred to an appendix.

We will work with an object language $\mathcal{L}$ whose elements (sentences) are denoted by lower-case Greek letters and whose subsets are denoted by upper-case Latin letters. $\mathcal{L}$ contains the usual n-ary truth-functional operators $\bot$ and $\top$ $(n = 0)$, $\neg$ $(n = 1)$, $\vee, \wedge, \rightarrow$, and $\leftrightarrow$ $(n = 2)$. The logic is represented by a Tarskian consequence operation Cn which includes classical propositional logic, is reflexive, idempotent, monotonous, compact, and satisfies the deduction theorem. We write $M \vdash \phi$ for $\phi \in \mathrm{Cn}(M)$, $\phi \vdash \psi$ and $\psi \dashv \phi$ for $\psi \in \mathrm{Cn}(\{\phi\})$, $\vdash \phi$ for $\phi \in \mathrm{Cn}(\varnothing)$, and $M \vdash_{\exists} N$ for $\mathrm{Cn}(M) \cap N \neq \varnothing$. We write $M \vdash_{min} \phi$ if $M \vdash \phi$, but $M' \nvdash \phi$ for all proper subsets M' of M. By K and variants like K', $K \dot{-} \phi$ etc., we denote a theory in $\mathcal{L}$, i.e., a subset of $\mathcal{L}$ that is closed under Cn.

[6] Compare the comments by Alchourrón and Makinson (1986), Sect. 2.1, on "Dominance and Maximality", and by Sen (1997, Sect. 5) on "Maximization and Optimization". For a short discussion of the advantages of using strict relations, cf. Rott (1992b, pp. 50–52). The non-strict AGM relations $\sqsubseteq$ are, more correctly speaking, of the "less-than-or-equal-to-or-incomparable-with" type.

2 Hierarchies and Their Properties

The following properties of a hierarchy were introduced by Alchourrón and Makinson:

(H0) If $\phi \prec \psi$, $\phi \dashv\vdash \phi'$ and $\psi \dashv\vdash \psi'$, then $\phi' \prec \psi'$ (Intersubstitutivity)[7]
(H1) If $\psi_1 \prec \cdots \prec \psi_n$, then not $\psi_n \prec \psi_1$ (Acyclicity)[8]
(H2$^\uparrow$) If $\phi \prec \psi$ and $\psi \vdash \chi$, then $\phi \prec \chi$ (Continuing up)
(H2$^\downarrow$) If $\phi \prec \psi$ and $\chi \vdash \phi$, then $\chi \prec \psi$ (Continuing down)
(H3) If $\phi \prec \psi$, then $\phi \prec \chi$ or $\chi \prec \psi$ (Virtual connectivity)

Let us note that Acyclicity implies Asymmetry, which in turn implies Irreflexivity. Irreflexivity and Transitivity taken together imply acyclicity. Asymmetry and Virtual connectivity taken together imply Transitivity.

Virtual connectivity, (H3), is sometimes called Almost-connectedness, Negative transitivity or Modularity. Virtual connectivity for $\prec$ implies that $\not\prec$, the relation such that $\phi \not\prec \psi$ if and only if not $\phi \prec \psi$, is transitive. If an irreflexive relation $\prec$ satisfies Virtual connectivity, then its symmetrical converse, $\not\lessgtr$ defined such that $\phi \not\lessgtr \psi$ holds if and only if neither $\phi \prec \psi$ nor $\psi \prec \phi$, is an equivalence relation.

Definition 3. *A* hierarchy *over the belief set K is a relation over K that satisfies (H0) and (H1). A hierarchy $\prec$ is said to be* regular *over K if and only if it also satisfies* (H2$^\uparrow$) *and* (H2$^\downarrow$)*, i.e., if and only if it continues up and down $\vdash$ over K.*

If both (H2$^\uparrow$) and (H2$^\downarrow$) hold, then so does (H0). Furthermore, it is easy to check that in the presence of (H1) and (H3), (H2$^\downarrow$) and (H2$^\uparrow$) are equivalent.

3 Safe Contraction and AGM-Style Postulates for Contraction

3.1 *The AGM Postulates*

Alchourrón, Gärdenfors and Makinson (1985) put forward a set of eight postulates for theory contraction that have become widely known as the *AGM postulates*.

($K \dot{-} 1$) if K is a theory, then $K \dot{-} \phi$ is a theory (Closure)
($K \dot{-} 2$) $K \dot{-} \phi \subseteq K$ (Inclusion)
($K \dot{-} 3$) if $\phi \notin K$, then $K \dot{-} \phi = K$ (Vacuity)
($K \dot{-} 4$) if $\phi \notin \mathrm{Cn}(\emptyset)$, then $\phi \notin K \dot{-} \phi$ (Success)
($K \dot{-} 5$) $K \subseteq \mathrm{Cn}((K \dot{-} \phi) \cup \{\phi\})$ (Recovery)
($K \dot{-} 6$) if $\mathrm{Cn}(\phi) = \mathrm{Cn}(\psi)$, then $K \dot{-} \phi = K \dot{-} \psi$ (Extensionality)
($K \dot{-} 7$) $K \dot{-} \phi \cap K \dot{-} \psi \subseteq K \dot{-} (\phi \wedge \psi)$ (Conjunctive overlap)

---

[7] Intersubstitutivity was not mentioned in Alchourrón and Makinson (1985), but introduced, under the name "normality", in Alchourrón and Makinson (1986).

[8] Acyclicity is also known as non-circularity.

(K$\dot{-}$ 8) if $\phi \notin K \dot{-} (\phi \wedge \psi)$, then $K \dot{-} (\phi \wedge \psi) \subseteq K \dot{-} \phi$ (Conjunctive inclusion)

Justifications for these postulates are given in Gärdenfors (1988, pp. 61–65). Postulates (K$\dot{-}$ 1) – (K$\dot{-}$ 6) have been called the *basic* AGM postulates, and postulates (K$\dot{-}$ 7) – (K$\dot{-}$ 8) have been called the *supplementary* AGM postulates. As shown by Alchourrón, Gärdenfors and Makinson (1985), the basic postulates exactly characterize the class of partial meet contractions, and postulates (K$\dot{-}$ 1)–(K$\dot{-}$ 8) taken together exactly characterize the class of transitively relational partial meet contractions.

We consider a number of relevant variations and weakenings of the supplementary postulates that we will use later on. First, we consider two variants of (K$\dot{-}$ 7).

(K$\dot{-}$ 7P) $K \dot{-} \phi \cap \mathrm{Cn}(\phi) \subseteq K \dot{-} (\phi \wedge \psi)$ (Partial antitony)

(K$\dot{-}$ 7p) If $\phi \in K \dot{-} (\phi \wedge \psi)$, then $\phi \in K \dot{-} (\phi \wedge \psi \wedge \chi)$ (Conjunctive trisection)

Theorem 1. *If the basic postulates hold, then (K$\dot{-}$ 7), (K$\dot{-}$ 7P), and (K$\dot{-}$ 7p) are all equivalent.*

Secondly, we consider some weakenings of (K$\dot{-}$ 8).

(K$\dot{-}$ 8c) If $\psi \in K \dot{-} (\phi \wedge \psi)$, then $K \dot{-} (\phi \wedge \psi) \subseteq K \dot{-} \phi$

(K$\dot{-}$ 8r) $K \dot{-} (\phi \wedge \psi) \subseteq \mathrm{Cn}(K \dot{-} \phi \cup K \dot{-} \psi)$ (Weak conjunctive inclusion)

(K$\dot{-}$ 8r′) If $\chi \in K \dot{-} (\phi \wedge \psi \wedge \chi)$, then there are ξ and ρ such that $\chi \dashv\vdash \xi \wedge \rho$ and $\xi \in K \dot{-} (\phi \wedge \chi)$ and $\rho \in K \dot{-} (\psi \wedge \chi)$

(K$\dot{-}$ 8wd) If $\chi \in K \dot{-} (\phi \wedge \psi)$, then either $\psi \vee \chi \in K \dot{-} \phi$ or $\phi \vee \chi \in K \dot{-} \psi$

(K$\dot{-}$ 8p) If $\phi \in K \dot{-} (\phi \wedge \psi \wedge \chi)$, then $\phi \in K \dot{-} (\phi \wedge \psi)$ or $\phi \in K \dot{-} (\phi \wedge \chi)$

(K$\dot{-}$ 8d) $K \dot{-} (\phi \wedge \psi) \subseteq K \dot{-} \phi \cup K \dot{-} \psi$

(K$\dot{-}$ 8d′) $K \dot{-} (\phi \wedge \psi) \subseteq K \dot{-} \phi$ or $K \dot{-} (\phi \wedge \psi) \subseteq K \dot{-} \psi$ (Conjunctive covering)

Given the basic AGM postulates, these are all weakenings of (K$\dot{-}$ 8). (K$\dot{-}$ 8c) plays a central role in Rott (1992b). (K$\dot{-}$ 8r) was called (K$\dot{-}$ 8vwd) in Rott (2001). Even given the postulates (K$\dot{-}$ 1) – (K$\dot{-}$ 7), (K$\dot{-}$ 8r) is logically independent of (K$\dot{-}$ 8c).[9] Given the basic postulates, (K$\dot{-}$ 8wd) and (K$\dot{-}$ 8d) are both non-Horn conditions that are stronger than (K$\dot{-}$ 8r). Again given the basic postulates, (K$\dot{-}$ 8wd) is equivalent to (K$\dot{-}$ 8p), which seems to be a new condition. In the choice-theoretic interpretation of Rott (2001), (K$\dot{-}$ 7p) and (K$\dot{-}$ 8p) correspond fairly directly to Sen's property α and (a finite version of) Sen's property γ, respectively. If a selection function takes all and only the finite subsets of some domain as arguments, then properties α and γ are necessary and jointly sufficient for the selection function to be relationalizable (or "rationalizable").[10] (K$\dot{-}$ 8d) is somewhat stronger than (K$\dot{-}$ 8wd); it corresponds to Disjunctive rationality in non-monotonic reasoning.[11] (K$\dot{-}$ 8d′) is the "covering" condition of AGM (1985). It seems to be stronger than (K$\dot{-}$ 8d), but they are in fact equivalent. Given the basic postulates, (K$\dot{-}$ 8c) and (K$\dot{-}$ 8wd) taken

---

[9] Cf. Rott (1993, p. 1438).

[10] Cf. Sen (1986, p. 1097).

[11] See Rott (2001, especially p. 104).

together imply (K$\dot{-}$ 8d). There is no analogous implication starting from (K$\dot{-}$ 8c) and (K$\dot{-}$ 8r). The more interesting of these observations are summarized in Theorem 2 and Fig. 1.

Theorem 2. *(i) If $\dot{-}$ satisfies (K$\dot{-}$ 1), (K$\dot{-}$ 2), (K$\dot{-}$ 5), and (K$\dot{-}$ 6), then it satisfies (K$\dot{-}$ 8r) if and only if it satisfies (K$\dot{-}$ 8r′).*

(ii) If $\dot{-}$ satisfies (K$\dot{-}$ 1), (K$\dot{-}$ 2), (K$\dot{-}$ 5), and (K$\dot{-}$ 6), then it satisfies (K$\dot{-}$ 8wd) if and only if it satisfies (K$\dot{-}$ 8p).

(iii) If $\dot{-}$ satisfies (K$\dot{-}$ 1), then it satisfies (K$\dot{-}$ 8d) if and only if it satisfies (K$\dot{-}$ 8d′).

(iv) If $\dot{-}$ satisfies (K$\dot{-}$ 1), (K$\dot{-}$ 2), (K$\dot{-}$ 5), (K$\dot{-}$ 6), (K$\dot{-}$ 8c) and (K$\dot{-}$ 8wd), then it satisfies (K$\dot{-}$ 8d).

(v) Even if $\dot{-}$ satisfies (K$\dot{-}$ 1)–(K$\dot{-}$ 7), (K$\dot{-}$ 8c) and (K$\dot{-}$ 8r), it does not necessarily satisfy (K$\dot{-}$ 8wd).

Taken together with two of the basic postulates, (K$\dot{-}$ 7) and (K$\dot{-}$ 8c) imply a transitivity and an acyclicity condition for contractions:

Theorem 3. *(i) Let $\dot{-}$ satisfy (K$\dot{-}$ 1)–(K$\dot{-}$ 6) and (K$\dot{-}$ 7). Then it also satisfies*

(K$\dot{-}$ Acyc) If $\phi_{i+1} \in K \dot{-} (\phi_i \wedge \phi_{i+1})$ and $\nvdash \phi_i \wedge \phi_{i+1}$ for all $i = 1, \ldots, n-1$, then not both $\phi_1 \in K \dot{-} (\phi_n \wedge \phi_1)$ and $\nvdash \phi_n \wedge \phi_1$.

(ii) Let $\dot{-}$ satisfy (K$\dot{-}$ 1)–(K$\dot{-}$ 6), (K$\dot{-}$ 7) and (K$\dot{-}$ 8c). Then it also satisfies

(K$\dot{-}$ Trans) If $\psi \in K \dot{-} (\phi \wedge \psi)$ and $\nvdash \phi \wedge \psi$ and $\chi \in K \dot{-} (\psi \wedge \chi)$ and $\nvdash \psi \wedge \chi$ then $\chi \in K \dot{-} (\phi \wedge \chi)$ and $\nvdash \phi \wedge \chi$.

It is a plausible conjecture that all safe contractions satisfy (K$\dot{-}$ Acyc), but we have no proof for this claim.[12]

3.2 Every Safe Contraction is a Partial Meet Contraction

The following fundamental result was reported already in the paper that introduced safe contraction:

Theorem 4. *(Alchourrón and Makinson 1985, Observation 3.2) Any safe contraction function over a belief set K satisfies (K$\dot{-}$ 1) – (K$\dot{-}$ 6).*

Since Alchourrón, Gärdenfors and Makinson (1985) showed that (K$\dot{-}$ 1) – (K$\dot{-}$ 6) exactly characterize the partial meet contractions, this immediately implies the following result:

---

12 Makinson (1989, p. 16) finitistic "loop condition" for non-monotonic reasoning is similar in spirit to (K$\dot{-}$ Trans) and (K$\dot{-}$ Acyc). Translated into the language of contractions, it reads thus: If for all $i \in \{i_1, \ldots i_{n-1}\}$, $\phi_{i+1} \to \phi_i \in K \dot{-} \phi_i$ and furthermore $\phi_1 \to \phi_n \in K \dot{-} \phi_n$, then $\phi_n \to \phi_1 \in K \dot{-} \phi_1$.

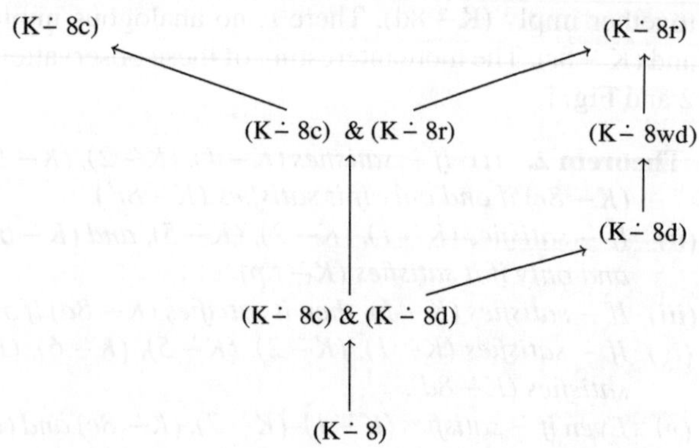

Fig. 1 A Hasse diagram for combinations of weakenings of (K∸ 8) over a belief set K, given contraction postulates (K∸ 1) – (K∸ 7).

Theorem 5. *Any safe contraction function over a belief set K can be represented as a partial meet contraction function.*

The converse of Theorem 5, however, does not hold.

Theorem 6. *It is not the case that all partial meet contractions are safe contractions.*

The idea behind the counterexample used in the proof of Theorem 6 is simple[13]: There are basic AGM contraction functions that intuitively involve cycles, but cycles are not permitted in the hierarchies used in the construction of safe contractions.

At the time of writing, we do not know an axiomatic characterization of safe contraction. Neither do we know an axiomatic characterization of those safe contractions that are based on transitive relations. However, axiomatic characterizations of other interesting classes of safe contraction can be obtained.

3.3 Properties Induced by Logic-Respecting Hierarchies

Reflecting on the meaning of hierarchies as specifying comparative vulnerability or exposure, it is natural to assume that logically stronger sentences cannot be less vulnerable or exposed than logically weaker sentences. This very plausible intuition is captured in the conditions Continuing up and Continuing down to which we will now turn. Fundamental results about both of them have been reported by Alchourrón and Makinson:

Theorem 7. *Any safe contraction over a belief set K generated by a hierarchy $\prec$ that continues either up or down $\vdash$ over K satisfies* (K∸ 7).

This two-pronged theorem was first presented and proved in Observations 4.3 and 5.3 of Alchourrón, Gärdenfors and Makinson (1985). Although the properties of

[13] For a similar example, compare Rott (2003, p. 268, see Footnote 20).

Continuing up and down appear to be similar, the respective proof methods for the two parts of this theorem are remarkably different. We will meet different consequences of Continuing up and Continuing down below.

The part of the following theorem that concerns (K$\dot{-}$8r) is new. For the case of regular safe contraction functions over a logically finite theory K, i.e., a theory that has only finitely many elements modulo Cn, it is an immediate consequence of Theorem 1 of Alchourrón and Makinson (1986, p. 193), taken together with an observation of Rott (1993, pp. 1440–1442, Theorem 2 and Corollary 2). We prove a somewhat more general result here:

Theorem 8. *Every safe contraction $\dot{-}$ over a belief set K generated by a hierarchy $\prec$ that continues down $\vdash$ over K satisfies (K$\dot{-}$ 7) and (K$\dot{-}$ 8r).*

As already pointed out, (K$\dot{-}$ 8r) is a weakening of the AGM postulate (K$\dot{-}$ 8). Safe contractions based on transitive hierarchies that continue up $\vdash$ over K satisfy (K$\dot{-}$ 8c), which is another weakening of (K$\dot{-}$ 8).

Theorem 9. *Every safe contraction $\dot{-}$ over a belief set K generated by a transitive hierarchy $\prec$ that continues up $\vdash$ over K satisfies (K$\dot{-}$ 7) and (K$\dot{-}$ 8c).*

3.4 Representation Theorems

The following recipe for retrieving a hierarchy from a safe contraction can be used to prove representation theorems for various classes of safe contractions.

Definition 4. *Assuming that a contraction function $\dot{-}$ over K is given, we define a hierarchy $\prec$ over K from it in the following way:*

(CH) $\quad \phi \prec \psi$ *iff* $\psi \in K \dot{-} (\phi \wedge \psi)$ *and* $\nvdash \phi$

The hierarchy obtained by (CH) has a number of interesting properties:

Lemma 10. *Let the contraction function $\dot{-}$ satisfy (K$\dot{-}$ 1)–(K$\dot{-}$ 6), and let $\prec$ be defined from $\dot{-}$ with the help of (CH). Then*

(i) $\prec$ *satisfies Intersubstitutivity, Irreflexivity and*
$\phi \prec \psi \wedge \chi$ *iff* $(\phi \wedge \psi \prec \chi$ *and* $\phi \wedge \chi \prec \psi)$ *(Choice)*
(ii) *if* $\dot{-}$ *satisfies (K$\dot{-}$ 7), then* $\prec$ *satisfies Continuing down*
(iii) *if* $\dot{-}$ *satisfies (K$\dot{-}$ 8c), then* $\prec$ *satisfies Continuing up*
(iv) *if* $\dot{-}$ *satisfies (K$\dot{-}$ 7) and (K$\dot{-}$ 8c), then* $\prec$ *is transitive*
(v) *if* $\dot{-}$ *satisfies (K$\dot{-}$ Trans), then* $\prec$ *is transitive*
(vi) *if* $\dot{-}$ *satisfies (K$\dot{-}$ 8r), then* $\prec$ *satisfies*
If $\phi \wedge \psi \prec \chi$ *and* $\nvdash \chi$*, then there are* ξ *and* ρ *such that* $\chi \dashv\vdash \xi \wedge \rho$
and $\phi \wedge \chi \prec \xi$ *and* $\psi \wedge \chi \prec \rho$ *(EII$^-$)*[14]

[14] If $\prec$ satisfies Intersubstitutivity and Choice, then (EII$^-$) is equivalent with the following property: If $\phi \wedge \psi \prec \chi$ and $\nvdash \chi$, then there are ξ and ρ such that $\chi \dashv\vdash \xi \wedge \rho$ and $\phi \wedge \rho \prec \xi$ and $\psi \wedge \xi \prec \rho$.

(vii) *if* $\dot{-}$ *satisfies (K$\dot{-}$ 8wd), then* $\prec$ *satisfies*

If $\phi \wedge \psi \prec \chi$, *then either* $\phi \prec \chi$ *or* $\psi \prec \chi$ *(EII)*

(viii) *if* $\dot{-}$ *satisfies (K$\dot{-}$ 8d), then* $\prec$ *satisfies*

If $\phi \wedge \psi \prec \chi \wedge \xi$ *and* $\nvdash \chi$ *and* $\nvdash \xi$, *then either* $\phi \prec \chi$ *or* $\psi \prec \xi$ *(EII$^+$)*

(ix) *if* $\dot{-}$ *satisfies (K$\dot{-}$ 8), then* $\prec$ *satisfies*

If $\phi \wedge \psi \prec \chi$, *then either* $\phi \prec \chi$ *or* $\psi \prec \phi \wedge \chi$

(K$\dot{-}$ 8) does not exactly give Virtual connectivity, but rather the condition mentioned in part (ix) of Lemma 10. Taken together with Intersubstitutivity, Irreflexivity, Choice and Continuing down, however, this condition implies Virtual connectivity.[15] As will become clear in Sect. 4, the main message of this Lemma can be seen in the fact that the hierarchies obtained from (CH) are epistemic entrenchment relations (cf. Definition 5 below).

We give the following representation result in two versions, one for the general case and one for the case of a logically finite belief set K.

Theorem 11. *(i) Every contraction* $\dot{-}$ *over a belief set K satisfying the AGM postulates (K$\dot{-}$ 1) – (K$\dot{-}$ 7) and (K$\dot{-}$ 8d) can be represented as a safe contraction function generated by a hierarchy* $\prec$ *that continues down* $\vdash$ *over K. If (K$\dot{-}$ 8c) is satisfied as well, then* $\prec$ *is regular and transitive.*

(ii) Every contraction $\dot{-}$ *over a logically finite belief set K satisfying the AGM postulates (K$\dot{-}$ 1) – (K$\dot{-}$ 7), as well as* either *(K$\dot{-}$ 8d)* or *both (K$\dot{-}$ 8c) and (K$\dot{-}$ 8r), can be represented as a safe contraction function generated by a hierarchy* $\prec$ *that continues down* $\vdash$ *over K. If (K$\dot{-}$ 8c) is satisfied as well, then* $\prec$ *is regular and transitive.*

Comment on the general case of Theorem 11: Besides the postulates (K$\dot{-}$ 1) – (K$\dot{-}$ 7), we have used (K$\dot{-}$ 8wd) for one direction of the proof, and either (K$\dot{-}$ 8d) or (K$\dot{-}$ 8c) for the other direction. Notice that we know from Theorem 2 (iii) that the weaker (K$\dot{-}$ 8wd) implies the stronger (K$\dot{-}$ 8d), if (K$\dot{-}$ 8c) is present as well. So we have found no way of avoiding the use of (K$\dot{-}$ 8d), This is why we have formulated part (i) of Theorem 11 in a non-disjunctive way. If (K$\dot{-}$ 8c) is present, this just guarantees the regularity and transitivity of the hierarchy used, by Lemma 10 (iii) and (iv). But it is not necessary for the representation itself if we have (K$\dot{-}$ 8d) anyway.

Comment on the finite case of Theorem 11: There are two alternative formulations of part (ii) of Theorem 11. This is because in the proof of this part, we have used (K$\dot{-}$ 8r) for one direction, and either (K$\dot{-}$ 8d) or (K$\dot{-}$ 8c) for the other direction. But (K$\dot{-}$ 8r) does *not* imply the stronger postulate (K$\dot{-}$ 8d), even when (K$\dot{-}$ 1) – (K$\dot{-}$ 7) and (K$\dot{-}$ 8c) are present as well. This we know from Theorem 2 (iv). So this representation theorem for the finite case goes through if *either* (K$\dot{-}$ 8d) *or* both (K$\dot{-}$ 8c) and (K$\dot{-}$ 8r) are given.

(Footnote 14 continued)
Notice that given (CH), this condition, as well as (EII$^-$) itself, is a transcription of the contraction postulate (K$\dot{-}$ 8r$'$) into the language of entrenchments.

[15] See Rott (2001), Lemma 52 (viii).

However, this result for the finite case can further be improved. While (CH) has the advantages of being simple and being applicable to the case of a belief set that is infinite modulo Cn, it has, as far as we can see, the disadvantage of requiring ($K\dot{-}$ 8c). But it can be shown that ($K\dot{-}$ 8c) is not necessary after all. In order to see this, we plug together two earlier observations. First, there is a central result of Alchourrón and Makinson's that we have already appealed to:

Theorem 12. *(Alchourrón and Makinson 1986, Theorem 1) Let K be a logically finite belief set. A contraction function $\dot{-}$ over K is a safe contraction function generated by a regular hierarchy $\prec$ iff it is a relational partial meet contraction (based on a strict preference relation over the maximal consistent subsets of K).*[16]

In proving this theorem, Alchourrón and Makinson use the following somewhat more complex strategy. First they retrieve from the contraction function $\dot{-}$ an intersubstitutive and acyclic[17] relation $\lhd$ over the co-atoms (the weakest non-tautologous elements) of the belief set K. Then, and this is their crucial idea, they extend $\lhd$ to a hierarchy $\prec$ over the whole of K by putting

($\exists\forall$) $\phi \prec \psi$ iff there is some co-atom χ of K with $\phi \vdash \chi$ such that for all co-atoms ξ of K with $\psi \vdash \xi$, $\chi \lhd \xi$

As Alchourrón and Makinson point out, this construction (their function f_4) extends $\lhd$, preserves Intersubstitutivity and Acyclicity, and the hierarchy $\prec$ so obtained (trivially) satisfies Continuing up and down. It is easy to check that this construction also preserves Transitivity: If $\lhd$ is transitive over the co-atoms of K, so is $\prec$ over K (this will be needed for part (ii) of Theorem 15). It is equally easy to see that $\prec$ satisfies (EII). Having noted the latter, we can immediately supplement Theorem 12 by the following

Theorem 13. *Let K be a logically finite belief set. A contraction function $\dot{-}$ over K is a safe contraction function generated by a regular hierarchy $\prec$ satisfying (EII) iff it is a relational partial meet contraction (based on a strict preference relation over the maximal consistent subsets of K).*

The following variant of Alchourrón and Makinson's construction ($\exists\forall$) can serve the same purpose, but has some different properties:

($\forall\exists$) $\phi \prec \psi$ iff for all co-atoms ξ of K with $\psi \vdash \xi$ there is a co-atom χ of K with $\phi \vdash \chi$ such that $\chi \lhd \xi$

Like ($\exists\forall$), this alternative construction extends $\lhd$, preserves Intersubstitutivity, Acyclicity and Transitivity, and $\prec$ so obtained (trivially) satisfies Continuing up and down. The hierarchy $\prec$ does not necessarily satisfy (EII) (for this $\lhd$ would have to be virtually connected), but it does satisfy Choice provided that $\lhd$ is transitive. In

16 For the formulation in terms of *strict* preference relations, compare Alchourrón and Makinson (1986, Sect. 2.1).

17 The acyclicity of relations involved in relational partial meet contractions is guaranteed by the requirement that choice functions select a non-empty choice set whenever the menu is non-empty.

contrast, (∃∀) leads to a hierarchy satisfying Choice only if $\lhd$ is virtually connected. Transitivity is a weaker and more natural requirement than Virtual connectivity. For this reason, the decision whether to prefer (∃∀) or (∀∃) will depend on whether one prefers to have $\prec$ satisfy (EII) or Choice.

Having Alchourrón and Makinson's result in place, we make use of the fact that there is a characterization of (transitively) relational partial meet contractions over a finite belief set K:

Theorem 14. *(Rott 1993, Corollary 2) Let K be a logically finite belief set. Then*

(i) A contraction function $\dot{-}$ over K is a partial meet contraction based on a strict preference relation over the maximal consistent subsets of K iff it satisfies (K$\dot{-}$ 1) – (K$\dot{-}$ 7) and (K$\dot{-}$ 8r).

(ii) A contraction function $\dot{-}$ over K is a partial meet contraction based on a transitive strict preference relation over the maximal consistent subsets of K iff it satisfies (K$\dot{-}$ 1) – (K$\dot{-}$ 7), (K$\dot{-}$ 8c) and (K$\dot{-}$ 8r).[18]

Putting Theorems 12 and 14 together, we obtain

Theorem 15. *Let K be a logically finite belief set. Then*

(i) Every contraction function $\dot{-}$ over K satisfying the AGM postulates (K$\dot{-}$ 1) – (K$\dot{-}$ 7) and (K$\dot{-}$ 8r) can be represented as a safe contraction function generated by a regular hierarchy $\prec$ over K.

(ii) Every contraction function $\dot{-}$ over K satisfying the AGM postulates (K$\dot{-}$ 1) – (K$\dot{-}$ 7), (K$\dot{-}$ 8c) and (K$\dot{-}$ 8r) can be represented as a safe contraction function generated by a transitive regular hierarchy $\prec$ over K.

This theorem can be proven directly by applying (CH) only to the co-atoms of the logically finite belief set K and subsequently applying Alchourrón and Makinson's transitivity-preserving construction (∃∀). The construction produces a hierarchy that is in general different from the one obtained by applying (CH) over the whole of K. For instance, while (∃∀) gives Continuing up for $\prec$ for free, (CH) needs (K$\dot{-}$ 8c) to validate Continuing up for $\prec$.

What have we achieved so far? We have identified the logic of safe contractions based on regular hierarchies in the finite case. It consists of (K$\dot{-}$ 1) – (K$\dot{-}$ 7) and (K$\dot{-}$ 8r). But this is not very surprising, given the results that have been available for a long time in Alchourrón and Makinson (1986) and Rott (1993). We have also identified the logic of safe contractions based on transitive regular hierarchies in the finite case which in addition has (K$\dot{-}$ 8c). All this follows from the results of Sect. 3.3 and Theorems 11 and 15.

Unfortunately, we have not been able to identify the logic of safe contractions based on regular hierarchies for infinite belief sets K. We know that safe contractions based on regular hierarchies satisfy (K$\dot{-}$ 1) – (K$\dot{-}$ 7) and (K$\dot{-}$ 8r), and if the hierarchy is in addition transitive, we also get (K$\dot{-}$ 8c) (Theorems 8 and 9). We

[18] For the formulation in terms of *strict* preference relations, please compare Rott (1993, p. 1430, and in particular Footnote 1).

know that contraction functions satisfying $(K\dot{-}1)-(K\dot{-}7)$ and $(K\dot{-}8d)$ can be represented as safe contractions based on regular hierarchies, and if they in addition satisfy $(K\dot{-}8c)$, the hierarchies are also transitive (Theorem 11 (i)). But there remains the gap between the weaker condition $(K\dot{-}8r)$ and the stronger condition $(K\dot{-}8d)$. And this gap—indeed, already the gap between $(K\dot{-}8r)$ and $(K\dot{-}8wd)$—is wide, even in the finite case, as our example for Theorem 2 (v) makes clear. The same example teaches us that the gap could only be filled from the side of the representation theorem, since we know that safe contractions based on transitive regular hierarchies need not satisfy $(K\dot{-}wd)$, even in the logically finite case. But we do not know how to fill this gap.

It remains to have a look at what happens if hierarchies satisfy the very strong condition Virtual connectivity.

Theorem 16. *(Alchourrón and Makinson 1985, Observation 6.2) Every safe contraction over a belief set K generated by a hierarchy $\prec$ that is virtually connected and continues up or down $\vdash$ over K satisfies $(K\dot{-}8)$.*

The following representation theorem is due to Alchourrón and Makinson (1986, Theorem 2) for the finite case. It was generalized to the infinite case in Rott (1992a, Corollary to Theorem 4):

Theorem 17. *Every contraction $\dot{-}$ over a belief set K satisfying the AGM postulates $(K\dot{-}1)-(K\dot{-}8)$ can be represented as a safe contraction function generated by a regular and virtually connected hierarchy $\prec$ over K.*

3.5 Overview of the Connections

Based on the the above-mentioned results, the relationships between safe contraction and some other major forms of contraction on belief sets are depicted in Fig. 2. Information about axiomatic characterizations is summarized in Table 1. In view of our decision to focus on relations of the "less-than" type, we refer to the converse of complements of Alchourrón, Gärdenfors and Makinson's non-strict relations and consequently rename their "transitively relational partial meet contraction functions" into "virtually connected relational partial meet contractions".[19]

4 Every Safe Contraction is an Entrenchment-Based Contraction

The concept of entrenchment was developed by Gärdenfors (1988) and Gärdenfors and Makinson (1988). Like hierarchies, entrenchment relations are relations between

[19] Compare Footnote 6 and the surrounding text.

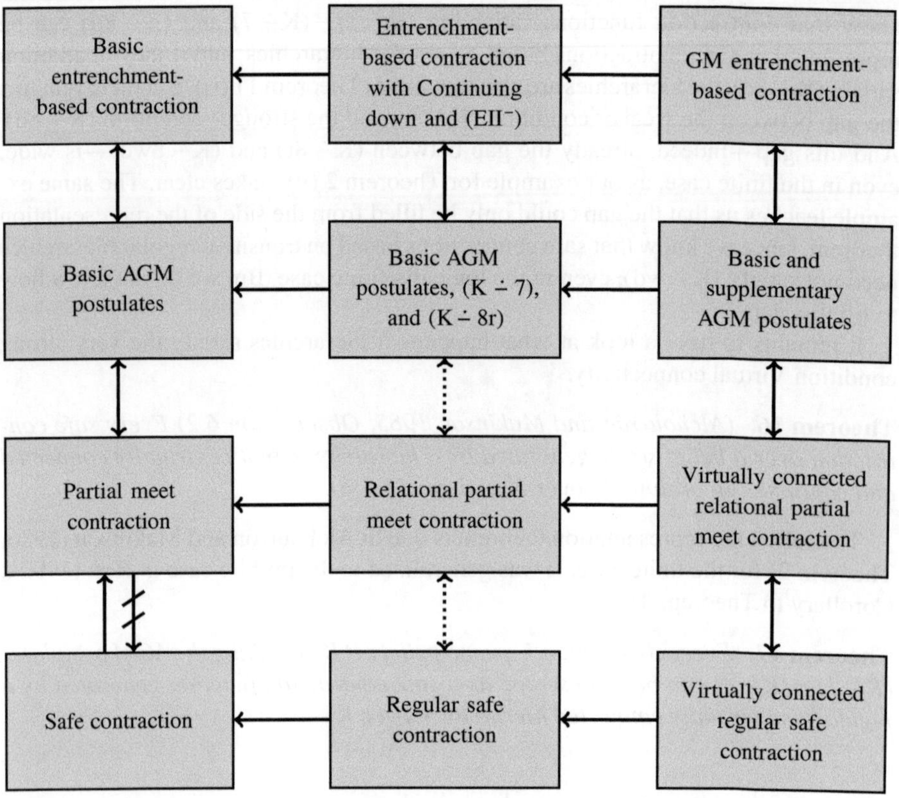

Fig. 2 Logical relationships between some major operations of contraction on belief sets. The dotted arrows represent equivalences that hold for logically finite belief sets. It is not known what relations hold here in the general case. (The proofs of most of these logical relationships are given in Sect. 3. The references for the relations involving the middle box in the top row are essentially Lemma 10, Parts (ii) and (vi), Rott (2001, Observation 68, line 4), Theorem 2 (i) and footnote 14 above.)

sentences. Entrenchment is an interesting concept of doxastic preference which allows for several quite diverse interpretations. Suppose, for instance, that a belief set K is generated from a finite and weakly ordered *belief base* H, i.e., $K = \mathrm{Cn}(H)$, and that there is an asymmetric and virtually connected (and, hence, transitive) *base hierarchy* $\lhd$ over H specifying the strength of the basic beliefs. Then an element ϕ of K can be called more entrenched than another element ψ of K if for every subset G of H implying ψ there is a subset G' of H implying ϕ that has only elements that are $\lhd$-stronger than the $\lhd$-weakest element in G. This is a *positive* concept of entrenchment, saying intuitively that there is a way of deriving ϕ that is 'better' (has a stronger base) than any way of deriving ψ. But an alternative idea makes equally good sense: An element ϕ of K can be called more entrenched than another element ψ of K if for every subset G of H not implying ϕ there is another subset G' of H not

Table 1 Some major operations of contraction on belief sets that are mentioned in this paper, with sources for their definitions and axiomatic characterizations

Type of operation	Defined	Axiomatically characterized
Partial meet contraction	Alchourrón et al. (1985)	Alchourrón et al. (1985)
Relational partial meet contraction	Alchourrón et al. (1985)	Rott (1993) (under some restrictions)
Transitively relational partial meet contraction	Alchourrón et al. (1985)	Rott (1993) (under some restrictions)
Virtually connected relational partial meet contraction	Alchourrón et al. (1985)	Alchourrón et al. (1985)
Gärdenfors-Makinson entrenchment-based contraction	Gärdenfors (1988), Gärdenfors and Makinson (1988)	Gärdenfors (1988), Gärdenfors and Makinson (1988)
Basic entrenchment-based contraction	Rott (2003)	Rott (2003)
Safe contraction	Alchourrón and Makinson (1985)	Unsolved problem
Regular safe contraction	Alchourrón and Makinson (1985)	This paper (the finite case); unsolved problem (the general case)
Transitive regular safe contraction	This paper	This paper (the finite case); unsolved problem (the general case)
Virtually connected regular safe contraction	Alchourrón and Makinson (1985)	Alchourrón and Makinson (1986) (the finite case); Rott (1992a) (the general case)
Kernel contraction	Hansson (1994)	Hansson (1994)
Base-generated kernel contraction	Hansson (1994)	Hansson (1994)

implying ψ which is 'better' than G in the following sense: There is an element of H that is lost in G but not in G' and is $\lhd$-stronger than any element of H that is lost in G' but not in G. This is a *negative* concept of entrenchment, saying intuitively that there are 'better' (less costly) ways of getting rid of ψ than there are ways of getting rid of ϕ.[20] Yet another, much more general but less constructive way is to interpret entrenchment as a *revealed preference relation* that can be reconstructed from binary choices between beliefs: ϕ is more entrenched in K than ψ if the contraction of K with respect to $\phi \wedge \psi$ (which comes down to the contraction of K with respect to at least one of ϕ and ψ) contains ϕ but does not contain ψ. A bit more precisely and in symbols:

---

[20] It can be shown that negative entrenchment is a refinement of positive entrenchment. Negative entrenchment introduces new distinctions within levels of positive entrenchment. While all beliefs are comparable in terms of the less discriminating positive entrenchment relation, negative entrenchment introduces incomparabilities. For this, see Rott (2000).

$$\text{(CE) } \phi < \psi \text{ iff } \psi \in K \dot{-} \phi \wedge \psi \text{ and } \nvdash \phi.$$

The fact that (CE) coincides with the above-mentioned recipe (CH) for retrieving a hierarchy from a contraction function indicates that there is a close relationship between hierarchies and entrenchment relations. It was proposed in Rott (1992a) that entrenchment relations are in fact particularly well-behaved hierarchies. (Cf. also Rott (2001), in particular pp. 229, 263.)

Clearly, the use of (CE) is re-constructive rather than constructive, since it presumes that a method of belief contraction is already given. But it captures the intuitive meaning of the term 'entrenchment' very well, and it does not assume that the belief set K is generated from a particular (prioritized) belief base.

It is this last concept of entrenchment that can be utilized in the context of safe contractions. Our aim is to transform an arbitrary hierarchy $\prec$ over a belief set K into an entrenchment relation $<$ over the whole of $\mathcal{L}$ such that this entrenchment relation, applied in the way entrenchment relations are applied for belief contraction, leads to the same result as the original safe contraction based on $\prec$.

For that purpose, we need to introduce the notion of epistemic entrenchment in a more precise way. We distinguish between three versions of this notion with increasing logical strength, introduced and motivated in Rott (2003, 1992b) and Gärdenfors and Makinson (1988), respectively.

Definition 5. *(a) A* basic entrenchment relation $<$ *satisfies the conditions*

(E0) If $\phi \dashv\vdash \phi'$ *and* $\psi \dashv\vdash \psi'$, *then* $\phi < \psi$ *if and only if* $\phi' < \psi'$ *(Intersubstitutivity)*

(E1) Not $\phi < \phi$ *(Irreflexivity)*

(E2) $\phi < \psi \wedge \chi$ *iff* $(\phi \wedge \psi < \chi$ *and* $\phi \wedge \chi < \psi)$ *(Choice)*

(E3) If $\nvdash \phi$, *then* $\phi < \psi$ *for some* ψ *(Maximality)*

(b) A generalized entrenchment relation $<$ *satisfies Irreflexivity, Choice, Maximality, as well as Continuing up and Continuing down for* $<$.[21]

(c) A GM entrenchment relation $<$ *satisfies Irreflexivity, Choice, Maximality, Continuing up and Continuing down as well as Virtual connectivity for* $<$ *and*

(E4) If K is consistent, then ϕ *is in K if and only if* $\psi < \phi$ *for some* ψ *(Minimality).*

Evidently, all GM entrenchment relations are generalized entrenchment relations, and all generalized entrenchment relations are basic entrenchment relations. Basic entrenchment relations need not be acyclic, but generalized entrenchment relations are (Rott 2003).

Lemma 18. *(i) Intersubstitutivity, Irreflexivity and Choice jointly imply Asymmetry, i.e., if* $\phi < \psi$ *then not* $\psi < \phi$.

(ii) Intersubstitutivity, the right-to-left direction of Choice and Asymmetry jointly entail

If $\phi \vdash \psi$, *then not* $\psi < \phi$ *(GM-dominance)*

[21] Recall that Continuing up and Continuing down jointly imply Intersubstitutivity.

and
Not both $\phi \wedge \psi < \phi$ *and* $\phi \wedge \psi < \psi$ *(GM-conjunctiveness)*

(iii) Intersubstitutivity and the right-to-left direction of Choice entail
If $\phi \wedge \psi < \psi$, *then* $\phi < \psi$ *(Conjunction down)*

(iv) Continuing down and the right-to-left direction of Choice entail
If $\phi < \psi$ *and* $\phi < \chi$, *then* $\phi < \psi \wedge \chi$ *(Conjunction up)*

(v) Continuing up and Continuing down entail the left-to-right direction of Choice

(vi) Continuing down, Conjunction up and Conjunction down entail the right-to-left direction of Choice.

(vii) Irreflexivity, the right-to-left direction of Choice, Continuing up and Continuing down taken together entail that $<$ *is transitive and thus also acyclic.*

(viii) The right-to-left direction of Choice and Continuing down entail the following condition:

If $\{\alpha_1, \ldots, \alpha_n\}$ *and* $\{\beta_1, \ldots, \beta_m\}$ *are finite sets of co-atoms of* K, *then:*

$$\bigwedge \alpha_i < \bigwedge \beta_j \text{ if for all } \beta_j \text{ there is an } \alpha_i \text{ such that } \alpha_i < \beta_j.$$

Intersubstitutivity, the right-to-left direction of Choice, Continuing up and (EII$^-$*) entail its converse:*

If $\{\alpha_1, \ldots, \alpha_n\}$ *and* $\{\beta_1, \ldots, \beta_m\}$ *are finite sets of co-atoms of* K, *then:*

$$\bigwedge \alpha_i < \bigwedge \beta_j \text{ only if for all } \beta_j \text{ there is an } \alpha_i \text{ such that } \alpha_i < \beta_j.$$

Part (viii) of Lemma 18 vindicates the (∀∃) construction of the previous section whenever the properties mentioned are satisfied. We consider the properties of the first part of (viii) to be much less demanding than those mentioned in the second part.

Having defined some important notions of entrenchment, we can now return to the question of how to obtain entrenchment relations from hierarchies. Feeding in the construction recipe of safe contractions into (CE), we get

(i) $\phi < \psi$ iff there is some $M \subseteq K$ with $M \vdash \psi$ such that for all $N \subseteq K$ with $N \vdash_{min} \phi \wedge \psi$, N is non-empty and for every χ in $M \cap N$ there is a ξ in N such that $\xi \prec \chi$.

Rott (1992a) used the following much simpler idea:

(ii) $\phi < \psi$ iff there is some $M \subseteq K$ with $M \vdash \psi$ such that for all $N \subseteq K$ with $N \vdash \phi$, N is non-empty and for every χ in M there is a ξ in N such that $\xi \prec \chi$.

Condition (ii) is similar to the positive interpretation of entrenchment generated by belief bases, but it does not depend on $\prec$ being virtually connected.

Lemma 19. *Conditions (i) and (ii) are equivalent if* $\prec$ *is transitive and continues down* $\vdash$ *over* K.

Theorem 20. *(Rott 1992a) Let K be a belief set and $\prec$ a hierarchy that continues down $\vdash$ over K. Let $<$ be the entrenchment relation constructed from $\prec$ according to (ii). Then the safe contraction based on $\prec$ is identical with the entrenchment-based contraction based on $<$ which is defined by*

$$\text{(EC)} \quad K \dot{-} \phi = \begin{cases} \{\psi \in K : \phi < (\phi \vee \psi)\} & \text{if } \nvdash \phi \\ K \dot{-} \phi = K & \text{otherwise} \end{cases}$$

The construction of *entrenchment-based contractions* by means of (EC) has been standard since Gärdenfors (1988) and Gärdenfors and Makinson (1988). It is remarkable, however, that it can be applied even in contexts in which the entrenchment relation satisfies only much weaker conditions than those envisaged by Gärdenfors and Makinson. Theorem 20 implies that any safe contraction function based on a hierarchy $\prec$ that continues down $\vdash$ over K can be represented as an entrenchment-based contraction function. As it turns out, this result can indeed further be improved, effectively by using (i) rather than (ii).

Theorem 21. *Any safe contraction function over a belief set K can be represented as an entrenchment-based contraction function.*

Since basic entrenchment relations need not in general be acyclic, they cannot always be applied as hierarchies for use in safe contractions. But the following question is interesting. Under what conditions can an entrenchment relation be used in a safe contraction, that is, as specified in Definition 2, and gives the same result as when used in (EC)? (Rott 1992a, Theorem 4 (ii)) showed that GM-entrenchment relations are suitable, but this result can be sharpened.

Theorem 22. *Let $<$ be a generalized entrenchment relation that satisfies*

(EII) *If $\phi \wedge \psi < \chi$, then either $\phi < \chi$ or $\psi < \chi$*

Then $<$ can be used as a hierarchy in safe contraction and as an entrenchment relation in an entrenchment-based contraction with the same result.

Theorem 22 shows that it is not necessary to have a GM-entrenchment in order to get the same result in safe contraction as in entrenchment-based contraction. A generalized entrenchment relation is sufficient, provided that it also satisfies (EII). (EII) is weaker than Virtual connectivity.[22] It was shown in Rott (2001) that it corresponds to the contraction postulate (K$\dot{-}$ 8wd).

[22] Suppose that Virtual connectivity is given and (EII) is violated. Concerning the latter, suppose that $\phi \wedge \psi < \chi$ and $\phi \not< \chi$ and $\psi \not< \chi$. Then by Virtual connectivity, $\phi \wedge \psi < \phi$ and $\phi \wedge \psi < \psi$. Then by the right-to-left direction of Choice, $\psi < \phi$ and $\phi < \psi$, contradicting Asymmetry.

5 Iterated Safe Contraction

As we noted in Sect. 1, operators of partial meet contraction, as defined by Alchourrón, Gärdenfors, and Makinson have a very limited range of application: each such operator can only be applied to one belief set. This makes iterated contraction, such as $K \dot{-} \phi_1 \dot{-} \phi_2 \ldots \dot{-} \phi_n$, impossible except in the trivial case when $K \dot{-} \phi_1 = K$ etc. The reason for this is that in order to obtain $K \dot{-} \top = K$ we need to have $\gamma(K \bot \top) = \{K\}$, i.e., $\gamma(\varnothing) = \{K\}$. In order to obtain $K' \dot{-} \top = K'$ for some $K' \neq K$ we similarly need a selection function γ' such that $\gamma'(\varnothing) = \{K'\}$. It follows that $\gamma \neq \gamma'$. For the same reason, iterated revision such as $K * \phi_1 * \phi_2 \cdots * \phi_n$, is impossible except in the trivial case when $K * \phi_1 = K$ etc. This limitation in partial meet contraction, as conventionally defined, has been a major driving force behind the extensive discussion in the belief revision literature on the construction of contraction and revision operators that can be used repeatedly. (For a brief summary of this literature, see Fermé and Hansson (2011, pp. 307–309)..)

For safe contraction this is different. Let $\prec$ be a hierarchy that is defined for the whole language ($\mathcal{L}$). Then $\prec$ is formally viable for use to contract any set of sentences. This formal advantage of safe contraction was noted already by Alchourrón and Makinson (1985).

However, this difference does not seem to reflect a fundamental difference between the two operations. As was noted for instance in Hansson (1992), it is fairly easy to redefine partial meet contraction so that it becomes globally applicable. All we need to do is to treat the limiting case of contraction by tautologies separately, without reference to the selection function.[23] This gives rise to the following definition:

Definition 6. *(Hansson 1992) A global partial meet contraction is an operation* $\dot{-}$ *that is based on a function* Γ *that assigns a selection function* $\Gamma(K) = \gamma_K$ *to each logically closed subset of* $\mathcal{L}$*, such that* $K \dot{-} \phi = \bigcap \gamma_K(K \bot \phi)$ *whenever* $K \bot \phi \neq \varnothing$*, and otherwise* $K \dot{-} \phi = K$.

A global partial meet contraction is unified if and only if it is based on a function Γ *such that* $\Gamma(K) = \Gamma(K')$ *for all belief sets* K *and* K'.

Expressed more simply, a unified global partial meet contraction is an operation $\dot{-}$ that is based on a global selection function γ, such that $K \dot{-} \phi = \bigcap \gamma(K \bot \phi)$ whenever $K \bot \phi \neq \varnothing$, and otherwise $K \dot{-} \phi = K$.

---

23 Another way to achieve a similar result is to apply partial meet contraction primarily to the inconsistent belief set, i.e., the whole language, and then define contraction on consistent sets as restrictions in the following way:

$$K \dot{-} \phi = \begin{cases} K \cap (\mathcal{L} \dot{-} \phi) & \text{if} \phi \in K \\ K & \text{otherwise} \end{cases}$$

This idea was briefly discussed for entrenchment-based contractions in Rott (1992b, Sect. 7) and more thoroughly analyzed for all AGM contraction models by Areces and Becher (2001).

However, perhaps surprisingly, the property of being unified is an empty condition for global partial meet contraction[24]:

Theorem 23. *(Hansson 2012) Let $\mathcal{L}$ be infinite and let $\dot{-}$ be a global operator on the belief sets in $\mathcal{L}$. Then $\dot{-}$ is a global partial meet contraction if and only if it is a unified global partial meet contraction.*

Analogously, we can distinguish between unified and non-unified safe contraction:

Definition 7. *The operator $\dot{-}$ on belief sets is a global safe contraction if and only if there is a function $\mathfrak{S}$ on beliefs sets such that for each belief set K, $\mathfrak{S}(K)$ is a hierarchy on K and for all belief sets K and sentences ϕ:*

$$K \dot{-} \phi = K \dot{-}_{\mathfrak{S}(K)} \phi$$

where $\dot{-}_{\mathfrak{S}(K)}$ is the safe contraction based on $\mathfrak{S}(K)$.

A global safe contraction is unified if and only if it is based on a function $\mathfrak{S}$ such that $\mathfrak{S}(K) = \mathfrak{S}(K')$ for all belief sets K and K'.

It is an open question whether or not global safe contraction and unified global safe contraction coincide. (It was shown in Hansson (2012, p. 153) that the corresponding kernel contractions coincide.) An axiomatic characterization of unified safe contraction is also lacking.

Obviously, the logical convenience of unified global safe contraction does not prove it to be plausible. There are indeed strong reasons to believe that it is not. The unified construction presupposes that the comparative retractability of sentences is left unchanged in all contractions. However, it is easy to find examples showing for instance that contraction by a sentence can weaken the agent's belief in some other sentence without leading her to remove it (Rott 1999, pp. 393–394, 2001, pp. 95–97, 219–222). It appears to be more reasonable to see unified global safe contraction as a limiting case, rather than as a fully realistic account of iterated contraction. Such a limiting case is important not least since logical principles that do not hold in the limiting case should expectedly also be invalid in the general case (Hansson 2012).

6 The Choice-Functional Generalization

One of the most conspicuous differences between safe contraction and partial meet contraction is that the former is defined as relational, whereas the latter is defined primarily as non-relational. But this is a non-essential difference since the basic idea of safe contraction can be reconstructed in a more general, choice-functional fashion.

---

[24] This applies to belief sets. As was shown in Hansson (1993), for belief bases it is a non-empty condition. A global partial meet contraction on belief bases is unified if and only if it satisfies:

If $\phi \notin \mathrm{Cn}(\varnothing)$ and each element of Z implies ϕ, then $B \dot{-} \phi = (B \cup Z) \dot{-} \phi$ (Redundancy).

This was done in Hansson (1994) under the name of kernel contraction. It is based on a choice mechanism that operates on kernel sets and—just like a hierarchy—removes at least one element from each ϕ-kernel, where ϕ is the sentence to be removed. Since such a function makes an incision into each ϕ-kernel, it is called an incision function:

Definition 8. *(Hansson 1994) An* incision function *for a belief set K is a function σ such that for all ϕ:*

(a) $\sigma(K \perp\!\!\!\perp \phi) \subseteq \bigcup(K \perp\!\!\!\perp \phi)$
(b) If $\varnothing \neq X \in K \perp\!\!\!\perp \phi$, *then* $X \cap \sigma(K \perp\!\!\!\perp \phi) \neq \varnothing$.

The kernel contraction *on K that is based on σ is the operation $\dot{-}$ such that for all $\phi \in \mathcal{L}$:*

$$K \dot{-} \phi = K \backslash \sigma(K \perp\!\!\!\perp \phi).$$

There are two ways to ensure that kernel contraction satisfies the closure postulate (K$\dot{-}$ 1):

Definition 9. *An incision function σ for K is* smooth *if and only if it holds for all subsets X of K that if $X \vdash \psi$ and $\psi \in \sigma(K \perp\!\!\!\perp \phi)$, then $X \cap \sigma(K \perp\!\!\!\perp \phi) \neq \varnothing$. A kernel contraction is smooth if and only if it is based on a smooth incision function.*

Definition 10. *Let σ be an incision function for K. The* saturated kernel contraction *on K that it gives rise to is the operation $\dot{-}$ such that for all $\phi \in \mathcal{L}$:*

$$K \dot{-} \phi = K \cap \mathrm{Cn}(K \backslash \sigma(K \perp\!\!\!\perp \phi))$$

It makes no difference which of the two methods to obtain closure we choose. Saturated and smooth kernel contractions are two ways to construct one and the same class of operations.

Theorem 24. *(Hansson 1994) An operator $\dot{-}$ for a belief set K is a saturated kernel contraction if and only if it is a smooth kernel contraction.*[25]

Furthermore, for belief sets these operations coincide with partial meet contraction.

Theorem 25. *(Hansson 1994) Let K be a belief set. Then an operation $\dot{-}$ is a smooth kernel contraction for K if and only if it is a partial meet contraction for K.*

Close connections between the incision function of a kernel contraction and the selection function of its corresponding partial meet contraction have been shown by Falappa et al. (2006).

Only one direction of the previous theorem can be extended to belief bases. All partial meet contractions on belief bases are smooth kernel contractions, but there are smooth kernel contractions on belief bases that are not partial meet contractions.

---

25 This also holds for belief bases.

This distinction resists logical closure. By a *base-generated kernel contraction* on K is meant an operation $\dot{-}$ such that $K \dot{-} \phi = \text{Cn}(B \simeq \phi)$ for all ϕ, where B is a set of sentences such that $\text{Cn}(B) = K$ and $\simeq$ is a kernel contraction on B. The base-generated kernel contractions on K do not coincide with the (direct) kernel contractions on K, and neither do they coincide with the correspondingly defined base-generated partial meet contractions on K (Hansson 1994).

The connection between kernel contraction and safe contraction is preferably expounded in two steps, with cumulative kernel contractions as an intermediate stage:

Definition 11. *(Hansson 1999a) A* kernel selection function *for a belief set K is a function s such that for all X such that $X \in K \perp\!\!\!\perp \phi$ for some ϕ:*

1. *$s(X) \subseteq X$*
2. *$s(X) \neq \varnothing$ if $X \neq \varnothing$.*

Two sets X and X' are sententially equivalent' *if and only if for each $x \in X$ there is some $x' \in X'$ with $x \leftrightarrow x' \in \text{Cn}(\varnothing)$, and vice versa. A kernel selection function s for K is* extensional *if and only if for all sententially equivalent X and X', if $x \in s(X)$, then $x' \in s(X')$ for all sentences $x' \in X'$ that are logically equivalent with x.*

An incision function σ is cumulative *if and only if there is some kernel selection function s for K such that for all ϕ:*

$$\sigma(K \perp\!\!\!\perp \phi) = \bigcup\{s(X) : X \in K \perp\!\!\!\perp \phi\}$$

An operator of kernel contraction is cumulative if and only if it is based on a cumulative incision function.

In the next step, we identify the relational kernel selection functions:

Definition 12. *A kernel selection function s for the belief set K is* relational *if and only if there is a relation $\prec$ on K such that whenever $X \in K \perp\!\!\!\perp \phi$:*

$\phi \in s(X)$ if and only if $\phi \in X$ and there is no $\rho \in X$ such that $\rho \prec \phi$.

s is acyclically relational *if and only if it is relational with respect to an acyclic relation $\prec$. An incision function is relational (acyclically relational) if and only if it is the cumulation of some relational (acyclically relational) kernel selection function.*

It will be seen from this definition and the above definition of safe contraction that safe contraction is the same as saturated kernel contraction based on some extensional and acyclically relational incision function. The interesting intermediate construction is cumulative kernel contraction which is, by the way, another operation in need of an axiomatic characterization.

7 Other Versions of Safe Contraction

Finally, we will briefly introduce three additional developments of safe contraction, all of which are worth much more detailed investigation than what they have as yet received.

7.1 *Multiple Safe Contraction*

Most of the literature on belief revision has been devoted to changes by single sentences as inputs. This applies also to the two seminal papers from 1985 in which partial meet contraction and safe contraction were introduced. However, there is also a small part of the literature in which the inputs for contractions and revisions are taken to be sets of sentences, rather than single sentences (Fuhrmann and Hansson 1994). This more general case is usually called "multiple contraction". There are two major types of multiple contraction. In package contraction, the success criterion is that all elements of the contractee are removed. In choice contraction, the success criterion is that at least one of its elements is removed. Choice contraction by a finite set can be reduced to single-sentence contraction by the conjunction of all its elements, since a contraction outcome $K \dot{-} A$ fails to imply all elements of A if and only if it fails to imply their conjunction. For package contraction no credible reduction to single-sentence contraction is available even in the finite case, and therefore package contraction is the operation type that has attracted most attention.

Several operations of belief change have been generalized to multiple change. This applies for instance to partial meet contraction (Hansson 1989; Fuhrmann and Hansson 1994; Li 1998), kernel contraction (Fermé et al. 2003), specified meet contraction (Hansson 2010), ranking based contraction (Spohn 2010) and Grove-style revision in sphere systems (Fermé and Reis 2012). As far as we can see, multiple safe contraction has not been treated in the literature (although the need to investigate it was mentioned in Fermé et al. 2003). It is, however, quite trivial to extend the definition of safe contraction to multiple inputs. For the package contraction variant we generalize kernel sets so that $K \perp\!\!\!\perp A$ is the set of inclusion-minimal subsets of K that imply some element of A.[26] Hierarchies are defined in the same way as for singeton (single sentence) safe contraction. A sentence ψ is $\prec$-safe with respect to A if and only if it holds for all $M \in K \perp\!\!\!\perp A$ that if $\psi \in M$, then there is some $\chi \in M$ with $\chi \prec \psi$. The associated safe contraction $\dot{-}$ is given by $K \dot{-} A = \mathrm{Cn}(K/A)$, where K/A is the set of sentences in K that are $\prec$-safe with respect to A. For the choice contraction variant we define $K \veebar A$ as the set of inclusion-minimal subsets of K that imply all elements of A.[27] Incisions functions and (choice-)safe contraction are defined as in the previous variant. It remains to investigate the properties of these constructions.

---

26 Thus $X \in K \perp\!\!\!\perp A$ if and only if (1) $X \subseteq K$, (2) $X \vdash_{\exists} A$, and (3) $X' \nvdash_{\exists} A$ for all $X' \subset X$.

27 Thus $X \in K \veebar A$ if and only if (1) $X \subseteq K$, (2) $A \subseteq \mathrm{Cn}(X)$, and (3) $A \nsubseteq \mathrm{Cn}(X')$ for all $X' \subset X$.

7.2 Non-prioritized Safe Contraction

By a non-prioritized contraction is meant an operation that (intuitively speaking) removes some but not all contingent input sentences. The rationale behind this type of operation is that some non-tautologous elements of the belief set may be so strongly believed that they cannot be removed in a single contraction step, i.e., there may be non-tautologous ϕ for which $K \dot{-} \phi \vdash \phi$.

Correspondingly, an operator of non-prioritized revision is one that does not always incorporate an input sentence even if it can be consistently adopted if some previous beliefs are removed. The rationale is of course that previously held beliefs may sometimes have higher priority for preservation than the input.

Quite a few operators of non-prioritized belief change have been developed, most of them through generalizations of partial meet contraction or some of its equivalent formulations. (Makinson 1997; Hansson 1999b; Hansson et al. 2001) As far as we know, no non-prioritized version of safe contraction has been proposed, but there is no difficulty involved in making such a construction. As was mentioned above, the reason why the relation $\prec$ in the definition of safe contraction is required to satisfy acyclicity (H1) is that this guarantees satisfaction of the success postulate (K$\dot{-}$ 4). A non-prioritized safe contraction can therefore only be constructed if we give up non-cyclicity.

7.3 Safe Revision

In the AGM framework revision, i.e., consistency-preserving incorporation of new information, is constructed from contraction through the Levi identity:

$$K * \phi = \mathrm{Cn}((K \dot{-} \neg\phi) \cup \{\phi\})$$

The rationale behind this construction is that a sentence ϕ can be added to a set without giving rise to inconsistency if and only if that set does not imply $\neg\phi$. Provided that ϕ is not itself inconsistent, we can therefore prepare a set K for consistent addition of ϕ by contracting $\neg\phi$ from it. Furthermore, if $*$ is a revision obtained in this way (a partial meet revision), then we can use the Harper identity to obtain an operator of contraction from it:

$$K \dot{-} \phi = K \cap (K * \neg\phi)$$

Gärdenfors (1982) was the first to note the close interdependence between the two identities, and in an important paper, Makinson (1987) further clarified their relationship. For every operator $\dot{-}$ of contraction, let $\mathbb{R}(\dot{-}\,)$ be the operator generated from $\dot{-}$ via the Levi identity. Furthermore, for every operator $*$ of revision, let $\mathbb{C}(*)$ be the operator generated from $*$ via the Harper identity. Then:

Theorem 26. *If $\dot{-}$ is an operator on K that satisfies (K$\dot{-}$ 1) – (K$\dot{-}$ 4), and (K$\dot{-}$ 6), then $\mathbb{C}(\mathbb{R}(\dot{-}))$ satisfies (K$\dot{-}$ 1) – (K$\dot{-}$ 6), and $\mathbb{R}(\mathbb{C}(\mathbb{R}(\dot{-}))) = \mathbb{R}(\dot{-})$.*

It follows from this and a theorem by (Alchourrón and Makinson 1985, p. 409, Observation 3.2) that the revision operator obtainable from a safe contraction can alternatively be obtained from a partial meet contraction. However, the converse relationship does not hold.

Theorem 27. *It is not the case that all partial meet revisions are safe revisions.*

This theorem leaves us with an interesting open issue, namely the axiomatic characterization of safe revision.

8 In Conclusion: Ten Open Issues

It has been our goal to rekindle the interest in the method of safe contraction. We summarized known results on safe contraction and have begun a fresh attempt at exploring it in the light of results that have been obtained for other prominent models of belief change. In particular, we have shown that the class of safe contractions is narrower than the classes of partial meet contraction functions and of basic entrenchment functions. Not every contraction function that satisfies AGM's basic set of postulates can be rendered as a safe contraction. We have somewhat deepened the results of Alchourrón and Makinson by isolating the effects of the Continuing down and the Continuing up conditions more distinctly. It turned out to be essential to distinguish a number of weakenings of the (very strong) eighth of the famous AGM postulates. Using (K$\dot{-}$ 8r) and (K$\dot{-}$ 8c), we have been able to give a logical characterization for safe contractions in the finite case that are based on regular hierarchies, as well as for those that are based on transitive regular hierarchies. We explored the possibilities and the limits of a unified method for representation results that brings hierarchies more in line with the idea of epistemic entrenchment. We have also surveyed what is known about iterated safe contraction, about kernel contraction, i.e., the choice-theoretic generalization of safe contraction, and about other variants of safe contraction.

In addition to reviewing and to some extent extending the available results on safe contraction we have identified ten interesting open issues concerning safe contraction:

1. Safe contraction has not been axiomatically characterized, not even in the finite case.
2. In particular, it remains to find out whether all safe contractions satisfy the acyclicity condition (K$\dot{-}$ Acyc), and whether this condition is sufficient for the representability of a contraction function as a safe contraction.
3. Safe contraction based on transitive hierarchies has not been axiomatically characterized, not even for logically finite belief sets, and the same comment applies

to safe contraction based on transitive regular hierarchies for logically infinite belief sets.
4. Safe contraction based on a regular hierarchy has not been axiomatically characterized other than for logically finite belief sets.
5. Unified global safe contraction has not been axiomatically characterized. It is also unknown whether it coincides with global (not necessarily unified) safe contraction.
6. Cumulative kernel contraction has not been axiomatically characterized.
7. Relational kernel contraction has not been axiomatically characterized.
8. Multiple safe contraction remains to be investigated.
9. Non-prioritized safe contraction (perhaps better called "unsafe contraction") has not been investigated either.
10. Safe revision has not been axiomatically characterized, not even in the finite case.

It should also be noted that we have mostly considered various modifications of safe contraction, such as non-relational, multiple, non-prioritized, global etc. variants, one at a time. Much remains to be discovered through systematic studies of the many combinations of these modifications that are possible. In conclusion, we believe that safe contraction, broadly conceived, is an equally fruitful area of study as partial meet contraction, broadly conceived. We are convinced that it is worth much more attention than it has as yet received.[28]

Appendix: Proofs

Proof of Theorem 1. It was shown by Alchourrón, Gärdenfors and Makinson (1985, Observation 3.3) that (K$\dot{-}$ 7) and (K$\dot{-}$ 7P) are equivalent, and by Rott (1992b, Lemma 2) that (K$\dot{-}$ 7) and (K$\dot{-}$ 7p) are equivalent. QED

Proof of Theorem 2. (i) (K$\dot{-}$ 8r) implies (K$\dot{-}$ 8r′): By (K$\dot{-}$ 6) and (K$\dot{-}$ 8r), we get that $K \dot{-} (\phi \wedge \psi \wedge \chi) = K \dot{-} ((\phi \wedge \chi) \wedge (\psi \wedge \chi)) \subseteq \mathrm{Cn}(K \dot{-} (\phi \wedge \chi) \cup K \dot{-} (\psi \wedge \chi))$. Now let $\chi \in K \dot{-} (\phi \wedge \psi \wedge \chi)$. Thus, by the compactness of Cn, we have finite sets $M \subseteq K \dot{-} (\phi \wedge \chi)$ and $N \subseteq K \dot{-} (\psi \wedge \chi)$ such that $M \cup N \vdash \chi$. Now define ξ to be $(\bigwedge M) \vee \chi$ and ρ to be $(\bigwedge N) \vee \chi$. By (K$\dot{-}$ 1), ξ is in $K \dot{-} (\phi \wedge \chi)$ and ρ is in $K \dot{-} (\psi \wedge \chi)$. Furthermore, we have $\chi \dashv\vdash \xi \wedge \rho$, as required by (K$\dot{-}$ 8r′).

(K$\dot{-}$ 8r′) implies (K$\dot{-}$ 8r): Let $\chi \in K \dot{-} (\phi \wedge \psi)$. By (K$\dot{-}$ 2), $\chi \in K$, so by (K$\dot{-}$ 5) and (K$\dot{-}$ 1), $\phi \rightarrow \chi \in K \dot{-} \phi$ and $\psi \rightarrow \chi \in K \dot{-} \psi$. Thus $\mathrm{Cn}(K \dot{-} (\phi \wedge \chi) \cup K \dot{-} (\psi \wedge$

[28] We would like to dedicate this paper to David Makinson, whom we admire as a researcher and who has been, in many and various ways, a long-term friend and mentor to both of us. Thank you so much, David! Our paper is the outcome of a truly joint undertaking. However, unless otherwise indicated, the results reported in Sects. 3 and 4 are Hans Rott's and the results reported in Sects. 5, 6 and 7 are Sven Ove Hansson's. We are grateful to Eduardo Fermé, David Makinson and in particular Richard Booth for very helpful comments on an earlier version of this paper.

$\chi))$ contains $\neg(\phi \vee \psi) \vee \chi$. In order to show that it contains χ, it is thus sufficient to show that it also contains $\phi \vee \psi \vee \chi$. By (K$\dot{-}$ 1), $\phi \vee \psi \vee \chi \in K \dot{-} (\phi \wedge \psi)$, and by (K$\dot{-}$ 6), $\phi \vee \psi \vee \chi \in K \dot{-} (\phi \wedge \psi \wedge (\phi \vee \psi \vee \chi))$. Now we are ready to apply (K$\dot{-}$ 8r$'$) and get that there are ξ and ρ such that $(\phi \vee \psi \vee \chi) \dashv\vdash \xi \wedge \rho$ and $\xi \in K \dot{-} (\phi \wedge (\phi \vee \psi \vee \chi))$ and $\rho \in K \dot{-} (\psi \wedge (\phi \vee \psi \vee \chi))$. That is, by (K$\dot{-}$ 1), $\xi \in K \dot{-} \phi$ and $\rho \in K \dot{-} \psi$. Hence, $\xi \wedge \rho \in \text{Cn}(K \dot{-} \phi \cup K \dot{-} \psi)$ and, finally, $\phi \vee \psi \vee \chi \in \text{Cn}(K \dot{-} \phi \cup K \dot{-} \psi)$, as desired.

(ii) (K$\dot{-}$ 8wd) implies (K$\dot{-}$ 8p): Let $\phi \in K \dot{-} (\phi \wedge \psi \wedge \chi) = K \dot{-} ((\phi \wedge \psi) \wedge (\phi \wedge \chi))$. Then by (K$\dot{-}$ 8wd), $\phi \vee (\phi \wedge \chi) \in K \dot{-} (\phi \wedge \psi)$ or $\phi \vee (\phi \wedge \psi) \in K \dot{-} (\phi \wedge \chi)$, which simplifies to $\phi \in K \dot{-} (\phi \wedge \psi)$ or $\phi \in K \dot{-} (\phi \wedge \chi)$.

(K$\dot{-}$ 8p) implies (K$\dot{-}$ 8wd): Let $\chi \in K \dot{-} (\phi \wedge \psi)$. Then by (K$\dot{-}$ 1) and (K$\dot{-}$ 6) $\phi \vee \psi \vee \chi \in K \dot{-} ((\phi \vee \psi \vee \chi) \wedge (\phi \wedge \psi))$. Thus by (K$\dot{-}$ 8p) $\phi \vee \psi \vee \chi \in K \dot{-} ((\phi \vee \psi \vee \chi) \wedge \phi)$ or $\phi \vee \psi \vee \chi \in K \dot{-} ((\phi \vee \psi \vee \chi) \wedge \psi)$. Which reduces by (K$\dot{-}$ 6) to $\phi \vee \psi \vee \chi \in K \dot{-} \phi$ or $\phi \vee \psi \vee \chi \in K \dot{-} \psi$. By (K$\dot{-}$ 2) and (K$\dot{-}$ 5), $\phi \rightarrow (\psi \vee \chi) \in K \dot{-} \phi$ and $\psi \rightarrow (\phi \vee \chi) \in K \dot{-} \psi$. By (K$\dot{-}$ 1), it follows that $\psi \vee \chi \in K \dot{-} \phi$ or $\phi \vee \chi \in K \dot{-} \psi$.

(iii) For the non-obvious direction, suppose to the contrary that (K$\dot{-}$ 8d) holds but not (K$\dot{-}$ 8d$'$). Since (K$\dot{-}$ 8d$'$) does not hold there are α, β, ϕ, and ψ such that $\alpha \in K \dot{-} (\phi \wedge \psi)$, $\alpha \notin K \dot{-} \phi$, $\beta \in K \dot{-} (\phi \wedge \psi)$, and $\beta \notin K \dot{-} \psi$. It follows from (K$\dot{-}$ 1) that $\alpha \wedge \beta \in K \dot{-} (\phi \wedge \psi)$, $\alpha \wedge \beta \notin K \dot{-} \phi$, and $\alpha \wedge \beta \notin K \dot{-} \psi$, contrary to (K$\dot{-}$ 8d).

(iv) Suppose that $\chi \in K \dot{-} \phi \wedge \psi$. We want to show that either $\chi \in K \dot{-} \phi$ or $\chi \in K \dot{-} \psi$. By (K$\dot{-}$ 1), $(\phi \wedge \psi) \vee \chi \in K \dot{-} \phi \wedge \psi$, and by (K$\dot{-}$ 6), $(\phi \wedge \psi) \vee \chi \in K \dot{-} \phi \wedge \psi \wedge ((\phi \wedge \psi) \vee \chi)$. So by (K$\dot{-}$ 8p) which we already proved to be relatively equivalent with (K$\dot{-}$ 8wd), either $(\phi \wedge \psi) \vee \chi \in K \dot{-} \phi \wedge ((\phi \wedge \psi) \vee \chi)$ or $(\phi \wedge \psi) \vee \chi \in K \dot{-} \psi \wedge ((\phi \wedge \psi) \vee \chi)$. By (K$\dot{-}$ 8c), this reduces to either $(\phi \wedge \psi) \vee \chi \in K \dot{-} \phi$ or $(\phi \wedge \psi) \vee \chi \in K \dot{-} \psi$. Using (K$\dot{-}$ 1), we get that either $\phi \vee \chi \in K \dot{-} \phi$ or $\psi \vee \chi \in K \dot{-} \psi$. But by (K$\dot{-}$ 2) and (K$\dot{-}$ 5), both $\phi \rightarrow \chi \in K \dot{-} \phi$ and $\psi \rightarrow \chi \in K \dot{-} \psi$. So by (K$\dot{-}$ 1) again, either $\chi \in K \dot{-} \phi$ or $\chi \in K \dot{-} \psi$, as desired.

(v) Figure 3 depicts a contraction function over $K = K_\perp$ that satisfies (K$\dot{-}$ 1) – (K$\dot{-}$ 7), (K$\dot{-}$ 8c) and (K$\dot{-}$ 8r), but not (K$\dot{-}$ 8wd). This contraction function derives from the (vacuously transitive) partial ordering $\leftarrow$ between the four possible worlds pq, $p\overline{q}$, $\overline{p}q$ and $\overline{p}\overline{q}$ that is fully specified by $pq \leftarrow \overline{p}\overline{q}$ and $p\overline{q} \leftarrow \overline{p}q$ (where the arrows point to the "more plausible" worlds). Having Grovean semantics for AGM in mind, it can be recognized as a partial meet contraction based on a transitive strict preference relation, and thus Theorem 14 (ii) applies (a similar example is given in Rott 2001, p. 164).[29] The violation of (K$\dot{-}$ 8wd) is this: We have $p \in K \dot{-} ((p \wedge q) \wedge (p \wedge \neg q)) = K \dot{-} \perp = \text{Cn}(p)$ but neither $p \vee (p \wedge q) \in K \dot{-} (p \wedge \neg q) = \text{Cn}(q)$ nor $p \vee (p \wedge \neg q) \in K \dot{-} (p \wedge q) = \text{Cn}(\neg q)$. QED

---

[29] Those who are worried about the fact that the contraction function operates over the absurd belief set $K_\perp$ may want to modify the example by introducing a third atom r, replacing $K_\perp$ by $\text{Cn}(r)$ and replacing $\binom{\phi}{\psi}$ by $\binom{r \vee \phi}{r \vee \psi}$.

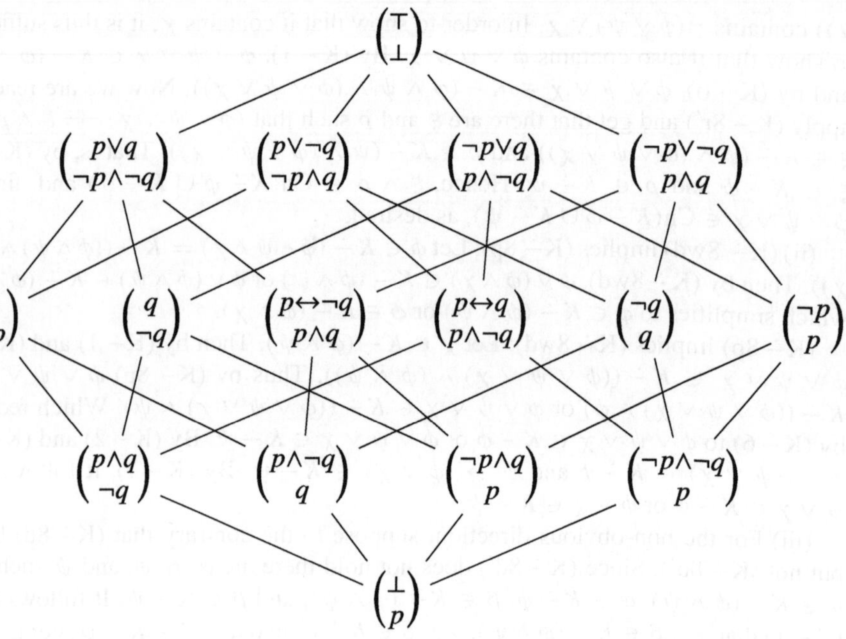

Fig. 3 The Hasse diagram for a contraction function over $K = K_\bot$ that satisfies (K$\dot{-}$ 1) – (K$\dot{-}$ 7), (K$\dot{-}$ 8c) and (K$\dot{-}$ 8r), but not (K$\dot{-}$ 8wd). Read "$\binom{\phi}{\psi}$" as "$K \dot{-} \phi = \text{Cn}(\psi)$"

Proof of Theorem 3. (i) To prove (K$\dot{-}$ Acyc), suppose for reductio that $\phi_{i+1} \in K \dot{-} (\phi_i \wedge \phi_{i+1})$ and $\nvdash \phi_i \wedge \phi_{i+1}$ for all $i = 1, \ldots, n-1$, and that $\phi_1 \in K \dot{-} (\phi_n \wedge \phi_1)$ and $\nvdash \phi_n \wedge \phi_1$. By (K$\dot{-}$ 7p) which is equivalent with (K$\dot{-}$ 7), it follows that $\phi_{i+1} \in K \dot{-} (\phi_1 \wedge \cdots \wedge \phi_n)$ for all $i = 1, \ldots, n-1$, and $\phi_1 \in K \dot{-} (\phi_1 \wedge \cdots \wedge \phi_n)$. Thus, by (K$\dot{-}$ 1), $\phi_1 \wedge \cdots \wedge \phi_n \in K \dot{-} (\phi_1 \wedge \cdots \wedge \phi_n)$. But since clearly $\nvdash \phi_1 \wedge \cdots \wedge \phi_n$, this contradicts (K$\dot{-}$ 4).

(ii) To prove (K$\dot{-}$ Trans), we assume that $\psi \in K \dot{-} (\phi \wedge \psi)$, $\nvdash \phi \wedge \psi$, $\chi \in K \dot{-} (\psi \wedge \chi)$, and $\nvdash \psi \wedge \chi$. Applying (K$\dot{-}$ 7p) and (K$\dot{-}$ 6) to $\psi \in K \dot{-} (\phi \wedge \psi)$ we obtain $\psi \in K \dot{-} ((\phi \wedge \chi) \wedge \psi)$, to which we apply (K$\dot{-}$ 8c) and (K$\dot{-}$ 6) to obtain $K \dot{-} (\phi \wedge \psi \wedge \chi) \subseteq K \dot{-} (\phi \wedge \chi)$. We also have $\chi \in K \dot{-} (\psi \wedge \chi)$ which with (K$\dot{-}$ 7p) and (K$\dot{-}$ 6) yields $\chi \in K \dot{-} (\phi \wedge \psi \wedge \chi)$. We can finally combine this with $K \dot{-} (\phi \wedge \psi \wedge \chi) \subseteq K \dot{-} (\phi \wedge \chi)$ and obtain $\chi \in K \dot{-} (\phi \wedge \chi)$ as desired. It remains to be shown that $\nvdash \phi \wedge \chi$. Suppose for reductio that $\vdash \phi \wedge \chi$. Then we can conclude from $\vdash \phi$ and (K$\dot{-}$ 6) that $K \dot{-} (\phi \wedge \psi) = K \dot{-} \psi$. In combination with the assumption $\psi \in K \dot{-} (\phi \wedge \psi)$ this yields $\psi \in K \dot{-} \psi$. Then (K$\dot{-}$ 4) yields $\vdash \psi$. Thus $\vdash \phi \wedge \psi$, contradicting one of our assumptions. QED

Proof of Theorem 6. We give an example of a contraction function satisfying (K$\dot{-}$ 1) – (K$\dot{-}$ 6) that cannot be represented as a safe contraction. Consider the language $\mathcal{L}$ that has only two propositional variables, p and q. Let $K = \text{Cn}(\{p, q\})$.

Consider now a basic AGM contraction function with the following values:

$$K \dot{-} p = \text{Cn}(q)$$
$$K \dot{-} q = \text{Cn}(p \leftrightarrow q)$$
$$K \dot{-} (p \leftrightarrow q) = \text{Cn}(p)$$

There is nothing in (K$\dot{-}$ 1) – (K$\dot{-}$ 6) that prevents a contraction function from having these values. The values for other sentences are irrelevant. We now show that such a function cannot be generated through safe contraction.

Consider first $K \dot{-} p$. Besides the singleton p-kernels $\{p \wedge q\}$ and $\{p\}$, there are four p-kernels of K (modulo logical equivalence), viz.,

$$\{p \vee q, p \vee \neg q\}, \{p \vee q, p \leftrightarrow q\},$$
$$\{q, p \vee \neg q\} \quad , \quad \{q, p \leftrightarrow q\}$$

Suppose that the contraction function is a safe contraction associated with a hierarchy $\prec$. Notice that $\neg p \vee q$, being a consequence of $\neg p$, is $\prec$-safe with respect to p anyway. Notice also that since $K \dot{-} p = \text{Cn}(q)$, $p \vee \neg q$ and $p \leftrightarrow q$ must not be $\prec$-safe with respect to p. Since $K \dot{-} p = \text{Cn}(q)$, either $p \vee q$ or q itself has to be $\prec$-safe with respect to p. Inspecting the four kernels, we realize that this means that

(A1) $p \vee \neg q \prec p \vee q$ and $p \leftrightarrow q \prec p \vee q$ *or*
(A2) $p \vee \neg q \prec q$ and $p \leftrightarrow q \prec q$

Second, we consider $K \dot{-} q = \text{Cn}(p \leftrightarrow q)$. By exactly the same argument that we have just applied, we get that

(B1) $p \vee q \prec \neg p \vee q$ and $p \prec \neg p \vee q$ *or*
(B2) $p \vee q \prec p \leftrightarrow q$ and $p \prec p \leftrightarrow q$

Third, we consider $K \dot{-} (p \leftrightarrow q) = \text{Cn}(p)$. Applying the same argument a third time, we get that

(C1) $\neg p \vee q \prec p \vee \neg q$ and $q \prec p \vee \neg q$ *or*
(C2) $\neg p \vee q \prec p$ and $q \prec p$

Assume, for the first part, that (A1) is true. Then, by the asymmetry of $\prec$, (B2) cannot be true. So (B1) is true. But then, again by the asymmetry of $\prec$, (C2) cannot be true. So (C1) is true as well. But from (A1), (B1) and (C1), we get the cycle

$$p \vee \neg q \prec p \vee q \prec \neg p \vee q \prec p \vee \neg q$$

So assume for the first part that (A2) is true. Then, by the asymmetry of $\prec$, (C1) cannot be true. So (C2) is true. But then, again by the asymmetry of $\prec$, (B1) cannot be true. So (B2) is true. But from (A2), (C2) and (B2), we get the cycle

$$p \leftrightarrow q \prec q \prec p \prec p \leftrightarrow q$$

So either way, we have found a cycle, which is forbidden by the assumption that $\prec$ is a hierarchy. We have found a contradiction and proved that $\dot{-}$ cannot be a safe contraction associated with $\prec$. QED

Proof of Theorem 8. That $\dot{-}$ satisfies (K$\dot{-}$ 7) is proven in Observation 5.3 of Alchourrón and Makinson (1985). It remains to verify (K$\dot{-}$ 8r). Let χ be in $K \dot{-} (\phi \wedge \psi)$. That is, there is a set M of $\phi \wedge \psi$-safe elements such that $M \vdash \chi$. In order to show that χ is in $\mathrm{Cn}(K \dot{-} \phi \cup K \dot{-} \psi)$, it suffices to show that there are sets N_1 of ϕ-safe and N_2 of ψ-safe elements such that $N_1 \cup N_2 \vdash \chi$. But since $\prec$ continues down $\vdash$ over K, each $\phi \wedge \psi$-safe element is either ϕ-safe or ψ-safe, by Corollary 5.4 of Alchourrón and Makinson (1985). So M can easily be split up into sets N_1 and N_2 as desired. QED

Proof of Theorem 9. That $\dot{-}$ satisfies (K$\dot{-}$ 7) is proven in Observation 4.3 of Alchourrón and Makinson (1985). It remains to verify (K$\dot{-}$ 8c). Let ψ and χ be in $K \dot{-} (\phi \wedge \psi)$. That is, there are sets M and N of $\phi \wedge \psi$-safe elements such that $M \vdash \psi$ and $N \vdash \chi$. Let M and N be chosen minimal, so that they are finite, by compactness. We want to show that χ is in $K \dot{-} \phi$ as well, that is, that there is a set P of ϕ-safe elements such that $P \vdash \chi$. Put $P = N$. Now suppose that there is a set R such that $R \vdash_{min} \phi$. By compactness, R is finite. Let $\xi \in N$ be in R. We need to show that there is a ρ in R such that $\rho \prec \xi$. Clearly, $M \cup R \vdash \phi \wedge \psi$. Define $M' = \{(\bigwedge R) \rightarrow \sigma : \sigma \in M\}$. Then clearly, $M' \cup R \vdash \phi \wedge \psi$. Since the elements of M are $\phi \wedge \psi$-safe and $\prec$ continues up $\vdash$ over K, the elements of M' are $\phi \wedge \psi$-safe as well, by Lemma 4.1 of Alchourrón and Makinson (1985). Take a subset S of $M' \cup R$ that minimally implies $\phi \wedge \psi$. Notice that $R \subseteq S$, because if some α from R were missing from S, then S could not imply ψ. For suppose that $(R - \{\alpha\}) \cup M' \vdash \phi \wedge \psi$. Then, since $\neg\phi$ is logically stronger than $\neg \bigwedge R$ and thus than M', we get that $(R - \{\alpha\}) \cup \{\neg\phi\} \vdash \phi \wedge \psi$, and thus $R - \{\alpha\} \vdash \phi$, contradicting the minimality of R.

Now since $S \vdash_{min} \phi \wedge \psi$, we can apply the fact that $\xi \in N$ is $\phi \wedge \psi$-safe and conclude that there is a ρ^0 in S such that $\rho^0 \prec \xi$. Now this ρ^0 is either in R or in M'. But if it is in M', we can, by the fact that the elements of M' are $\phi \wedge \psi$-safe, find another sentence ρ^1 with $\rho^1 \prec \rho^0$ which again is either in R or in M'. If ρ^1 is in M', we repeat the process again, and so on. Since M' is finite and $\prec$ is acyclic and transitive, there must finally be a ρ^n with $\rho^n \prec \xi$ which is not in M' but in R. This is the ρ we have been looking for. QED

Proof of Lemma 10. In the presence of the basic AGM postulates including (K$\dot{-}$ 4), Definition 4 above is equivalent to Definition 17 of Rott (2001, p. 229). Thus, the hierarchies obtained by (CH) are relations of epistemic entrenchment. With the exception of (v) and (vi), the claims of Lemma 10 are all listed in part (ii) of Observation 68 of Rott (2001, p. 263), although (viii) is a slightly strengthened form for the present context. (v) is trivial. We prove (vi). Suppose that $\phi \wedge \psi \prec \chi$ and

$\nvdash \chi$. This means, by (CH), that $\chi \in K \dot{-} (\phi \wedge \psi \wedge \chi)$ and $\nvdash \phi \wedge \psi$. By (K$\dot{-}$ 8r), it follows that $\chi \in \mathrm{Cn}(K \dot{-} (\phi \wedge \chi) \cup K \dot{-} (\psi \wedge \chi))$. By the compactness of Cn, there are finite sets $M \subseteq K \dot{-} (\phi \wedge \chi)$ and $N \subseteq K \dot{-} (\phi \wedge \chi)$ such that $M \cup N \vdash \chi$. By (K$\dot{-}$ 1), we have $(\bigwedge M) \vee \chi \in K \dot{-} (\phi \wedge \chi)$, $(\bigwedge N) \vee \chi \in K \dot{-} (\psi \wedge \chi)$ and $\chi \dashv\vdash ((\bigwedge M) \vee \chi) \wedge ((\bigwedge N) \vee \chi)$. Let ξ be $(\bigwedge M) \vee \chi$ and ρ be $(\bigwedge N) \vee \chi$. Since both ξ and ρ are in $\mathrm{Cn}(\chi)$, we have, by (K$\dot{-}$ 6), $\xi \in K \dot{-} (\phi \wedge \chi \wedge \xi)$ and $\rho \in K \dot{-} (\psi \wedge \chi \wedge \rho)$. Thus, since $\nvdash \phi \wedge \chi$ and $\nvdash \psi \wedge \chi$, $\phi \wedge \chi \prec \xi$ and $\psi \wedge \chi \prec \rho$, as desired. QED

Proof of Theorem 11. The hierarchy $\prec$ defined by (CH) can be used to reconstruct the given contraction function $\dot{-}$ as a safe contraction, provided that $\dot{-}$ has the properties mentioned. The properties of $\prec$ are listed in Lemma 10.

The main claim is:

$\psi \in K \dot{-} \phi$ iff

there is a set M elements in K that are of $\prec$-safe with respect to ϕ and jointly entail ψ iff

there is an $M \subseteq K$ such that $M \vdash \psi$ and for all $N \subseteq K$ such that $N \vdash_{min} \phi$ and all $\chi \in N \cap M$ there is a $\xi \in N$ such that $\xi \prec \chi$ iff

there is an $M \subseteq K$ such that $M \vdash \psi$ and for all $N \subseteq K$ such that $N \vdash_{min} \phi$ and all $\chi \in N \cap M$ there is a $\xi \in N$ such that $\chi \in K \dot{-} \xi \wedge \chi$ and $\nvdash \xi$

This claim is trivially satisfied if $\phi \notin K$, since then (K$\dot{-}$ 3) and the fact that K is logically closed entail that both the left-hand side and the right-hand side are equivalent to $\psi \in K$. So we suppose that $\phi \in K$ in the following.

(i) From left to right, the general case. Suppose that $\psi \in K \dot{-} \phi$. Take the set $M = \{\phi \vee \psi, \neg\phi \vee \psi\}$. By (K$\dot{-}$ 1) and (K$\dot{-}$ 2), $M \subseteq K$. As a consequence of $\neg\phi$, $\neg\phi \vee \psi$ is not in any minimal set entailing ϕ, so it is $\prec$-safe with respect to ϕ. It remains to check $\phi \vee \psi$. Suppose $N \vdash_{min} \phi$ with $\phi \vee \psi$ in $N \subseteq K$. We want to show that there is a ξ in N such that $\phi \vee \psi \in K \dot{-} \xi \wedge (\phi \vee \psi)$ and $\nvdash \xi$. (Notice that the minimality condition will not be used again for this direction.)

By (K$\dot{-}$ 1), we can infer from $\psi \in K \dot{-} \phi$ that $\phi \vee \psi \in K \dot{-} \phi$. So by the Partial antitony condition (K$\dot{-}$ 7P) which is equivalent with (K$\dot{-}$ 7), we get $\phi \vee \psi \in K \dot{-} ((\bigwedge N) \wedge \phi)$, or, by (K$\dot{-}$ 6), $\phi \vee \psi \in K \dot{-} ((\bigwedge N) \wedge (\phi \vee \psi))$. By (K$\dot{-}$ 8p) which is equivalent with (K$\dot{-}$ 8wd), $\phi \vee \psi \in K \dot{-} (\xi \wedge (\phi \vee \psi))$ for some ξ in N. And by the minimality of N, $\nvdash \xi$, as desired.

From right to left. Suppose that there is an $M \subseteq K$ such that $M \vdash \psi$ and for all $N \subseteq K$ such that $N \vdash_{min} \phi$ and all $\chi \in N \cap M$ there is a $\xi \in N$ such that $\chi \in K \dot{-} \xi \wedge \chi$ and $\nvdash \xi$. Notice that ψ is in K, by (K$\dot{-}$ 1).

In the limiting case $\vdash \phi$, we have $N \vdash_{min} \phi$ if and only if $N = \emptyset$. So the supposition reduces to $K \vdash \psi$ or, by the fact that K is logically closed, $\psi \in K$. But by (K$\dot{-}$ 5), $K \subseteq \mathrm{Cn}(K \dot{-} \phi \cup \{\phi\}) = \mathrm{Cn}(K \dot{-} \phi) = K \dot{-} \phi$, so $\psi \in K \dot{-} \phi$, as desired. So let us now turn to the principal case in which $\nvdash \phi$.

Let M' be a subset of M such that $M' \vdash_{min} \phi \vee \psi$. From the supposition it follows that for all $N \subseteq K$ such that $N \vdash_{min} \phi$ and all $\chi \in N \cap M'$ there is a $\xi \in N$ such that $\chi \in K \dot{-} \xi \wedge \chi$ and $\nvdash \xi$. By Partial antitony (K$\dot{-}$ 7P), we get that for all $N \subseteq K$ such that $N \vdash_{min} \phi$ and all $\chi \in N \cap M'$, $\chi \in K \dot{-} \bigwedge N$. In short, by (K$\dot{-}$ 1), for all $N \subseteq K$ such that $N \vdash_{min} \phi$, $\bigwedge(N \cap M') \in K \dot{-} \bigwedge N$.

Suppose for reductio that $M' \vdash \phi$. Then, since $\nvdash \phi$, also $\nvdash \bigwedge M'$. It follows from $M' \vdash_{min} \phi \vee \psi$ that $M' \vdash_{min} \phi$, and we get $\bigwedge(M' \cap M') \in K \dot{-} \bigwedge M'$, contradicting (K$\dot{-}$ 4). So $M' \nvdash \phi$.

Now consider the set $N_0 = M' \cup \{\psi \rightarrow \phi\}$. Since K is logically closed and $\phi \in K$, $N_0 \subseteq K$. We show that $N_0 \vdash_{min} \phi$. Since $M' \vdash_{min} \phi \vee \psi$, clearly $N_0 \vdash \phi$. Since $M' \nvdash \phi$, $\psi \rightarrow \phi$ is not redundant. But neither is any element ρ of M' redundant. For suppose that $(M' - \{\rho\}) \cup \{\psi \rightarrow \phi\} \vdash \phi$. Then $M' - \{\rho\} \vdash (\psi \rightarrow \phi) \rightarrow \phi$, i.e., $M' - \{\rho\} \vdash \phi \vee \psi$, contradicting the minimality of M'.

From the above, we get $\bigwedge(N_0 \cap M') \in K \dot{-} \bigwedge N_0$. Now we have $M' \subseteq N_0$, thus $\bigwedge(N_0 \cap M') = \bigwedge M'$ and $\bigwedge M' \in K \dot{-} \bigwedge N_0$. Equivalently, by (K$\dot{-}$ 6), $\bigwedge M' \in K \dot{-} (\bigwedge M') \wedge \phi$. So by (K$\dot{-}$ 4) and (K$\dot{-}$ 8d), or alternatively by (K$\dot{-}$ 8c), $\bigwedge M' \in K \dot{-} \phi$, and by (K$\dot{-}$ 1) $\phi \vee \psi \in K \dot{-} \phi$. Finally, by (K$\dot{-}$ 5) and (K$\dot{-}$ 1), $\psi \in K \dot{-} \phi$, as desired.

(ii) The case of a logically finite belief set K, from left to right. Suppose that $\psi \in K \dot{-} \phi$. Let $\text{Coatoms}_K(\phi \vee \psi)$ denote the set of co-atoms of K, i.e., the logically weakest non-tautological elements of K, that are implied by $\phi \vee \psi$, and take the set $M = \text{Coatoms}_K(\phi \vee \psi) \cup \{\neg\phi \vee \psi\}$. As a consequence of $\neg\phi$, $\neg\phi \vee \psi$ is not in any minimal set entailing ϕ, so it is $\prec$-safe with respect to ϕ. It remains to check for every co-atom α in $\text{Coatoms}_K(\phi \vee \psi)$ that it is $\prec$-safe with respect to ϕ. Suppose $N \vdash_{min} \phi$ with α in N. We want to show that there is a ξ in N such that $\alpha \in K \dot{-} \xi \wedge \alpha$ and $\nvdash \xi$.

By (K$\dot{-}$ 1), we can infer from $\psi \in K \dot{-} \phi$ that $\alpha \in K \dot{-} \phi$. So by $\phi \vdash \alpha$ and Partial antitony (K$\dot{-}$ 7P) which is equivalent with (K$\dot{-}$ 7), we get $\alpha \in K \dot{-} ((\bigwedge N) \wedge \phi)$, or, by (K$\dot{-}$ 6), $\alpha \in K \dot{-} \bigwedge\{\xi \wedge \alpha : \xi \in N\}$. By (K$\dot{-}$ 8r), $\alpha \in \text{Cn}(\bigcup\{K \dot{-} (\xi \wedge \alpha) : \xi \in N\})$. Since α is a co-atom of K, $\alpha \in \text{Cn}(K \dot{-} (\xi \wedge \alpha)) = K \dot{-} (\xi \wedge \alpha)$ for some ξ in N, by Boolean considerations.

The right-to-left direction of the case of a logically finite belief set K, is proven literally as in the general case. QED

Proof of Lemma 18. (i) Suppose that $\phi < \psi$ and $\psi < \phi$. Then, by Intersubstitutivity, $\phi < (\psi \wedge \psi)$ and $\psi < (\phi \wedge \phi)$, so by the left-to-right direction of Choice $(\phi \wedge \psi) < \psi$ and $(\psi \wedge \phi) < \phi$. By Intersubstitutivity again, $((\phi \wedge \psi) \wedge \phi) < \psi$ and $((\phi \wedge \psi) \wedge \psi) < \phi$. By the right-to-left direction of Choice, $(\phi \wedge \psi) < (\phi \wedge \psi)$. But this contradicts Irreflexivity. (ii)–(vi) are easy. (vii) For transitivity, let $\phi < \psi$ and $\psi < \chi$. Then, by Continuing down, $\phi \wedge \chi < \psi$ and $\phi \wedge \psi < \chi$. By the right-to-left direction of Choice, it follows that $\phi < \psi \wedge \chi$. So by Continuing up, $\phi < \chi$. Acyclicity follows from Transitivity and Irreflexivity. (viii) For the first part, assume that for all β_j there is an α_i such that $\alpha_i < \beta_j$. By Continuing down, for all β_j, it holds that $\bigwedge \alpha_i < \beta_j$. So by Conjunction up, which follows from Continuing down and the right-to-left direction of Choice (see part (iv) of this lemma), $\bigwedge \alpha_i < \bigwedge \beta_j$, as desired. For the second part, assume that $\bigwedge \alpha_i < \bigwedge \beta_j$. By Continuing up, $\bigwedge \alpha_i < \beta_j$ for all β_j. Take an arbitrary such β_j. Now repeated application of (EII$^-$) tells us that there are sentences $\phi_j^1, \ldots, \phi_j^k$ such that $\beta_j \dashv\vdash \phi_j^1 \wedge \cdots \wedge \phi_j^k$ and for all $i = 1, \ldots, n$, $\alpha_i \wedge \beta_j < \phi_j^i$. But since β_j is a co-atom of K, it follows from Boolean considerations that each ϕ_j^i is either a tautology or equivalent with β_j and that in fact at least one of the ϕ_j^i's is equivalent with β_j. It follows from Intersubstitutivity that $\alpha_i \wedge \beta_j < \beta_j$ for at least one i, and the right-to-left direction of Choice and Intersubstitutivity once more that $\alpha_i < \beta_j$ for at least one α_i. But this is what we wanted to prove. QED

Proof of Lemma 19. That (ii) implies (i) is trivial. Take M from (ii) and notice that every N that minimally implies $\phi \wedge \psi$ obviously implies ϕ.

Now we show that (i) implies (ii), given Transitivity and Continuing down for $\prec$. Take M from (i) and let M' be a subset of M that minimally implies ψ. Obviously M' also satisfies (i), and by compactness, M' is finite. Suppose that $N \vdash \phi$. Put $N' = \{(\bigwedge M') \rightarrow \xi : \xi \in N\}$. Since $M' \cup N \vdash \phi \wedge \psi$, also $M' \cup N' \vdash \phi \wedge \psi$. Let S be a minimal subset of $M' \cup N'$ that implies $\phi \wedge \psi$. First notice that $M' \subseteq S$, because if some χ from M' were missing from S, then S could not imply ϕ. For suppose that $(M' - \{\chi\}) \cup N' \vdash \phi \wedge \psi$. Then, since $\neg\psi$ is logically stronger than $\neg \bigwedge M'$ and thus than N', we get that $(M' - \{\chi\}) \cup \{\neg\psi\} \vdash \phi \wedge \psi$, and thus $M' - \{\chi\} \vdash \psi \vee (\phi \wedge \psi)$, contradicting the minimality of M'. So since $S \vdash_{min} \phi \wedge \psi$, we can apply (i) and find that S is non-empty and for every χ in $M' \cap S = M'$ there is a ξ in S such that $\xi \prec \chi$. Now this ξ is either in M' or in N'. But if it is in M', we can apply (i) again and find another sentence ξ^1 with $\xi^1 \prec \xi$ which again is either in M' or in N'. If ξ^1 is in M', we repeat the process again, and so on. Since M' is finite and $\prec$ is acyclic and transitive, there must finally be a ξ^n with $\xi^n \prec \chi$ which is not in M' but in N'. This ξ^n is of the form $(\bigwedge M') \rightarrow \rho$ for some ρ in N. Since, finally, this ρ implies ξ^n and $\prec$ continues down $\vdash$ over K, $\rho \prec \chi$, as desired. QED

Proof of Theorem 21. By Theorem 4, every safe contraction function $\dot{-}$ over K satisfies (K$\dot{-}$ 1) – (K$\dot{-}$ 6). We now show that (EC) follows from (CE), provided that $\dot{-}$ satisfies (K$\dot{-}$ 1), (K$\dot{-}$ 2), (K$\dot{-}$ 5) and (K$\dot{-}$ 6).[30] (CE) gives us:

$$\phi < \phi \vee \psi \text{ iff } \phi \vee \psi \in K \dot{-} (\phi \wedge (\phi \vee \psi)) \text{ and } \nvdash \phi$$

By (K$\dot{-}$ 6), this is equivalent with

$$(\dagger) \quad \phi < \phi \vee \psi \text{ iff } \phi \vee \psi \in K \dot{-} \phi \text{ and } \nvdash \phi$$

For (EC), we need to show that

$\psi \in K \dot{-} \phi$ iff $\psi \in K$ and either $\phi < \phi \vee \psi$ or $\vdash \phi$

Let $\psi \in K \dot{-} \phi$. By (K$\dot{-}$ 2), $\psi \in K$. By (K$\dot{-}$ 1), $\phi \vee \psi \in K \dot{-} \phi$. Suppose that $\nvdash \phi$. Then, by (†), $\phi < \phi \vee \psi$, as desired.

For the converse, let $\psi \in K$ and either $\phi < \phi \vee \psi$ or $\vdash \phi$. If firstly $\phi < \phi \vee \psi$, then by (†) $\phi \vee \psi \in K \dot{-} \phi$. Since $\psi \in K$, we get $\phi \rightarrow \psi \in K \dot{-} \phi$, by (K$\dot{-}$ 5) and

30 Compare Rott (2001, Observation 64).

(K$\dot{-}$ 1). Thus by (K$\dot{-}$ 1) $\psi \in K \dot{-} \phi$, as desired. If secondly $\vdash \phi$, then $K \subseteq K \dot{-} \phi$, by (K$\dot{-}$ 5), so again $\psi \in K \dot{-} \phi$, as desired. QED

Proof of Theorem 22. Since $<$ is a generalized entrenchment relation, it is acyclic, by Lemma 18 (vii), and can thus be used as a hierarchy for a safe contraction. Let ψ be in K. The case when $\vdash \psi$ is trivial so we assume that $\nvdash \psi$. Then ψ is in $K \dot{-} \phi$ acccording to the entrenchment-based contraction (EC) iff

(i) $\phi < \phi \vee \psi$
ψ is in $K \dot{-} \phi$ acccording to the safe contraction defined in Definition 2 iff
(ii) there is a set M of ϕ-safe elements such that $M \vdash \psi$

First, we show that (i) implies (ii). Suppose that $\phi < \phi \vee \psi$. From this we get, by Intersubstitutivity and the right-to-left direction of Choice, $\phi \vee \neg\psi < \phi \vee \psi$. Now consider the set $M = \{\phi \vee \psi, \neg\phi \vee \psi\}$. We show that this set is suitable for (ii). $\neg\phi \vee \psi$ is ϕ-safe anyway, because it is a consequence of $\neg\phi$ and thus will not be included in any minimal set implying ϕ. It remains to show that $\phi \vee \psi$ is ϕ-safe as well. Suppose that $\phi \vee \psi$ is in N for some N such that $N \vdash_{min} \phi$. Notice that N is finite and that $N - \{\phi \vee \psi\} \vdash \phi \vee \neg\psi$. Thus, by Continuing down, $\bigwedge(N - \{\phi \vee \psi\}) < \phi \vee \psi$, and by repeated application of (EII), we get that there is some χ in $N - \{\phi \vee \psi\}$ such that $\chi < \phi \vee \psi$. Thus $\phi \vee \psi$ is ϕ-safe, and this proves (ii).

Second, we show that (ii) implies (i). Pick an M from (ii). If $N \vdash_{min} \phi$ and χ is in $M \cap N$, then there is a ξ in N such that $\xi < \chi$. Take some subset $M' \subseteq M$ such that $M' \vdash_{min} \phi \vee \psi$. Then, if $N \vdash_{min} \phi$ and χ is in $M' \cap N$, there is a ξ in N such that $\xi < \chi$. By Continuing down, $\bigwedge N < \chi$. By repeated application of Conjunction up, which follows from Continuing down (by Lemma 18 (iv)), $\bigwedge N < \bigwedge(M' \cap N)$.

Now we show that $M' \nvdash \phi$. Suppose for reductio that $M' \vdash \phi$. Then, since $\nvdash \phi$, also $\nvdash \bigwedge M'$. It follows from $M' \vdash_{min} \phi \vee \psi$ that $M' \vdash_{min} \phi$, and we get $\bigwedge(M' \cap M') \in K \dot{-} \bigwedge M'$, contradicting (K$\dot{-}$ 4).

Now consider the set $N_0 = M' \cup \{\psi \to \phi\}$. It follows from $M' \vdash \phi \vee \psi$ that $N_0 \vdash \phi$. Since $M' \nvdash \phi$, $\psi \to \phi$ is not redundant. Suppose that some $\rho \in M'$ is redundant. Then $M' - \{\rho\} \cup \{\psi \to \phi\} \vdash \phi$ and consequently $M' - \{\rho\} \vdash \phi \vee \psi$, contrary to $M' \vdash_{min} \phi \vee \psi$. We therefore have $N_0 \vdash_{min} \phi$.

This gives us $\bigwedge N_0 < \bigwedge(M' \cap N_0)$ and since $N_0 = M' \cup \{\psi \to \phi\}$, $\bigwedge N_0 < \bigwedge M'$. By the right-to-left direction of Choice, $\psi \to \phi < \bigwedge M'$, and by Continuing up and Continuing down $\phi < \phi \vee \psi$, i.e. we have (i) as desired. QED

Proof of Theorem 27. Let the language have only the two propositional variables p and q, and let $K = \text{Cn}(p \wedge q)$. Let $*$ be a revision function on K such that $K * \neg p = Cn(\neg p \wedge q)$, $K * \neg q = Cn(\neg p \wedge \neg q)$, and $K * \neg(p \leftrightarrow q) = Cn(p \wedge \neg q)$.

$K \bot p = \{\text{Cn}(q), \text{Cn}(p \leftrightarrow q)\}$, and therefore $K \dot{-} p$ is either $\text{Cn}(q)$, $\text{Cn}(p \leftrightarrow q)$, or $\text{Cn}(p \to q)$. Of these only $\text{Cn}(q)$ satisfies the Levi identity $(K \dot{-} p) + \neg p = \text{Cn}(\neg p \wedge q)$. Thus $K \dot{-} p = \text{Cn}(q)$.

$K \bot q = \{\text{Cn}(p), \text{Cn}(p \leftrightarrow q)\}$, and therefore $K \dot{-} q$ is either $\text{Cn}(p)$, $\text{Cn}(p \leftrightarrow q)$, or $\text{Cn}(q \to p)$. Of these only $\text{Cn}(p \leftrightarrow q)$ satisfies the Levi identity $(K \dot{-} q) + \neg q = \text{Cn}(\neg p \wedge \neg q)$. Thus $K \dot{-} q = \text{Cn}(p \leftrightarrow q)$.

$K \perp (p \leftrightarrow q) = \{\text{Cn}(p), \text{Cn}(q)\}$, and therefore $K \dot{-} (p \leftrightarrow q)$ is either $\text{Cn}(p)$, $\text{Cn}(q)$, or $\text{Cn}(p \vee q)$. Of these only $\text{Cn}(p)$ satisfies the Levi identity $(K \dot{-} p \leftrightarrow q) + \neg(p \leftrightarrow q) = \text{Cn}(p \wedge \neg q)$. Thus $K \dot{-} (p \leftrightarrow q) = \text{Cn}(p)$.

Thus we have $K \dot{-} p = \text{Cn}(q)$, $K \dot{-} q = \text{Cn}(p \leftrightarrow q)$, and $K \dot{-} (p \leftrightarrow q) = \text{Cn}(p)$. It was shown in the proof of Theorem 6 that this contraction function is not reconstructible as a safe contraction, and consequently $*$ is not reconstructible as a safe revision. QED

References

Alchourrón, C., Gärdenfors, P., & Makinson, D. (1985). On the logic of theory change: Partial meet contraction and revision functions. *Journal of Symbolic Logic*, *50*, 510–530.

Alchourrón, C. E., & Makinson, D. (1985). On the logic of theory change: Safe contraction. *Studia Logica*, *44*, 405–422.

Alchourrón, C. E., & Makinson, D. (1986). Maps between some different kinds of contraction function: The finite case. *Studia Logica*, *45*, 187–198.

Areces, C., & Becher, V. (2001). Iterable AGM functions. In M.-A. Williams & H. Rott (Eds.), *Frontiers in Belief Revision* (pp. 261–277). Dordrecht: Kluwer.

Besnard, P., & Hunter, A. (2008). *Elements of Argumentation*. Cambridge, MA: MIT Press.

Carnota, R., & Rodríguez, R. (2011). AGM theory and artificial intelligence. In E. J. Olsson & S. Enqvist (Eds.), *Belief Revision meets Philosophy of Science* (pp. 1–42). Dordrecht: Springer.

Falappa, M., Fermé, E., & Kern-Isberner, G. (2006). On the Logic of Theory Change: Relations Between Incision and Selection Functions. In G. Brewka et al. (eds.), *ECAI'2006, 17th European Conference on Artificial Intelligence* (pp. 402–406). IOS Press.

Fermé, E., & Hansson, S. O. (2011). AGM 25 years of research in belief change. *Journal of Philosophical Logic*, *40*, 295–331.

Fermé, E., & Reis, M. (2012). System of spheres-based multiple contractions. *Journal of Philosophical Logic*, *41*, 29–52.

Fermé, E., Saez, K., & Sanz, P. (2003). Multiple kernel contraction. *Studia Logica*, *73*, 183–195.

Fuhrmann, A., & Hansson, S. O. (1994). A survey of multiple contractions. *Journal of Logic, Language, and Information*, *3*, 39–76.

Gärdenfors, P. (1982). Rules for rational changes of belief. In T. Pauli (Ed.), ⟨320311⟩: *Philosophical Essays Dedicated to Lennart Åqvist on His Fiftieth Birthday*, Vol. 34 (pp. 88–101). Uppsala: University of Uppsala Philosophical Studies.

Gärdenfors, P. (1988). *Knowledge in flux: Modeling the dynamics of epistemic states*. Cambridge, MA: Bradford Books, MIT Press.

Gärdenfors, P., & Makinson, D. (1988). Revisions of knowledge systems using epistemic entrenchment. In M. Vardi (Ed.), *TARK'88: Proceedings of the Second Conference on Theoretical Aspects of Reasoning About Knowledge* (pp. 83–95). Los Altos, CA: Morgan Kaufmann.

Hansson, S. O. (1989). New operators for theory change. *Theoria*, *55*, 114–132.

Hansson, S. O. (1992). A dyadic representation of belief. In P. Gärdenfors (Ed.), *Belief Revision* (pp. 89–121). Cambridge: Cambridge University Press.

Hansson, S. O. (1993). Reversing the Levi identity. *Journal of Philosophical Logic*, *22*, 637–669.

Hansson, S. O. (1994). Kernel contraction. *Journal of Symbolic Logic*, *59*, 845–859.

Hansson, S. O. (1999a). *A Textbook of Belief Dynamics. Theory Change and Database Updating*. Dordrecht: Kluwer.

Hansson, S. O. (1999b). A survey of non-prioritized belief revision. *Erkenntnis*, *50*, 413–427.

Hansson, S. O. (2010). Multiple and iterated contraction reduced to single-step single-sentence contraction. *Synthese*, *173*, 153–177.

Hansson, S. O. (2012). Global and iterated contraction and revision: An exploration of uniform and semi-uniform approaches. *Journal of Philosophical Logic*, *41*, 143–172.

Hansson, S. O., Fermé, E., Cantwell, J., & Falappa, M. (2001). Credibility-limited revision. *Journal of Symbolic Logic*, *66*, 1581–1596.

Li, J. (1998). A note on partial meet package contraction. *Journal of Logic, Language and Information*, *7*, 139–142.

Makinson, D. (1987). On the status of the postulate of recovery in the logic of theory change. *Journal of Philosophical Logic*, *16*, 383–394.

Makinson, D. (1989). General theory of cumulative inference. In M. Reinfrank, J. de Kleer, M. Ginsberg, & E. Sandewall (Eds.), *Non-Monotonic Reasoning—Proceedings of the 2nd International Workshop Grassau, FRG, June 13–15, 1988* (pp. 1–18). Berlin: Springer.

Makinson, D. (1997). Screened revision. *Theoria*, *63*, 14–23.

Reis, M., & Fermé, E. (2012). Possible worlds semantics for partial meet multiple contraction. *Journal of Philosophical Logic*, *41*, 7–28.

Rott, H. (1992a). On the logic of theory change: More maps between different kinds of contraction functions. In P. Gärdenfors (Ed.), *Belief Revision* (pp. 122–141). Cambridge: Cambridge University Press.

Rott, H. (1992b). Preferential belief change using generalized epistemic entrenchment. *Journal of Logic, Language and Information*, *1*, 45–78.

Rott, H. (1993). Belief contraction in the context of the general theory of rational choice. *Journal of Symbolic Logic*, *58*, 1426–1450.

Rott, H. (1999). Coherence and conservatism in the dynamics of belief. Part I: Finding the right framework. *Erkenntnis*, *50*, 387-412.

Rott, H. (2000). Just because: Taking belief bases seriously. In S. R. Buss, P. Hájek, & P. Pudlák (Eds.), *Logic Colloquium '98—Proceedings of the Annual European Summer Meeting of the Association for Symbolic Logic held in Prague, Urbana* (pp. 387–408). Urbana, Ill: Association for Symbolic Logic.

Rott, H. (2001). *Change, choice and inference: A study of belief revision and nonmonotonic reasoning*. Oxford: Oxford University Press.

Rott, H. (2003). Basic entrenchment. *Studia Logica*, *73*, 257–280.

Sen, A. (1986). Social choice theory. In K. J. Arrow & M. D. Intriligator (Eds.), *Handbook of Mathematical Economics* (Vol. III, pp. 1073–1181). Amsterdam: Elsevier/North-Holland.

Sen, A. (1997). Maximization and the act of choice. *Econometrica*, *65*, 745–779.

Spohn, W. (2010) Multiple contraction revisited. In M. Suárez, M. Dorato & M. Rédei (Eds.), *EPSA Epistemology and Methodology of Science. Launch of the European Philosophy of Science Association*, (Vol. 1, pp. 279–288). Dordrecht: Springer.

A Panorama of Iterated Revision

Pavlos Peppas

Abstract In 1985, David Makinson, together with Carlos Alchourron and Peter Gärdenfors published an article, the now renowned "AGM paper", that gave rise to an entire new area of research: Belief Revision. The AGM paper set the stage for studying belief revision and provided the first fundamental results in the area. There was however one aspect of belief revision that was not addressed in the AGM paper: iterated belief revision. Since 1985, there have been numerous attempts to tackle this problem. In this chapter, we shall review some of the most influential approaches to the problem of iterated belief revision, and discuss their strengths and shortcomings.

Keywords Belief revision · Iterated revision · Possible worlds

1 Introduction

Almost three decades ago, David Makinson, together with Carlos Alchourron and Peter Gärdenfors published their classical work on the logic of theory change, the renowned "AGM paper" Alchourron et al. (1985). From this paper the area of Belief Revision was born and hundreds of future publications drew inspiration from it and contributed to the development of the field. Yet in this survey we will not focus on what the AGM did do, but rather on what it left out: *iterated* belief revision.

Let us set the stage with an example. Vasiliki is an archeologist participating in an excavation of an ancient temple near Athens. One day her trowel hits on an ancient Greek vase. The vase is a typical third century BC Greek vase. The design, the painting, the archeological layer it was discovered in, all confirm without doubt that the vase was made in Greece some 2300 years ago. The big surprise however came

P. Peppas (✉)
Department of Business Administration, University of Patras, Patra, Greece
e-mail: pavlos@upatras.gr

P. Peppas
Faculty of Engineering and IT, University of Technology, Sydney, Australia

S. O. Hansson (ed.), *David Makinson on Classical Methods for Non-Classical Problems*, Outstanding Contributions to Logic 3, DOI: 10.1007/978-94-007-7759-0_5,

when Vasiliki looked inside the vase. A small stone tablet with Maya script inscribed on it was lying inside! The discovery was mind-blowing. To the young enthusiastic archeologist, eager to make a name for herself, this was without doubt concrete proof that Columbus was not the first European to reach America. The ancient Greeks had beat him by 1800 years! History books had to be rewritten! Vasiliki spent the night thinking of all the changes that had to be made in our theories about ancient Greeks (and perhaps ancient Mayas).

The theory developed by Alchourron, Gärdenfors, and Makinson, was designed to address precisely these kind of scenarios. A rational agent receives new reliable information φ that (in principle) contradicts her initial belief set K; the agent is thus forced to move to a new belief set $K * \varphi$ containing φ. Moreover, the new belief set $K * \varphi$ has to be a *rational* change to K given φ. The inner workings of the transition from K to $K * \varphi$ are studied and formalized in the AGM paper, setting the stage for a plethora of exciting results to follow over the next three decades.

Yet, Vasiliki's story doesn't end here. The next day, Vasiliki tells her friend Margarita of her discovery. Margarita was more cautious in her reaction. She tells Vasiliki that the local museum is hosting an exhibition of ancient Mayan artifacts, and that this may have something to do with her tablet. Still, Vasiliki refuses to make the connection, and holds on to her theory of an ancient Greek Columbus. Yet her theory survived only another few hours. On the 8 o'clock news that evening Vasiliki finds out that the police had arrested a man who had confessed to having stolen an ancient Greek vase and a Maya tablet from the local museum, both of which he had hidden in a hole he dug at the very spot Vasiliki made her discovery.

Surprising as it may sound, the AGM paper doesn't cater for Vasiliki's later revision. At first, this seems very strange. Since the AGM framework formalizes correctly one-step belief revision, why can't we use the AGM models for a sequence of revisions, simply by treating such a sequence as a series of one-step revisions? The short answer is that by doing so, we are essentially assuming that each one-step revision in the series is *independent* from the rest, thus failing to capture the fact that all these one-step revisions are performed by *the same rational agent*, and therefore have to be related. The way that these one-step revisions are related to one-another is a hot topic that has led to many interesting proposals, the most influential of which we shall review in this survey.[1]

2 Preliminaries

Let us first fix some notation and terminology. Throughout this article we shall be working with a propositional language L in which the agent's beliefs as well as the new information will be expressed.[2] Sometimes we shall refer to L as *the object language*.

---

[1] There is some overlap between this article and an earlier survey of ours on belief revision Peppas (2008).

[2] We note that this assumption is made only to simplify the exposition; many of the approaches discussed herein can work, at least technically, in a more general setting.

For a set of sentences Γ of L, we denote by $Cn(\Gamma)$ the set of all logical consequences of Γ, i.e. $Cn(\Gamma) = \{\varphi \in L\colon \Gamma \vdash \varphi\}$. A theory K of L is any set of sentences of L closed under $\vdash$, i.e. $K = Cn(K)$. We shall denote the set of all theories of L by $\mathbb{K}_L$. A theory K of L is complete iff for all sentences $\varphi \in L$, $\varphi \in K$ or $\neg\varphi \in K$. We shall denote the set of all consistent complete theories of L by $\mathbb{M}L$. For a set of sentences Γ of L, $[\Gamma]$ denotes the set of all consistent complete theories of L that contain Γ. Often we shall use the notation $[\varphi]$ for a sentence $\varphi \in L$, as an abbreviation of $[\{\varphi\}]$. For a theory K and a set of sentences Γ of L, we shall denote by $K + \Gamma$ the closure under $\vdash$ of $K \cup \Gamma$, i.e. $K + \Gamma = Cn(K \cup \Gamma)$. For a sentence $\varphi \in L$ we shall often write $K + \varphi$ as an abbreviation of $K + \{\varphi\}$. Finally, the symbols $\top$ and $\bot$ will be used to denote an arbitrary (but fixed) tautology and contradiction of L respectively.

3 The AGM Postulates for Belief Revision

In the AGM paradigm the process of belief revision is modeled as a function $*$ mapping a theory K and a sentence φ to a new theory $K * \varphi$. Of course certain constraints need to be imposed on $*$ in order for it to capture the notion of *rational belief revision* correctly. A guiding intuition in formulating these constraints has been the *principle of minimal change* according to which a rational agent ought to change her beliefs *as little as possible* in order to (consistently) accommodate the new information.

In Gärdenfors (1984), a set of eight postulates, known as the *AGM postulates for belief revision*,[3] which are now widely regarded to have captured much of what is the essence of rational belief revision:

(K $*$ 1) $K * \varphi$ is a theory of L.
(K $*$ 2) $\varphi \in K * \varphi$.
(K $*$ 3) $K * \varphi \subseteq K + \varphi$.
(K $*$ 4) If $\neg\varphi \notin K$ then $K + \varphi \subseteq K * \varphi$.
(K $*$ 5) If φ is consistent then $K * \varphi$ is also consistent.
(K $*$ 6) If $\vdash \varphi \leftrightarrow \psi$ then $K * \varphi = K * \psi$.
(K $*$ 7) $K * (\varphi \wedge \psi) \subseteq (K * \varphi) + \psi$.
(K $*$ 8) If $\neg\psi \notin K * \varphi$ then $(K * \varphi) + \psi \subseteq K * (\varphi \wedge \psi)$.

Any function $* : \mathbb{K}_L \times L \mapsto \mathbb{K}_L$ satisfying the AGM postulates for revision $(K*1)$–$(K*8)$ is called an *AGM revision function* . The first six postulates $(K*1)$–$(K*6)$ are known as the *basic* AGM postulates (for revision), while $(K*7)$–$(K*8)$ are called the *supplementary* AGM postulates.

Postulate $(K * 1)$ says that the agent, being an ideal reasoner, remains logically omniscient after she revises her beliefs. Postulate $(K * 2)$ says that the new infor-

[3] Although these postulates where first proposed by Gärdenfors alone, they were extensively studied in collaboration with Alchourron and Makinson (1985); thus their name.

mation φ should *always* be included in the new belief set. $(K * 2)$ places enormous faith on the reliability of φ. The new information is perceived to be so reliable that it prevails over all previous conflicting beliefs, no matter what these beliefs might be. Postulates $(K * 3)$ and $(K * 4)$ viewed together state that whenever the new information φ does not contradict the initial belief set K, there is no reason to remove any of the original beliefs at all; the new belief state $K * \varphi$ will contain the whole of K, the new information φ, and whatever follows from the logical closure of K and φ (and nothing more). Essentially $(K * 3)$ and $(K * 4)$ express the notion of minimal change in the limiting case where the new information is consistent with the initial beliefs. $(K * 5)$ says that the agent should aim for consistency at any cost; the *only* case where it is "acceptable" for the agent to fail is when the new information in itself is inconsistent (in which case, because of $(K * 2)$, the agent can't do anything about it). $(K * 6)$ is known as the *irrelevance of syntax postulate*. It says that the syntax of the new information has no effect on the revision process; all that matters is its content (i.e. the proposition it represents). Hence, logically equivalent sentences φ and ψ change a theory K in the same way.

Finally, postulates $(K * 7)$ and $(K * 8)$ are best understood taken together. They say that for any two sentences φ and ψ, if in revising the initial belief set K by φ one is lucky enough to reach a belief set $K * \varphi$ that is consistent with ψ, then to produce $K * (\varphi \wedge \psi)$ all that one needs to do is to expand $K * \varphi$ with ψ; in symbols $K * (\varphi \wedge \psi) = (K * \varphi) + \psi$. The motivation for $(K * 7)$ and $(K * 8)$ comes again from the principle of minimal change. The rationale is (loosely speaking) as follows: $K * \varphi$ is a minimal change of K to include φ and therefore there is no way to arrive at $K * (\varphi \wedge \psi)$ from K with "less change". In fact, because $K * (\varphi \wedge \psi)$ also includes ψ one might have to make further changes apart from those needed to include φ. If however ψ is consistent with $K * \varphi$, these further changes can be limited to simply adding ψ to $K * \varphi$ and closing under logical implications—no further withdrawals are necessary.

As already mentioned, the AGM postulates are widely accepted as sound properties of rational belief revision. Notice however that all postulates refer to a *single* belief set K; no constraints are placed to relate future revisions with past ones.[4] We will have more to say about this in Sect. 5.

For the sake of readability from now on we shall restrict our attention only to revision by *consistent* epistemic input φ; i.e. we assume $\nvdash \neg\varphi$. We note however that most approaches discussed herein can also deal with the limiting case of revising by an inconsistent sentence.

4 Epistemic Entrenchment

Suppose that two different rational agents hold the same beliefs K and they receive the same information φ. Will the two agents revise K in the same way? The answer in general is no. The reason is that extra-logical factors come into play that may lead

[4] Although $(K * 7)$ and $(K * 8)$ seem to relate different belief sets, as will become apparent from the constructive models later on, this is not really the case.

the agents to respond differently to φ. Hence the AGM postulates do not determine a *single* rational revision function but a whole family of them. The choice of the "right" function for a particular scenario depends on the extra-logical factors mentioned above (see Gärdenfors and Makinson (1988), Peppas (2008) for details). These extra-logical factors essentially assign an *epistemic value* to the agent's individual beliefs which determine their fate during revision.

Considerations like these led Gärdenfors and Makinson (1988) to introduce the notion of *epistemic entrenchment* as a means of encoding the extra-logical factors that are relevant to belief revision. We note that originally epistemic entrenchment was introduced as a constructive model for another type of belief change called, *belief contraction*. Nevertheless because of the close connection between belief contraction and belief revision (see Gärdenfors and Makinson (1988), Peppas (2008) for details), epistemic entrenchment can also be regarded as a constructive model of belief revision and this is how we shall treat it herein.

Intuitively, the epistemic entrenchment of a belief ψ is the degree of resistance that ψ exhibits to change: the more entrenched ψ is, the less likely it is to be swept away during revision by some other belief φ. Formally, epistemic entrenchment is defined as a preorder $\leq$ on L satisfying the following axioms:

(EE1) If $\varphi \leq \psi$ and $\psi \leq \chi$ then $\varphi \leq \chi$.
(EE2) If $\varphi \vdash \psi$ then $\varphi \leq \psi$.
(EE3) $\varphi \leq \varphi \wedge \psi$ or $\psi \leq \varphi \wedge \psi$.
(EE4) When K is consistent, $\varphi \notin K$ iff $\varphi \leq \psi$ for all $\psi \in L$.
(EE5) If $\psi \leq \varphi$ for all $\psi \in L$, then $\vdash \varphi$.

Axiom (EE1) states that $\leq$ is transitive. (EE2) says that the stronger a belief is logically, the less entrenched it is. At first this may seem counter-intuitive. A closer look however will convince us otherwise. Consider two beliefs φ and ψ both of them members of a belief set K, and such that $\varphi \vdash \psi$. Then clearly, if one decides to give up ψ one will also have to remove φ (for otherwise logical closure will bring ψ back). On the other hand it is possible to give up φ and retain ψ. Hence giving up φ produces less epistemic loss than giving up ψ and therefore the former should be preferred whenever a choice exists between the two. Thus axiom (EE2). For axiom (EE3) notice that, again because of logical closure, one cannot give up $\varphi \wedge \psi$ without removing at least one of the sentences φ or ψ. Hence either φ or ψ (or even both) is at least as vulnerable as $\varphi \wedge \psi$ during revision. We note that from (EE1)–(EE3) it follows that $\leq$ is *total*; i.e. for any two sentences $\varphi, \psi \in L$, $\varphi \leq \psi$ or $\psi \leq \varphi$.

The final two axioms deal with the two ends of this total preorder $\leq$, i.e. with its minimal and its maximal elements. In particular, axiom (EE4) says that in the principal case where K is consistent, all non-beliefs (i.e. all the sentences that are not in K) are minimally entrenched; and conversely, all minimally entrenched sentences are non-beliefs. At the other end of the entrenchment spectrum we have all tautologies, which according to (EE5) are the only maximal elements of $\leq$ and therefore the hardest to remove.

Given a belief set K and an epistemic entrenchment $\leq$ associated with K (which encodes all the extra-logical factors that are relevant to belief revision), we can fully

Fig. 1 A System of Spheres

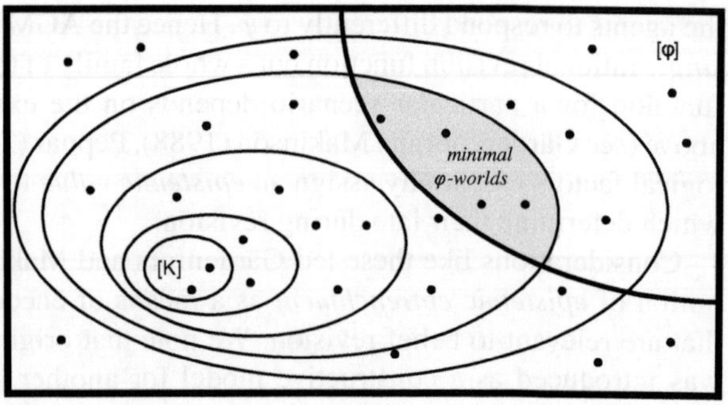

determine the new belief set $K * \varphi$ for any epistemic input φ by means of the following condition:

(E*) $\psi \in K * \varphi$ iff $(\varphi \to \neg\psi) < (\varphi \to \psi)$.

Loosely speaking, (E*) can be interpreted as follows: the presence of a belief ψ in $K * \phi$ is fully determined by its epistemic entrenchment relative to its negation, *under the assumption* φ. That is, $\psi \in K * \phi$ iff *assuming* φ, the belief ψ is more entrenched than its negation.

From the results obtained by Gärdenfors and Makinson in (1988) it follows that the class of revision functions induced from epistemic entrenchments via (E*) corresponds precisely to the family of AGM revision functions (see also Rott (1991); Peppas and Williams (1995)). In other words, epistemic entrenchment is a sound and complete model of the extra-logical factors relevant to AGM revision.

4.1 System of Spheres

Building on earlier work by Lewis (1973), Grove (1988), introduced another constructive model for belief revision based on a structure called *system of spheres*. Like an epistemic entrenchment, a system of sphere is essentially a preorder. However the objects being ordered are no longer sentences but consistent complete theories.

Given an initial belief set K a *system of spheres centered on* $[K]$ is formally defined as a collection S of subsets of $\mathbb{M}L$, called *spheres*, satisfying the following conditions (see Fig. 1)[5]:

(S1) S is totally ordered with respect to set inclusion; that is, if $V, U \in S$ then $V \subseteq U$ or $U \subseteq V$.

(S2) The smallest sphere in S is $[K]$; that is, $[K] \in S$, and if $V \in S$ then $[K] \subseteq V$.

(S3) $\mathbb{M}L \in S$ (and therefore $\mathbb{M}L$ is the largest sphere in S).

[5] Recall that $\mathbb{M}L$ is the set of all consistent complete theories of L, and for a theory K of L, $[K]$ is the set of all consistent complete theories that contain K.

(S4) For every $\varphi \in L$, if there is any sphere in S intersecting $[\varphi]$ then there is also a smallest sphere in S intersecting $[\varphi]$.

Intuitively a system of spheres S centered on $[K]$ represents the *relative plausibility* of consistent complete theories, which in this context play the role of possible worlds: the closer a consistent complete theory is to the center of S, the more plausible it is. Conditions (S1)–(S4) are then read as follows. (S1) says that any two worlds in S are always comparable in terms of plausibility. Condition (S2) tells us that the most plausible worlds are those compatible with the agent's initial belief set K. Condition (S3) says that all worlds appear somewhere in the plausibility spectrum. Finally condition (S4), also known as the *Limit Assumption*, is of a more technical nature. It guarantees that for any consistent sentence φ, if one starts at the outermost sphere $\mathbb{M}L$ (which clearly contains a φ-world) and gradually progresses towards the center of S, one will eventually meet the *smallest* sphere containing φ-worlds.[6] The smallest sphere in S intersecting $[\varphi]$ is denoted $c(\varphi)$. In the limiting case where φ is inconsistent, $c(\varphi)$ is defined to be equal to $\mathbb{M}L$.

Suppose now that we want to revise K by a sentence φ. Intuitively, the rational thing to do is to select the most plausible φ-worlds and define through them the new belief set $K * \varphi$:

(S*) $K * \varphi = \bigcap(c(\varphi) \cap [\varphi])$

Condition (S*) is precisely what Grove proposed as a means of constructing a revision function $*$ from a system of spheres S. Moreover Grove proved that the functions so constructed coincide with the functions satisfying the AGM postulates. Hence, like an epistemic entrenchment, a system of spheres is also a sound and complete model of the extra-logical factors relevant to belief revision.

Notice that there is an obvious connection between systems of spheres and preorders on possible worlds. In particular, let us call a preorder $\leq$ in $\mathbb{M}L$ *inductive* iff every non-empty subset of $\mathbb{M}L$ has a minimal element. For any belief set K and system of spheres S centered on $[K]$ we can derive an inductive total preorder on possible worlds $\leq$ as follows:

($S\leq$) $r \leq r'$ iff every sphere of S containing r' also contains r.

If $*$ is the revision function (at K) induced from S, then it is not hard to verify that

($\leq *$) $K * \varphi = min([\varphi], \leq)$.[7]

Conversely, for any theory K and inductive total preorder $\leq$ whose minimal worlds are all the K-worlds, we can construct by means of the condition ($\leq S$) below, a system of spheres S centered on $[K]$ such that the revision function $*$ induced from S at K is identical to the revision function induced from $\leq$ at K (via condition ($\leq *$)).

[6] We note that the Limit Assumption is similar to *smoothness* as defined by Kraus et al. (1990), also known as *stoppering* in Makinson (1994).

[7] For a set of possible worlds V, $min(V, \leq)$ denotes the set of *minimal* worlds in V with respect to $\leq$; i.e. $min(V, \leq) = \{r \in V : \text{for all } r' \in V, \text{if } r' \leq r \text{ then } r \leq r'\}$.

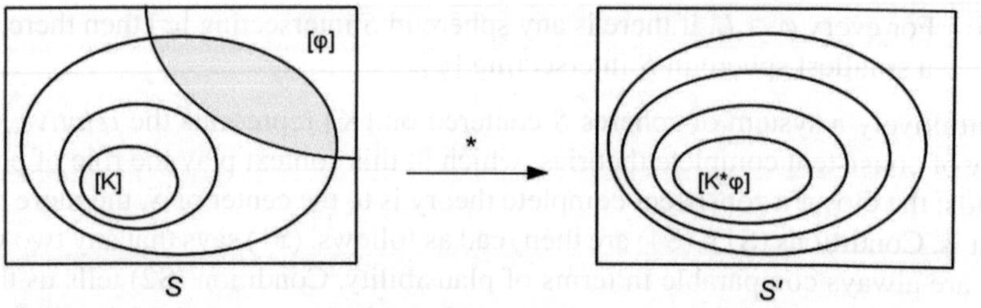

Fig. 2 What we need for Iterated Belief Revision

$(\leq S)$ $\quad V \in S$ iff there is an $r \in \mathbb{M}L$ such that $V = \{r' \in \mathbb{M}L : r' \leq r\}$.

Given the inter-definability of systems of spheres and inductive total preorders, in the rest of the article we shall switch back and forth between the two as the need arises. Moreover, for the rest of the article, all preorders are assumed to be inductive, unless explicitly stated otherwise.

5 The Problem of Iterated Revision

As already noted, when a rational agent performs a sequence of revisions, the individual steps in this sequence must somehow be related. Returning to the introductory example, the fact that Vasiliki responded to her discovery of a Maya tablet in an ancient Greek vase by adopting the belief of an ancient-Greek-Colombus (rather than, say, that the tablet and the vase were fake), may very well be related to her subsequent decision to hold on to this belief even when faced with the information of a local museum displaying Maya artifacts. Unfortunately, the AGM postulates tell us very little about the relationship between the individual steps of a sequence of revisions.

The problem can be expressed more formally as follows. Consider a theory K coupled with a structure encoding extra-logical information relevant to belief change, say, a system of spheres S centered on $[K]$. Suppose that we now receive new information φ, such that $\varphi \notin K$, thus leading us to the new theory $K * \varphi$. Notice that $K * \varphi$ is *fully determined* by S and φ, and moreover, as Grove has shown, the transition from the old to the new belief set satisfies the AGM postulates. So far, so good. This however is as far as the classical AGM paradigm can take us. If at this point we receive further evidence ψ that is inconsistent with $K * \varphi$ (but not self-contradictory), we have no means of producing $K * \varphi * \psi$. What we are missing is a new system of spheres S' associated with $K * \varphi$ that would determine our revision policy at this point (see Fig. 2).[8]

Presumably, the new system of spheres S' would be a *rational* offspring of S and φ. Even if S' in not *fully determined* by S and φ, it should at least be constrained

[8] Notice that the original system of spheres S is centered on $[K]$, not on $[K * \varphi]$, and therefore cannot be used to direct further revisions.

by them. More generally, the problem of iterated revision is the problem of formulating constraints that capture the dynamics of the structure used to encode one-step revision policies (be it a system of spheres, an epistemic entrenchment, or any other mathematical object with a similar function). The rest of this article reviews (some of) the best known proposals.

6 Iterated Revision with Enriched Epistemic Input

Spohn (1988), was one of the first to address the problem of iterated belief revision, and the elegance of his solution has influenced most of the proposals that followed. This elegance however comes with a price; to produce the new preference structure from the old one, Spohn requires as input not only the new information φ, but also the *degree of firmness* by which the agent accepts the new information. Let us take a closer look at Spohn's solution (to simplify discussion, in this section we shall consider only revision by *consistent* sentences on *consistent* theories).

To start with, Spohn uses a richer structure than a system of spheres to represent the preference information related to a belief set K. He calls this structure an *ordinal conditional function* (OCF). Formally, an OCF κ is a function from the set $\mathbb{M}L$ of possible worlds to the class of ordinals such that at least one world is assigned the ordinal 0. Intuitively, κ assigns a plausibility grading to possible worlds: the larger $\kappa(r)$ is for some world r, the less plausible r is.[9] This plausibility grading can easily be extended to sentences: for any consistent sentence φ, we define $\kappa(\varphi)$ to be the κ-value of the most plausible φ-world; in symbols, $\kappa(\varphi) = \min(\{\kappa(r) : r \in [\varphi]\})$.

Clearly, the most plausible worlds of all are those whose κ-value is zero. These worlds define the belief set that κ is related to. In particular, we shall say that the belief set K is related to the OCF κ iff $K = \bigcap\{r \in \mathbb{M}L\colon \kappa(r) = 0\}$. Given a theory K and an OCF κ related to it, Spohn can produce the revision of K by any sentence φ, *as well as* the new ordinal conditional function related to $K * \varphi$. The catch is, as mentioned earlier, that apart from φ, its degree of firmness d is also needed as input. The new OCF produced from κ and the pair $\langle \varphi, d \rangle$ is denoted $\kappa * \langle \varphi, d \rangle$ and it is defined as follows[10]:

$$\text{(CON)} \quad \kappa * \langle \varphi, d \rangle(r) = \begin{cases} \kappa(r) - \kappa(\varphi) & \text{if } r \in [\varphi] \\ \kappa(r) - \kappa(\neg\varphi) + d & \text{otherwise} \end{cases}$$

Essentially condition (CON) works as follows. Starting with κ, all φ-worlds are shifted "downwards" against all $\neg\varphi$-worlds until the most plausible of them hit the bottom of the rank; moreover, all $\neg\varphi$-worlds are shifted "upwards" until the most

9 In this sense an ordinal conditional function κ is quite similar to a system of spheres S: both are formal devices for ranking possible worlds in terms of plausibility. However κ not only tells us which of any two worlds is more plausible; it also tells us by *how much* is one world more plausible than the other.

10 The left subtraction of two ordinals α, β such that $\alpha \geq \beta$, is defined as the unique ordinal γ such that $\alpha = \beta + \gamma$.

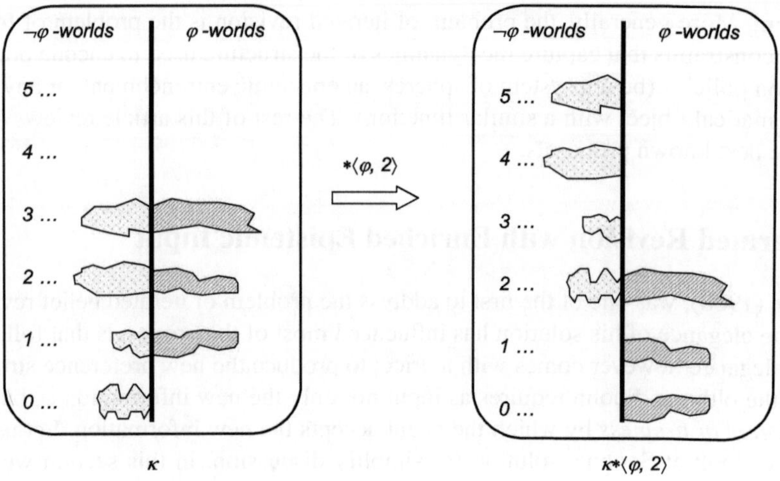

Fig. 3 Spohn's Conditionalization

plausible of them are at distance d from the bottom (see Fig. 3). Spohn calls this process *conditionalization* (more precisely, the $\langle\varphi, d\rangle$-*conditionalization* of κ) and argues that is the right process for revising OCFs.

Conditionalization is indeed intuitively appealing and has many nice formal properties, including compliance with the AGM postulates[11] (see Spohn 1988, Gärdenfors 1988, Williams 1994). Moreover notice that the restriction of κ to $[\varphi]$ and to $[\neg\varphi]$ remains unchanged during conditionalization, hence in this sense the principle of minimal change is observed not only for transitions between belief sets, but also for their associated OCFs.

There are however other ways of interpreting minimal change in the context of iterated revision. Williams (1994) proposes the process of *adjustment* as an alternative to conditionalization. Given an OCF κ, Williams defines the $\langle\varphi, d\rangle$—*adjustment of* κ, which we denote by $\kappa\circ\langle\varphi, d\rangle$, as follows:

$$\text{(ADJ)} \quad \kappa\circ\langle\varphi, d\rangle(r) = \begin{cases} 0 & \text{if } r \in [\varphi],\ d > 0, \text{ and } \kappa(r) = \kappa(\varphi) \\ d & \text{if } r \in [\neg\varphi], \text{ and } \kappa(r) = \kappa(\neg\varphi) \text{ or } \kappa(r) \leq d \\ \kappa(r) & \text{otherwise} \end{cases}$$

Adjustment minimizes changes to the grades of possible worlds in *absolute* terms. To see this, notice that in the principal case where $\kappa(\varphi) > 0$ and $d > 0$,[12] the only φ-worlds that change grades are the most plausible ones (wrt κ), whose grade becomes

[11] That is, given an OCF κ and any $d > 0$, the function $*$ defined as $K * \varphi = \bigcap\{r \in \mathbb{M}L : \kappa * \langle\varphi, d\rangle(r) = 0\}$ satisfies the AGM postulates $(K * 1)$–$(K * 8)$.

[12] This is the case where the new information φ contradicts the original belief set (since $\kappa(\varphi) > 0$, the agent originally believes $\neg\varphi$).

zero. Moreover, the only $\neg\varphi$-worlds that change grades are those with grades smaller that d, or, if no such world exists, the minimal $\neg\varphi$-worlds whose grade becomes d. Like conditionalization, adjustment satisfies all AGM postulates for revision. The process of adjustment was further developed in Williams (1996) and Benferhat et al. (2004), with the introduction of *maxi-adjustment* and *disjunctive maxi-adjustment* respectively.

The entire apparatus of OCFs and their dynamics (conditionalization or adjustment) can be reproduced using sentences rather than possible worlds as building blocks. To this end, Williams (1994) defined the notion of *ordinal epistemic entrenchments functions* (OEF) as a special mapping from sentences to ordinals, intended to encode the resistance of sentences to change: the higher the ordinal assigned to sentence, the higher the resistance of the sentence. As the name suggests, an OEF is an enriched version of an epistemic entrenchment (in the same way that an OCF is an enriched version of a system of spheres). Williams formulated the counterparts of conditionalization and adjustment for OEF and proved their equivalence with the corresponding operation on OCFs.

In Nayak (1994), Nayak took this line of work one step further. Using the original epistemic entrenchment model to encode sentences resistance to change, he considers the general problem of epistemic entrenchment dynamics. The novelty in Nayak's approach is that the epistemic input is no longer a simple sentence as in AGM, or even a sentence coupled with a degree of firmness as in OCF dynamics, but rather another epistemic entrenchment; i.e. an initial epistemic entrenchment $\leq$ is revised by another epistemic entrenchment $\leq'$, producing a new epistemic entrenchment $\leq * \leq'$. Notice that because of (EE4) (see Sect. 4), an epistemic entrenchment uniquely determines the belief set it relates to; we shall call this set the *content* of an epistemic entrenchment. Hence epistemic entrenchment revision should be interpreted as follows. The initial epistemic entrenchment $\leq$ represents both the original belief set K (defined as its content) as well as the preference structure related to K. The input $\leq'$ represents prioritized evidence: the content K' of $\leq'$ describes the new information, while the ordering on K' is related (but not identical) to the relative strength of acceptance of the sentences in K'. Finally, $\leq * \leq'$ encodes both the posterior belief set as well as the preference structure associated with it.

The construction of $\leq * \leq'$ is motivated by what is now known as *lexicographic* revision, defined in terms of systems of spheres as follows.[13] Consider two systems of spheres S and S', with the former representing the initial belief state, and the latter the epistemic input. Let $B1, B2, B3, \ldots$ be the bands of S, and $A1, A2, \ldots$ the bands of S'.[14] The revision of S by S' is defined to be the system of spheres composed by the following bands (after eliminating the empty sets): $A1 \cap B1$, $A1 \cap B2$, $A1 \cap B3, \ldots$, $A2 \cap B1$, $A2 \cap B2$, $A2 \cap B3, \ldots$

[13] There is a well known connection between a system of spheres S and an epistemic entrenchment $\leq$. In particular, the latter can easily be constructed from the former (while preserving the induced revision function) as follows: $\varphi \leq \psi$ iff $c(\neg\varphi) \subseteq c(\neg\psi)$, for all contingent $\varphi, \psi \in L$.

[14] Loosely speaking, the bands of a system of spheres are the sets of worlds between successive spheres.

Nayak's induced operators satisfy (a generalized version of) the AGM postulates for revision. Compared to Williams' OEFs dynamics, Nayak's work is closer to the AGM tradition (both use epistemic entrenchments to represent belief states and plausibility is represented in relative rather than absolute terms). On the other hand however, when it comes to the modeling epistemic input, Nayak departs even further than Williams from the AGM paradigm; an epistemic entrenchment (used by Nayak) is a much more complex structure than a weighted sentence (used by Williams), which in turn is richer than a simple sentence (used in the original AGM paradigm).

7 Iterated Revision with Simple Epistemic Input

This raises the question of whether a solution to iterated revision can be produced using only the apparatus of the original AGM framework; that is, using epistemic entrenchments (or systems of spheres or selection functions) to model belief states, and simple sentences to model epistemic input.

7.1 The DP Approach and its Sequels

One of the most influential proposals to this end is the work of Darwiche and Pearl ("DP" for short), (1997). The first important feature of this work is that, contrary to the original approach of Alchourron, Gärdenfors and Makinson (but similarly to Spohn (1988), Williams (1994), Nayak (1994)), revision functions operate on *belief states*, not on *belief sets*. In the present context a belief state (also referred to as an *epistemic state*) is defined as a belief set coupled with a structure that encodes relative plausibility (e.g., an epistemic entrenchment, a system of spheres, etc.). Clearly a belief state is a richer model that a belief set. Hence it could well be the case that two belief states agree on their belief content (i.e. their belief sets), but behave differently under revision because of differences in their preference structures. For ease of presentation, and although this is not required by Darwiche and Pearl, in the rest of this section we shall identify belief states with systems of spheres; note that given a system of spheres S we can easily retrieve its belief content–simply notice that $c(\top)$ is the smallest sphere of S and therefore $\bigcap c(\top)$ is the belief set associated with S.[15] We shall denote this belief set by $B(S)$; i.e. $B(S) = \bigcap c(\top)$. We may sometimes abuse notation and write for a sentence φ that $\varphi \in S$ instead of $\varphi \in B(S)$.:

[15] Recall that for any sentence ψ, $c(\psi)$ denotes the smallest sphere in S intersecting $[\psi]$.

With these conventions, $*$ becomes a function that maps a system of spheres S and a sentence φ, to a new system of spheres $S * \varphi$. Darwiche and Pearl reformulated the AGM postulates accordingly to reflect the shift from belief sets to belief states[16]:

(S*1) $S * \varphi$ is a system of spheres.
(S*2) $\varphi \in B(S * \varphi)$.
(S*3) $B(S * \varphi) \subseteq B(S + \varphi)$.
(S*4) If $\neg\varphi \notin B(S)$ then $B(S + \varphi) \subseteq B(S * \varphi)$.
(S*5) If φ is consistent then $B(S * \varphi)$ is also consistent.
(S*6) If $\vdash \varphi \leftrightarrow \psi$ then $B(S * \varphi) = B(S * \psi)$.
(S*7) $B(S * (\varphi \wedge \psi)) \subseteq B((S * \varphi) + \psi)$.
(S*8) If $\neg\psi \notin B(S * \varphi)$ then $B((S * \varphi) + \psi) \subseteq B(S * (\varphi \wedge \psi))$.

With this background, Darwiche and Pearl introduced four additional postulates to regulate iterated revisions[17]:

(DP1) If $\varphi \vdash \chi$ then $(S * \chi) * \varphi = S * \varphi$.
(DP2) If $\varphi \vdash \neg\chi$ then $(S * \chi) * \varphi = S * \varphi$.
(DP3) If $\chi \in S * \varphi$ then $\chi \in B((S * \chi) * \varphi)$.
(DP4) If $\neg\chi \notin S * \varphi$ then $\neg\chi \notin B((S * \chi) * \varphi)$.

Postulate (DP1) says that if the subsequent evidence φ is logically stronger than the initial evidence χ then φ overrides whatever changes χ may have made. (DP2) says that if two contradictory pieces of evidence arrive sequentially one after the other, it is the latter that will prevail. Notice that according to (DP1) and (DP2), the latter piece of evidence φ prevails in a very *strong* sense: φ fully determines, not just the next belief set, but the entire next belief state (alias system of spheres), overriding any effects that the former piece of evidence may have had in either of them. (DP3) says that if revising S by φ causes χ to be accepted in the new belief state, then revising first by χ and then by φ cannot possibly block the acceptance of χ. Finally, (DP4) captures the intuition that "no evidence can contribute to its own demise" Darwiche and Pearl (1997); if the revision of S by φ does not cause the acceptance of $\neg\chi$, then surely this should still be the case if S is first revised by χ before revised by φ.

Apart from their simplicity and intuitive appeal, postulates (DP1)–(DP4) also have a nice characterization in terms of systems-of-spheres dynamics. First however some more notation: for a system of spheres S, we shall denote by $\leq_S$ the preorder induced from S by $(S\ \leq)$. Moreover $<_S$ denotes the strict part of $\leq_S$.

Darwiche and Pearl proved that there is a one-to-one correspondence between (DP1)–(DP4) and the following constraints on system-of-spheres dynamics:

(DPS1) If $r, r' \in [\varphi]$ then $r \leq_{S*\phi} r'$ iff $r \leq_S r'$.

---

[16] It should be noted that Darwiche and Pearl use different notation, and as already mentioned, they leave open the representation of a belief state (it is not necessarily represented as a system of spheres).

[17] Like with (S*1)–(S*8), the original formulation of (DP1)–(DP4) is slightly different. Herein we have rephrased the Darwiche and Pearl postulates in AGM notation.

(DPS2) If $r, r' \in [\neg\varphi]$ then $r \leq_{S*\phi} r'$ iff $r \leq_S r'$.
(DPS3) If $r \in [\varphi]$ and $r' \in [\neg\varphi]$ then $r <_S r'$ entails $r <_{S*\phi} r'$.
(DPS4) If $r \in [\varphi]$ and $r' \in [\neg\varphi]$ then $r \leq_S r'$ entails $r \leq_{S*\phi} r'$.

Theorem 1 (Darwiche and Pearl 1997). *Let S be a belief state and $*$ a revision function satisfying the (DP-modified) AGM postulates. Then $*$ satisfies (DP1)–(DP4) iff it satisfies (DPS1)–(DPS4) respectively.*

In a way, Darwiche and Pearl were forced to make the shift from belief sets to belief states, for otherwise, as pointed out by Lehmann (1995), (DP2) would have conflicted with the AGM postulates.[18] For instance, let p, q be propositional variables, and define $K = Cn(\emptyset)$ and $K' = Cn(\{p\})$. From $(K * 1)$–$(K * 8)$ it follows that $K' * \neg q = K * (p \wedge \neg q)$. Therefore, $(K' * \neg q) * q = (K * (p \wedge \neg q)) * q$. On the other hand, from (DP2) we derive that $(K' * \neg q) * q = K' * q$, and similarly, $(K * (p \wedge \neg q)) * q = K * q$. Moreover from (K*1)–(K*8) it follows that $K' * q = K' + q = Cn(\{p, q\})$, whereas, $K * q = K + q = Cn(\{q\})$. Hence, $(K' * \neg q) * q \neq (K * (p \wedge \neg q)) * q$ Contradiction.

Nayak et al. (2003), proposed another way to reconcile (DP2) with the AGM postulates that does not require moving away from belief sets. It does however require two other changes to the original formulation of belief revision. Firstly, $*$ is defined as a *unary* rather than a binary function, mapping sentences to theories. That is, each theory K is assigned its own revision function which for any sentence φ produces the revision of K by φ. We shall denote the unary revision function assigned to K by $*_K$ and the result of revising K by φ as $*_K(\varphi)$. This change in notation will serve as a reminder of the unary nature of revision functions adopted in Nayak et al. (2003) Notice that this reformulation of revision functions does not require any modification to the AGM postulates, since all of them refer only to a single theory K.

The second modification to revision functions proposed in Nayak et al. (2003) is that they are *dynamic*; i.e. they could change as new evidence arrives. The implications of this modification are best illustrated in the following scenario. Consider an agent whose belief set at time t_0 is K_0, and who receives a sequence of new evidence $\varphi_1, \varphi_2, \ldots, \varphi_n$ and performs the corresponding n revisions that take him at time t_n to the belief set K_n. Suppose now that it so happens that $K_n = K_0$; i.e. after incorporating all the new evidence, the agent ended up with the theory she started with. Because of the dynamic nature of revision functions in Nayak et al. (2003), it is possible that the revision function assigned to K_0 at time t_0 is different from the one assigned to it at time t_n. Hence although the evidence $\varphi_1, \varphi_2, \ldots, \varphi_n$ did not change the agent's beliefs, they did alter her attitude towards new epistemic input.

These two modifications to revision functions take care of the inconsistency between (DP2) and the AGM postulates when applied to belief sets. There is however another problem with (DP1)–(DP4) identified in Nayak et al. (2003). Nayak et al. argue that (DP1)–(DP4) are also too permissive; i.e. there are revision functions that comply with both the AGM and DP postulates and nevertheless lead to

[18] Although it should be noted that Darwiche and Pearl argue that this shift is not necessitated by technical reasons alone; conceptual considerations also point the same way.

counter-intuitive results. Moreover, an earlier proposal by Boutilier (1993, 1996), which strengthens (DP1)–(DP4) still fails to block the unintended revision functions (and introduces some problems of its own–see Darwiche and Pearl (1997)). Hence Nayak *et al.* proposed the following addition to (DP1)–(DP4) instead, called the *Conjunction Postulate* :

(CNJ) If $\chi \wedge \varphi \nvdash \bot$, then $*^{\chi}_{*_K(\chi)}(\varphi) = *_K(\chi \wedge \varphi)$.

Some comments on the notation in (CNJ) are in order. As usual, K denotes the initial belief set, and $*_K$ the unary revision function associated with it. When K is revised by a sentence χ, a new theory $*_K(\chi)$ is produced. This however is not the only outcome of the revision of K by χ; a new revision function associated with $*_K(\chi)$ is also produced. This new revision function is denoted in (CNJ) by $*^{\chi}_{*_K(\chi)}$. The need for the superscript χ is due to the dynamic nature of $*$ (as discussed earlier, along a sequence of revisions, the same belief set may appear more than once, each time with a different revision function associated to it, depending on the input sequence).

Postulate (CNJ) essentially says that if two pieces of evidence χ and φ are consistent with each other, then it makes no difference whether they arrive sequentially or simultaneously; in both cases the revision of the initial belief set K produces the same theory.

Nayak et al. show that (CNJ) is consistent with both AGM and DP postulates, and it blocks the counterexamples known at the time. In fact (CNJ) is strong enough to *uniquely* determine (together with (K*1)–(K*8) and (DP1)–(DP4)) the new revision function $*^{\chi}_{*_K(\chi)}$. A construction of this new revision function from $*_K$ and χ is given in Nayak et al. (2003)

Yet, some authors have argued, Zhang (2004), Jin and Thielscher (2005), that while (DP1)–(DP4) are too permissive, the addition of (CNJ) is too radical (at least in some cases). Accordingly, Jin and Thielscher proposed a weakening of (CNJ), which they call the *Independence postulate* Jin and Thielscher (2005, 2007). The Independence postulate is formulated within the DP framework; that is, it assumes that belief states rather than belief sets are the primary objects of change[19]:

(Ind) If $\neg\chi \notin S * \varphi$ then $\chi \in B((S * \chi) * \varphi)$.

The Independence postulate, apart from performing well in indicative examples (see Jin and Thielscher (2005)), also has a nice characterization in terms of system of spheres dynamics:

(IndR) If $r \in [\varphi]$ and $r' \in [\neg\varphi]$ then $r \ \leq_S r'$ entails $r \ <_{S*\phi} r'$.

Theorem 2 (Jin and Thielscher 2005). *Let S be a belief state and $*$ a revision function satisfying the (DP-modified) AGM postulates. Then $*$ satisfies (Ind) iff it satisfies (IndR).*

---

[19] It should be noted that (Ind) was also studied independently by Booth et al. (2005), Booth and Meyer (2006).

The Independence postulate can be shown to be weaker than (CNJ) and in view of Theorems 1, 2, it is clearly stronger than (DP3) and (DP4). Jin and Thielscher show that (Ind) is consistent with the AGM and DP postulates combined.

7.2 Conflicts

Despite the popularity of the DP approach and the remedies introduced to fix its initial problems, their is still some controversy surrounding the DP postulates. One of the latest criticisms comes in the form of a result proving that the DP postulates are in conflict with another important aspect of belief revision; namely, the desideratum that beliefs that are not *relevant* to the new information should be immune to the revision process.

As noted by Parikh in (1999), the AGM postulates are too liberal in their treatment of relevance, and hence he proposed an extra postulate called (P) to supplement them presented below. First some notation: for any sentence x, L_x denotes the (unique) smallest language in which x can be expressed:

(P) If $K = Cn(x \wedge y)$, where x, y are sentences of disjoint languages L_x, L_y respectively, and moreover $\varphi \in L_x$, then $K * \varphi = (Cn(x) \circ \varphi) + y$, where $\circ$ is a revision function over the language L_x.

Hence (P) essentially says that, if the initial belief set can be split into two parts x and y of disjoint languages L_x and L_y, then the revision of K by any epistemic input in L_x leaves the y-part of K unaffected. Although (P) may not tell the whole story of relevance-sensitive belief revision, it is surely an intuitive first step.

At first glance postulate (P) appears to be unrelated to iterated revision and consequently to the DP postulates. Yet, in Peppas et al. (2008), using the semantics of (P) developed in Peppas et al. (2004), proved that (P) is in conflict with each one of the DP postulates:

Theorem 3 (Peppas et al. 2008). *In the presence of the AGM postulates, postulate (P) is inconsistent with each one of the postulates (DP1)–(DP4).*

It should be noted that according to the above result, the DP postulates are in conflict with (P) not only as a whole, but also in isolation. Of course Theorem 3 can be interpreted as a liability for axiom (P) rather than the DP postulates. Nevertheless, until the issue is settled, the DP postulates (and axiom (P)) remain contestable.

7.3 Enriched Epistemic States

Next we turn to two approaches to Iterated Belief Revision that, while keeping the epistemic input simple, use richer structures to model epistemic states.

Fig. 4 Distances Between Possible Worlds

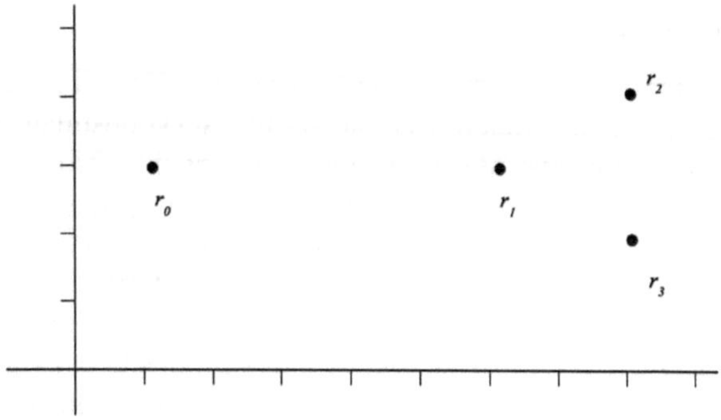

The first is a *distance-based approach* by Lehmann, Magidor, and Schlechta (or LMS for short), Lehmann et al. (2001). The LMS proposal is based on a simple and very intuitive idea: introduce an *priori* notion of *distance d* between possible worlds, and use d to derive the preorders associated with the initial belief set K *as well as* with all future belief sets resulting from K via iterated revisions. Formally, d is a function that maps any two possible worlds r and r' to a real number $d(r, r')$[20] which is to be interpreted as a measure of how far away r' looks from the standpoint of r.

Let us take a concrete example to illustrate the LMS approach. Suppose that the object language L is built from two propositional variables p and q, that give rise to four possible worlds $r_0 = Cn(\{p, q\})$, $r_1 = Cn(\{p, \overline{q}\})$, $r_2 = Cn(\{\overline{p}, q\})$, and $r_3 = Cn(\{\overline{p}, \overline{q}\})$.[21] Moreover assume that the distance d between these worlds is the Euclidean distance between the corresponding points in Fig. 4.

Suppose that the initial belief state is r_0. Then according to Fig. 4, the world closest to r_0 is r_1, followed by the worlds r_2, r_3 which are equidistant from r_0. Hence the preorder associated with r_0 is $r_0 \leq r_1 \leq r_2, r_3$. Future preorders can likewise be derived. If for example r_0 is revised by $\overline{p}$, the new preorder $\leq'$ associated with $r_0 * \overline{p}$ is $r_2, r_3 \leq' r_1 \leq' r_0$.[22] More generally, the preorder $\leq$ derived from a distance d which is associated with a belief set K, can be defined as follows:

(d$\leq$) $r \leq r'$ iff there is a $w \in [K]$ such that, for all $w' \in [K]$, $d(w, r) \leq d(w', r')$.

LMS assume very little about the properties of the distance function d. In fact they assume even less that what are generally considered reasonable properties of distance, committing themselves only to the following assumption:

(d1) $d(r, r') = 0$ iff $r = r'$.

If in addition to (d1), d satisfies the property (d2) below, it is called *symmetric*:

---

20 Distance between possible worlds does not have to be expressed in terms of real numbers; this is an assumption made herein for simplicity.

21 For any propositional variable x, by $\overline{x}$ we denote the negation of x.

22 The distance between a world r and a *set* of worlds V can be defined as the smallest distance between r and a world in V. Hence, according to Fig. 4, the closest world to $\{r_2, r_3\}$ is r_1 followed by r_0.

(d2) $d(r, r') = d(r', r)$.

It can be shown that under property (d1) alone, all revision functions induced from distance functions d satisfy the AGM postulates. The converse however is not true; there exist AGM revision functions that cannot be constructed by any distance function. Lehmann, Magidor, and Schlechta provided an exact characterization of the class of AGM functions that can be constructed from distance functions. Herein we shall present only the representation result related to symmetric distance functions, and only for the special case of finitary propositional languages (i.e. propositional language L built from *finitely many* propositional variables). For similar results on more general cases, the reader is referred to Lehmann et al. (2001).

When the object language L is finitary, any theory K can be represented as the logical closure of a single sentence (i.e. all theories are finitely axiomatizable). Hence for condition (d*) below, we shall abuse notation and extend the application of the disjunction operator to theories with the understanding that for any two theories K and K', $K \vee K'$ denotes the disjunction of sentences χ and χ' whose logical closure equals K and K' respectively.

(d *) If K_0 is consistent with $K_1 * (K_0 \vee K_2)$, K_1 is consistent with $K_2 * (K_1 \vee K_3)$, ..., and, K_{n-1} is consistent with $K_n * (K_{n-1} \vee K_0)$, then K_1 is consistent with $K_0 * (K_n \vee K_1)$

Theorem 4 (Lehmann et al. 2001). *An AGM revision function $*$ can be constructed from a symmetric distance function d over possible worlds, iff $*$ satisfies (d*).*

The key feature of the LMS approach is the use of a belief-set-independent meta-structure, namely a distance function d, to derive all preorders necessary to guide present and future revisions. Recently, Booth and Meyer (2011) also developed such a meta-structure with similar aims, which however is quite different from a distance function.

The basic idea in Booth and Meyer (2011), is to construct for each possible world r, a *"positive"* and *"negative"* clone, denoted r^+ and r^- respectfully. The set of all such "signed" possible worlds is denoted by $\mathbb{M}L^{\pm}$; moreover, $\mathbb{M}L^+$ and $\mathbb{M}L^-$ denote the sets of positively and negatively signed possible worlds respectfully (and hence $\mathbb{M}L^{\pm} = \mathbb{M}L^+ \cup \mathbb{M}L^-$).

The intuition behind signed possible worlds is given by the following passage from Booth and Meyer (2011):

> ... when we compare two different worlds x and y according to preference, often we are able to imagine different contingencies, according to whether all goes well in x and y or not. Our idea is to associate to each world x two abstract objects x^+ and x^-, with the intuition that x^+ represents x in positive circumstances, while x^- represents x in negative circumstances.

Against this background, Booth and Meyer introduced a preorder $\preccurlyeq$ over signed worlds, which is the meta-structure alluded to above. In particular $\preccurlyeq$ is defined as a total preorder in $\mathbb{M}L^{\pm}$ such that for all possible worlds r, w,

($\preccurlyeq$ 1) $r^+ \prec r^-$

Fig. 5 Signed Possible Worlds

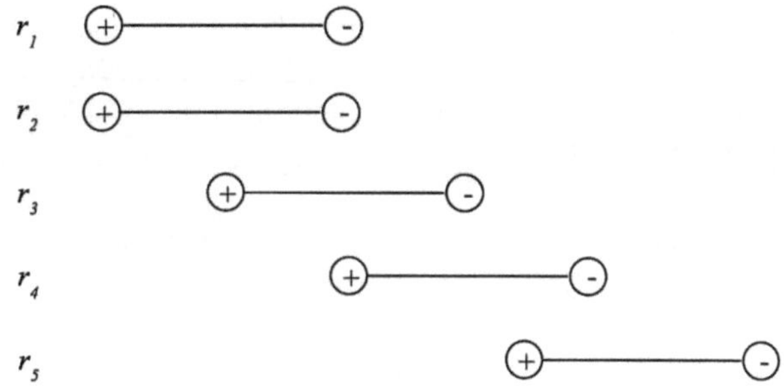

$(\preccurlyeq 2)$ $\quad r^+ \preccurlyeq w^+$ iff $r^- \preccurlyeq w^-$

Let us call a total preorder in $\mathbb{M}L^{\pm}$ that satisfies $(\preccurlyeq 1)$–$(\preccurlyeq 2)$ a *signed preorder*.[23] Signed preorders provide the model proposed by Booth and Meyer for epistemic states.

The best way to understand a signed preorder $\preccurlyeq$ and its role in iterated revision is through a graphical representation proposed in Booth and Meyer (2011). Suppose that each possible world r is represented by a "stick" of a fixed length whose left and right endpoints correspond to r^+ and r^- respectively (Fig. 5). The sticks are positioned in such a way as to reflect the preorder $\preccurlyeq$ over the endpoints. For instance, from Fig. 5 it follows that $r_1^+ \approx r_2^+ \prec r_3^+ \prec r_4^+ \prec r_5^+$, and $r_1^- \approx r_2^- \approx r_4^+ \prec r_3^-$.

From the preorder $\preccurlyeq$ over signed worlds, one can extract a preorder $\leq$ over plain worlds, essentially by taking the projection of $\preccurlyeq$ over $\mathbb{M}L^+$. More formally, $\leq$ is defined as total preorder in $\mathbb{M}L$, constructed from $\preccurlyeq$ as follows:

(Def-$\leq$) $\quad$ For all $r, w \in \mathbb{M}L$, $r \leq w$ iff $r^+ \preccurlyeq w^+$

The preorder $\leq$ extracted from $\preccurlyeq$ plays the role of a system of spheres.[24] Hence, the minimal worlds wrt $\leq$ define the agent's current beliefs K, and for any epistemic input φ, the revision of K by φ is defined as $K * \varphi = \bigcap(min([\varphi], \leq))$. As for the rest of $\preccurlyeq$, it is used to determine the preorder associated with the revised belief set $K * \varphi$. In particular, depending on whether a world r satisfies the new evidence φ or not, its plausibility relative to the new belief set $K * \varphi$ is identified with the plausibility of its positive or negative clone respectively. More formally, let $\vartheta_\varphi(r)$ denote r^+ if $r \vdash \varphi$, and r^- otherwise. Then the preorder $\leq'$ associated with the revised belief set $K * \varphi$ is defined as follows:

(Def-$\leq'$) $\quad$ For all $r, w \in \mathbb{M}L$, $r \leq' w$ iff $\vartheta_\varphi(r) \preccurlyeq \vartheta_\varphi(w)$.

There are two features of the above approach to iterated revision, let us call it the *BM approach*, that should be noted.[25] Firstly, the new information φ is not always

---

23 This is not the name used in Booth and Meyer (2011). In fact the overall exposition herein is slightly different from Booth and Meyer (2011), but the essence remains the same.

24 As already noted, a system of spheres is just another way of representing a total preorder on possible worlds.

25 The BM approach has also been characterized axiomatically in Booth and Meyer (2011).

included in the revised belief set $K * \varphi$; this places the BM approach in the realm of non-prioritized belief revision Hansson (1998). Secondly, the BM approach still comes short of a *complete* solution to the problem of iterated belief revision. In particular, notice that while the signed preorder $\preccurlyeq$ associated with the initial belief set K, fully determines the system of spheres related to K, *as well as* the system of spheres associated with $K * \varphi$ (for any input φ), it doesn't go any further than that; systems of spheres associated with future belief sets $K * \varphi * \psi$ remain unknown. What is needed to complete the picture is a method of cascading the signed preorder $\preccurlyeq$ to future belief sets. Booth and Meyer have already made important steps in this direction, Booth and Meyer (2011).

8 Other Approaches

The models discussed so far are among the most influential in the iterated revision literature. Other important works on the subject are briefly presented in this section (although the list is far from complete).

Ferme's and Rott's approach to iterated belief revision, Ferme and Rott (2004), can loosely speaking be described as follows. Let $\leq$ be a total preorder on possible worlds representing the agent's initial belief state, and let φ be the new information received by the agent. Like Spohn, Ferme and Rott require information about the firmness of φ, in addition to φ itself. Unlike Spohn however, firmness is not specified as an ordinal number (which Ferme and Rott argue is intuitively unclear), but through a *reference sentence* ψ. In particular, the epistemic input comes as a pair of sentences (φ, ψ) which essentially instructs the agent to revise the initial belief state $\leq$ in such a way so that at the new belief state $\leq'$, not only is φ accepted, but it is accepted with the same firmness as ψ (i.e. φ, ψ are equally entrenched). This can be achieved as follows. Let $r_{\neg\psi}$ denote any minimal $\neg\psi$-world wrt $\leq$. Ferme and Rott construct $\leq'$ from $\leq$ by moving all $\neg\varphi$-worlds to the left of $r_{\neg\psi}$, at the same layer as $r_{\neg\psi}$.[26] The approach is intuitively appealing and has many nice features. At the same time, it also has a number of shortcomings (for example, the preorders tend to become coarser, which among other things entails that one "can never build up an informative belief state from the state of complete ignorance" Ferme and Rott (2004)).

In Delgrande and Schaub (2003), propose a constructive approach to belief revision that includes a solution to the problem of iterated revision as a by-product. In particular, for a theory K and a sentence φ, the revision of K by φ is constructed as follows. Firstly, a new theory K' is built from K by replacing every atom p with a new atom p'. Observe that φ is consistent with K' (even if it is inconsistent with K) since the two are expressed in terms of disjoint languages. Starting with the (consistent) set $K' \cup \{\varphi\}$, one goes through the list of original atoms p and corresponding new atoms p', progressively adding assertions of the form $p \equiv p'$ for as long as

[26] That is, all $\neg\varphi$-worlds that are initially strictly more plausible than $r_{\neg\psi}$, are placed at the same level as $r_{\neg\psi}$.

consistency is maintained. Let us denote by EQ a maximal set of such assertions $p \equiv p'$ that can be consistently added to $K' \cup \{\varphi\}$. Delgrande and Schaub define the revision of K by φ as the projection of $Cn(K' \cup \{\varphi\} \cup \{EQ\})$ to the original language of K. If there is more than one maximal set of assertions EQ that is consistent with $K' \cup \{\varphi\}$, we can either use one of them (chosen by a selection function), or take the intersection of $Cn(K' \cup \{\varphi\} \cup \{EQ\})$ for all such EQs. The former is called *choice revision* and the latter *skeptical revision*.[27] Observe that since $K * \varphi$ is defined for every K and φ, this approach provides a solution both for one-step and for iterated belief revision.

In Delgrande et al. (2006), take a different approach to iterated revision. Given an initial belief base φ_0, and a sequence of observations $\varphi_1; \varphi_2; \ldots; \varphi_n$, the revision of φ_0 by the sequence is defined as the *prioritized merging* of the multiset $\{(\varphi_0, r_0), (\varphi_1, r_1), \ldots, (\varphi_n, r_n)\}$, where each r_i represents the reliability degree of the corresponding sentence φ_i. Observe that in the special case where $r_0 < r_1 < \cdots < r_n$, prioritized merging conforms with an assumption widely used in the iterated revision literature: recent observations are assumed to have priority over earlier ones. Delgrande, Dubois and Lang prove that most of the well known postulates for iterated revision (except (DP2)) are consequences of the postulate they propose for prioritized merging.

All the works discussed so far assume that the agent under consideration is situated in a *static* world. Recently, iterated belief revision has also been studied in the context of a *dynamic environment*. In particular, consider an agent operating in a world that changes due to actions taken by the agent herself or by other agents. Let us denote by K the agent's beliefs about the initial state of the world. If the agent is informed of an action that brought about φ, she has to modify K accordingly via a process known as *belief update*.[28] Moreover, in addition to actions that change the state of the world (*ontic* actions), there are also *sensing* actions, through which the agent acquires information about the world without changing it. The effects of sensing actions are modeled in terms of belief revision. We note that some sensing actions can indirectly reveal information about *past* world states, leading to an interesting interplay between alternating revisions and updates. Such scenarios have been studied by Hunter and Delgrande (2005, 2007), as well as by Shapiro et al. (2000) (the latter in the context of *situation calculus*).

In a different direction, considerations related to iterated belief revision are starting to be studied in a *dynamic epistemic logic* setting Baltag and Smets (2009). Finally, there is also important work on *iterated contraction* (see for example Chopra et al. (2008), Ramachandran et al. (2012)).

---

[27] It should be noted that the selection function employed in choice revision, does not depend on the initial belief set K.

[28] See Katsuno and Mendelzon (1991) for a formal treatment of belief update and its difference from belief revision.

9 Conclusion

Important steps have been made towards a general solution to the problem of iterated revision. On the other hand, there is still some controversy over the appropriateness of the proposed approaches and no signs of convergence towards a "standard" model (like AGM is for one-step revision).

Part of the reason for this has to do with the lack of consensus over the "right" *input* for iterated revision. As we have seen in this survey, input can vary from being a plain sentence, to a sentence coupled with an ordinal number, to an entire preorder on sentences.[29]

A second source of difficulty in the way that most proposals in the literature approach the problem of iterated revision is the following. To solve the problem of iterated revision (regardless of the type of input we employ) one will ultimately have to specify a relationship between the meta-structure $\mathcal{U}$ that guides one-step revisions at the initial time-point, with the meta-structure $\mathcal{U}'$ resulting after revision.[30] Any solution that *uniquely determines* $\mathcal{U}'$ from $\mathcal{U}$ and the epistemic input, runs the risk of not being general enough. On the other hand, if $\mathcal{U}'$ in *not* fully determined from $\mathcal{U}$ and the epistemic input, then a *meta-meta-structure* is needed to select between the different options for $\mathcal{U}'$. This however leads to a vicious circle since one would then have to develop a model for the dynamics of the meta-meta-structures (with similar problems to confront).

Clearly, there are still important pieces missing from the puzzle of iterated belief revision. One should treat this as an opportunity rather than a liability: exciting new results on the subject are waiting to be discovered!

Acknowledgments I am grateful to the anonymous reviewers for their valuable comments, and to the editor of this book, Sven Ove Hansson, for his excellent work in coordinating our joint efforts. Since this book is devoted to David Makinson's work, I would like to take this opportunity to express my gratitude to David for co-founding the area of Belief Revision that has constantly fueled my intellectual curiosity for over 20 years. More importantly though, I would like to thank David for setting such an inspiring example of a true scholar of the very best in academic traditions.

References

Alchourron, C., Gärdenfors, P., & Makinson, D. (1985). On the logic of theory change: Partial meet functions for contraction and revision. *Journal of Symbolic Logic*, *50*, 510–530.

Alchourron, C., & Makinson, D. (1985). On the logic of theory change: Safe contractions. *Studia Logica*, *44*, 405–422.

[29] Perhaps the best way to explain this is that there is no "right" input as such; different types of input are appropriate for different kinds of iterated revision scenarios.

[30] Depending on the model, $\mathcal{U}, \mathcal{U}'$ can be systems of spheres, ordinal conditional functions, preorders on possible worlds, etc.

Baltag, A., & Smets, S. (2009). Group belief dynamics under iterated revision: Fixed points and cycles of joint upgrades. *Proceedings of the* 20th *Conference on Theoretical Aspects of Rationality and Knowledge (TARK-09).*

Benferhat, S., Kaci, S., Le Berre, D., & Williams, M.-A. (2004). Weakening conflicting information for iterated revision and knowledge integration. *Artificial Intelligence, 153*(1–2), 339–371.

Booth, R., Chopra, S., & Meyer, T. (2005). Restrained revision. 6th *Workshop on Nonmonotonic Reasoning, Action, and Change (NRAC-05)*, Edinburgh.

Booth, R., & Meyer, T. (2006). Admissible and restrained revision. *Journal of artificial intelligence research, 26*, 127–151.

Booth, R., & Meyer, T. (2011). How to revise a total preorder. *Journal of Philosophical Logic, 40*, 193–238.

Boutilier, C. (1993). Revision sequences and nested conditionals. *Proceedings of the* 13th *International Joint Conference in, Artificial Intelligence*. (pp. 519–525).

Boutilier, C. (1996). Iterated revision and minimal change of conditional beliefs. *Journal of Philosophical Logic, 25*(3), 262–305.

Chopra, S., Ghose, A., Meyer, T., & Wong, K.-S. (2008). Iterated belief change and the recovery axiom. *Journal of Philosophical Logic, 37*(5), 501–520.

Darwiche, A., & Pearl, J. (1997). On the logic of iterated belief revision. *Artificial Intelligence, 89*, 1–29.

Delgrande, J., & Schaub, T. (2003). A consistency-based approach for belief change. *Artificial Intelligence, 151*, 1–41.

Delgrande, J., Dubois, D., & Lang, J. (2006). *Iterated revision as prioritized merging:In Proceedings of the 10th International Conference on Principles of Knowledge Representation and Reasoning* (pp. 210–220). Sa Fransico: Morgan Kauffman.

Ferme, E., & Rott, H. (2004). Revision by comparison. *Artificial Intelligence, 157*, 139–202.

Gärdenfors, P. (1984). Epistemic importance and minimal changes of belief. *Australasian Journal of Philosophy, 62*, 136–157.

Gärdenfors, P. (1988). *Knowledge in flux:Modeling the dynamics of epistemic states*. Cambridge: MIT Press.

Gärdenfors, P., & Makinson, D. (1988). Revisions of knowledge systems using epistemic entrenchment. *Proceedings of Theoretical Aspects of Reasoning about Knowledge* (pp. 83–95). USA:Morgan-Kaufmann.

Grove, A. (1988). Two modellings for theory change. *Journal of Philosophical Logic, 17*, 157–170.

Hansson, S.-O. (1999). *A textbook of belief dynamics: Theory change and database updating*. USA: Kluwer Academic Publishers.

Hansson, S.-O. (1998). A survey of non-prioritized belief revision. *Erkenntnis, 50*, 413–427.

Hunter, A., & Delgrande, J. (2005). *Iterated belief change: A Transition System Approach Proceedings of the* 19th *International Joint Conference in Artificial Intelligence* (pp 460–465).

Hunter, A., & Delgrande, J. (2007). An action description language for iterated belief change. *Proceedings of the* 20th *International Joint Conference in Artificial Intelligence* (pp. 2498–2503). Hyderabad.

Jin, Y., & Thielscher, M. (2005). Iterated revision, revised. *Proceedings of the 19th International Joint Conference in Artificial Intelligence* (pp 478–483), Edinburgh.

Jin, Y., & Thielscher, M. (2007). Iterated revision, revised. *Artificial Intelligence, 171*(1), 1–18.

Katsuno, H., & Mendelzon, A. (1991a). On the difference between updating a knowledge base and revising it. *Proceedings of the 2nd International Conference on Principles of Knowledge Representation and Reasoning* (pp. 387–394). USA: Morgan Kauffman.

Katsuno, H., & Mendelzon, A. (1991b). Propositional knowledge base revision and minimal change. *Artificial Intelligence, 52*, 263–294.

Konieczny, S., & Perez, R. P. (2000). A framework for iterated revision. *Journal of Applied Non-Classical Logics, 10*, 339–367.

Kraus, S., Lehmann, D., & Magidor, M. (1990). Nonmonotonic reasoning: Preferential models and cumulative logics. *Artificial Intelligence, 44*, 167–207.

Lehmann, D. (1995). Belief revision, revised. *Proceedings of the* 14th *International Joint Conference on Artificial Intelligence* (pp. 1534–1540). Montreal.

Lehmann, D., Magidor, M. K., & Schlechta, K. (2001). Distance semantics for belief revision. *Journal of Symbolic Logic*, *66*(1), 295–317.

Lewis, D. (1973). *Counterfactuals*. Cambridge, USA: Harvard University Press.

Makinson, D. (1994). General patterns in nonmonotonic reasoning. *Handbook of Logic in Artificial Intelligence and Logic Programming*, *3*, 35–110.

Nayak, A. (1994). Foundational belief change. *Journal of Philosophical Logic*, *23*, 495–533.

Nayak, A. (1994). Iterated belief change based on epistemic entrenchment. *Erkenntnis*, *41*, 353–390.

Nayak, A., Pagnucco, M., & Peppas, P. (2003). Dynamic belief revision operators. *Artificial Intelligence*, *146*, 193–228.

Parikh, R. (1999). Beliefs, belief revision, and splitting languages. In J. L. Moss, & M. de Rijke, (Eds.), *Logic, language, and computation* (Vol. 2, pp. 266–268). CSLI Lecture Notes No. 96, CSLI Publications.

Peppas, P., & Williams, M. (1995). Constructive modelings for theory change. *Notre Dame Journal of Formal Logic*, *36*(1), 120–133.

Peppas, P., Chopra, S., & Foo, N. (2004). Distance semantics for relevance-sensitive belief revision. *Proceedings of the 9th International Conference on the Principles of Knowledge Representation and Reasoning (KR2004)*, Canada: Whistler.

Peppas, P. (2008). Belief revision, In F. van Harmelen, V. Lifschitz, & B. Porter, (Eds). Handbook of Knowledge Representation. Elsevier.

Peppas, P., Fotinopoulos, A., & Seremetaki, S. (2008). Conflicts between relevance-sensitive and iterated belief revision. *Proceedings of the* 18th *European Conference on Artificial Intelligence (ECAI08)*. Greece: Patras.

Ramachandran, R., Nayak, A., & Orgun, M. (2012). Three approaches to iterated belief contraction. *Journal of Philosophical Logic*, *41*(1), 115–142.

Rott, H. (1991). Two methods of constructing contractions and revisions of knowledge systems. *Journal of Philosophical Logic*, *20*, 149–173.

Shapiro, S., Pagnucco, M., Lesperance, Y., & Levesque, H. (2000). Iterated belief change in the situation calculus. *Proceedings of the* 10th *International Conference on Principles of Knowledge Representation and Reasoning (KR'00)*, Morgan Kaufmann.

Spohn, W. (1988). Ordinal conditional functions: A dynamic theory of epistemic states, in Causation in Decision, Belief Change and Statistics, *2*, 105–134.

Williams, M.-A. (1994). Transmutations of knowledge systems. *Proceedings of the* 4th *International Conference on Principles of Knowledge Representation and Reasoning*, *29*, 619–629.

Williams, M.-A. (1996). A Practical approach to belief revision: Reason-based change. *Proceedings of the* 5th *International Conference on Principles of Knowledge Representation and Reasoning* (pp. 412–421). Trento, Italy.

Zhang, D. (2004). Properties of iterated multiple belief revision. *Proceedings of the* 7th *International Conference on Logic Programming and Nonmonotonic Reasoning (LPNMR)* (pp. 314–325). Austria: Springer.

AGM, Ranking Theory, and the Many Ways to Cope with Examples

Wolfgang Spohn

Abstract The paper first explains how the ranking-theoretic belief change or conditionalization rules entail all of the standard AGM belief revision and contraction axioms. Those axioms have met a lot of objections and counter-examples, which thus extend to ranking theory as well. The paper argues for a paradigmatic set of cases that the counter-examples can be well accounted for with various pragmatic strategies while maintaining the axioms. So, one point of the paper is to save AGM belief revision theory as well as ranking theory. The other point, however, is to display how complex the pragmatic interaction of belief change and utterance meaning may be; it should be systematically and not only paradigmatically explored.

Keywords Ordinal conditional function · Ranking theory · AGM · Success postulate · Preservation postulate · Superexpansion postulate · Intersection postulate · Recovery postulate

1 Introduction[1]

Expansions, revisions, and contractions are the three kinds of belief change intensely studied by AGM belief revision theory and famously characterized by the standard eight revision and eight contraction axioms. Even before their canonization in Alchourrón et al. (1985), ranking theory and its conditionalization rules for belief change (Spohn 1983, Sect. 5.3) generalized upon the AGM treatment. I always took the fact that these conditionalization rules entail the standard AGM axioms (as

[1] I am grateful to Paul Arthur Pedersen for discussing his example in Sect. 6 with me, to David Makinson and Brian Leahy for various hints and corrections, and to two anonymous referees for further valuable remarks considerably improving the paper.

W. Spohn (✉)
Department of Philosophy, University of Konstanz, 78457 Konstanz, Germany
e-mail: wolfgang.spohn@uni-konstanz.de

S. O. Hansson (ed.), *David Makinson on Classical Methods for Non-Classical Problems*, Outstanding Contributions to Logic 3, DOI: 10.1007/978-94-007-7759-0_6,

first observed in Spohn (1988), footnote 20, and in Gärdenfors (1988), Sect. 3.7) as reversely confirming ranking theory.

As is well known, however, a vigorous discussion has been going on in the last 20 years about the adequacy of those axioms, accumulating a large number of plausible counter-examples, which has cast a lot of doubt on the standard AGM theory and has resulted in a host of alternative axioms and theories. Via the entailment just mentioned these doubts extend to ranking theory; if those axioms fall, ranking theory falls, too. Following Christian Morgenstern's saying "weil nicht sein kann, was nicht sein darf", this paper attempts to dissolve those doubts by providing ranking-theoretic ways of dealing with those alleged counter-examples, which avoid giving up the standard AGM axioms. So, this defense of the standard AGM axioms is at the same time a self-defense of ranking theory.

This is the obvious goal of this paper. It is a quite restricted one, insofar as it exclusively focuses on those counter-examples. No further justification of AGM or ranking theory, no further comparative discussion with similar theories is intended; both are to be found extensively, if not exhaustively in the literature.

There is, however, also a mediate and no less important goal: namely to demonstrate the complexities of the pragmatic interaction between belief change and utterance meaning. I cannot offer any account of this interaction. Instead, the variety of pragmatic strategies I will be proposing in dealing with these examples should display the many aspects of that interaction that are hardly captured by any single account. So, one conclusion will be that this pragmatics, which has been little explored so far, should be systematically studied. And the other conclusion will be that because of those complexities any inference from such examples to the shape of the basic principles of belief change is premature and problematic. Those principles must be predominantly guided by theoretical considerations, as they are in both AGM and ranking theory in well-known ways.

The plan of the paper is this: I will recapitulate the basics of ranking theory in Sect. 2 and its relation to AGM belief revision theory in Sect. 3, as far as required for the subsequent discussion. There is no way of offering a complete treatment of the problematic examples having appeared in the literature. I have to focus on some paradigms, and I can only hope to have chosen the most important ones. I will first attend to revision axioms: Sect. 4 will deal with the objections against the Success Postulate, Sect. 5 with the Preservation Postulate, and Sect. 6 with the Superexpansion Postulate. Then I will turn to contraction axioms: Sect. 7 will be devoted to the Intersection Postulate, and Sect. 8 to the Recovery Postulate, perhaps the most contested one of all. Section 9 will conclude with a brief moral.

2 Basics of Ranking Theory

AGM belief revision theory is used to work with sentences of a given language **L** —just a propositional language; quantifiers and other linguistic complications are rarely considered. For the sake of simplicity let us even assume **L** to be finite, i.e., to have only finitely many atomic sentences. **L** is accompanied by some logic as specified in the consequence relation *Cn*, which is usually taken to be the classical compact Tarskian entailment relation. I will assume it here as well (although there are variations we need not go into). A *belief set* is a deductively closed set of sentences of **L**, usually a consistent one (since there is only one inconsistent belief set). Belief change then operates on belief sets. That is, expansion, revision, and contraction by $\varphi \in \mathbf{L}$ operate on belief sets; they carry a given belief set into a, respectively, expanded, revised, or contracted belief set.

By contrast, ranking theory is used to work with a Boolean algebra (or field of sets) $\mathcal{A}$ of propositions over a space W of possibilities. Like probability measures, ranking functions are defined on such an algebra. Let us again assume the algebra $\mathcal{A}$ to be finite; the technical complications and variations arising with infinite algebras are not relevant for this paper (cf. Huber 2006; Spohn 2012, Chap. 5). Of course, the two frameworks are easily intertranslatable. Propositions simply are truth conditions of sentences, i.e., sets of valuations of **L** (where we may take those valuations as the possibilities in W). And if $T(\varphi)$ is the truth condition of φ, i.e., the set of valuations in which φ is true, then $\{T(\varphi) | \varphi \in \mathbf{L}\}$ is an algebra—indeed a finite one, since we have assumed **L** to be finite.

I have always found it easier to work with propositions. For instance, logically equivalent sentences, which are not distinguished in belief revision theory, anyway (due to its extensionality axiom), reduce to identical propositions. And a belief set may be represented by a single proposition, namely as the intersection of all the propositions corresponding to the sentences in the belief set. The belief set is then recovered as the set of all sentences corresponding to supersets of that intersection in the algebra (since the classical logical consequence between sentences simply reduces to set inclusion between propositions).

Let me formally introduce the basic notions of ranking theory before explaining their standard interpretation:

Definition 1: κ is a *negative ranking function* for $\mathcal{A}$ iff κ is a function from $\mathcal{A}$ into $\mathbf{N}^+ = \mathbf{N} \cup \{\infty\}$ such that for all $A, B \in \mathcal{A}$

(1) $\kappa(W) = 0$,
(2) $\kappa(\emptyset) = \infty$, and
(3) $\kappa(A \cup B) = \min\{\kappa(A), \kappa(B)\}$.

Definition 2: β is a *positive ranking function* for $\mathcal{A}$ iff β is a function from $\mathcal{A}$ into $\mathbf{N}^+$ such that for all $A, B \in \mathcal{A}$

(4) $\beta(\emptyset) = 0$,
(5) $\beta(W) = \infty$, and
(6) $\beta(A \cap B) = \min\{\beta(A), \beta(B)\}$.

Negative and positive ranking functions are interdefinable via the equations:

(7) $\beta(A) = \kappa(\bar{A})$ and $\kappa(A) = \beta(\bar{A})$. A further notion that is often useful is this:

Definition 3: τ is a *two-sided ranking function* for $\mathscr{A}$ (corresponding to κ and/or β) iff

(8) $\tau(A) = \kappa(\bar{A}) - \kappa(A) = \beta(A) - \kappa(A)$.

The axioms immediately entail the *law of negation*:

(9) either $\kappa(A) = 0$ or $\kappa(\bar{A}) = 0$, or both (for negative ranks), and
(10) either $\beta(A) = 0$ or $\beta(\bar{A}) = 0$, or both (for positive ranks), and
(11) $\tau(\bar{A}) = -\tau(A)$ (for two-sided ranks).

Definition 4: Finally, the *core* of a negative ranking function κ or a positive ranking function β is the proposition

(12) $C = \bigcap\{A \mid \kappa(\bar{A}) > 0\} = \bigcap\{A \mid \beta(A) > 0\}$.

Given the finiteness of $\mathscr{A}$ (or a strengthening of axioms (3) and (6) to infinite disjunctions or, respectively, conjunctions), we obviously have $\beta(C) > 0$.

The standard interpretation of these notions is this: *Negative ranks express degrees of disbelief.* (Thus, despite being non-negative numbers, they express something negative and are therefore called negative ranks.) To be a bit more explicit, for $A \in \mathscr{A}$ $\kappa(A) = 0$ says that A is not disbelieved, and $\kappa(A) = n > 0$ says that A is disbelieved (to degree n). Disbelieving is taking to be false and believing is taking to be true. Hence, *belief in A* is the same as disbelief in $\bar{A}$ and thus expressed by $\kappa(\bar{A}) > 0$. Note that we might have $\kappa(A) = \kappa(\bar{A}) = 0$, so that A is neither believed nor disbelieved.

Positive ranks express degrees of belief directly. That is, $\beta(A) = 0$ iff A is not believed, and $\beta(A) = n > 0$ iff A is *believed* or *taken to be true* (to degree n). This interpretation of positive and negative ranks entails, of course, their interdefinability as displayed in (7).

Because of the axioms (1) and (4) beliefs are consistent; not everything is believed or disbelieved. Because of the axioms (3) and (6) beliefs are deductively closed. And the *core* of κ or β represents all those beliefs, by being their conjunction and entailing all of them and nothing else.

Finally, *two-sided ranks* are useful because they represent belief and disbelief in a single function. Clearly, we have $\tau(A)>0$, <0, or $= 0$, iff, respectively, A is believed, disbelieved, or neither. However, a direct axiomatization of two-sided ranks is clumsy; this is why I prefer to introduce them via Def. 3. Below I will freely change between negative, positive, and two-sided ranks.

As already indicated, ranks represent not only belief, but also degrees of belief; the larger $\beta(A)$, the firmer your degree of belief in A. So, they offer an alternative model of such degrees. The standard model is probability theory, of course. However, it is very doubtful whether probabilities are able to represent beliefs, as the huge discussion triggered by the lottery paradox shows. (The lottery paradox precisely shows that axiom (c) of Def. 2 cannot be recovered in probabilistic terms.) So I consider it an advantage of ranking theory that it can represent both, beliefs and degrees of belief. (For all this see Spohn 2012, Chaps. 5 and 10.)

Indeed, these degrees are cardinal, not ordinal (like Lewis' similarity spheres or AGM's entrenchment ordering), and they are accompanied by a measurement theory, which proves them to be measurable on a ratio scale (cf. Hild and Spohn 2008; Spohn 2012, Chap. 8). (Probabilities, by contrast, are usually measured on an absolute scale.)

I should perhaps mention that there are some formal variations concerning the range of ranking functions, which might consist of real or ordinal numbers instead of natural numbers; indeed, the measurement theory just mentioned works with real-valued ranking functions. In the infinite case, there is also some freedom in choosing the algebraic framework and in strengthening axioms (3) and (6). Here, we need not worry about such variations; it suffices to consider only the finite case and integer-valued ranking functions.

The numerical character of ranks is crucial for the next step of providing an adequate notion of *conditional belief*. This is generated by the notion of conditional ranks, which is more naturally defined in terms of negative ranking functions:

Definition 5: The *negative conditional rank* $\kappa(B|A)$ of $B \in \mathscr{A}$ *given* or *conditional on* $A \in \mathscr{A}$ (provided $\kappa(A) < \infty$) is defined by:

(13) $\kappa(B|A) = \kappa(A \cap B) - \kappa(A)$.

This is equivalent to the *law of conjunction*:

(14) $\kappa(A \cap B) = \kappa(A) + \kappa(B|A)$.

This law is intuitively most plausible: How strongly do you disbelieve $A \cap B$? Well, A might be false; then $A \cap B$ is false as well; so take $\kappa(A)$, your degree of disbelief in A. But if A should be true, B must be false in order $A \cap B$ to be false. So add $\kappa(B|A)$, your degree of disbelief in B given A.

The positive counterpart is the *law of material implication*:

(15) $\beta(A \to B) = \beta(B|A) + \beta(\bar{A})$—where $A \to B = \bar{A} \cup B$ is (set-theoretic) material implication and where the *positive conditional rank* $\beta(B|A)$ of B *given* A is defined in analogy to (7) by:

(16) $\beta(B|A) = \kappa(\bar{B}|A)$.

(15) is perhaps even more plausible: Your degree of belief in $A \to B$ is just your degree of belief in its vacuous truth, i.e., in $\bar{A}$, plus your conditional degree of belief in B given A. This entails that your conditional rank and your positive rank of the material implication coincide if you don't take A to be false, i.e., $\beta(\bar{A}) = 0$.

It should be obvious, though, that conditional ranks are much more tractable in negative than in positive terms. In particular, despite the interpretational differences there is a far-reaching formal analogy between ranks and probabilities. However, this analogy becomes intelligible only in terms of negative ranks and their axioms (1)–(3) and (13). This is why I have always preferred negative ranks to their positive counterparts.

Conditional ranks finally entail a notion of conditional belief:

(17) B is *conditionally believed given* A iff $\beta(B|A) = \kappa(\bar{B}|A) > 0$.

One further definition will be useful:

Definition 6: The negative ranking function κ is *regular* iff for all $A \in \mathcal{A}$ with $A \neq \varnothing\ \kappa(A) < \infty$.

Hence, in a regular ranking function only the contradiction is maximally firmly disbelieved, and only the tautology is maximally firmly believed. And conditional ranks are universally defined except for the contradictory condition. This corresponds to the probabilistic notion of regularity.

There is no space for extensive comparative observations. Just a few remarks: Ranking functions have ample precedent in the literature, at least in Shackle's (1961) functions of potential surprise, Rescher's (1964) hypothetical reasoning, and Cohen's (1970) functions of inductive support. All these predecessors arrived at the Baconian structure of (1)–(3) or (4)–(6), as it is called by Cohen (1980). However, none of them has an adequate notion of conditional ranks as given by (13) or (16); this is the crucial advance of ranking theory (cf. Spohn 2012, Sect. 11.1).

AGM belief revision theory seems to adequately capture at least the notion of conditional belief. However, in my view it founders at the problem of iterated belief revision. The point is that conditional belief is there explained only via the ordinal notion of an entrenchment ordering, but within these ordinal confines no convincing account of iterated revision can be found. (Of course, the defense of this claim would take many pages.) The iteration requires the cardinal resources of ranking theory, in particular the cardinal notion of conditional ranks (cf. Spohn 2012, Chaps. 5 and 8).

Finally, ranking theory is essentially formally equivalent to possibility theory as suggested by Zadeh (1978), fully elaborated in Dubois and Prade (1988), and further expanded in many papers; the theories are related by an exponential (or logarithmic) scale transformation. However, while ranking theory was determinately intended to capture the notion of belief, possibility theory was and is less determinate in my view. This interpretational indecision led to difficulties in defining conditional degrees of possibility, which is not an intuitive notion, anyway, and therefore formally explicable in various ways, only one of which corresponds to (13) (cf. Spohn 2012, Sect. 11.8). AGM unambiguously talk about belief, and therefore I continue my discussion in terms of ranking theory, which does the same.

Above I introduced my *standard interpretation* of ranking theory, which I then extended to conditional belief. However, one should note that it is by no means mandatory. On this interpretation, there are many degrees of belief, many degrees of disbelief, but only one degree of unopinionatedness, namely the two-sided rank 0. This looks dubious. However, we are not forced to this interpretation. We might as well take some threshold value $z > 0$ and say that only $\beta(A) > z$ expresses belief. Or in terms of two-sided ranks: $\tau(A) > z$ is belief, $-z \leq \tau(A) \leq z$ is unopinionatedness, and $\tau(A) < -z$ is disbelief. Then, the basic laws of belief are still preserved, i.e., belief sets are always consistent and deductively closed. It's only that the higher the threshold z, the stricter the notion of belief. I take this to account for the familiar vagueness of the notion of belief; there is only a vague answer to the question: How firmly do you have to believe something in order to believe it?

Still, the Lockean thesis ("belief is sufficient degree of belief") can be preserved in this way, while it must be rejected if degrees of belief are probabilities. Of course, the vagueness also extends to conditional belief. However, the ranking-theoretic apparatus underneath is entirely unaffected by that reinterpretation.

Let us call this the *variable interpretation* of ranking theory. Below, the standard interpretation will be the default. But at a few places, which will be made explicit, the variable interpretation will turn out to be useful.

3 AGM Expansion, Revision, and Contraction as Special Cases of Ranking-Theoretic Conditionalization

The notion of conditional belief is crucial for the next point. How do we change belief states as represented by ranking functions? One idea might be that upon experiencing A we just move to the ranks conditional on A. However, this means treating experience as absolutely certain (since $\beta(A|A) = \infty$); nothing then could cast any doubt on that experience. This is rarely or never the case; simple probabilistic conditionalization suffers from the same defect. This is why Jeffrey (1965/1983, Chap. 11) proposed a more general version of conditionalization, and in Spohn (1983, Sect. 5.3, 1988, Sect. 5) I proposed to transfer this idea to ranking theory:

Definition 7: Let κ be a negative ranking function for $\mathscr{A}$ and $A \in \mathscr{A}$ such that $\kappa(A)$, $\kappa(\bar{A}) < \infty$, and $n \in \mathbf{N}^+$. Then the *$A \to n$-conditionalization* $\kappa_{A\to n}$ of κ is defined by

(18) $\kappa_{A\to n}(B) = \min\{\kappa(B|A), \kappa(B|\bar{A}) + n\}$.

The $A \to n$-conditionalization will be called *result-oriented*.

It is easily checked that

(19) $\kappa_{A\to n}(A) = 0$ and $\kappa_{A\to n}(\bar{A}) = n$.

Thus, the parameter n specifies the posterior degree of belief in A and hence the result of the belief change; this is why I call it result-oriented. It seems obvious to me that learning must be characterized by such a parameter; the learned can be learned with more or less certainty. Moreover, for any B we have $\kappa_{A\to n}(B|A) = \kappa(B|A)$ and $\kappa_{A\to n}(B|\bar{A}) = \kappa(B|\bar{A})$. In sum, we might describe $A \to n$-conditionalization as shifting the A-part and the $\bar{A}$-part of κ in such a way that A and $\bar{A}$ get their intended ranks and as leaving the ranks conditional on A and on $\bar{A}$ unchanged. This was also the crucial rationale behind Jeffrey's generalized conditionalization (cf. also Teller 1976).

However, as just noticed, the parameter n specifies the effect of experience, but does not characterize experience by itself. This objection was also raised against Jeffrey—by Field (1978), who proposed quite an intricate way to meet it. In ranking terms the remedy is much simpler:

Definition 8: As before, let κ be a negative ranking function for $\mathscr{A}$, $A \in \mathscr{A}$ such that $\kappa(A), \kappa(\bar{A}) < \infty$, and $n \in \mathbf{N}^+$. Then the $A\uparrow n$-conditionalization $\kappa_{A\uparrow n}$ of κ is defined by

(20) $\kappa_{A\uparrow n}(B) = \min\{\kappa(A \cap B) - m, \kappa(\bar{A} \cap B) + n - m\}$, where $m = \min\{\kappa(A), n\}$.

The $A\uparrow n$-conditionalization will be called *evidence-oriented*.

The effect of this conditionalization is that, whatever the prior ranks of A and $\bar{A}$, the posterior rank of A improves by exactly n ranks in comparison to the prior rank of A. This is most perspicuous in the easily provable equation

(21) $\tau_{A\uparrow n}(A) - \tau(A) = n$

for the corresponding two-sided ranking function. So, now the parameter n indeed characterizes the nature and the strength of the evidence by itself—whence the name.

Of course, the two kinds of conditionalization are interdefinable; we have:

(22) $\kappa_{A\rightarrow n} = \kappa_{A\uparrow m}$, where $m = \tau(\bar{A}) + n$.

Result-oriented conditionalization is also called Spohn conditionalization, since it was the version I proposed, whereas evidence-oriented conditionalization is also called Shenoy conditionalization, since it originates from Shenoy (1991). There are, moreover, generalized versions of each, where either the direct effect of learning or the experience itself is characterized by some ranking function on some partition of the given possibility space (not necessarily a binary partition $\{A, \bar{A}\}$), as already proposed by Jeffrey (1965/1983, Chap. 11) for probabilistic learning. This generalized conditionalization certainly provides the most general and flexible learning rule in ranking terms. However, there is no need to formally introduce it; below I will refer only to the simpler rules already stated. (For more careful explanations of this material see Spohn 2012, Chap. 5.)

All of this is directly related to AGM belief revision theory. First, these rules of conditionalization map a ranking function into a ranking function. Then, however, they also map the associated belief sets (= set of all propositions entailed by the relevant core). Thus, they do what AGM expansions, revisions, and contractions do. The latter may now easily be seen to be special cases of result-oriented conditionalization. At least, the following explications seem to fully capture the intentions of these three basic AGM movements:

Definition 9: Expansion by A simply is $A\rightarrow n$-conditionalization for some $n > 0$, provided that $\tau(A) \geq 0$; that is, the prior state is does not take A to be false, and the posterior state believes or accepts A with some firmness n.

Definition 10: Revision by A is $A\rightarrow n$-conditionalization for some $n > 0$, provided that $-\infty < \tau(A) < 0$; that is, the prior state disbelieves A and the posterior state is forced to accept A with some firmness n. In the exceptional case where $\tau(A) = -\infty$ no $A\rightarrow n$-conditionalization and hence no revised ranking function is defined. In this case we stipulate that the associated belief set is the inconsistent one. With this stipulation, ranking-theoretic revision is as generally defined as AGM revision.

For both, expansion and revision by A, it does not matter how large the parameter n is, as long as it is positive. Although the posterior ranking function varies with different n, the posterior belief set is always the same; a difference in belief sets could only show up after iterated revisions.

As to *contraction by* A: $A \to 0$-conditionalization amounts to a two-sided contraction either by A or by $\bar{A}$ (if one of these contractions is substantial, the other one must be vacuous); whatever the prior opinion about A, the posterior state then is unopinionated about A. Hence, we reproduce AGM contraction in the following way:

Definition 11: Contraction by A is $A \to 0$-conditionalization in case A is believed, but not maximally, i.e., $\infty > \tau(A) > 0$, and as no change at all in case A is not believed, i.e., $\tau(A) \leq 0$. In the exceptional case where $\tau(A) = \infty$, no $A \to 0$-conditionalization and hence no contracted ranking function is defined. In this case we stipulate that the contraction is vacuous and does not change the belief set. Thereby ranking-theoretic contraction is also as generally defined as AGM contraction.

It should be clear that these three special cases do not exhaust conditionalization. For instance, there is also the case where evidence directly weakens, though does not eliminate the disbelief in the initially disbelieved A. Moreover, evidence might also speak against A; but this is the same as evidence in favor of $\bar{A}$.

The crucial observation for the rest of the paper now is that revision and contraction thus ranking-theoretically defined entail all eight AGM revision and all eight AGM contractions axioms, (K $*$ 1) – (K $*$ 8) and (K $\div$ 1) – (K $\div$ 8)—*provided* we restrict the ranking-theoretic operations to regular ranking functions. The effect of this assumption is that $\varnothing$ is the only exceptional case for revision and W the only exceptional case for contraction.

For most of the axioms this entailment is quite obvious (for full details see Spohn 2012, Sect. 5.5). In the sequel, I move to and fro between the sentential framework (using greek letters and propositional logic) and the propositional framework (using italics and set algebra). This should not lead to any misunderstanding. K is a variable for belief sets, $K * \varphi$ denotes the revision of K by $\varphi \in \mathbf{L}$ and $K \div \varphi$ the contraction of K by φ. Finally $A = T(\varphi)$ and $B = T(\psi)$.

(K$*$1), *Closure*, says: $K * \varphi = Cn(K * \varphi)$. It is satisfied by Definiton 10, because, according to each ranking function, the set of beliefs is deductively closed.

(K $*$ 2), *Success*, says in AGM terms: $\varphi \in K * \varphi$. With Def. 10 this translates into: $\kappa_{A \to n}(\bar{A}) > 0 \quad (n > 0)$. This is true by definition (where we require regularity entailing that $\kappa_{A \to n}$ is defined for all $A \neq \emptyset$).

(K $*$ 3), *Expansion 1*, says in AGM terms: $K * \varphi \subseteq Cn(K \cup \{\varphi\})$.

(K $*$ 4), *Expansion 2*, says: if $\neg \varphi \notin K$, then $Cn\ (K \cup \{\varphi\}) \subseteq K * \varphi$. Together, $(K * 3)$ and $(K * 4)$ are equivalent to $K * \varphi = Cn\ (K \cup \{\varphi\})$, provided that $\neg \varphi \notin K$. With Def. 10 this translates into: if $\kappa\ (A) = 0$ and if C is the core of κ, then the core of $\kappa_{A \to n}$ $(n > 0)$ is $C \cap A$. This is obviously true.

(K ∗ 5), *Consistency Preservation*, says in AGM terms: if $\perp \notin Cn(K)$ and $\perp \notin Cn(\varphi)$, then $\perp \notin K * \varphi$ ($\perp$ is some contradictory sentence). This holds because, if κ is regular, $\kappa_{A\to n}(n > 0)$ is regular, too, and both have consistent belief sets.

(K ∗ 6), *Extensionality*, says in AGM terms: if $Cn(\varphi) = Cn(\psi)$, then $K * \varphi = K * \psi$. And in ranking terms: $\kappa_{A\to n} = \kappa_{A\to n}$. It is built into the propositional framework.

(K ∗ 7), *Superexpansion*, says in AGM terms: $K * (\varphi \wedge \psi) \subseteq Cn((K * \varphi) \cup \{\psi\})$.

(K ∗ 8), *Subexpansion*, finally says: if $\neg\psi \notin K * \varphi$, then $Cn\ ((K * \varphi) \cup \{\psi\}) \subseteq K * (\varphi \wedge \psi)$. In analogy to $(K * 3)$ and $(K * 4)$, the conjunction of $(K * 7)$ and $(K * 8)$ translates via Def. 10 into: if $\kappa\ (B|A) = 0$ and if C is the core of $\kappa_{A\to n}$ $(n > 0)$ then the core of $\kappa_{A\cap B\to n}$ is $C \cap B$. This is easily seen to be true. The point is this: Although Rott (1999) is right in emphasizing that (K ∗ 7) and (K ∗ 8) are not about iterated revision, within ranking theory they come to that, and they say then that (K ∗ 3) and (K ∗ 4) hold also after some previous revision; and, of course, (K ∗ 3) and (K ∗ 4) hold for any ranking function.

Similarly for the contraction axioms:

(K ÷ 1), *Closure*, says: $K \div \varphi = Cn(K \div \varphi)$. It holds as trivially as (K ∗ 1).

(K ÷ 2), *Inclusion*, says in AGM terms: $K \div \varphi \subseteq K$. And via Definition 11 in ranking terms: the core of κ is a subset of the core of $\kappa_{A\to 0}$. This is indeed true by definition.

(K ÷ 3), *Vacuity*, says in AGM terms: if $\varphi \notin K$, then $K \div \varphi = K$. And in ranking terms: If $\kappa(\bar{A}) = 0$, then $\kappa_{A\to 0} = \kappa$. This is true by Definition 11.

(K ÷ 4), *Success*, says in AGM terms: $\varphi \notin K \div \varphi$, unless $\varphi \in Cn(\emptyset)$. And in ranking terms: if $A \neq W$, then $\kappa_{A\to 0}(A) = 0$. Again this is true by Def. 11, also because $\kappa_{A\to 0}$ is defined for all $A \neq W$ due to the regularity of κ.

(K ÷ 5), *Recovery*, says in AGM terms: $K \subseteq Cn((K \div \varphi) \cup \{\varphi\})$. With Def. 11 this translates into ranking terms: if C is the core of κ and C' the core of $\kappa_{A\to 0}$, then $C' \cap A \subseteq C$. This holds because $C \subseteq C'$ and $C' - C \subseteq \bar{A}$.

(K ÷ 6), *Extensionality*, says: if $Cn(\varphi) = Cn(\psi)$, then $K \div \varphi = K \div \psi$. It is again guaranteed by our propositional framework.

(K ÷ 7), *Intersection*, says in AGM terms: $(K \div \varphi) \cap (K \div \psi) \subseteq K \div (\varphi \wedge \psi)$.

(K ÷ 8), *Conjunction*, finally says: if $\varphi \notin K \div (\varphi \wedge \psi)$, then $K \div (\varphi \wedge \psi) \subseteq K \div \varphi$. Both translate via Def. 11 into the corresponding assertions about the cores of the ranking functions involved. I spare myself showing their ranking-theoretic validity, also because of the next observation. (But see Spohn 2012, p. 90.)

As to the relation between AGM revision and contraction, I should add that the *Levi Identity* and the *Harper Identity* also hold according to the ranking-theoretic account of those operations:

The *Levi Identity* says in AGM terms: $K * \varphi = Cn\ ((K \div \neg\, \varphi) \cup \{\varphi\})$. And in ranking terms: if C' is the core of $\kappa_{A\to n}$ $(n > 0)$ and C'' the core of $\kappa_{\bar{A}\to 0}$, then $C' = C'' \cap A$. It thus reduces revision to contraction (and expansion) and is immediately entailed by Defs. 10–11.

The *Harper Identity* says in AGM terms: $K \div \varphi = K \cap (K * \neg\, \varphi)$. And in ranking terms: if C is the core of κ, C' is the core of $\kappa_{\bar{A}\to n}$ $(n > 0)$, and C'' the core of $\kappa_{A\to 0}$, then $C'' = C \cup C'$. It thus reduces contraction to revision and holds again because of

Def. 10–11. Moreover, since the Harper Identity translates the eight revision axioms (K $*$ 1) – (K $*$ 8) into the eight contraction axioms (K $\div$ 1) – (K $\div$ 8) and since ranking-theoretic revision satisfies (K $*$ 1) – (K $*$ 8), as shown, ranking-theoretic contraction must satisfy (K $\div$ 1) – (K $\div$ 8); so, this proves (K $\div$ 7) – (K $\div$ 8).

I should finally add that the picture does not really change under the variable interpretation introduced at the end of the previous section. Only the variants of conditionalization increase thereby. I have already noted that expansion and revision are unique only at the level of belief sets, but not at the ranking-theoretic level. Under the variable interpretation, contraction looses its uniqueness as well, because under this interpretation there are also many degrees of unopinionatedness. However, rank 0 preserves its special status, since it is the only rank n for which possibly $\tau(A) = \tau(\bar{A}) = n$. Hence, the unique contraction within the standard interpretation may now be called *central contraction*, which is still special.

The problem I want to address in this paper is now obvious. If many of the AGM revision and contraction postulates seem objectionable or lead to unintuitive results, then the above ranking-theoretic explications of AGM revision and contraction, which entail those postulates, must be equally objectionable. Hence, if I want to maintain ranking theory, I must defend AGM belief revision theory against these objections. This is what I shall do in the rest of this paper closely following Spohn (2012, Sect. 11.3), and we will see that ranking theory helps enormously with this defense. I cannot cover the grounds completely. However, if my strategy works with the central objections to be chosen, it is likely to succeed generally.

4 The Success Postulate for AGM-Revision

Let me start with three of the AGM postulates for revision. A larger discussion originated from the apparently undue rigidity of the *Success* postulate ($K * 4$) requiring that

(23) $\varphi \in K * \varphi$,

i.e., that the new evidence must be accepted. Many thought that "new information is often rejected if it contradicts more entrenched previous beliefs" (Hansson 1997, p. 2) or that if new information "conflicts with the old information in K, we may wish to weigh it against the old material, and if it is … incredible, we may not wish to accept it" (Makinson 1997, p. 14). Thus, belief revision theorists tried to find accounts for what they called non-prioritized belief revision. Hansson (1997) is a whole journal issue devoted to this problem.

The idea is plausible, no doubt. However, the talk of weighing notoriously remains an unexplained metaphor in belief revision theory; and the proposals are too ramified to be discussed here. Is ranking theory able to deal with non-prioritized belief revision?

Yes. After all, ranking theory is made for the metaphor of weighing (cf. Spohn 2012, Sect. 6.3). So, how do we weigh new evidence against old beliefs? Above I explained revision by A as result-oriented $A \rightarrow n$-conditionalization for some $n > 0$ (as far as belief sets were concerned, the result was the same for all $n > 0$). And thus *Success* was automatically satisfied. However, I also noticed that evidence-oriented $A \uparrow n$-conditionalization may be a more adequate characterization of belief dynamics insofar as its parameter n pertains only to the evidence. Now we can see that this variant conditionalization is exactly suited for describing non-prioritized belief revision.

If we assume that evidence always comes with the same firmness $n > 0$, then $A \uparrow n$-conditionalization of a ranking function κ is sufficient for accepting A if $\kappa(A) < n$ and is not sufficient for accepting A otherwise. One might object that the evidence A is here only weighed against the prior disbelief in A. But insofar as the prior disbelief in A is already a product of a weighing of reasons (as described in Spohn 2012, Sect. 6.3), the evidence A is also weighed against these old reasons. It is not difficult to show that $A \uparrow n$-conditionalization with a fixed n is a model of screened revision as defined by Makinson (1997, p. 16). And if we let the parameter n sufficient for accepting the evidence vary with the evidence A, we should also be able to model relationally screened revision (Makinson 1997, p. 19).

Was this a defense of *Success* and thus of AGM belief revision? Yes and no. The observation teaches the generality and flexibility of ranking-theoretic conditionalization. We may define belief revision within ranking theory in such a way as to satisfy *Success* without loss. But we also see that ranking theory provides other kinds of belief change which comply with other intuitive desiderata and which we may, or may not, call belief revision. In any case, ranking-theoretic conditionalization is broad enough to cover what has been called non-prioritized belief revision.

5 The Preservation Postulate

Another interesting example starts from the observation that (K * 4), *Expansion* 2, is equivalent to the *Preservation* postulate, given (K * 2), *Success*:

(24) if $\neg \varphi \notin K$, then $K \subseteq K * \varphi$

Preservation played an important role in the rejection of the unrestricted Ramsey test in Gärdenfors (1988, Sect. 7.4). Later on it became clear that *Preservation* is wrong if applied to conditional sentences φ (cf. Rott 1989) or, indeed, to any kind of auto-epistemic or reflective statements. Still, for sentences φ in our basic language **L**, *Preservation* appeared unassailable.

Be this as it may, even *Preservation* has met intuitive doubts. Rabinowicz (1996) discusses the following simple story: Suppose that given all my evidence I believe that Paul committed a certain crime (= ψ); so $\psi \in K$. Now a new witness turns up producing an alibi for Paul (= φ). Rabinowicz assumes that φ, though surprising, might well be logically compatible with K; so $\neg \varphi \notin K$. However, after the testimony I no longer believe in Paul's guilt, so $\psi \notin K * \varphi$, in contradiction to *Preservation*.

κ	ψ	$\neg\psi$
φ	3	6
$\neg\varphi$	0	9

κ'	ψ	$\neg\psi$
φ	0	3
$\neg\varphi$	6	15

Fig. 1 A Counter-example to *Preservation?*

Prima facie, Rabinowicz' assumptions seem incoherent. If I believe Paul to be guilty, I thereby exclude the proposition that any such witness will turn up; the appearance of the witness is a surprise initially disbelieved. So, we have $\neg\varphi \in K$ after all, and *Preservation* does not apply and holds vacuously.

Look, however, at the following negative ranking function κ and its $\varphi\rightarrow$6- or $\varphi\uparrow$9-conditionalization κ' (again, forgive me for mixing the sentential and the propositional framework) (Fig. 1).

As it should be, the witness is negatively relevant to Paul's guilt according to κ (and vice versa); indeed, Paul's being guilty (ψ) is a necessary and sufficient reason for assuming that there is no alibi ($\neg\varphi$)—in the sense that $\neg\varphi$ is believed given ψ and φ is believed given $\neg\psi$. Hence, we have $\kappa(\neg\psi) = 6$, i.e., I initially believe in Paul's guilt, and confirming our first impression, $\kappa(\varphi) = 3$, i.e., I initially disbelieve in the alibi.

However, I have just tacitly assumed the standard interpretation in which negative rank > 0 is the criterion of disbelief. We need not make this assumption. I emphasized at the end of Sect. 2 that we might conceive disbelief more strictly according to the variable interpretation, say, as negative rank > 5. Now note what happens in our numerical example: Since $\kappa(\neg\psi) = 6$ and $\kappa(\varphi) = 3$, I do initially believe in Paul's guilt, but not in the absence of an alibi (though one might say that I have positive inclinations toward the latter). Paul's guilt is still positively relevant to the absence of the alibi, but neither necessary nor sufficient for believing the latter. After getting firmly informed about the witness, I change to $\kappa'(\neg\varphi) = 6$ and $\kappa(\psi) = 3$; that is, I believe afterwards that Paul has an alibi (even according to our stricter criterion of belief) and do not believe that he has committed the crime (though I am still suspicious).

By thus exploiting the vagueness of the notion of belief, we have found a model that accounts for Rabinowicz' intuitions. Moreover, we have described an operation that may as well be called belief revision, even though it violates *Preservation*. Still, this is not a refutation of *Preservation*. If belief can be taken as more or less strict, belief revision might mean various things and might show varying behavior. And the example has in fact confirmed that, under our standard interpretation (with disbelief being rank > 0), belief revision should conform to preservation.

This raises an interesting question: What is the logic of belief revision (and contraction) under the variable interpretation of belief within ranking theory just used? I don't know; I have not explored the issue. What is clear is only that the logic of central contraction (cf. the end of Sect. 3) is the same as the standard logic of contraction, because central contraction *is* contraction under the standard interpretation.

6 The Superexpansion Postulate

As already noticed by Gärdenfors (1988, p. 57), (K $*$ 7), *Superexpansion*, is equivalent to the following assertion, given (K $*$ 1) – (K $*$ 6):

(25) $K * \varphi \cap K * \psi \subseteq K * (\varphi \vee \psi)$.

Arthur Paul Pedersen has given the following very plausible example that is at least a challenge to that assertion (quote from personal communication):

Tom is president of country *X*. Among other things, Tom believes

$\neg\varphi$: Country *A* will not bomb country *X*.
$\neg\psi$: Country *B* will not bomb country *X*.

Tom is meeting with the chief intelligence officer of country *X*, who is competent, serious, and honest.

Scenario 1: The intelligence officer informs Tom that country *A* will bomb country *X* (φ). Tom accordingly believes that country *A* will bomb country *X*, but he retains his belief that country *B* will not bomb country *X* ($\neg\psi$). Because Tom's beliefs are closed under logical consequence, Tom also believes that either country *A* or country *B* will not bomb country *X* ($\neg\varphi \vee \neg\psi$).

So $\varphi, \neg\psi, \neg\varphi \vee \neg\psi$ are in $K * \varphi$.

Scenario 2: The intelligence officer tells Tom that country *B* will bomb country *X* (ψ). Tom accordingly believes that country *B* will bomb country *X*, but he retains his belief that country *A* will not bomb country *X* ($\neg\varphi$). Because Tom's beliefs are closed under logical consequence, Tom also believes that either country *A* or country *B* will not bomb country *X* ($\neg\varphi \vee \neg\psi$).

So $\psi, \neg\varphi, \neg\varphi \vee \neg\psi$ are in $K * \psi$.

Scenario 3: The intelligence officer informs Tom that country *A* or country *B* or both will bomb country *X* ($\varphi \vee \psi$). In this scenario, Tom does not retain his belief that country *A* will not bomb country *X* ($\neg\varphi$). Nor does Tom retain his belief that country *B* will not bomb country *X* ($\neg\psi$). Furthermore, Tom does not retain his belief that either country *A* or country *B* will not bomb country *X* ($\neg\varphi \vee \neg\psi$)—that is to say, his belief that it is not the case that both country *A* and country *B* will bomb country *X*—for he now considers it a serious possibility that both country *A* and country *B* will bomb country *X*. Accordingly, Tom accepts that country *A* or country *B* or both will bomb country *X* ($\varphi \vee \psi$), but Tom retracts his belief that country *A* will not bomb country *X* ($\neg\varphi$), his belief that country *B* will not bomb country *X* ($\neg\psi$), and his belief that either country *A* or country *B* will not bomb country *X* ($\neg\varphi \vee \neg\psi$).

So $\varphi \vee \psi$ is in $K * (\varphi \vee \psi)$.
Importantly, $\neg\varphi \vee \neg\psi$ is not in $K * (\varphi \vee \psi)$!

One can understand the reason for the retraction of $\neg\varphi \vee \neg\psi$ in Scenario 3 as follows: If after having learned that either country *A* or country *B* will bomb country *X* Tom learns that country *A* will bomb country *X*, for him it is not settled whether country *B* will bomb country *X*. Yet if Tom were to retain his belief that either country *A* or country *B* will not

κ	ψ	$\neg\psi$
φ	2	1
$\neg\varphi$	1	0

Initial State

κ_1	ψ	$\neg\psi$
φ	1	0
$\neg\varphi$	2	1

revision of κ *by* φ

κ_2	ψ	$\neg\psi$
φ	1	2
$\neg\varphi$	0	1

revision of κ *by* ψ

κ_3	ψ	$\neg\psi$
φ	1	0
$\neg\varphi$	0	1

revision of κ *by* $\varphi \vee \psi$

κ_4	ψ	$\neg\psi$
φ	0	0
$\neg\varphi$	0	1

contraction of κ_3 *by* $\neg\varphi \vee \neg\psi$

Fig. 2 A Counter-example to *Superexpansion?*

> bomb country *X*, this issue would be settled for Tom, for having learned that country *A* will bomb country *X*, Tom would be obliged to believe that country *B* will not bomb country *X*—and this is unreasonable to Tom.

Obviously (K $*$ 7), or the equivalent statement (25), is violated by this example.

Still, I think we may maintain (K $*$ 7). Figure 2 below displays a plausible initial epistemic state κ. Scenarios 1 and 2 are represented by κ_1 and κ_2, which are, more precisely, the $\varphi \rightarrow 1$- and the $\psi \rightarrow 1$-conditionalization of κ. However, more complicated things are going on in scenario 3. Pedersen presents the intelligence officer's information that "country *A* or country *B* or both will bomb country *X*" in a way that suggests that its point is to make clear that the "or" is to be understood inclusively, not exclusively. If the information had been that "either country *A* or country *B* (and not both) will bomb country *X*", there would be no counter-example, and the supplementary argument in the last paragraph of the quote would not apply; after learning that country *A* will bomb country *X*, Tom would indeed be confirmed in believing that country *B* will not bomb country *X*.

However, the communicative function of "or" is more complicated. In general, if I say "*p* or *q*", I express, according to Grice's maxim of quantity, that I believe that *p* or *q*, but do not believe *p* and do not believe *q*, and hence exclude neither *p* nor *q*; otherwise my assertion would have been misleading. And according to Grice's maxim of quality, my evidence is such as to justify the disjunctive belief, but not any stronger one to the effect that *p*, non-*p*, *q*, or non-*q*.

So, if the officer says "φ or ψ or both", the only belief he expresses is indeed the belief in $\varphi \vee \psi$, but he also expresses many non-beliefs, in particular that he excludes neither φ, nor ψ, *nor* $\varphi \wedge \psi$. And if Tom trusts his officer, he adopts the officer's doxastic attitude, he revises by $\varphi \vee \psi$, *and* he contracts by $\neg \varphi \vee \neg \psi$, in order not to exclude $\varphi \wedge \psi$. Given the symmetry between φ and ψ, the other attitudes concerning φ and ψ then follow. That is, if Grice's conversational maxims are correctly applied, there is not only a revision going in scenario 3, but also a contraction. And then, of course, there is no counter-example to *Superexpansion*. This is again displayed in Fig. 2, where κ_3 is the $\varphi \vee \psi \rightarrow 1$-conditionalization of the initial κ (in which $\neg \varphi \vee \neg \psi$ is still believed) and κ_4 is the $\neg \varphi \vee \neg \psi \rightarrow 0$-conditionalization of κ_3 (in which $\neg \varphi \vee \neg \psi$ is no longer believed).

Note that these tables assume a symmetry concerning φ and ψ, concerning the credibility of the attacks of country *A* and country *B*. We might build in an asymmetry instead, and then the situation would change.

To confirm my argument above, suppose that in scenario 1 the officer informs Tom that country *A* will bomb country *X* *or* both countries will. The belief thereby expressed is the same as that in the original scenario 1. But why, then, should the officer choose such a convoluted expression? Because he thereby expresses different non-beliefs, namely that he does not exclude that both countries will bomb country *X*. And then, Tom should again contract by $\neg\varphi \vee \neg\psi$. In the original scenario 1, by contrast, the officer does not say anything about country *B*, and hence Tom may stick to his beliefs about country *B*, as Pedersen has assumed.

We might change scenario 3 in a converse way and suppose that the officer only says that country *A* or country *B* will bomb country *X*, without enforcing the inclusive reading of "or" by adding "or both". Then the case seems ambiguous to me. Either Tom might read "or" exclusively and hence stick to his belief that not both countries, *A* and *B*, will bomb country *X*. Or Tom might guess that the inclusive reading is intended; but then my redescription of the case holds good. Either way, no counter-example to *Superexpansion* seems to be forthcoming.

7 The Intersection Postulate for AGM-Contraction

Let me turn to some of the AGM contraction postulates, which have, it seems, met even more doubt. And let me start with the postulate (K $*$ 7), *Intersection*, which says:

(26) $(K \div \varphi) \cap (K \div \psi) \subseteq K \div (\varphi \wedge \psi)$.

This corresponds to the revision postulate (K $*$ 7) just discussed. Sven Ove Hansson has been very active in producing (counter-)examples. In (1999, p. 79) he tells a story also consisting of three scenarios and allegedly undermining the plausibility of *Intersection*:

> I believe that Accra is a national capital (φ). I also believe that Bangui is a national capital (ψ) As a (logical) consequence of this, I also believe that either Accra or Bangui is a national capital ($\varphi \vee \psi$).
>
> *Case 1*: 'Give the name of an African capital' says my geography teacher.
>
> 'Accra' I say, confidently.
>
> The teacher looks angrily at me without saying a word. I lose my belief in φ. However, I still retain my belief in ψ, and consequently in $\varphi \vee \psi$.
>
> *Case 2*: I answer 'Bangui' to the same question. The teacher gives me the same wordless response. In this case, I lose my belief in ψ, but I retain my belief in φ and consequently my belief in $\varphi \vee \psi$.
>
> *Case 3*: 'Give the names of two African capitals' says my geography teacher.
>
> 'Accra and Bangui' I say, confidently.
>
> The teacher looks angrily at me without saying a word. I lose confidence in my answer, that is, I lose my belief in $\varphi \wedge \psi$. Since my beliefs in φ and in ψ were equally strong, I cannot choose between them, so I lose both of them.

After this, I no longer believe in $\varphi \vee \psi$.

κ	ψ	$\neg\psi$
φ	0	1
$\neg\varphi$	1	2

Initial State

κ_1	ψ	$\neg\psi$
φ	0	1
$\neg\varphi$	0	1

contraction of κ *by* φ

κ_2	ψ	$\neg\psi$
φ	0	0
$\neg\varphi$	1	1

contraction of κ *by* ψ

κ_3	ψ	$\neg\psi$
φ	0	0
$\neg\varphi$	0	1

contraction of κ *by* $\varphi \wedge \psi$ *or by* $[\varphi, \psi]$

κ_4	ψ	$\neg\psi$
φ	0	0
$\neg\varphi$	0	0

contraction of κ *first by* φ *and then by* ψ

Fig. 3 A Counter-example to *Contraction?*

At first blush, Hansson's response to case 3 sounds plausible. I suspect, however, this is so because the teacher's angry look is interpreted as, respectively, φ and ψ being *false*. So, if case 1 is actually a revision by $\neg\varphi$, case 2 a revision by $\neg\psi$, and case 3 a revision by $\neg\varphi \wedge \neg\psi$, Hansson's intuitions concerning the retention of $\varphi \vee \psi$ come out right. It is not easy to avoid this interpretation. The intuitive confusion of inner and outer negation—in this case of disbelief and non-belief—is ubiquitous. And the variable interpretation of (dis)belief would make the confusion even worse.

Still, let us assume that the teacher's angry look just makes me insecure so that we are indeed dealing only with contractions. Fig. 3 then describes all possible contractions involved. κ_1 and κ_2 represent the contractions in case 1 and case 2. These cases are unproblematic.

However, I think that case 3 is again ambiguous. The look might make me uncertain about the whole of my answer. So I contract by $\varphi \wedge \psi$, thus give up φ as well as ψ (because I am indifferent between them) and retain $\varphi \vee \psi$. This is represented by κ_3 in Fig. 3.

It is more plausible, though, that the look makes me uncertain about both parts of my answer. So I contract by φ and by ψ. This may be understood as what Fuhrmann and Hansson (1994) call package contraction by $[\varphi, \psi]$, in which case I still retain $\varphi \vee \psi$ (according to Fuhrmann and Hansson (1994), and according to my ranking-theoretic reconstruction of multiple and in particular package contraction in Spohn (2010)—for details see there). The result is also represented by κ_3 in Fig. 3. The sameness is accidental; in general, single contraction by $\varphi \wedge \psi$ and package contraction $[\varphi, \psi]$ fall apart.

Or it may be understood as an iterated contraction; I first contract by φ and then by ψ (or the other way around). Then the case falls into the uncertainties of AGM belief revision theory *vis-à-vis* iterated contraction (and revision). Ranking-theoretic contraction, by contrast, can be iterated (for the complete logic of iterated contraction see Hild and Spohn (2008)). And it says that by first contracting by φ and then by ψ one ends up with no longer believing $\varphi \vee \psi$ (at least if φ and ψ are doxastically independent in the ranking-theoretic sense, as may be plausibly assumed in Hanssons's example). This is represented by κ_4 in Fig. 3.

Again, these results depend on the built-in symmetries between φ and ψ and their independence and thus on the prior state κ and its acquisition. If it were different, the contractions might have different results.

Thus, I have offered two different explanations of Hansson's intuition without the need to reject *Intersection*. In this case, I did not allude to maxims of conversation as in the previous section (since the teacher does not say anything). The effect, however, is similar. Plausibly, other or more complicated belief changes are going on in this example than merely single contractions. Therefore it does not provide any reason to change the postulates characterizing those single contractions.

8 The Recovery Postulate

Finally, I turn to the most contested of all contraction postulates, *Recovery* (K $\div$ 5), which asserts:

(27) $K \subseteq Cn((K \div \varphi) \cup \{\varphi\})$

Hansson (1999, p. 73) presents the following example: Suppose I am convinced that George is a murderer ($= \psi$) and hence that George is a criminal ($= \varphi$); thus φ, $\psi \in K$. Now I hear the district attorney stating: "We have no evidence whatsoever that George is a criminal." I need not conclude that George is innocent, but certainly I contract by φ and thus also lose the belief that ψ. Next, I learn that George has been arrested by the police (perhaps because of some minor crime). So, I accept that George is a criminal, after all, i.e., I expand by φ. Recovery then requires that $\psi \in Cn((K \div \varphi) \cup \{\varphi\})$, i.e., that I also return to my belief that George is a murderer. I can do so only because I must have retained the belief in $\varphi \rightarrow \psi$ while giving up the belief in φ and thus in ψ. But this seems absurd, and hence we face a clear counter-example against *Recovery*.

This argument is indeed impressive—but not unassailable. First, let me repeat that the ranking-theoretic conditionalization rules are extremely flexible; any standard doxastic movement you might want to describe can be described with them. The only issue is whether the description is natural. However, that is the second point: what is natural is quite unclear. Is the example really intended as a core example of contraction theory, such that one must find a characterization of contraction that directly fits the example? Or may we give more indirect accounts? Do we need, and would we approve of, various axiomatizations of contraction operations, each fitting at least one plausible example? There are no clear rules for this kind of discussion, and as long as this is so the relation between theory and application does not allow any definite conclusions.

Let us look more closely at the example. Makinson (1997) observes (with reference to the so-called filtering condition of Fuhrmann (1991), p. 184) that I believe φ (that George is a criminal) *only because* I believe ψ (that George is a murderer). Hence I believe $\varphi \rightarrow \psi$, too, *only because* I believe ψ, so that by giving up φ and hence ψ the belief in $\varphi \rightarrow \psi$ should disappear as well. This implicit appeal

κ	ψ	$\neg\psi$
φ	0	1
$\neg\varphi$	∞	2

Initial State

κ_1	ψ	$\neg\psi$
φ	0	1
$\neg\varphi$	∞	0

contraction of κ *by* φ

κ_3	ψ	$\neg\psi$
φ	0	0
$\neg\varphi$	∞	1

contraction of κ *by* $\varphi \wedge \psi$

κ_4	ψ	$\neg\psi$
φ	0	0
$\neg\varphi$	∞	0

contraction of κ *first by* ψ *and the n by* φ

Fig. 4 A Counter-example to *Recovery*?

to justificatory relations captures our intuition well and might explain the violation of *Recovery* (though the "only because" receives no further explication). However, I find the conclusion of Makinson (1997, p. 478) not fully intelligible:

> Examples such as those above … show that even when a theory is taken as closed under consequence, recovery is still an inappropriate condition for the operation of contraction when the theory is seen as comprising not only statements but also a relation or other structural element indicating lines of justification, grounding, or reasons for belief. As soon as contraction makes use of the notion "*y* is believed only because of *x*", we run into counterexamples to recovery … But when a theory is taken as "naked", i.e. as a bare set of statements closed under consequence, then recovery appears to be free of intuitive counterexamples.

I would have thought that the conclusion is that it does not make much sense to consider "naked" theories, i.e., belief states represented simply as sets of sentences, in relation to contraction, since the example makes clear that contraction is governed by further parameters not contained in that simple representation. This is exactly the conclusion elaborated by Haas (2005, Sect. 2.10).

I now face a dialectical problem, though. A ranking function is clearly not a naked theory in Makinson's sense. It embodies justificatory relations; whether it does so in a generally acceptable way, and whether it can specifically explicate the "only because", does not really matter. (I am suspicious of the "only because"; we rarely, if ever, believe things only for one reason.) Nevertheless, it is my task to defend *Recovery*. Indeed, my explanation for our intuitions concerning George is a different one.

First, circumstances might be such that recovery is absolutely right. There might be only one crime under dispute, a murder, and the issue might be whether George has committed it, and not whether George is a more or less dangerous criminal. Thus, I might firmly believe that he is either innocent or a murderer so that, when hearing that the police arrested him, my conclusion is that he is a murderer, after all.

These are special circumstances, though. The generic knowledge about criminals to which the example appeals is different. In my view, we are not dealing here with two sentences or propositions, φ and ψ, of which one, ψ, happens to entail the other, φ. We are rather dealing with a single scale or variable which, in this simple case, takes only three values: "murderer", "criminal, but not a murderer", and "not criminal". (See Fig. 4, where φ and ψ generate a 2×2 matrix. However, one field is impossible and receives negative rank ∞; one can't be an innocent murderer. So, you should rather read the remaining three fields as a single, three-valued scale.)

The default for such scales or variables is that a distribution of degrees of belief over the scale is *single-peaked*. In the case of negative ranks this means that the distribution of negative ranks over the scale has only one local minimum; so, the distribution should rather be called 'single-dented'.

In the present example, the default means: For each person, there is one degree of criminality which is most credible (where credibility is measured here by two-sided ranks, but the default as well applies to other kinds of credibility like probabilities), and other degrees of criminality are the less credible, the further away they are from the most credible degree, i.e., they decrease in a weakly monotonous way. This default is obeyed in my initial doxastic state κ displayed in Fig. 4, in which I believe George to be a murderer; there negative ranks take their minimum at the value "murderer" and then increase.

Now, a standard AGM contraction by φ (or a $\varphi \rightarrow 0$-conditionalization), as displayed in the second matrix of Fig. 4, produces a two-peaked or 'two-dented' distribution: both "not criminal" and "murderer" receive negative rank 0 and only the middle value ("criminal, but not a murderer" receives a higher negative rank (and remains thus disbelieved). This just reflects the retention of $\varphi \rightarrow \psi$. Thus, AGM contraction violates the default of single-peakedness (or 'single-dentedness').

Precisely for this reason we do not understand the district attorney's message as an invitation for a standard contraction. Rather, I think the message "there is no evidence that George is a criminal" is tacitly supplemented by "let alone a murderer", in conformity to Grice's maxim of quantity. That is, we understand it as an invitation to contract not by $\varphi \wedge \psi$ (as displayed in the third matrix of Fig. 4), but by ψ (George is a murderer), and then, if still necessary, by φ or, what comes to the same, by $\varphi \wedge \neg \psi$ (as displayed in the fourth matrix of Fig. 4). In other words, we understand it as an invitation to perform a mild contraction by φ in the sense of Levi (2004, p. 142f.), after which no beliefs about George are retained. Given this reinterpretation there is no conflict between *Recovery* and the example.

Levi (2004, p. 65f.) finds another type of example to be absolutely telling against *Recovery* (see also his discussion of still another example in Levi (1991), p. 134ff.). Suppose you believe that a certain random experiment has been performed ($= \varphi$), say, a coin has been thrown, and furthermore you believe in a certain outcome of that experiment ($= \psi$), say, heads. Now, doubts are raised as to whether the experiment was at all performed. So, you contract by φ and thereby give up ψ as well. Suppose, finally, that your doubts are dispelled. So, you again believe in φ. Levi takes it to be obvious that, in this case, it should be entirely open to you whether or not the random ψ obtains—another violation of *Recovery*.

I do not find this story so determinate. Again, circumstances might be such that *Recovery* is appropriate. For instance, the doubt might concern the correct execution of the random experiment; it might have been a fake. Still, there is no doubt about its result, if the experiment is counted as valid. In that case *Recovery* seems mandatory.

However, I agree with Levi that this is not the normal interpretation of the situation. But I have a different explanation of the normal interpretation. In my view, the point of the example is not randomness, but presupposition. ψ presupposes φ (in the formal linguistic sense); one cannot speak of the result of an experiment unless the experiment has been performed. And then it seems to be a pragmatic rule that, if the requirement is to withdraw a presupposition, then one has to withdraw the item depending on this presupposition explicitly, and not merely as an effect of giving up the presupposition.

Let us look at the situation a bit more closely. Of course, the issue depends on which formal account of presuppositions to accept. We may say that q (semantically) presupposes p if both q and $\neg q$ logically entail p, although p is not logically true; since Strawson (1950) this is standard as a first attempt at semantic presupposition. Then, however, it is clear that our propositional framework, or the sentential framework with its consequence relation Cn, is not suited for formally dealing with presuppositions. Or we may treat presuppositions within dynamic semantics. But again, our framework is not attuned to such alternatives. Hence we have to be content with an informal discussion; it will be good enough.

To begin with, it seems that any argument and hence any belief change concerning q leaves the presupposition p untouched. For instance, if we argue about, and take various attitudes towards, whether or not Jim quit smoking, or whether or not John won the race, all this takes place on the background of the presupposition that he did smoke in the past, or, respectively, that there was a race.

What happens, though, if we argue about the presupposition p itself? I think we may distinguish two cases then, instantiated by the two examples just given. Let us look at the first example and suppose that I believe that Jim quit smoking and hence smoked in the past. Now doubts are raised that Jim smoked in the past, and maybe I accept these doubts. What happens then to my belief that Jim quit smoking? Well, why did I have this belief in the first place? Presumably, because I haven't seen Jim smoking for quite a while and because I thought to remember to have often seen him smoking in the past. It is characteristic of this example that "Jim quit smoking" can be decomposed into two logically independent sentences "Jim smoked in the past" and "Jim does not smoke now". Hence, if I am to give up that Jim smoked in the past, I have to give up "Jim quit smoking" as well, but I will retain "Jim does not smoke now". This entails, however, that, if the doubts are dispelled and I return to my belief that Jim smoked in the past, I will also return to my belief that Jim quit smoking, since I retained the belief that Jim does not smoke now. And so we have a case of *Recovery*.

However, this characteristic does not always hold. Let us look at a second example, where q = "John won the race", which presupposes p = "there was a race". Again, assume that I believe both and that doubts are raised about the presupposition. The point now is "John won the race" is not decomposable in the way above. It is usually very unclear what John is supposed to have done if there was no race at all, what it is apart from the presupposition that is correctly described as John's winning the race (with the help of the presupposition). So, in this case doubts about the presupposition are at the same time doubts about John's having done anything that could be described as winning the race in the case there should have been a race. If so, the withdrawal of the presupposition p must be accompanied by an explicit withdrawal of q, so that the material implication $p \to q$ is lost as well. Again, we have no counter-example against *Recovery*; *Recovery* does not apply at all, because a more complex doxastic change has taken place in the second example. And it seems to me that, at least under the normal interpretation, Levi's example of the random experiment is of the second characteristic. If the coin has not been thrown at all, there is no behavior of the coin that could be described as the coin's showing head in case it had been thrown.

So, the pragmatic rule stated above seems to apply at least to the second kind of example characterized by the non-decomposability of presupposition and content. This pragmatic rule is quite different from my above observation about scales. The pragmatic effect, however, is the same. And again this effect agrees with Levi's mild contraction. Note, by the way, that what I described as special circumstances in the criminal and the random example above can easily be reconciled with mild contraction; informational loss is plausibly distributed under these circumstances in such a way that mild contraction and AGM contraction arrive at the same result.

Hence I entirely agree with Levi on the description of the examples. I disagree on their explanation. Levi feels urged to postulate another kind of contraction operation governed by different axioms, and Makinson has the hunch that taking account of justificatory relations will lead to such a different contraction operation. By contrast, I find AGM contraction sufficient on the theoretical level and invoke various pragmatic principles explaining why more complex things might be going on in certain situations than single AGM contractions.

9 Conclusion

All in all, I feel justified in repeating the conclusions already sketched in the introduction. First, ranking-theoretic conditionalization includes expansion, revisions, and contraction as special cases. And since the latter can plausibly be explicated by ranking theory only in the way specified in Sect. 3, this entails that the standard AGM postulates $(K * 1) - (K * 8)$ and $(K \div 1) - (K \div 8)$ must hold for revisions and contractions. However, because of its much larger generality (which in turn is due to the additional structure assumed in ranking theory) ranking-theoretic conditionalization has resources to cope with other kinds of examples and with more kinds of belief change than the standard AGM theory. On a theoretical level ranking-theoretic conditionalization is all we need.

The second conclusion is more important. I did not, and did not attempt to, offer any systematic account for dealing with all kinds of examples. On the contrary, I intentionally used a variegated bunch of pragmatic and interpretational strategies for coping with the examples. I believe that all these strategies, and certainly more, are actually applied. So there is no reasonable hope for a unified treatment of the examples. Rather, we must study all the pragmatic and interpretational ways in systematic detail. (Cf., e.g., Merin 1999, 2003a, b, who has made various interesting and relevant observations concerning the formal pragmatics of presuppositions and scale phenomena, though not in direct connection to belief revision.) And we must study the interaction of those strategies. I see here a potentially very rich, but so far little explored research field at the interface between linguistics and formal epistemology. In a way, the gist of the paper was at least to point at this large research field.

And the third conclusion is immediate: If this large research field interferes, there can be no direct argument from intuitions about examples to the basic axioms of belief change; there is always large space for alternative explanations of the intuitions within this interfering field. Hence, I have little sympathy for experimenting with these basic axioms. Rather, these axioms have theoretical justifications, which are amply provided within ranking theory (see Spohn 2012, Chaps. 5 and 8). These theoretical justifications are the important ones, and hence I stand by the standard AGM axioms unshaken.

References

Alchourrón, C. E., Gärdenfors, P., & Makinson, D. (1985). On the logic of theory change: Partial meet functions for contraction and revision. *Journal of Symbolic Logic*, *50*, 510–530.
Cohen, L. J. (1970). *The implications of induction*. London: Methuen.
Cohen, L. J. (1980). Some historical remarks on the Baconian conception of probability. *Journal of the History of Ideas*, *41*, 219–231.
Dubois, D., & Prade, H. (1988). *Possibility theory: An approach to computerized processing of uncertainty*. New York: Plenum Press.
Field, H. (1978). A note on Jeffrey conditionalization. *Philosophy of Science*, *45*, 361–367.
Fuhrmann, André. (1991). Theory contraction through base contraction. *Journal of Philosophical Logic*, *20*, 175–203.
Fuhrmann, A., & Hansson, S. O. (1994). A survey of multiple contractions. *Journal of Logic, Language, and Information*, *3*, 39–76.
Gärdenfors, P. (1988). *Knowledge in flux*. Cambridge, MA: MIT Press.
Haas, G. (2005). *Revision und Rechtfertigung*. Eine Theorie der Theorieänderung, Heidelberg: Synchron Wissenschaftsverlag der Autoren.
Hansson, S. O. (1997). Special issue on non-prioritized belief revision. *Theoria*, *63*, 1–134.
Hansson, S. O. (1999). Theory change and database updating. *A Textbook of Belief Dynamics*. Dordrecht: Kluwer.
Hild, M., & Spohn, W. (2008). The measurement of ranks and the laws of iterated contraction. *Artificial Intelligence*, *172*, 1195–1218.
Huber, F. (2006). Ranking functions and rankings on languages. *Artificial Intelligence*, *170*, 462–471.
Jeffrey, R. C. (1965/1983). The logic of decision (2nd ed.). Chicago: University of Chicago Press.
Levi, I. (1991). *The fixation of belief and its undoing*. Cambridge: Cambridge University Press.
Levi, I. (2004). *Mild contraction: Evaluating loss of information due to loss of belief*. Oxford: Oxford University Press.
Makinson, D. (1997a). Screened revision. *Theoria*, *63*, 14–23.
Makinson, D., et al. (1997b). On the force of some apparent counterexamples to recovery. In E. G. Valdés (Ed.), *Normative systems in legal and moral theory* (pp. 475–481). Festschrift for Carlos Alchourrón and Eugenio Bulygin, Berlin: Duncker & Humblot.
Merin, A. (1999). Information, relevance and social decision-making: Some principles and results of decision-theoretic semantics. In L. S. Moss, J. Ginzburg, & M. De Rijke (Eds.), *Logic, Language, and Computation* (Vol. 2, pp. 179–221). Stanford, CA: CSLI Publications (Online at http://www.semanticsarchive.net)

Merin, A. (2003a). *Replacing horn scales by act-based relevance orderings to keep negation and numerals meaningful*. Forschungsberichte der DFG-Forschergruppe Logik in der Philosophie, No. 110, University of Konstanz. (Online at http://www.semanticsarchive.net)

Merin, A. (2003b). *Presuppositions and practical reason: A study in decision-theoretic semantics*. Forschungsberichte der DFG-Forschergruppe Logik in der Philosophie, No. 114, University of Konstanz. (Online at http://www.semanticsarchive.net)

Rabinowicz, W. (1996). Stable revision, or is preservation worth preserving?. In A. Fuhrmann & H. Rott (Eds.), *Logic, action, and information* (pp. 101–128). de Gruyter: Essays on Logic in Philosophy and Artificial Intelligence, Berlin.

Rescher, N. (1964). *Hypothetical reasoning*. Amsterdam: North-Holland.

Rott, H. (1989). Conditionals and theory change: Revisions expansions and additions. *Synthese, 81*, 91–113.

Rott, H. (1999). Coherence and conservatism in the dynamics of belief Part I: Finding the right framework. *Erkenntnis, 50*, 387–412.

Shackle, G. L. S. (1961). *Decision, order, and time in human affairs* (2nd ed.). Cambridge: Cambridge University Press.

Shenoy, P. P. (1991). On Spohn's rule for revision of beliefs. *International Journal of Approximate Reasoning, 5*, 149–181.

Spohn, W. (1983). Eine Theorie der Kausalität, unpublished Habilitationsschrift, Universität München, pdf-version at: http://www.uni-konstanz.de/FuF/Philo/Philosophie/philosophie/files/habilitation.pdf

Spohn, W. (1988). Ordinal conditional functions. A dynamic theory of epistemic states. In W. L. Harper & B. Skyrms (Eds.), *Causation in decision, belief change, and statistics* (Vol. II, pp. 105–134). Dordrecht: Kluwer.

Spohn, W. (2010). Multiple contraction revisited. In M. Suárez, M. Dorato, & M. Rédei (Eds.), *EPSA epistemology and methodology of science* (Vol. 1, pp. 279–288).Launch of the European Philosophy of Science Association, Dordrecht: Springer.

Spohn, W. (2012). *The laws of belief*. Ranking Theory and Its Philosophical Applications, Oxford: Oxford University Press.

Strawson, P. F. (1950). On referring. *Mind, 59*, 320–344.

Teller, P. (1976). Conditionalization, observation and change of preference. In W. L. Harper & C. A. Hooker (Eds.), *Foundations of Probability Theory* (pp. 205–259). Reidel, Dordrecht: Statistical Inference and Statistical Theories of Science.

Zadeh, Lofti A. (1978). Fuzzy sets as a basis for a theory of possibility. *Fuzzy Sets and Systems, 1*, 3–28.

Liars, Lotteries, and Prefaces: Two Paraconsistent Accounts of Belief Change

Edwin Mares

Abstract In this paper two closely related systems of paraconsistent belief change are presented. A paraconsistent system of belief change is one that allows for the non-trivial treatment of inconsistent belief sets. The aim in this paper is to provide theories that help us investigate responses to paradoxes. According to the strong system of paraconsistent belief change, if an agent accepts an inconsistent belief set, he or she has to accept the conjunctions of all the propositions in it. The weak system, on the other hand, allows people to have beliefs without believing their conjunctions. The weak system seems to deal better than with the lottery and preface paradoxes than does strong system, although the strong system has other virtues.

Keywords Paraconsistent belief change · Paraconsistent consequence relation · Relevance logic · Lottery paradox · Preface paradox · Liar paradox · Russell's paradox · Reject set

1 Introduction

This paper sets out two systems of belief change. Both of these systems allow one both to hold inconsistent beliefs and to be thought of as epistemically rational. By this I mean that the systems can exhibit a rationale for accepting an inconsistent belief set. One of these systems is a streamlined form of the view of Mares (2002) and the other is a modification of this first view to treat the lottery paradox and the preface paradox.

E. Mares (✉)
Centre for Logic, Language, and Computation, Victoria University of Wellington, Wellington, New Zealand
e-mail: Edwin.Mares@vuw.ac.nz

S. O. Hansson (ed.), *David Makinson on Classical Methods for Non-Classical Problems*, Outstanding Contributions to Logic 3, DOI: 10.1007/978-94-007-7759-0_7,

The themes in this paper intersect with David Makinson's work at various points. The first and most obvious connection between them is that the systems[1] of belief change presented in this paper are variations on the AGM account of belief change. The second point of interesection is almost as obvious. In his well known paper, "The Paradox of the Preface" Makinson (1965b),[2] Makinson presents an example of an agent who is rationally motivated to hold an inconsistent set of beliefs. I employ this example both to motivate the overall project and to support the second system that I present. The third point of overlap concerns the logic I use in order to define the closure of belief sets. This is the weak relevant logic, B. Early in his career, Makinson worked on another weak relevant logic, the system of first degree entailments (FDE) Makinson (1965a) and he discusses this logic further in Makinson (1973). These logics are closely linked and I could use FDE in the current systems, with minor modifications.[3]

I use a relevant logic because relevant logics are *paraconsistent*. A paraconsistent logic is a logic that does not contain every instance of the *rule of explosion*:

$$\begin{array}{c} \neg A \\ A \\ \hline \therefore\ B \end{array}$$

The reason for rejecting explosion, in the study of belief revision, is to be able to treat inconsistent belief sets in a non-trivial manner. There are two possible motivations for wanting to do this. First, we might want to deal with contradictory beliefs that an agent may happen to have, without her intending to accept a contradiction. Second, we might want to study the conditions under which it might be rational intentionally to accept inconsistent sets of propositions. Although the systems that I explore in this paper can be used to study belief sets that just happen to be inconsistent, my focus is on the conditions under which it is rational knowingly to accept inconsistent beliefs.

A situation in which it seems to be rational for an agent to accept inconsistent beliefs is given by Makinson (1965b). Suppose that an agent writes a non-fiction book—a book purporting to state facts about some subject. Let $A_1, \ldots, A_n$ be the statements that she makes in the main text of the book. In the preface, being both modest and sincere, the agent says 'not every statement I make in this book is true'. She believes every statement she makes, but she knows that, being a substantial book, it is extremely unlikely that every one of the statements of that book is true. So, she believes all of $A_1, \ldots, A_n$ but also believes that at least one of $A_1, \ldots, A_n$ is false.

---

[1] I use 'systems' of belief change throught this paper rather than the more usual 'theories' because I use 'theory' to refer to sets of formulas that are closed under conjunction and logical consequence. In the original draft of this paper I used 'theory' in both of these senses, producing a rather confusing mess (as a referee pointed out).

[2] In a recent paper Makinson (2012), Makinson returns to the preface paradox and compares it with the lottery paradox.

[3] Makinson has returned much more recently to work relevant logic. Recently, he is investigating the relationship between relevant and classical logic, in particular in his contribution to this volume.

Clearly, the agent has inconsistent beliefs, but it also seems that she is rational. We need a system that can account for this.

Discussing the preface paradox in slightly more depth allows us to understand a distinction that is central to this paper. An agent might accept contradictory beliefs in one of two senses. She might believe a single proposition that is by itself a contradiction or she might believe more than one proposition that together are inconsistent. When one accepts a contradiction, then we say that she has adopted a *dialetheic solution* to a problem (presumably no one rational would accept a contradiction without being faced with a problem of some sort). When one accepts an inconsistent set of propositions but does not accept any inconsistent conjunctions of these propositions in order to avoid a problem we say that she has adopted a *non-adjunctive solution* to that problem.[4]

Perhaps the most familiar motivations for dialetheic solutions involve the liar paradox and Russell's paradox. The naïve account of truth leads to the liar paradox and the naïve notion of a set as an arbitrary collection leads to Russell's paradox. There are good grounds, however, to accept naïve theories: they are conceptually simpler and less contrived than the alternatives. On the other hand, consider the lottery paradox Kyburg (1961). Suppose that there is a fair lottery with one million tickets. Each ticket has an equal chance of winning—$1/1{,}000{,}000$. Given this, for each ticket, it is rational to believe that it will not win. But it is not rational to believe that no ticket will win and, moreover, it is rational to believe that some ticket will win. Let W_i represent the belief that ticket i will win. Thus, a rational person can believe $\neg W_1, \ldots, \neg W_{1,000,000}$ and $W_1 \vee \ldots \vee W_{1,000,000}$ which are, taken together, inconsistent. But it is not rational to believe that $\neg W_1 \wedge \ldots \wedge \neg W_{1,000,000}$ and so a non-adjunctive paraconsistent solution seems more reasonable for the lottery paradox.

In this paper, I first set out a system that allows only for adjunctive solutions to paradoxes. An adjunctive solution is one in which the agent's belief set is fully closed under conjunction. One sort of adjunctive solution is a dialetheic solution, that is, a solution in which we accept statements that are themselves inconconsistent. Another sort of adjunctive solution is to reject a source of the paradox and have a consistent belief set that is closed under conjunction. The second system that I present is a modification of the first to allow non-adjunctive solutions to paradoxes. I call the system that allows only belief sets that are closed under conjunction, the *strong system*, and the system that does not require this, the *weak system*. The reason for these names will be made clear soon.

2 Relevant Consequence Relations

At the heart of any system of belief revision is a consequence operator, or family of consequence operators. The ones that the present systems utilize are, of course, paraconsistent consequence operators. The particular logic that I use to define them

---

4 This terminology is common in the literature. See Arruda (1977); Priest and Routley (1989).

is a relevant logic. I could use a non-relevant paraconsistent logic, and this would require only minor modifications to my view.

I define two consequence operators based on the weak relevant logic **B**. I pick this logic because it is weak and so makes very few assumptions about the nature of logical consequence. But, without any real modifications to what I say later, I could choose any relevant logic between **B** and **RM3**.[5] What is needed is a consequence operator Cn that has the following properties. Where X is a set of formulas,

- $X \subseteq Cn(X)$ (inclusion);
- $Cn(Cn(X)) = Cn(X)$ (idempotence);
- $X \subseteq Y$ implies $Cn(X) \subseteq Cn(Y)$ (monotonicity);
- if $A \in Cn(X)$, then there is a finite $Y \subseteq X$, $A \in Cn(Y)$ (compactness).

Let us now turn to the formal details. The language is a standard propositional language with a non-empty set of propositional variables $p, q, r, \ldots$, parentheses (,), binary connectives $\rightarrow$, $\wedge$, $\circ$ and a unary connective $\neg$. The connective $\circ$ is called *fusion*. It is an intensional conjunction. It plays no direct role in the formulation of the systems of belief revision but is included to make possible the formulation of the relevant consequence relation given below. I also use two defined connectives:

$$A \vee B =_{Df} \neg(\neg A \wedge \neg B)$$

$$A \leftrightarrow B =_{Df} (A \rightarrow B) \wedge (B \rightarrow A)$$

The Logic B

Axioms:

1. $A \rightarrow A$	2. $A \rightarrow (A \vee B)$
3. $B \rightarrow (A \vee B)$	4. $(A \wedge B) \rightarrow A$
5. $(A \wedge B) \rightarrow B$	6. $(A \rightarrow C) \rightarrow ((A \wedge B) \rightarrow C)$
7. $((A \rightarrow C) \wedge (B \rightarrow C))$ $\leftrightarrow ((A \vee B) \rightarrow C)$	8. $(A \wedge (B \vee C))$ $\rightarrow ((A \wedge B) \vee (A \wedge C))$
9. $\neg\neg A \leftrightarrow A$	

Rules:

Name	Rule
Modus Ponens	$\vdash A \rightarrow B, \vdash A \Longrightarrow \vdash B$
Adjunction	$\vdash A, \vdash B \Longrightarrow \vdash A \wedge B$
Suffixing	$\vdash A \rightarrow B \Longrightarrow \vdash (B \rightarrow C) \rightarrow (A \rightarrow C)$
Prefixing	$\vdash B \rightarrow C \Longrightarrow \vdash (A \rightarrow B) \rightarrow (A \rightarrow C)$
Contraposition	$\vdash A \rightarrow B \Longrightarrow \vdash \neg B \rightarrow \neg A$
Importation/Exportation	$\vdash A \rightarrow (B \rightarrow C) \Longleftrightarrow \vdash (A \circ B) \rightarrow C$

[5] For a survey of the various relevant logics, their properties, and their semantics, see Routley et al. (1983).

This logic determines several different consequence relations. I distinguish between three of these here. The first is the *relevant consequence relation*, $\vdash$. This relation holds between a *structure* of formulas on the left and a single formula on the right. I borrow the notion of a structure of formulas from Slaney (1990) and Restall (2000). The set of structures is the smallest set such that if A is a formula, then A is a structure and if X and Y are structures, then so is the ordered pair (X, Y). We say that

$$X \vdash B \text{ if and only if } \vdash \mathring{X} \rightarrow B$$

where $\mathring{X}$ results from replacing all occurrences of the comma in X with fusion. So, in particular, $A \vdash B$ if and only if $\vdash A \rightarrow B$. The relevant consequence relation forms the basis for the other two relations. This relation, as opposed to the other two, is singular on the right; it holds between structures of formulas on the left and single formulas on the right.

The relation $\vdash$ does not itself give yield a consequence operator, so we move on to develop a second consequence relation—the *conjunctive consequence relation*, $\Vdash$. This is a relation between sets of formulas. For finite non-empty sets it is defined as follows:

$$\{A_1, ..., A_n\} \Vdash \{B_1, ..., B_m\} \text{ iff } A_1 \wedge ... \wedge A_n \vdash B_1 \vee ... \vee B_m,$$

where $n, m > 0$. The conjunction and disjunction used here are extensional. The conjunction is not the same as fusion. One important difference is that exensional conjunction does not residuate with implication, that is we cannot infer in every case from $\{A, B\} \Vdash \{C\}$ to $\{A\} \Vdash \{B \rightarrow C\}$. In addition, unlike the relevant consequence relation defined in terms of fusion, $\vdash$, we have $\{...A...\} \Vdash \Delta$, if and only if we also have $\{...A, A...\} \Vdash \Delta$. The same holds for the conclusion set. In fact, $\Vdash$ obeys weakening, that is, if $\Gamma \Vdash \Delta$, then $\Gamma' \cup \Gamma \Vdash \Delta$ and $\Gamma \Vdash \Delta \cup \Delta'$.

The definition of $\Vdash$ is extended to cover infinite sets as follows:

$$X \Vdash Y \text{ iff there are finite } X' \subseteq X \text{ and } Y' \subseteq Y \text{ such that } X' \Vdash Y'.$$

Note that the empty set does not entail anything under $\Vdash$ and no set entails the empty set. Where no confusion will result, I drop the set brackets and write '$A_1, ..., A_n \Vdash B_1, ..., B_m$' instead of '$\{A_1, ..., A_n\} \Vdash \{B_1, ..., B_m\}$'.

The third consequence relation is the single entailment consequence relation, $\Vvdash$. It is a relation between two non-empty sets of formulas, X and Y such that

$$X \Vvdash Y \text{ iff there is a wff } A \in X \text{ and a } B \in Y \text{ such that } A \vdash B.$$

This definition holds for both finite and infinite sets. As we can see from the definition, the empty set does not entail any other set under $\Vvdash$ nor does any set entail the empty set.

I use the consequence relations $\Vdash$ and $\Vvdash$ to define closure operations. In the usual way, the relations can be used to define the consequence closure of a set X:

$$Cn_{\Vdash}(X) = \{A : X \Vdash A\}$$

The closure $Cn_{\Vdash}(X)$ of X is called a *theory*. It is closed both under the provable entailments of the logic B and under conjunction. In one version of the current view, I use theories to represent belief sets.

It is easy to show that for any set of formulas X, $Cn_{\Vdash}$ has the following properties:

- $X \subseteq Cn_{\Vdash}(X)$ (inclusion);
- $Cn_{\Vdash}(Cn_{\Vdash}(X)) = Cn_{\Vdash}(X)$ (idempotence);
- $X \subseteq Y$ implies $Cn_{\Vdash}(X) \subseteq Cn_{\Vdash}(Y)$ (monotonicity).

From the definition of $Cn_{\Vdash}$ it follows that it is compact, that is,

- If $A \in Cn_{\Vdash}(X)$, then there is some finite subset Y of X such that $A \in Cn_{\Vdash}(Y)$.

In addition to theories, I need sets that are duals of theories, called "co-theories" or "reject sets", or merely *rejects*. From an algebraic point of view, a theory is a filter on the Lindenbaum algebra of the logic. A reject is an ideal on the Lindenbaum algebra.[6] A reject is a set of formulas closed under disjunction and is closed downwards under entailment. I introduce the operator $Do_{\Vdash}$ to represent this form of closure:

$$Do_{\Vdash}(X) = \{A : A \Vdash X\}$$

$Do_{\Vdash}$ is called a *downward consequence operator*. A set of formulas Δ is a reject iff and only if $\Delta = Do_{\Vdash}(\Delta)$. For each formula in a reject, the reject also includes all the formulas that entail it and for all formulas A, B in the reject, $A \vee B$ is also in the reject. The downward consequence operator has some similar properties to the upward consequence operator:

- $X \subseteq Do_{\Vdash}(X)$ (inclusion);
- $Do_{\Vdash}(Do_{\Vdash}(X)) = Do_{\Vdash}(X)$ (idempotence);
- $X \subseteq Y$ implies $Do_{\Vdash}(X) \subseteq Do_{\Vdash}(Y)$ (monotonicity);
- if $A \in Do_{\Vdash}(X)$, then there is some finite subset Y of X such that $A \in Do_{\Vdash}(Y)$ (compactness).

I also use consequence and downwards consequence operators based on $\Vvdash$:

$$Cn_{\Vvdash}(X) = \{A : X \Vvdash A\} \; Do_{\Vvdash}(X) = \{A : A \Vvdash X\}$$

We call a set Γ closed under $Cn_{\Vvdash}$ a *semi-theory* and a set Δ closed under $Do_{\Vvdash}$ a *semi-reject*. Both $Cn_{\Vvdash}$ and $Do_{\Vvdash}$ satisfy inclusion, idempotence, monotonicity, and compactness.

---

[6] Strictly speaking, a filter or ideal on the Lindenbaum algebra is not a set of formulas, but a set of equivalence classes of formulas.

Every semi-theory is the union of a set of theories, in particular, where Γ is a semi-theory

$$\Gamma = \bigcup_{A \in \Gamma} Cn_{||\text{-}}(\{A\}).$$

Similarly, every semi-reject is the union of a set of rejects. Where Δ is a semi-reject,

$$\Delta = \bigcup_{A \in \Delta} Do_{||\text{-}}(\{A\}).$$

These facts will be used in Sect. 10 below.

3 Belief and Denial

The AGM account treats contraction, expansion, and revision as operations on sets of formulas (or propositions or actions). These are belief sets. When we move to discussing the possibility of one's accepting contradictory beliefs, there are good reasons why we should move away from belief sets to consider slightly more complicated structures. Consider the version of the AGM account based on epistemic entrenchment (Gärdenfors and Makinson 1988). When we revise a belief set, we remove all those formulas that are least entrenched that contradict the new formula. The motivating principles behind this view are conservativeness about relinquishing beliefs and the principle that we should maintain the consistency of our belief sets. Now consider what happens if we merely drop the principle that we should maintain consistency. Conservativeness alone would direct us not to abandon any beliefs when we integrate new ones. This is clearly inadequate.

We could instead merely take it for granted that we would not consider identifying revision with expansion as adequate and treat revision as a primitive notion. Then we could see which of the AGM postulates is appropriate when we relinquish consistency as a condition of belief sets. Some researchers have done just this (see Priest (2001) and especially Tanaka (2005)).[7] But I wish to present a systems that expose more of the way in which people working with inconsistent theories understand what they are doing.[8]

When people working in inconsistent mathematics, for example, discuss how they have created their theories, they often talk about the fragments of classical mathematics that they wish to keep and about those formulas that they wish to avoid proving. Moreover, there are various degrees to which one might wish to avoid certain formulas. For example, everyone wants to avoid proving $0 = 1$, since having it in

---

[7] For a good survey of AGM-like theories based on non-classical logics, see Wassermann (2011).

[8] Restall and Slaney (1995) also try to expose more detail in how people do or should work with inconsistencies in belief revision, but they are concerned with the elimination of inconsistencies, not with when we should accept them.

a system of arithmetic will threaten to allow us to prove every equation. But in an inconsistent set theory, one might dither about whether to accept or deny principles such as the axiom of choice and the continuum hypothesis. One might accept those principles if he or she needs them to prove something else and cannot see a way to do so that is as elegant without them. So our rejection of a formula can come in different strengths.

This notion of rejection is called *denial*. A denial can be either an overt speech act or a propositional attitude. The idea behind my view of belief revision is to treat the process of revision as a trade-off between retaining beliefs and retaining the sentences that we deny. I represent the content of an agent's belief and denials as a pair of sets of formulas, (Γ, Δ). I call such pairs, *contents*. In this paper, I discuss two sorts of contents: strong contents and weak contents.[9]

Definition 1 A strong content *is a pair* (Γ, Δ) *of sets of formulas such that* (i) Γ *is a theory*, (ii) Δ *is a reject, and* (iii) $\Gamma \cap \Delta = \emptyset$.

Definition 2 A weak content *is a pair* (Γ, Δ) *of sets of formulas such that* (i) Γ *is a semi-theory*, (ii) Δ *is a semi-reject, and* (iii) $\Gamma \cap \Delta = \emptyset$.

A strong content is what (Restall 2013), calls a "bitheory". In strong contents, the theory represents an agent's belief set and the reject represents his or her denial set. In weak contents, the semi-theory represents an agent's beliefs and the semi-reject represents his or her denials.

Contents are consistent in a sense. Beliefs and denials do not overlap. This is a specific case of what I call pragmatic consistency or *p-consistency*.[10] A pair of sets of formulas (Γ, Δ) is strongly p-consistent if and only if $\Gamma \not\Vdash \Delta$. It is weakly p-consistent if and only if $\Gamma \not\Vvdash \Delta$.

In this paper, I set out two systems of belief revision. In the first, revision, contraction and expansion are operations on strong contents. I call this the *strong system*. In the second, they are operations on weak contents and, correspondingly, I call it the *weak system*.

4 A Semi-Lattice of Strong Contents

In setting out my two systems of belief revision, I employ an ordering on contents. Here I set out the ordering on strong contents, $\sqsubseteq$. It is defined as follows:

$$(\Gamma, \Delta) \sqsubseteq (\Gamma', \Delta') \quad \text{iff} \quad \Gamma \subseteq \Gamma' \text{ and } \Delta \subseteq \Delta'$$

I also utilize a intersection operator on contents:

---

[9] There are other theories of belief revision that use rejection sets, see, e.g., Alchourón and Bulygin (1981) and Gomolinska (1998).

[10] In fact the term used in Mares (2002) is "coherence", but was changed in Mares (2004) to "pragmatic consistency". I now prefer the latter term.

$$(\Gamma, \Delta) \sqcap (\Gamma', \Delta') =_{Df} (\Gamma \cap \Gamma', \Delta \cap \Delta')$$

where $SCon$ is the set of strong contents, the triple $(SCon, \sqsubseteq, \sqcap)$ is a meet semilattice. $SCon$ is not just closed under finite intersections. Rather, where X is an arbitrary subset of $SCon$, the ordered pair, $\sqcap X = (\cap X_1, \cap X_2)$ is a strong content, where X_1 is the set of belief sets of members of X and X_2 is the set of denial sets of members of X.

The bottom element of the semi-lattice is $(\emptyset, \emptyset)$. There is no unique top member. Every strong content (Γ, Δ) such that $\Gamma \cup \Delta$ is the set of formulas is a $\sqsubseteq$-maximal member of $SCon$, and the only maximal members are such that their belief and denial sets sum to the set of all formulas. This is easy to prove:

Proposition 3 *If a pair* (Γ, Δ) *is a* $\sqsubseteq$*-maximal strong content, then* $\Gamma \cup \Delta$ *is the set of formulas.*

Proof Suppose that there is no strong content that $\sqsubseteq$-contains (Γ, Δ). And suppose that there is some formula A such that $A \notin \Gamma$ and $A \notin \Delta$. From $\sqsubseteq$-maximality, we know that $\Gamma \cup \{A\} \Vdash \Delta$ and $\Gamma \Vdash \Delta \cup \{A\}$. We can derive from this that there is some formula $G \in \Gamma$ and some $D \in \Delta$ such that $\vdash G \rightarrow (A \vee D)$ and $\vdash (G \wedge A) \rightarrow D$. Since Γ is a theory, $A \vee D \in \Gamma$. But $G \in \Gamma$, hence $G \wedge (A \vee D) \in \Gamma$. The logic is distributive (by axiom 8), and so $(G \wedge A) \vee (G \wedge D) \in \Gamma$. As we have said $\vdash (G \wedge A) \rightarrow D$ and we know that $\vdash (G \wedge D) \rightarrow D$, hence by axiom 7, adjunction and modus ponens, we obtain $D \in \Gamma$. But if $D \in \Gamma$, then (Γ, Δ) is not strongly p-consistent and so is not a strong content. Thus, there is no formula that is in neither Γ nor Δ. ■

Note that this little proof requires that the logic be distributive. Thus it does not work for non-distributive logics, such as linear logic. Whether Proposition 3 can be proven for non-distributive logics I do not know. As we shall see in Sect. 8, a similar proposition is provable for weak contents.

5 Aside I: Strong Contents and Prime Theories

The present approach to belief revision is syntactic. It treats revision and the other AGM operations in terms of formulas contained in belief and denial sets. We might, however, be able to set out a Grove-like semantics for paraconsistent belief revision and extract from it the properties of the AGM operations. This has been done for views of belief revision that are rather different from mine by Restall and Slaney (1995), Tanaka (2005), Priest (2001). In Grove's semantics, which uses classical logic, a model is based on a set of maximally consistent sets of formulas Grove (1988). In these paraconsistent systems, models are based instead on *prime theories* of the logic that is being used. A theory Γ is prime if and only if for every disjunction $A \vee B$ in Γ at least one disjunct A or B is in Γ.

It might be advantageous to have a semantical version of the strong system. In the case of AGM, having the Grove semantics led to logical semantics that explicitly deal with iterated revision, such as dynamic doxastic logic Segerberg (1995) and dynamic epistemic logic.[11] Perhaps the same progression from a syntactic system to a form of dynamic semantics could occur for paraconsistent belief revision as well. The model theory for relevant logics are based on sets of indices, like the possible worlds of modal logic. The indices of relevant semantics are known as set-ups, (possible and impossible) worlds, or situations. In the canonical model for relevant logics, indices are represented by prime theories. The relationship between prime theories and the indices of the relevant logic on one hand and prime theories and strong contents on the other hand might allow us to create a logic of strong contents based on the model theory for relevant logic. A proponent of the strong system might follow what Segerberg, van Benthem, and others have done in creating a modal logic of belief revision but based on the Routley-Meyer model theory for relevant logic rather than the Kripke semantics for classical modal logic.

A belief set is modelled by a set of prime theories if and only if the intersection of that set is exactly the belief set. We can extend this idea to model strong contents in the following way.

Definition 4 *A strong content* (Γ, Δ) *is* strongly modelled *by a set* X *of prime theories if and only if* (i) $\Gamma = \cap X$ *and* (ii) $\Delta = \{A : A \notin \cup X\}$.

In other words, a strong content is strongly modelled by a set of prime theories X if its belief set is the set of formulas true in every prime theory in X and its denial set is the set of formulas that are in none of the prime theories in X.

It turns out that not every strong content is strongly modelled by a set of prime theories. Only those that are closed under the *saturation rules* are modelled in this sense. The saturation rules are the following:

$$\frac{\Gamma \cup \{D\} \Vdash \Delta}{\therefore\ D \in \Delta}.$$

$$\frac{\Gamma \Vdash \Delta \cup \{G\}}{\therefore\ G \in \Gamma}$$

Let us call all strong contents that satisfy the saturation rule, *saturated strong contents*. The first saturation rule is a sort of pragmatic form of modus tollens. It tells us that if we cannot add a particular formula to our stock of beliefs, then we should deny it. The second rule is a pragmatic form of disjunctive syllogism. It tells us, in effect, that if our beliefs entail a disjunction and all but one of those disjuncts is denied then we should believe that remaining disjunct. Whether these rules place reasonable constraints on contents is a topic that I will leave to another occasion.

[11] For an excellent narrative explaining the development of modal logics of belief revision from the AGM theory, see Fuhrmann (2011). For clear exposition of dynamic epsitemic logic and an up-to-date bibliography, see Benthem (2011).

Proposition 5 *Let X be a set of prime theories. Then the content strongly modelled by X is a saturated strong content.*

Proof Since X is a set of theories, $\Gamma = \cap X$ is also a theory. Now consider $\Delta = Fml - \cup X$, where Fml is the set of formulas. Assume that $A \in \Delta$ and that $\vdash B \rightarrow A$. Then $B \in \Delta$, for otherwise there would be at least one $x \in X$ such that $B \in x$. Since x is a theory, A would also be in x. Now suppose that A and B are both in Δ. I show that $A \vee B \in \Delta$. Since X is the set of prime theories, if $A \vee B$ were in one member x of X, then either A or B would also be in x. But if either A or B were in x, it would not be in Δ. So, $A \vee B \in \Delta$. Clearly, (Γ, Δ) is strongly p-consistent. Thus, (Γ, Δ) is a strong content.

Now, let us show that (Γ, Δ) is saturated. First, suppose that $\Gamma \cup \{D\} \Vdash \Delta$. By the definition of 'strong modelling', for every $x \in X$, $\Gamma \subseteq x$. Suppose that $D \in x$. Then, because x is a theory, some disjunction in Δ is also in x. But Δ is closed under disjunction, and so there is some formula in Δ that is also in x. But, by the definition of 'strong modelling' this is impossible. Thus, $D \notin x$. Generalizing, D is not in any member of X and therefore $D \in \Delta$.

Now, suppose that $\Gamma \Vdash \Delta \cup \{G\}$. Then, by the definition of $\Vdash$, there is some $B \in \Delta$ and some $A \in \Gamma$ such that $\vdash A \rightarrow (B \vee G)$. So, $B \vee G \in x$ for every $x \in X$ and, since x is prime, either B or G is in x. But B is in Δ and hence, by the construction of Δ, B is not in x. Thus, $G \in x$. Generalizing, $G \in \Gamma$. Therefore, (Γ, Δ) is saturated. ■

In order to prove the next proposition, I need a version of the Lindenbaum extension lemma. This version was originally proven by Nuel Belnap and independently by Gabbay (1974).[12]

Lemma 6 *Let Γ and Δ be sets of formulas such that $\Gamma \nVdash \Delta$. Then there is a prime theory Γ' such that $\Gamma \subseteq \Gamma'$ and $\Gamma' \nVdash \Delta$.*

Proposition 7 *Let (Γ, Δ) be a saturated strong content. There is a set of prime theories X that strongly models (Γ, Δ).*

Proof Let A be an arbitrary formula not in Δ. Since (Γ, Δ) is saturated, if $\Gamma \cup \{A\} \Vdash \Delta$, then $A \in \Delta$, so $\Gamma \cup \{A\} \nVdash \Delta$. By Lemma 6, there is a prime theory x extending Γ and containing A such that $x \nVdash \Delta$. Let us call the set of prime theories x extending Γ such that $x \nVdash \Delta$, X. Clearly, the set of formulas that belong to no member of X is Δ.

I now show that $\Gamma = \cap X$. We have stipulated that for each $x \in X$, $\Gamma \subseteq x$. Now let B be an arbitrary formula not in Γ. I show that there is some $x \in X$ such that $B \notin x$. Note that $\Gamma \nVdash \{B\} \cup \Delta$. For, if $\Gamma \Vdash \{B\} \cup \Delta$, by saturation, $B \in \Gamma$. Thus, by Lemma 6, there is some $x \in X$ such that $B \notin x$. ■

---

12 As far as I know, Belnap never published his proof.

6 Contraction, Expansion, and Revision in the Strong System

The strong system is a slightly modified version of the system of Mares (2002). According to that system, in revising one's beliefs, an agent adds the new proposition to the belief set, and perhaps contracts her belief set and/or her denial set to maintain p-consistency. Similarly, when an agent revises her denial set, she adds a new proposition to her denials and contracts her denials and/or her beliefs to maintain p-consistency.

The system has twice as many operations as does the standard AGM account. It has contraction, expansion and revision operations for beliefs and one of each of these for denials. Belief contraction is formalized using the symbol $-$ and denial contraction by $\ominus$. Similarly belief expansion is represented by $+$ and denial expansion by $\oplus$, belief revision by $*$, and denial revision by $\circledast$. To define expansions, contractions, and revisions I use a set of functions σ_A°, where $\circ$ is a variable ranging over operators from the set $\{+, -, *, \oplus, \ominus, \circledast\}$ and A is a formula.[13] Every σ_A° is a function from strong contents to sets of strong contents. Belief and denial expansion, contraction, and revision are all defined in the same way as partial meets[14]:

$$(\Gamma, \Delta) \circ A \;=\; \sqcap \sigma_A^\circ((\Gamma, \Delta)).$$

The differences between the operations are to be understood through the different constraints that we place on the various σ_A°.

Let us start with the two sorts of contraction. The idea is quite simple. When revising a belief set Γ by a formula A, an agent takes a set of subtheories of Γ that do not contain A and are most acceptable to that agent. The contraction of Γ is the intersection of these acceptable subtheories of A. Similarly, when contracting a denial set Δ by A, she takes a set of sub-rejects of Δ that are most acceptable to her and that do not contain A and takes the intersection of them. More formally, the constraints on belief contraction are that (i) if $(\Gamma', \Delta') \in \sigma_A^-((\Gamma, \Delta))$, then $\Gamma' \subseteq \Gamma$, (ii) $\Delta' = \Delta$, and (iii) $A \notin \Gamma'$. Similarly, (i) if $(\Gamma', \Delta') \in \sigma_A^\ominus((\Gamma, \Delta))$, then $\Delta' \subseteq \Delta$, (ii) $\Gamma' = \Gamma$, and (iii) $A \notin \Delta'$.

Unlike most other AGM-like systems, in this system expansion is more complicated than contraction. If we were just to add formulas to belief or desire sets, then we could violate strong p-consistency. When one expands her belief set, she may at the same have to contract her denial set and, similarly, when she expands her denial set she may have to contract her belief set. These contractions are difficult to describe in general. If we expand a belief set by A, we may have to contract formulas other than A from the denial set. Suppose that $B \wedge A$ is also in the denial set and B is in the belief set. We will have a violation of p-consistency when A is added to the beliefs unless $B \wedge A$ is removed from the denial set. The constraints on belief expansion are that (i) if $(\Gamma', \Delta') \in \sigma_A^+((\Gamma, \Delta))$, then $\Delta' \subseteq \Delta$ and (ii) $\Gamma' = Cn_{\Vdash}(\Gamma \cup \{A\})$. In

---

[13] I use 'σ' to connote 'strong' as in 'strong theories'.

[14] Partial meets were first introduced into the AGM theory as a way of treating contraction Alchourón et al. (1985). Here I generalize this approach to define all the AGM operations.

addition, (i) if $(\Gamma', \Delta') \in \sigma^{\oplus}_A((\Gamma, \Delta))$, then $\Gamma' \subseteq \Gamma$ and (ii) $\Delta' = Do_{\Vdash}(\Delta \cup \{A\})$. When there is no need to contract the denial set, we should not do so. Thus I add the constraint that if $\Gamma \cup \{A\} \not\Vdash \Delta$, then $\sigma^{+}_A((\Gamma, \Delta)) = \{(Cn_{\Vdash}(\Gamma \cup \{A\}), \Delta)\}$ and if $\Gamma \not\Vdash \Delta \cup \{A\}$, then $\sigma^{\oplus}_A((\Gamma, \Delta)) = \{(\Gamma, Do_{\Vdash}(\Delta \cup \{A\}))\}$.

Unfortunately, revision is not definable in terms of expansion and contraction in this setting. Revision is a little like expansion in that it may change both an agent's belief and denial sets. But unlike expansion, the idea behind revision is that the agent may trade off between relinquishing beliefs or denials in order to integrate a new belief or denial. For example, suppose that I come home to find a note that my partner has taken our dogs for a walk. It is natural for me to integrate this proposition into my set of beliefs by relinquishing the belief that she is at home rather than the denials that she and the dogs are invisible and that she is walking them inside our house.

The constraints on belief revision are that (i) if $(\Gamma', \Delta') \in \sigma^{*}_A((\Gamma, \Delta))$, then $A \in \Gamma'$, (ii) if $(\Gamma', \Delta') \in \sigma^{*}_A((\Gamma, \Delta))$, then $\Delta' \subseteq \Delta$, and (iii) if $(\Gamma', \Delta') \in \sigma^{*}_A((\Gamma, \Delta))$, then $\Gamma' \subseteq Cn_{\Vdash}(\Gamma \cup \{A\})$. Similarly, the constraints on denial revision are if $(\Gamma', \Delta') \in \sigma^{\circledast}_A((\Gamma, \Delta))$, then $A \in \Delta'$, (ii) if $(\Gamma', \Delta') \in \sigma^{*}_A((\Gamma, \Delta))$, then $\Gamma' \subseteq \Gamma$, and (iii) if $(\Gamma', \Delta') \in \sigma^{*}_A((\Gamma, \Delta))$, then $\Delta' \subseteq Do_{\Vdash}(\Gamma \cup \{A\})$. I place an additional constraint on revision. Where adding a formula does not cause any conflict between beliefs and desires, we should not otherwise modify the belief and desire sets. That is,

$$\text{where } \Gamma \cup \{A\} \not\Vdash \Delta,\ \sigma^{*}_A((\Gamma, \Delta)) = \{(Cn_{\Vdash}(\Gamma \cup \{A\}), \Delta)\}$$

and

$$\text{where } \Gamma \not\Vdash \Delta \cup \{A\},\ \sigma^{\circledast}_A((\Gamma, \Delta)) = \{(\Gamma, Do_{\Vdash}(\Delta \cup \{A\}))\}.$$

So, if $\Gamma \cup \{A\}$ and Δ are strongly p-consistent, then the belief revision of (Γ, Δ) by A is just the same as the belief expansion of that content by A and, similarly, if Γ and $\Delta \cup \{A\}$ are strongly p-consistent, then the denial revision by A of (Γ, Δ) is the same as its denial expansion by A.

7 The AGM Postulates

There are motivations for the use of relevant logic other than its paraconsistency. It is not the case that any formula belongs to all theories. In fact the empty set is itself a theory of **B** (and of every other relevant logic). Similarly, for every formula, there is at least one reject that does contain it. The set of formulas itself is a reject (and a theory). Moreover, for any formula there is at least one theory that contains it and at least one reject that fails to contain it. These facts allow the strong system to satisfy unrestricted forms of the AGM success postulate. The following all hold of the strong system:

$$\text{where } \circ \in \{+, *\},\ A \in_1 ((\Gamma, \Delta) \circ A) \text{ (positive success 1);}$$
$$\text{where } \circ \in \{\oplus, \circledast\},\ A \in_2 ((\Gamma, \Delta) \circ A) \text{ (positive success 2);}$$
$$A \notin_1 ((\Gamma, \Delta) - A) \text{ (negative success 1);}$$
$$A \notin_2 ((\Gamma, \Delta) \ominus A) \text{ (negative success 2).}$$

Here I use $A \in_1 (X, Y)$ to mean that $A \in X$ and $A \in_2 (X, Y)$ to mean that $A \in Y$.

It is also straightforward to show that the inclusion schemas hold universally in this system. That is, where $(\Gamma, \Delta) + A = (\Gamma', \Delta')$, $\Gamma \subseteq \Gamma'$ and where $(\Gamma, \Delta) \oplus A = (\Gamma', \Delta')$, $\Delta \subseteq \Delta'$. And, as in the AGM account, If $A \in_1 (\Gamma, \Delta)$, then $(\Gamma, \Delta) \circ A = (\Gamma, \Delta)$ where $\circ \in \{+, *\}$. Similarly, If $A \in_2 (\Gamma, \Delta)$, then $(\Gamma, \Delta) \circ A = (\Gamma, \Delta)$ where $\circ \in \{\oplus, \circledast\}$.

The logical equivalence postulate, if $\vdash A \leftrightarrow B$, then $(\Gamma, \Delta) \circ A = (\Gamma, \Delta) \circ B$, is easily satisfied, for all the operations. We need only set $\sigma^\circ_A = \sigma^\circ_B$ whenever $\vdash A \leftrightarrow B$.

The case concerning the special AGM postulates for conjunction are somewhat more complicated. Here they are translated into the idiom of my view:

- $(\Gamma, \Delta) * A \wedge B \sqsubseteq ((\Gamma, \Delta) * A) + B$;
- If $B \notin_2 (\Gamma, \Delta) * A$, then $((\Gamma, \Delta) * A) + B \sqsubseteq (\Gamma, \Delta) * A \wedge B$.

The problem with the first of these is to determine how Δ needs to be contracted in order to allow Γ to be modified to integrate $A \wedge B$ or to integrate A or to be expanded to contain B. It is not clear that, in $((\Gamma, \Delta) * A) + B$, Δ will be shrunk less than in $(\Gamma, \Delta) * A \wedge B$. A weaker postulate that we can enforce is is that the belief set in $(\Gamma, \Delta) * A \wedge B$ will always be a subset of the belief set in $((\Gamma, \Delta) * A) + B$. The second special postulate can be accepted if it altered slightly:

$$\text{If } \Gamma' \cup \{B\} \not\Vdash \Delta', \text{ then } ((\Gamma, \Delta) * A) + B \sqsubseteq (\Gamma, \Delta) * A \wedge B,$$

where Γ' is the belief set and Δ' the denial set in $(\Gamma, \Delta) * A$. The dual of the second special postulate is also of interest:

$$\text{If } \Gamma' \not\Vdash \Delta' \cup \{B\}, \text{ then } ((\Gamma, \Delta) \circledast A) \oplus B \sqsubseteq (\Gamma, \Delta) \circledast A \vee B,$$

where Γ' and Δ' are the belief set and denial set respectively of $(\Gamma, \Delta) \circledast A$. Satisfying either or both forms of the second special postulate requires fiddling with the selection functions in rather complicated ways, but it can be done. The question is whether these postulates are intuitive enough (or do enough work in the resulting system) to make complexity worthwhile. This is a question I leave to the reader to answer.

8 The Lottery and Preface Paradoxes and the Weak System

In this section I consider in more detail the lottery paradox and non-adjunctive solutions to it. My presentation closely follows that of Douven and Williamson (2006).

The lottery paradox, in effect, sets up an opposition between two principles. The first of these principles is what Douven and Williamson call the "sufficiency thesis" that we should believe any proposition the probability of which is sufficiently close to 1 and the second is what is usually known as the "adjunction principle", that our beliefs should be closed under conjunction. Some philosophers, such as Pollock (1995) and levi (1997), have suggested restricting the sufficiency thesis in reaction to the lottery paradox. Pollock says:

> Observing that the probability is only 0.000001 of a ticket's being drawn given that it is a ticket in the lottery, it seems reasonable to accept that the ticket will not win. The prima facie reason involved in this reasoning is statistical syllogism. By the same reasoning, it will be reasonable to believe, for each ticket, that it will not win. However, these conclusions conflict jointly with something else that we are justified in believing: that some ticket will win. We cannot be justified in believing each member of an explicitly contradictory set of propositions, and we have no way of choosing between them, so it follows intuitively that we are not justified in believing of any ticket that it will not win (Pollock 1995, p. 112).

Pollock thinks that the inconsistency of a set of potential beliefs rules out an ideal agent's accepting it. As a paraconsistent logician I disagree with this, but his idea can be translated into the idiom of the present view. Suppose that the agent denies that $\neg W_1 \wedge \ldots \wedge \neg W_{10^6}$ and that this denial is more entrenched than the belief that $\neg W_i$ for each i between 1 and 1,000,000 if she were to adopt that belief. Then, since she has the same evidence for each of the $\neg W_i$s, she should abandon all of them to maintain strong p-consistency. Thus, we can adopt Pollock-style reasoning and maintain the adjunction principle despite the lottery paradox.[15]

I am not sure, however, that I should adopt this sort of reasoning. In order to see why, let us turn to the preface paradox. We can think of the preface paradox as coming in two forms: a negation form and a denial form. First, there is the form in which the agent accepts the negative sentence, 'It is not the case that every statement in this book is true'. This is the negation form of the paradox. Second, there is the form of the paradox in which the agent denies the sentence, 'every statement in this book is true'. This latter version is the denial form of the paradox. In its denial form, the agent asserts a set of statements $A_1, \ldots, A_n$, and she denies a conjunction, $A_1 \wedge \ldots \wedge A_n$. This agent's content is not strongly p-consistent, but it does seem that the she is rational. Being both modest and realistic, the agent should believe that at least one of his beliefs are false. A paraconsistent account of belief change should be able to accommodate this sort of content. With regard to the preface paradox, there is no move available like restricting the sufficiency thesis. This is because the agent has no choice but to assert the statements in his own book. Thus, I think that there are cases in which we want to allow agents to have strongly p-inconsistent contents and so we need to move to a system that requires only weak p-consistency (or something like it).

Given this abandonment of strong p-consistency, it seems reasonable to adopt a view of the lottery paradox as well as the preface paradox that allows an agent to be strongly p-inconsistent.

---

15 Although this solution might run afoul of Douven and Williamson's generalizations of the lottery paradox Douven and Williamson (2006).

Before I go on, I would like to discuss briefly why a dialetheic approach to the lottery paradox is counter-intuitive. Suppose that an agent takes a dialetheic attitude towards the lottery. That is to say, before the lottery takes place, the agent believes $\neg W_i$ for each i between 1 and 1,000,000 and also believes that $\neg W_1 \wedge \neg W_2 \wedge ... \wedge \neg W_{10^6}$. Before the lottery is drawn, however, the agent knows that one of her beliefs of the form $\neg W_i$ is inaccurate, but she doesn't know which one it is. But she does know that the belief that $\neg W_1 \wedge \neg W_2 \wedge ... \wedge \neg W_{10^6}$ will turn out to be inaccurate no matter which ticket is drawn. So, the agent knows before the drawing that one of her $\neg W_i$s will need to be rejected but not which one, but she does know that no matter what, if the drawing takes place, that she will eventually have to remove $\neg W_1 \wedge \neg W_2 \wedge ... \wedge \neg W_{10^6}$ from her belief set. In terms of my view, this is a good reason for the agent to deny $\neg W_1 \wedge \neg W_2 \wedge ... \wedge \neg W_{10^6}$ from the start and not believe that $\neg W_1 \wedge \neg W_2 \wedge ... \wedge \neg W_{10^6}$. Thus, it would seem that we should reject dialetheic approaches to the lottery paradox.

With regard to the preface paradox the situation is slightly more complicated. As we have seen, the denial form requires a non-adjunctive solution. The negation form of the paradox, however, can be given either a non-adjunctive solution or a dialetheic solution. There is no formal or epistemological barrier to conjoining all of the beliefs other than the desire to maintain consistency. It may turn out that none of the statements in the body of the book are false and that the only false statement is the one in the preface. So, hedging one's bets by believing both the conjunction of every statement in the body of the book and the disjunction of their negation might be seen to be rational. In other words, it does not seem irrational to accept a dialethic solution to the negation form of the preface paradox.

The advantage of the weak system that I present below is that it allows an agent to employ dialethic solutions to some paradoxes at the same time as taking on non-adjunctive solutions to other paradoxes. The flexibility of the weak system in this regard is an important positive feature.

9 The Semi-Lattice of Weak Contents

The algebraic structure of the set of is weak contents is similar to that of the set of strong contents. The triple $(WCon, \sqsubseteq, \sqcap)$ is a semi-lattice, where $WCon$ is the set of weak contents. One might think that the quadruple $(WCon, \sqsubseteq, \sqcap, \sqcup)$ is a lattice, where $(\Gamma, \Delta) \sqcup (\Gamma', \Delta') = (\Gamma \cup \Gamma', \Delta \cup \Delta')$, since semi-theories and semi-rejects are closed under union. But this is not the case. The problem is that even if (Γ, Δ) and (Γ', Δ') are weak contents, $(\Gamma, \Delta) \sqcup (\Gamma', \Delta')$ might not be weakly p-consistent.

The following proposition gives us a partial picture of the nature of this semi-lattice. Where Fml is the set of formulas,

Proposition 8 *A weak content* (Γ, Δ) *is* $\sqsubseteq$*-maximal in the set* $WCon$ *if and only if* $\Gamma \cup \Delta = Fml$.

Proof $\Rightarrow$ Suppose that (Γ, Δ) is $\sqsubseteq$-maximal in the set $WCon$. Let A be a formula not in Γ. Then $\Gamma \cup \{A\} \Vdash \Delta$. Since $\Gamma \nVdash \Delta$, we know that $A \Vdash \Delta$, hence $A \in \Delta$. $\Leftarrow$ is obvious. ■

10 Conjunction/Disjunction Maximality

There is, however, a problem with the modification of the system of belief revision to use weak content rather than strong contents. The complete loss of closure under conjunction seems too radical. For the most part, we are justified in believing a conjunction when we believe its conjuncts. If I believe that I am cold and that it is a sunny warm day, then I am justified in believing that the weather is warm and that I feel cold. The dissonance between the conjuncts should be allowed to come together so that I can investigate why both of these are true at the same time. I might ask myself whether I have a virus and whether I should go to the doctor. A content should be *appropriately* closed under conjunction and disjunction. The problem is to determine what is appropriate closure under conjunction.

In this section, I start the search for the appropriate notion of closure by presenting and investigating the notion of a $\wedge\vee$-*maximal content* (pronounced 'conjunction/disjunction maximal content'). As a first approximation, the aim of our contractions, expansions, and revisions should be to construct a content in which the belief set is as closed under conjunction and the denial set is as closed under disjunction as is possible while maintaining weak p-consistency. Here is a formal definition:

Definition 9 *A weak content* (Γ, Δ) *is* $\wedge\vee$-*maximal if and only if* (i) *for every* $A, B \in \Gamma$ *if* $A \wedge B \notin \Gamma$, *then* $\Gamma \cup \{A \wedge B\} \Vdash \Delta$ *and* (ii) *for every* $C, D \in \Delta$ *if* $C \vee D \notin \Delta$ *then* $\Gamma \Vdash \Delta \cup \{C \vee D\}$.

Any weak content can be extended to one that is $\wedge\vee$-maximal. This extension, moreover, is unique. The operator Max defined below produces a $\wedge\vee$-maximal content when applied to any weak content.

Definition 10 *Where* (Γ, Δ) *is a weakly p-consistent pair of sets of formulas,* $Max((\Gamma, \Delta))$ *is a pair of sets* (Γ', Δ') *such that* (i) *for any formula* G, $G \in \Gamma'$ *if and only if there is some* $X \subseteq \Gamma$ *such that* $X \nVdash \Delta$ *and* $X \Vdash G$ *and* (ii) *for any formula* D, $D \in \Delta'$ *if and only if there is some* $Y \subseteq \Delta$ *such that* $\Gamma \nVdash Y$ *and* $D \Vdash Y$.

Theorem 11 shows that Max produces $\wedge\vee$-maximal contents from arbitrary weak contents.

Theorem 11 *Where* (Γ, Δ) *is a weakly p-consistent pair of sets of formulas,* $Max((\Gamma, \Delta))$ *is* (a) *a weak content and* (b) $\wedge\vee$-*maximal.*

Proof Let (Γ, Δ) be a weakly p-consistent pair of sets of formulas and let $Max((\Gamma, \Delta)) = (\Gamma', \Delta')$.

(a) It is clear that Γ' is a semi-theory and Δ' is a semi-reject. So it is sufficient to show that (Γ', Δ') is weakly p-consistent. Suppose that it is not weakly p-consistent, then $\Gamma' \Vdash \Delta'$. Thus, there are $G \in \Gamma'$ and $D \in \Delta'$ such that $G \vdash D$. But, by the definition of Max, there is some $X \subseteq \Gamma$ and some $Y \subseteq \Delta$ such that $X \nVdash \Delta$, $\Gamma \nVdash Y$, $X \Vdash G$, and $D \Vdash \Delta$. But then $X \Vdash G$, $G \vdash D$, and $D \Vdash \Delta$. So, $X \Vdash \Delta$, contrary to the construction of $Max((\Gamma, \Delta))$. Therefore, $Max((\Gamma, \Delta))$ is weakly p-consistent.

(b) Suppose that $A, B \in \Gamma'$ and $A \wedge B \nVdash \Delta'$. Then, $\{A, B\} \nVdash \Delta$. $\{A, B\} \Vdash A \wedge B$ hence $A \wedge B \in \Gamma'$. Now suppose that $A, B \in \Delta'$ and $\Gamma' \nVdash A \vee B$. Then, $\Gamma \nVdash \{A, B\}$ hence $A \vee B \in \Delta'$. Therefore, $Max((\Gamma, \Delta))$ is $\wedge\vee$-maximal. ■

In the cases in which (Γ, Δ) is strongly p-consistent, $Max((\Gamma, \Delta))$ is a strong content.

The following theorem will also be useful in the construction of the non-adjunctive system of belief revision.

Theorem 12 *If* (Γ, Δ) *is a* $\wedge\vee$*-maximal weak content, then* $(\Gamma, \Delta) = Max((\Gamma, \Delta))$. ■

Proof Suppose that (Γ, Δ) is a $\wedge\vee$-maximal weak content. By the definition of Max and the properties of $\Vdash$, $(\Gamma, \Delta) \sqsubseteq Max((\Gamma, \Delta))$.

I now show that $Max((\Gamma, \Delta)) \sqsubseteq (\Gamma, \Delta)$. Suppose that $A \in_1 Max((\Gamma, \Delta))$. Then, there is some $X \subseteq \Gamma$ such that $X \nVdash \Delta$ and $X \Vdash A$. Then, by the definition of $\Vdash$, there are some $G_1, ..., G_n \in X$ such that $G_1 \wedge ... \wedge G_n \vdash A$. (Γ, Δ) is $\wedge\vee$-maximal, so if $G_1 \wedge ... \wedge G_n \notin \Gamma$ then $\{G_1 \wedge ... \wedge G_n\} \Vdash \Delta$, hence $X \Vdash \Delta$, contrary to the assumption. So, $G_1 \wedge ... \wedge G_n \in \Gamma$ and since $G_1 \wedge ... \wedge G_n \Vdash A$, $A \in \Gamma$ as required.

Now suppose that $B \in_2 Max((\Gamma, \Delta))$. Then, there is a $Y \subseteq \Delta$ such that $\Gamma \nVdash Y$ and $B \Vdash Y$. Thus, there are $D_1, ..., D_m \in Y$ such that $B \Vdash D_1 \vee ... \vee D_m$. By the definition of $\wedge\vee$-maximality, if $D_1 \vee ... \vee D_m \notin \Delta$, then $\Gamma \Vdash D_1 \vee ... \vee D_m$, but then $\Gamma \Vdash Y$, contradiction the assumption. So, $D_1 \vee ... \vee D_m \in \Delta$, hence $B \in \Delta$. ■

11 Belief and Denial Fade

I do not claim that our contents should always be $\wedge\vee$-maximal. In the context of the lottery paradox, considered from the point of view of classical logic, Kyburg points out that even in some cases in which we maintain consistency of the agent's belief set conjoining can be unwarranted (Kyburg 1997, p. 118). Suppose that the core of the agent's belief set just contains $\neg W_i$ for each ticket i and $W_1 \vee ... \vee W_{10^6}$. If we add $\neg W_1 \wedge ... \wedge \neg W_{999,999}$ then we still have a consistent set, but that additional belief is unreasonable. Kyburg thinks that our beliefs should fade out or in with their probabilities. Even if we do not agree with the probabilistic point of view, we can accept the gist of what Kyburg says.[16] At some point in conjoining the $\neg W_i$s, the

[16] Makinson develops Kyburg's idea in a more formally percise way in terms of the notion of a "lossy rule". Conjunction introduction is lossy; it is unreliable in certain situations. In Makinson (2012), Makinson formulates rules that allow conjunction introduction to be "applied sparingly".

agent starts to produce statements that she does not believe. But she might not want to deny them either. She merely reach a point at which she wishes to withhold belief. It seems counterintuitive in these cases to demand that her contents be $\wedge\vee$-maximal.[17]

Thus, we need a system that encourages some closure of belief sets under conjunction and denial sets under disjunction, but stops short of demanding $\wedge\vee$-maximality. I suggest that we do this by defining the AGM operations by means of partial meets of $\wedge\vee$-maximal contents. The intersection of a set of $\wedge\vee$-maximal contents itself may not be $\wedge\vee$-maximal. Consider, for example, $Max(\{p, q\}, \emptyset) \sqcap Max(\{p, q\}, \{p \wedge q\}) = (Cn_{|||\vdash}\{p, q\}, \emptyset)$ and this is not $\wedge\vee$-maximal. But, if an agent's content is the intersection of just a few weak contents and they are very similar to one another, and they are $\wedge\vee$-maximal, then their intersection will be largely closed under conjunction, although perhaps not fully closed.

I use 'ω' to represent the various selection functions on the class of weak contents, because 'ω' looks rather like 'w'. To ensure that the operations deliver meets of $\wedge\vee$-maximal contents, I use the operation Max as follows. For all $\circ \in \{+, -, *, \oplus, \ominus, \circledast\}$,

$$(\Gamma, \Delta) \circ A \;=\; \sqcap\{Max((\Gamma', \Delta')) : (\Gamma', \Delta') \in \omega^{\circ}_{A}((\Gamma, \Delta))\}$$

In words, the meet corresponding to each $\circ$ and formula A is the intersection of the Max of each of the weak contents chosen by ω°_{A}. For example, in the case of the lottery paradox, one maximalized potential content may contain as a belief $\neg W_1 \wedge \ldots \wedge \neg W_{999,999}$ but another may contain it as a denial, and so in the intersection it appears in neither the belief or denial set.

In cases in which only strong contents seem right to an agent, the application of the operations will always yield a strong content.

Theorem 13 *If all* $(\Gamma', \Delta') \in \omega^{\circ}_{A}((\Gamma, \Delta))$ *are strong contents, then* $(\Gamma, \Delta) \circ A$ *is a strong content.*

Proof Suppose that every $(\Gamma', \Delta') \in \omega^{\circ}_{A}((\Gamma, \Delta))$ is a strong content. Then, by Theorem 12, $Max((\Gamma', \Delta')) = (\Gamma', \Delta')$. So, $(\Gamma, \Delta) \circ A = \sqcap\omega^{\circ}_{A}((\Gamma, \Delta))$ and the intersection of strong contents is a strong content. ■

---

[17] In Kyburg (1997), Kyburg attacks Ray Jennings and Peter Schotch's theory of forcing Jennings and Schotch (1984) as a means for dealing with the lottery paradox. Since my original idea in this paper was to adapt the forcing strategy to the AGM theory, it is appropriate for me to explain forcing briefly. In that theory, a (finite) premise set is partitioned into consistent subsets. The fewest number of sets such that their union is the original premise set is called the level of the premise set. Suppose that the level of a premise set X is n. Then we say that X forces a formula A (written $X[\vdash A$) if and only if in every partition of X into n subsets there is a set Y in that partition such that Y classically entails A. As a defender of forcing, Brown (1999) says that this method does not allow us to derive all the conjunctions that one would accept in lottery type cases. Rather, extra logical considerations usually come into play. This is true in the weak theory as well, since the formal requirements on the selection functions do not determine exactly which contents are selected, maximalized, and then intersected.

I impose the same constraints on the ω_A°s as I did on the σ_A°s to make the system obey the same postulates. The operations all satisfy the corresponding success and inclusion postulates and the same difficulties exist with the special postulates.

The treatment of expansion in the weak system is, however, somewhat interesting. Suppose that an agent has a content (Γ, Δ) and wishes to expand her belief set by A. The belief set of each member of $\omega_A^+((\Gamma, \Delta))$ is $Cn_{|||\vdash}(\Gamma \cup \{A\})$ and so $(\Gamma, \Delta) + A = \sqcap\{Max((Cn_{|||\vdash}(\Gamma \cup \{A\}), \Delta')) : (Cn_{|||\vdash}(\Gamma \cup \{A\}), \Delta') \in \omega_A^+((\Gamma, \Delta))\}$. What is different in the weak system from the strong system is that without knowing some contingent facts about the selection function, we cannot know what the belief sets of expansion are. The treatment of denial expansion is similar.

12 Aside II: Weak Contents and Prime Theories

In Sect. 5, we saw that a saturated strong content is strongly modelled by a set of prime theories. The relationship between weak theories and contents is not as elegant. A weak content is modelled in a similar way by a set of sets of prime theories. But, interestingly, unlike strong contents, we do not need any additional conditions, such as saturation, in order to model weak contents using prime theories.

Part of the interest in showing that saturated strong contents are modelled by sets of prime theories is that this fact might be useful in constructing a Kripke semantics for a modal logic of belief revision based on the strong system. The move to sets of sets of prime theories is not clearly linked to modal logic in the same way, but it *might* be that a logic based on a neighbourhood-like semantics could be constructed to capture the ideas behind the weak system. I am not sure that this can be done, but for the sake of comprehensiveness, I include the following discussion on models for weak contents.

First, I need a notion of weak modelling to parallel the concept of strong modelling that I used in Sect. 5:

Definition 14 *A weak content* (Γ, Δ) *is weakly modelled by a set of sets of prime theories* Σ *if and only if* (i) *for each formula* $G \in \Gamma$ *there is a* $X \in \Sigma$ *such that for all* $x \in X$, $G \in x$ *and* (ii) *for each formula* $D \in \Delta$ *there is some* $Y \in \Sigma$ *such that for all* $y \in Y$, $D \notin y$.

Not just any set of sets of prime theories weakly models a weak content. A further condition is needed to ensure that the pair of sets of formulas modelled is weakly p-consistent. This is the *overlap* condition. A set of sets of formulas X is said to be *overlapping* if and only if for any Y and Z in X, $Y \cap Z$ is non-empty. This is the only condition we need:

Proposition 15 *Every set of overlapping sets of prime theories weakly models a weak content.*

Proof Let Σ be a set of overlapping sets of prime theories. First, consider the set Γ of formulas A such that there is some X in Σ, for all $x \in X$, $A \in x$. Then, every formula B such that $A \Vdash B$ is also in every $x \in X$. Thus, Γ is a semi-theory. Second, consider the set Δ of formulas B such that there is some $Y \in \Sigma$ for all $y \in Y$, $B \notin y$. Clearly, this set is a semi-reject, since for any theory y if $B \notin y$ and $C \Vdash B$, then $C \notin y$.

Now, I show that (Γ, Δ) is weakly p-consistent. Suppose that $A \in \Gamma$ and $B \in \Delta$. Then there are $X, Y \in \Sigma$ such that for all $x \in X$, $A \in x$, and for all $y \in Y$, $B \notin y$. If $A \Vdash B$, then for all $y \in Y$, $A \notin y$. But, there is some prime theory in both X and Y and so this is not possible. Thus, (Γ, Δ) is weakly p-consistent. ■

The converse can be proven as well.

Proposition 16 *Every weak content is weakly modelled by some set of overlapping sets of prime theories.*

Proof Let (Γ, Δ) be a weak content.

Suppose that $A \in \Gamma$ and that X_A is the set of prime theories that contain A. Consider an arbitrary formula C such that $A \nVdash C$. By Lemma 6, is an $x \in X_A$ such that $C \notin x$. Thus, if $C \in x$ for all $x \in X_A$, then $A \Vdash C$ and hence $C \in \Gamma$.

Now suppose that $B \in \Delta$ and that Y_B is the set of prime theories that do not contain B. Let C be an arbitrary formula such that $C \nVdash B$. By Lemma 6, there is a prime theory y such that $C \in y$ but $B \notin y$. This y is in Y_B.

I now construct a set of sets of formulas that models (Γ, Δ), i.e., $\Sigma = \{X_A : A \in \Gamma\} \cup \{Y_B : B \in \Delta\}$. I still need to show that it does in fact model (Γ, Δ).

Let Z be a member of Σ. Assume first that $E \in z$ for all $z \in Z$. Case 1. Z is some X_A for some $A \in \Gamma$. Then, as I have shown, $A \Vdash E$, hence $E \in \Gamma$. Case 2. z is Y_B for some $B \in \Delta$. This is impossible, since $\emptyset \in Y_B$ for all $B \in \Delta$.

Second, assume that $E \notin z$ for all $z \in Z$. Case 1. Z is some X_A for some $A \in \Gamma$. This is impossible since Fml—the set of all formulas—is a member of all X_A for all $A \in \Gamma$. Case 2. z is Y_B for some $B \in \Delta$. As I have already shown, $E \Vdash B$, hence $E \in \Delta$.

We also need to show that the sets in Σ are overlapping. Consider two sets, W and Z in Σ. Case 1. There is an $A \in \Gamma$ and a $B \in \Delta$ such that $W = X_A$ and $Z = Y_B$. By hypothesis, $A \nVdash B$. So, by Lemma 6, there is a prime theory that contains A and does not contain B and this prime theory is in both W and Z. Case 2. There is an $A \in \Gamma$ and a $B \in \Gamma$ such that $W = X_A$ and $Z = X_B$. As I have already pointed out, the set of formulas is in X_A and X_B, so they overlap. Case 3. There is an $A \in \Delta$ and a $B \in \Delta$ such that $W = Y_A$ and $Z = Y_B$. As I have already pointed out, the empty set is in Y_A and Y_B, so they overlap.

Therefore, Σ weakly models (Γ, Δ). ■

13 Conclusion: Assessing the Strong and Weak Systems

In this paper, I sketch two paraconsistent systems of belief change. The first is the strong system, that takes belief sets to be theories and denial sets to be rejects. The mathematical relationships between theories and rejects is well understood. Theories correspond algebraically to filters in lattices. Rejects correspond to ideals. Filter-ideal pairs are elegant structures. Moreover, the modelling of a certain class of strong contents by sets of prime theories is attractive. The second system that I present is the weak system of belief revision. That account treats belief sets as semi-theories and denial sets as semi-rejects. These are not as widely used. Nor are they as elegant. Furthermore, they are modelled by much more complicated structures. On aesthetic grounds, the strong system seems to win.

But when we turn to applications, the weak system comes into its own. In order to provide a wider understanding of the rationality of accepting inconsistencies, we have to look both at cases in which it seems rational to accept a single formula that is a contradiction and at cases in which we have an inconsistent set of formulas but which are not by themselves contradictions. In the second sort of case, there seem to some instances in which it is desirable, all things considered, to advise an agent not to conjoin her beliefs or disjoin her denials. Thus, we need the weak system.

Acknowledgments David Makinson suggested to me that I write a paper for this volume and for this he is in part responsible for my rethinking my views on paraconsistency and belief revision. I am very grateful to David for his close reading of a draft of this paper and his extremely helpful suggestions. I am grateful to Rob Goldblatt, Patrick Girard, Jeremy Seligman, and Zach Weber for listening to and commenting on an earlier version of this paper. I am also indebted to Sven Ove Hansson and his two referees for very helpful comments. They saved me from some very embarrassing errors.

References

Alchourón, C., & Bulygin, E. (1981). The expressive conception of norms. In R. Hilpinen (Ed.), *New essays in deontic logic* (pp. 95–124). Dordrecht: Reidel.

Alchourón, C., Gärdenfors, P., & Makinson, D. (1985). On the logic of theory change: Partial meet contraction and revision functions. *The Journal of Symbolic Logic*, *50*, 510–530.

Arruda, A. (1977). A survey of paraconsistent logic. In A. I. Arruda, N. Da Costa, & R. Chuaqui (Eds.), *Non-classical logic* (pp. 1–41). Amsterdam: North Holland.

Brown, B. (1999). Adjunction and aggregation. *Noûs*, *33*, 273–283.

Douven, I., & Williamson, T. (2006). Generalizing the lottery paradox. *British Journal for the Philosophy of Science*, *57*, 755–779.

Fuhrmann, A. (2011). Theories of belief change. In S. Bernecker & D. Pritchard (Eds.), *Routledge companion to epistemology* (pp. 621–638). London: Routledge.

Gabbay, D. M. (1974). On second order intuitionist propositional calculus with full comprehension. *Archiv für Mathematische Logik und Grundlagenforschung*, *16*, 177–186.

Gärdenfors, P., & Makinson, D. (1998). Revisions of knowledge systems using epistemic entrenchment. In *Second Conference on Theoretical Aspects of Reasoning about Knowledge*, (pp. 83–95).

Gomolinska, A. (1998). On the logic of acceptance and rejection. *Studia Logica*, *60*, 233–251.

Grove, A. (1988). Two modellings for theory change. *Journal of Philosophical Logic, 17*, 157–170.
Jennings, R., & Schotch, P. (1984). The preservation of coherence. *Studia Logica, 43*, 89–106.
Kyburg, H. (1961). *Probability and the logic of rational belief*. Middletown: Wesleyan University Press.
Kyburg, H. (1997). The rule of adjunction and reasonable inference. *The Journal of Philosophy, 94*, 109–125.
Levi, I. (1997). *The covenant of reason: Rationality and the commitments of thought*. Cambridge: Cambridge University Press.
Makinson, D. (1965a). An alternative characterization of first-degree entailment. *Logique et Analyse, 8*, 305–311, Misprints corrected in *Logique et Analyse, 9*, 394, 1966.
Makinson, D. (1965b). The paradox of the preface. *Analysis, 25*, 205–207.
Makinson, D. (1973). *Topics in modern logic*. London: Methuen.
Makinson, D. (2012). Logical questions behind the lottery paradox: Lossy rules for uncertain inference. *Synthese, 186*, 511–529.
Mares, E. (2002). A paraconsistent theory of belief revision. *Erkenntnis, 56*, 229–246.
Mares, E. (2004). *Relevant logic: A philosophical interpretation*. Cambridge: Cambridge University Press.
Pollock, J. (1995). *Cognitive carpentry*. Cambridge: MIT Press.
Priest, G. (2001). Paraconsistent belief revision. *Theoria, 67*, 214–228.
Priest, G., & Routley, R. (1989). On paraconsistency. In G. Priest, R. Routley, & J. Norman (Eds.), *Paraconsistent logic*. Philosophia: Munich.
Restall, G. (2000). *Introduction to subsructural logics*. London: Routlege.
Restall, G. (2013). Assertion, denial, and non-classical theories. In K. Tanaka, F. Berto, E. Mares, & F. Paoli (Eds.), *Paraconsistency: Logic and applications* (pp. 81–99). Dordrecht: Springer.
Restall, G., & Slaney, J. (1995). Realistic belief revsion. In *Proceedings of the Second World Conference on the Fundamentals of Aritificial Intelligence* (pp. 367–378).
Routley, Richard, Brady, Ross, Meyer, Robert, & Routley, Val. (1983). Relevant Logics and Their Rivals (Vol. 1). Atascardero, CA: Ridgeview Press.
Segerberg, K. (1995). Belief revision from the point of view of doxastic logic. *Bulletin of the IGPL, 3*, 535–553.
Slaney, J. (1990). A general logic. *Australasian Journal of Philosophy, 68*, 7489.
Tanaka, K. (2005). The AGM theory and inconsistent belief change. *Logique et Analyse, 48*, 113–150.
van Benthem, J. (2011). *Logical dynamics of information and interaction*. Cambridge: Cambridge University Press.
Wassermann, R. (2011). On AGM for non-classical logics. *Journal of Philosophical Logic, 40*, 271–294.

Epistemic Reasoning in Life and Literature

Rohit Parikh

Abstract Epistemic Logic has recently acquired importance as a growing field with influence in Distributed Computing, Philosophy, Economics, and of late even in Social Science and Animal behavior. Here, on a relatively light note, we give examples of epistemic reasoning occurring in literature and used very effectively by writers like Shakespeare, Shaw, Arthur Conan Doyle, and O'Henry. For variety we also give an example of epistemic reasoning used by fireflies, although it is far fetched to suppose that fireflies use epistemic reasoning in any kind of a conscious way. Surely they are getting a lot of help from Darwin. It is this writer's hope that epistemic reasoning as a formal discipline will some day acquire importance comparable to that of Statistics. Hopefully these examples make part of the case.

Keywords Epistemic reasoning · Theory of mind · Multiagent epistemology · Logic in literature · Game theory in literature

1 Introduction

The celebrated AGM theory (Alchourron et al. 1985) does not quite tell us how to revise our beliefs when we receive information which contradicts what we believe already. But it does tell us that if we have a procedure for revising our beliefs, what logical properties that procedure (called *) should have. Now AGM do not provide a unique solution and indeed there are very many. In practice, however, when we read the newspaper, we constantly learn things which contradict what we had thought earlier. And most of the time our revision of existing beliefs is effortless. How is this so?

R. Parikh (✉)
Brooklyn College of the City University of New York, New York, USA
e-mail: rparikh@gc.cuny.edu

Graduate Center of the City University of New York, New York, USA

S. O. Hansson (ed.), *David Makinson on Classical Methods for Non-Classical Problems*, Outstanding Contributions to Logic 3, DOI: 10.1007/978-94-007-7759-0_8,

Parikh (1999, 2011); Kourousias and Makinson (2007) provide part of an answer. We divide our beliefs into subareas; beliefs about politics, about our children, about our teeth, and many others. When revising, we tend to revise only in some subarea. And this makes the problem easier, but the solution is still not unique. Even politics, or our children still give rise to large areas and we should expect computational difficulties. Generally we do not.

So the general issue is, *How far can logic—or formal reasoning—help us understand actual human behavior?*

In 'Formal models for real people' Lambalgen and Counihan (2008) address this issue. Their thoughts are by no means conclusive. But their general point, that logic has relevance even in thinking about real people seems valid. If this were not so we would be totally lost trying to understand other people. But in practice we are able to reduce our conjectures to small sets.

Someone has just poured a cup of coffee and is stretching out her hand. We can reasonably conclude that she is looking for milk or sugar or perhaps *Equal*. If we know her well, we can even be sure that it is not milk, to which she is allergic. So reasoning, and previously known facts form a backdrop which allows us to reduce the enormous complexity of the social world to a manageable size.

In this paper I want to talk about some examples from real life or from literature which show how logic does help us to reason about other people. I will then indicate some limitations of logic which may well need to be supplemented by facts or where perhaps the situation is simply hopeless.

2 Theory of Mind

How much do we know of what others are thinking? A striking experiment related by Wimmer and Perner (1983) shows that small children do not have an understanding of other people's minds.

In their experiment, a group of small children are told about a small boy called Maxi who comes back from grocery shopping with his mother and helps her to unload the groceries. Among the groceries is some chocolate which Maxi puts in a red cupboard and then goes out to play.

While Maxi is gone, mother takes the chocolate out of the red cupboard, uses some of it and puts the rest in the blue cupboard.

Now Maxi comes back from play and wants the chocolate.

The little children who are told this story are then asked, "Where will Maxi look for the chocolate?" Children at the age of five or more answer that he will look in the red cupboard where they know he put it. But younger children of three or four tend to say that he will look in the blue cupboard where they know the chocolate is.

The expression *Theory of Mind* was coined by Premack and Woodruff (1978) to describe the phenomenon where we represent another's mind, and perhaps even represent their representation of our minds.

While Wimmer and Perner talk about children, it is known from experiments, e.g. by Verbrugge and her colleagues See for instance, Verbrugge and Mol (2008) that adults are far from perfect and fail to reach levels of mind more than two or three.

Many examples of epistemic reasoning in real life and literature involve the theory of mind. Sometimes one may know someone well enough to know when they are *not* using the theory of mind but examples do not come readily to mind.

Sacks (1995) does give examples of people who are striking in their lack of a theory of mind. Temple Grandin who is both autistic and a well respected scholar (Grandin 2006) would probably be a source of examples, But surely it would not be ethical to take advantage of people who lack TOM and can easily be taken in.[1]

3 Examples

3.1 *Inducing Beliefs: Shakespeare's Much ado About Nothing*

At Messina, a messenger brings news that Don Pedro, a Spanish prince from Aragon, and his officers, Claudio and Benedick, have returned from a successful battle. Leonato, the governor of Messina, welcomes the messenger and announces that Don Pedro and his men will stay for a month.[2] Beatrice, Leonato's niece, asks the messenger about Benedick, and makes sarcastic remarks about his ineptitude as a soldier. Leonato explains that "There is a kind of merry war betwixt Signior Benedick and her".

Various events take place and Claudio wins the hand in marriage of Hero, Leonato's only daughter and the wedding is to take place in a week.

Don Pedro and his men, bored at the prospect of waiting a week for the wedding, hatch a plan to matchmake between Beatrice and Benedick who inwardly love each other but outwardly display contempt for each other.

According to this strategem, the men led by Don Pedro proclaim Beatrice's love for Benedick while knowing he is eavesdropping on their conversation. Thus we have, using b for Benedick, d for Don Pedro and E for the event of eavesdropping,

$$K_b(E),\ K_d(E) \text{ and } \neg K_b(K_d(E))$$

All these conditions are essential and of course the plot would be spoiled if we had $K_b(K_d(E))$ instead of $\neg K_b(K_d(E))$. Benedick would be suspicious and would not credit the conversation.

The women led by Hero carry on a similar charade for Beatrice.

---

1 Although Temple has great difficulty figuring out what adults and children are up to, she has enormous communication and empathy with animals. She has worked out methods to ease the last moments on earth of animals bound for slaughter.

2 Some of the examples in this section also occur in (Parikh et al.). Some are new.

Beatrice and Benedick are now convinced that their own love is returned, and hence decide to requite the love of the other.

The play ends with all four lovers getting married.

Beatrice's Decision problem

	Love	Nolove
Propose	100	−20
Nopropose	−10	0

Here *love* means "Beatrice loves me" and *nolove* the other possibility.

If Benedick believes *nolove* then *nopropose* is his best strategy. But this changes if he comes to learn *love*. Note also that the game played on Beatrice and Benedick does not lead to common knowledge that they love each other. Rather in each of them a *justified true belief* is produced that she (he) loves me. It is true since the person with the belief is indeed loved. And it is also justified since,while the belief originates in a charade put on for the benefit of the believer the believer is justified in taking it at face value. Nonetheless, as with the Gettier cases, it is not knowledge since the justification is defective.

Beatrice's Decision Problem

	Propose	Nopropose
Accept	100	−20
Ridicule	−10	10

Since Beatrice loves Benedick, if he does propose, she is best off accepting him. If he does not, then ridiculing him is at least a minor pleasure. And what can 'accept' mean in the context where he does not propose? We could take it to mean that she reveals her own love and willingness to marry him prematurely, thereby being embarrassed and earning a negative utility of −20.

3.2 Bernard Shaw's You Never Can Tell

In this play, a rich businessman Mr. Crampton is estranged from his wife (although being English they are polite to each other) and has not seen his three children for many years. The three children in question are Gloria, the oldest, and Dolly and Philip (who are what we might describe as good natured brats).

It so happens that a meeting is now being arranged in a restaurant between the mother and the three children on one side, and Mr. Crampton on the other. The conversation below takes place just before Mr. Crampton arrives. William is the sagacious waiter. McComas is a family friend.

PHILIP. Mr. Crampton is coming to lunch with us.

WAITER (puzzled). Yes, sir. (Diplomatically.) Don't usually lunch with his family, perhaps, sir?

PHILIP (impressively). William: he does not know that we are his family. He has not seen us for eighteen years. He won't know us. (To emphasize the communication he seats himself on the iron table with a spring, and looks at the waiter with his lips compressed and his legs swinging.)

DOLLY. We want you to break the news to him, William.

WAITER. But I should think he'd guess when he sees your mother, miss. (Philip's legs become motionless at this elucidation. He contemplates the waiter raptly.)

DOLLY (dazzled). I never thought of that.

PHILIP. Nor I. (Coming off the table and turning reproachfully on McComas.) Nor you.

It is obvious that what the waiter is saying makes perfect sense. And yet, obviously lacking logical omniscience, Dolly and Philip fail to make the inference.

Can we formalize William's inference and perhaps write a little program in *Prolog* to carry out this inference? Surely yes, though I am not sure what purpose that would serve.

A hint of the direction in which we might proceed occurs in Parikh (2008).[3]

> Suppose that Vikram believes it is not raining. In that case he will be in a belief state b such that $ch(b, \{U, \neg U\}) = (b', \neg U)$. Given a choice between taking an umbrella or not, he chooses not to take the umbrella, and goes into state b'.
>
> Suppose, however, that he looks out the window and sees drops of rain falling. Let r be the event of rain falling. Then the update operation $\rightarrow_e$ causes him to go into state c such that in state c he chooses U from $\{U, \neg U\}$. Thus $ch(c, \{U, \neg U\}) = (c', U)$
>
> Note that state c will also have other properties beside choosing to take an umbrella. It may also cause him to say to others, "You know, it is raining," or to complain, "Gosh, does it always rain in this city?" There is no such thing as *the state* of believing that it is raining. Every belief state has many properties, most of them unrelated to rain.

It is clear that Mr. Crampton will recognize his wife when he meets her and will guess that the three young adults are his own children. However, the imagined update (on seeing his wife) is not caused by a *sentence* (the case usually studied in the literature) but by an *event*. See Parikh and Ramanujam (2003) for a semantics. William makes the second order inference that Mr. Crampton will make this inference, and Dolly and Philip, less used to the world, fail to make it.

Another play by Shaw, *The Man of Destiny* also has some epistemic considerations. A letter has been received by Napoleon, presumably detailing his wife Josephine's infidelities. A pretty woman, presumably a friend of Josephine, appears before Napoleon reads the letter and tries to use her beauty and her intelligence to talk him out of reading it. At one stage Napoleon says:

---

[3] In the quoted paragraph, $ch(b, \{U, \neg U\}$ refers to Vikram's choice between the two actions, *take umbrella* and *not take umbrella* when Vikram is in state b. Note that we are taking a state as an *element* of the space of all possible states. *Believing that it is raining* will be a *subset* consisting of all states in which Vikram believes it is raining. Thus a state where Vikram believes it is raining and is standing will be distinct from a state where he believes it is raining and is sitting down. Neither state deserves to be called *the* state of believing that it is raining.

NAPOLEON (with coarse familiarity, treating her as if she were a vivandiere). Capital! Capital! (He puts his hands behind him on the table, and lifts himself on to it, sitting with his arms akimbo and his legs wide apart). Come: I am a true Corsican in my love for stories. But I could tell them better than you if I set my mind to it. Next time you are asked why a letter compromising a wife should not be sent to her husband, answer simply that the husband would not read it. Do you suppose, little innocent, that a man wants to be compelled by public opinion to make a scene, to fight a duel, to break up his household, to injure his career by a scandal, when he can avoid it all by taking care not to know?

Neyman (1991) suggests that Napoleon's worries are unjustified—that what the husband should fear is not that he himself knows but that others know he knows. However, Napoleon realizes this full well. He does read the letter but only after protesting that he will not read it, and then proceeds to read the letter while he is alone in the garden.

I will let the reader of *this paper* find out for herself what happens next.

3.3 Deception in Fireflies

This example is from Skyrms (2010).

Fireflies use their light for sexual signalling. In the western hemisphere, males fly over the meadows, flashing a signal. If a female on the ground gives the proper sort of answering flashes, the male descends and they mate. The flashing "code" is species-specific. Females and males in general use and respond to the pattern of flashes only of their own species.

There is, however, an exception. A female firefly of the genus Photuris, when she observes a male of the genus Photinus, may mimic the female signals of the male's species, lure him in, and eat him. She gets not only a nice meal, but also some useful protective chemicals that she cannot get in any other way. One species, *Photuris versicolor*, is a remarkably accomplished mimic—capable of sending the appropriate flash patterns of 11 Photinus species.

Here the female firefly of the genus Photuris is relying on a convention which exists in the Photinus species that a certain pattern of signals means, *A Photinus female is waiting for you*. Someone is indeed waiting for the poor Photinus male, but alas it is not a female of his species.[4]

3.4 Hamlet's Trick to Catch his Father's Killer

Of course everyone is familiar with the main story of the play but in brief, Hamlet arrives back in Denmark to find that his father is dead, ostensibly bitten by a poisonous snake as he slept in the garden. His father's brother Claudius is now king and has married Hamlet's mother.

Naturally Hamlet is not too pleased at these events and wonders if he has come to attend his father's funeral or his mother's wedding. At this, a ghost claiming to

[4] Note, however that we do not address false information, as conveyed by the Photuris in this paper. That will be the subject of a sequel.

be Hamlet's father appears on the ramparts of the castle Elsinore and tells Hamlet a rather different story of his father's death.

> Now, Hamlet, hear: 'Tis given out that, sleeping in my orchard, a serpent stung me; so the whole ear of Denmark is by a forged process of my death rankly abused: but know, thou noble youth, the serpent that did sting thy father's life now wears his crown.

Hamlet should revenge his father's death by killing his killer Claudius.

But can Hamlet trust the ghost? He says to himself,

> The spirit that I have seen may be the devil: and the devil hath power To assume a pleasing shape; yea, and perhaps out of my weakness and my melancholy, as he is very potent with such spirits, abuses me to damn me.

Thus comes the strategem which Hamlet devises.

> I have heard that guilty creatures sitting at a play have by the very cunning of the scene been struck so to the soul that presently they have proclaim'd their malefactions; for murder, though it have no tongue, will speak with most miraculous organ. I'll have these players play something like the murder of my father before mine uncle: I'll observe his looks; I'll tent him to the quick: if he but blench, I know my course...
>
> I'll have grounds More relative than this: the play's the thing wherein I'll catch the conscience of the king.

The king does indeed react with shock and dismay and the ghost is proved right.

Let G mean that the ghost is right, let N mean that Claudius acts normally and D mean that he acts with shock and dismay. Hamlet's theory consists of the two formulas,

$\neg G \to N$ and $G \to D$.

Since Hamlet observes D and hence $\neg N$, G follows, the ghost is proved right, and Hamlet's course is clear.

Alas, Hamlet continues to act with hesitation, refuses to kill Claudius when the latter is at prayer (for that would send him to heaven which he does not deserve) and the play ends in tragedy. Claudius does indeed die, but so does Hamlet himself, his mother Gertrude, and a few others who are innocent in the affair.

Hamlet's reasoning goes beyond pure epistemic logic into something deeper. Claudius's reaction is not that of his rational thoughtful sense who might have seen through Hamlet's trick and maintained a 'poker face'. It is the reaction of a man caught by surprise whose rational self is momentatily overcome.

3.5 *Sherlock Holmes in the Silver Blaze*

This particular story of Holmes is a veritable goldmine of epistemic reasoning. A horse, Silver Blaze, the favourite in a forthcoming race, is missing, and his trainer, Mr. Straker is found dead on the moor, killed by a blow from some blunt instrument. Among the contents of Straker's pocket is a receipt for a dress costing 22 guinees. But the receipt is in the name of a Mr. Darbyshire and 22 guinees is rather a high price in

Holmes's times. Wanting to find out if the dress is for Straker's wife, Holmes tricks her into revealing the fact that it isn't (a direct question might make her suspicious).[5]

> No, Mrs. Straker; but Mr. Holmes, here, has come from London to help us, and we shall do all that is possible.
>
> "Surely I met you in Plymouth, at a garden party, some little time ago, Mrs. Straker," said Holmes. "No, sir; you are mistaken."
>
> "Dear me; why, I could have sworn to it. You wore a costume of dove-coloured silk, with ostrich feather trimming."
>
> "I never had such a dress, sir," answered the lady.
>
> "Ah; that quite settles it," said Holmes; and, with an apology, he followed the Inspector outside.

Holmes now knows what he suspected, that Straker had another household under the name of Darbyshire.

Later he turns to the question of what has happened to the horse and even though Holmes clearly suffers from Asperger's syndrome, he nonetheless shows a mastery of second order epistemic reasoning, by putting himself in the horse's mind.

> "It's this way, Watson," he said at last. We may leave the question of who killed John Straker for the instant, and confine ourselves to finding out what has become of the horse. Now, supposing that he broke away during or after the tragedy, where could he have gone to? The horse is a very gregarious creature. If left to himself his instincts would have been either to return to King's Pyland, or go over to Mapleton.
>
> Why should he run wild upon the moor? He would surely have been seen by now. And why should gipsies kidnap him? These people always clear out when they hear of trouble, for they do not wish to be pestered by the police. They could not hope to sell such a horse. They would run a great risk and gain nothing by taking him. Surely that is clear.
>
> Where is he, then?
>
> I have already said that he must have gone to King's Pyland or to Mapleton. He is not at King's Pyland, therefore he is at Mapleton. Let us take that as a working hypothesis and see what it leads us to. This part of the moor, as the Inspector remarked, is very hard and dry. But it falls away towards Mapleton, and you can see from here that there is a long hollow over yonder, which must have been very wet on Monday night. If our supposition is correct, then the horse must have crossed that, and there is the point where we should look for his tracks.

Holmes does indeed find the horse's tracks on the other side of the hollow and also learns that the horse is in fact at Mapleton. He also forms a theory of how Straker came to die.

Having finished his inquiries, Holmes is now ready to leave the site of the crime and return to London.

> Colonel Ross still wore an expression which showed the poor opinion which he had formed of my companion's ability, but I saw by the Inspector's face that his attention had been keenly aroused.
>
> "You consider that to be important?" he asked.

[5] This incident is also discussed at length in Genot and Jacot (2012).

"Exceedingly so."

"Is there any other point to which you would wish to draw my attention?"

"To the curious incident of the dog in the night-time."

"The dog did nothing in the night-time."

"That was the curious incident," remarked Sherlock Holmes.

Holmes is pointing out that since the dog did not bark, no stranger was involved in the crime. It turns out that the criminal was Mr. Straker, who was well known to the dog, and had taken the horse out to the moor in order to injure him slightly and prevent him from winning the race. The horse, alarmed at the goings on, lashed out with his foot, thereby killing Mr. Straker. The horse, as anticipated by Holmes, headed for Mapleton (his original stable) and since they had their own horse in the race, they hid Silver Blaze. After they realize that the jig is up, they agree with Holmes to arrange for Silver Blaze to run and he wins the race.

Michael Chwe in his (Chwe) also discusses game theoretic issues arising in literature. But epistemic considerations do not play a major role. They do in his earlier book *Rational Ritual*, Chwe (2003).

3.6 Gift of the Magi

In this story by O'Henry (William Sidney Porter), James Dillingham Young (Jim) and his wife, Della, are a couple living in a modest flat. They each have one possession in which they take pride: Della's beautiful long, flowing hair and Jim's gold watch, which had belonged to his father and grandfather.

On Christmas Eve, with only $1.87 in hand, and desperate to find a gift for Jim, Della sells her hair for $20, and eventually finds a platinum fob chain for Jim's watch for $21. Happy to have found the perfect gift at last, she runs home and begins to prepare dinner.

When Jim comes home, he looks at his short haired wife with a strange expression. Della then admits to Jim that she sold her hair to buy him his Christmas present.

The dumbfounded Jim then gives Della her present, an array of expensive combs for her hair. Della shows Jim the chain she bought for him, and Jim responds that he sold his watch to get the money to buy her combs.

Jim and Della are now left with gifts that neither one can use, but they realize how far they are willing to go to show their love for each other.

The story ends with the narrator comparing the pair's mutually sacrificial gifts of love with those of the Biblical Magi.

In the following game Della is the row player and Jim the column player. The payoffs given are only the surface payoffs. The $(-10, -10)$ in the bottom right corner becomes (100, 100) once they realize how much each loves the other.

	Keep watch	Sell watch
Keep hair	0, 0	10, 5
Sell hair	5, 10	−10, −10

Note that in this case both Della and Jim have a default assumption. Della's default is that Jim *has* his watch, and with this assumption, the second column vanishes and selling her hair becomes a dominant strategy for Della which she carries out.

Similarly, Jim's default assumption is that Della *has* her long flowing hair and thus the bottom row disappears. Selling watch then becomes a dominat strategy for Jim.

This situation is somewhat analogous to the iterated elimination of dominated strategies, a technique used in game theory (Osborne and Rubinsein 1994). However, what we have here is not an elimination of dominated strategies, but an elmination of *non-default* strategies.

An investigation of this phenomenon would seem to require insights from both game theory and default logic.

4 Concluding Remarks

It seems clear that many writers from Shakespeare on[6] are using epistemic reasoning and game theoretic considerations in a very sophisticated way.

By comparison, purely formal thinking about epistemic reasoning, while occasionally sophisticated mathematically, is nonetheless rather primitive in terms of how well it relates to the actual world. It would be very valuable if professional logicians caught up with these brilliant literary minds in the formal work they do.

Here are some directions in which we could go.

1. Instead of concentrating on update via hearing a sentence, we could allow more general updates, for instance by witnessing an event.
2. We could take better account of default reasoning. For instance Della and Jim's actions are understandable in terms of their default assumptions. Grove spheres could be used as a handy mechanism in game theory as well.
3. We could take better account of inferring someone's beliefs in terms of how they act. It is Hamlet's uncle Claudius's dismay at watching the play put on by Hamlet which convinces Hamlet that what his father's ghost told him was true.
4. Instead of assuming common knowledge we could take better account of the limitations of our perception of the beliefs of other people.

6 For a still earlier example, see the incident of Yudhisthira and Drona in the battle of the *Mahabharata*, (Smith 2009; Wikipedia entry on Drona). Incidents of deception also occur in the *Hebrew Bible*, see Brams (2003), Sects. 6.2 and 6.3.

Acknowledgments Thanks to Steven Brams, Yang Liu, Carol Parikh, Prashant Parikh, Paul Pedersen, Hilary Putnam, Noson Yanofsky and Ruili Ye for comments. This paper was supported by a grant from PSC-CUNY under the *Faculty Research Assistance Program.*

References

Alchourron, C., Gärdenfors, P., & Makinson, D. (1985). On the logic of theory change: Partial meet contraction and revision functions. *Journal of Symbolic Logic*, *50*, 510–530.

Brams, S. (2011). *Game theory and the humanities: Bridging two worlds*. Cambridge: MIT press.

Brams, S. (2003). *Bibliocal games*. Cambridge: MIT press.

Chwe, M. S.-Y. (2013). *Jane Austen, Game theorist*. Princeton: Princeton University Press.

Chwe, M. S.-Y. (2003). *Rational ritual*. Princeton: Princeton University Press.

Wikipedia entry on Drona, http://en.wikipedia.org/wiki/Drona.

Genot, E., & Jacot, J. (2012). How can questions be informative before they are answered? Strategic information in interrogative games. *Episteme*, *9*(2), 189–204.

Grandin, T. (2006). *Thinking in pictures and other reports from my life with autism*. New York: Vintage Books.

Kourousias, G., & Makinson, D. (September 2007). Parallel interpolation, splitting, and relevance in belief change. *Journal of Symbolic Logic*, *72*, 994–1002.

van Lambalgen, M., & Counihan, M. (2008). Formal models for real people. *Journal of Logic Language and Information*, *17*, 385–389.

Neyman, A. (1991). The positive value of information. *Games and Economic Behavior*, *3*, 350–355.

Osborne, M., & Rubinsein, A. (1994). *A course in game theory*. Cambridge: MIT press.

Parikh, R. (1999). Belief revision and language splitting. In Proc. Logic, Language and Computation, Ed. Moss, Ginzburg and de Rijke, CSLI 1999, pp. 266–278 (earlier version appeared in 1996 in the prliminary proceedings).

Parikh, R. (March 2011). Beth definability, interpolation and language splitting. *Synthese*, *179*(2), 211–221.

Parikh, R. (2008). Sentences, belief and logical omniscience, or what does deduction tell us? *Review of Symbolic Logic*, *1*, 459–476.

Parikh, R., & Ramanujam, R. (2003). Journal of Logic, Language and Information. *A knowledge based semantics of messages*, *12*, 453–467.

Parikh, R., Tasdemir, C., & Witzel, A. (2011). The power of knowledge in games, http://cuny.academia.edu/RohitParikh/Papers/.

Premack, D. G., & Woodruff, G. (1978). Does the chimpanzee have a theory of mind? *Behavioral and Brain Sciences*, *1*(4).

Sacks, O. (1995). *An anthropologist on mars*. New York: Vintage Books.

Skyrms, B. (2010). *Signals: Evolution, learning and information*. New York: Oxford University press.

Smith, J. (2009). The mahabharata : An abridged translation (p. 477). New Delhi: Penguin Books.

Verbrugge, R., & Mol, L. (2008). Learning to apply theory of mind. *Journal of Logic Language and Information*, *17*(4), 489–511.

Wimmer, H., & Perner, J. (1983). Beliefs about beliefs: Representation and constraining function of wrong beliefs in young children's understanding of deception. *Cognition*, *13*(1), 103–128.

Part III
Uncertain Reasoning

New Horn Rules for Probabilistic Consequence: Is O+ Enough?

James Hawthorne

Abstract In our 2007 paper David and I studied consequence relations that correspond to conditional probability functions above thresholds, the *probabilistic consequence relations*. We showed that system O is a probabilistically sound system of Horn rules for the probabilistic consequence, and we conjectured that O might also provide a complete axiomatization of the set of finite premised Horn rules for probabilistic consequence relations. In a 2009 paper Paris and Simmonds provided a mathematically complex way to characterize all of the sound finite-premised Horn rules for the probabilistic consequence relations, and they established that the rules derivable from system O are insufficient. In this paper I provide a brief accounts of system O and the probabilistic consequence relations. I then show that O together with the probabilistically sound (Non-Horn) rule known as Negation Rationality implies an additional systematic collection of sound Horn rules for probabilistic consequence relations. I call O together with these new Horn rules 'O+'. Whether O+ is enough to capture all probabilistically sound finite premised Horn rules remains an open question.

Keywords Probabilistic consequence relation · Probability threshold · Horn rule · Nonmonotonic consequence

1 Introduction

In our 2007 paper, "The Quantitative/Qualitative Watershed for Rules of Uncertain Inference", David and I studied consequence relations that correspond to conditional probability functions above thresholds. That is, we studied the *probabilistic consequence relations* (hereafter the *ProbCR*s), defined as follows:

J. Hawthorne (✉)
University of Oklahoma, Norman, OK, USA
e-mail: hawthorne@ou.edu

S. O. Hansson (ed.), *David Makinson on Classical Methods for Non-Classical Problems*, Outstanding Contributions to Logic 3, DOI: 10.1007/978-94-007-7759-0_9,

Definition 1. Probabilistic Consequence Relations: Let p be any probability function defined on sentences of a language for sentential logic,[1] and let t be any real number such that $0 < t \leq 1$. The pair $< p, t >$ generates a probabilistic consequence relation $|\!\sim_{p,t}$ by the rule: $a|\!\sim_{p,t} x$ iff either $p(a) = 0$ or $p_a(x) \geq t$, where by definition $p_a(x) = p(x|a) = p(a \wedge x)/p(a)$ for $p(a) > 0$. The parameter t is called the *threshold* associated with p for the relation $|\!\sim_{p,t}$.

A leading idea of that paper was to show that a system of Horn rules called **O** lies at the cusp between probabilistically sound Horn rules and well-studied stronger qualitative systems that fail to be probabilistically sound. **O** consists of several familiar rules for nonmonotonic consequence relations together with some weakened versions of other familiar rules, all of them sound for the *ProbCR*s. Some other well-known rules, such as AND (if $a \mid\!\sim x$ and $a \mid\!\sim y$, then $a \mid\!\sim x \wedge y$) and OR (if $a \mid\!\sim x$ and $b \mid\!\sim x$, then $a \vee b \mid\!\sim x$), are unsound for the *ProbCR*s. However, **O** turns out to be strong enough that merely adding AND to **O** not only *jumps the divide* between probabilistically sound rules and stronger qualitative rules for uncertain inference, rather it takes the logic all the way over to the strong system called *preferential consequence relations*, which is usually characterized by the system of Horn rules called **P**. Nevertheless, strong as it is, **O** does not contain every rule that is sound for the *ProbCR*s. The well-know non-Horn rule NR (negation rationality: if $a \mid\!\sim x$, then either $a \wedge b \mid\!\sim x$ or $a \wedge \neg b \mid\!\sim x$) is also sound for *ProbCR*s. Adding NR to **O** results in the probabilistically sound system we call **Q**, also investigated in our 2007 paper.[2]

Our paper showed that no set of sound *finite-premised* Horn rules is *complete* for all *ProbCR*s.[3] For, we showed, there are *infinite-premised* Horn rules sound for *ProbCR*s that cannot be derived from any sound set of finite-premised Horn rules. This result left open the question of whether the sound finite-premised Horn rules we investigated (the system **O**) would suffice to derive all sound finite-premised Horn rules for *ProbCR*.

In the present paper, after summarizing the systems of sound rules for *ProbCR*, I'll extend some of the main ideas and results from our 2007 paper. These new results are based on some work David and I did after our 2007 paper was published. This new work was motivated by an exchange of email messages with Jeff Paris and Richard Simmonds, who contacted us after proving their important *completeness result* for a

---

[1] That is, p satisfies the usual classical probability axioms on sentence of a language for sentential logic: (1) $p(a) \geq 0$, (2) if $|\!- a$ (i.e. if a is a tautology), then $p(a) = 1$, (3) if $|\!- \neg(a \wedge b)$, then $p(a \vee b) = p(a) + p(b)$; and conditional probability is defined as $p(a \mid b) = p(a \wedge b)/p(b)$ whenever $p(b) > 0$. All of the other usual probabilistic rules follow from these axioms.

[2] The systems **O** and **Q**, and their probabilistic soundness, were first investigated in (Hawthorne 1996). The more recent paper, (Hawthorne and Makinson 2007), proves important new results about **O**, **Q**, and related systems.

[3] That is, any set of sound rules for *ProbCR*s that are in *Horn rule form* will be satisfied by some relations $|\!\sim$ on all pairs of sentences that are not in *ProbCR*s. A rule is in *Horn rule form* just when it is of form, "if $a_1|\!\sim x_1, \ldots, a_n|\!\sim x_n$, then $b|\!\sim y$" (with at most a finite number of premise conditions of form $a_1|\!\sim x_1, \ldots, a_n|\!\sim x_n$), and perhaps also containing *side conditions* about logical entailments among sentences.

characterization of the finite-premised Horn rules for *ProbCR*s. In their paper "**O** is Not Enough" (2009), Paris and Simmonds show how to capture all of the sound finite-premised Horn rules for *ProbCR*s, and they establish that the rules we investigated in our 2007 paper were *not enough*. Although Paris and Simmonds characterize a complete set of finite-premised Horn rules for *ProbCR*s, their characterization is fairly opaque. They provide an algorithm for generating all sound Horn rules (and prove that it does so). But the algorithm for generating the rules is complex enough that it's not at all easy to tell what the rules it generates will look like in advance of just cranking them out one at a time—and there are an infinite number of them to crank out.

Motivated by the Paris and Simmonds result, David and I have discovered an explicit infinite sequence of additional sound Horn rules. But it isn't clear whether these new rules suffice to derive all finite-premised Horn rules—i.e. to derive all of those rules generated by the Paris-Simmonds procedure. The present paper is devoted to specifying these additional rules and raising unsolved questions about *complete rules* for *ProbCR*s.

2 Probabilistic Consequence Relations and the O and Q Rules: A Quick Overview of Earlier Results

The family **O** of rules for consequence relations is defined as follows.

Definition 2. The family of rules **O**:

$a \mid\!\sim a$	(REFLEX: reflexivity)
If $a \mid\!\sim x$ and $x \mid\!- y$, then $a \mid\!\sim y$	(RW: right weakening)
If $a \mid\!\sim x$ and $a \mid\!- b$ and $b \mid\!- a$, then $b \mid\!\sim x$	(LCE: left classical equivalence)
If $a \mid\!\sim x \wedge y$, then $a \wedge x \mid\!\sim y$	(VCM: very cautious monotony)
If $a \wedge b \mid\!\sim x$ and $a \wedge \neg b \mid\!\sim x$, then $a \mid\!\sim x$	(WOR: weak OR)
If $a \mid\!\sim x$ and $a \wedge \neg y \mid\!\sim y$, then $a \mid\!\sim x \wedge y$	(WAND: weak AND).

These rules are sound for the *ProbCR*s (see Hawthorne and Makinson 2007; Hawthorne 1996). Furthermore, given the other rules, the WOR could be replaced by the following rule.

If $\mid\!- \neg(a \wedge)$ and $a \mid\!\sim x$ and $b \mid\!\sim x$, then $(a \vee b) \mid\!\sim x$ (XOR: exclusive OR).

That is, given the other rules, one can derive XOR from WOR, and *vice versa*. This is especially interesting because it turns out that the additional sound Horn rules we've discovered for *ProbCR* are extended versions of the XOR rule. I'll get to those in the next section.

The rules in **O** are quite similar to the rules in the family **P**, which is a sound and complete family of rules for the *preferential consequence relations*. These consequence relations are defined semantically in terms of *stoppered* (a.k.a. *smooth*) *preferential models* (see Krauss et al. 1990; Makinson 1989, 1994).

Definition 3. The family of rules **P**:
REFLEX, LCE, RW together with the following rules:

If $a|\!\sim x$ and $a|\!\sim y$, then $a \wedge x|\!\sim y$ (CM: cautious monotony)
If $a|\!\sim x$ and $b|\!\sim x$, then $a \vee b|\!\sim x$ (OR : disjunction in the premises)
If $a|\!\sim x$ and $a|\!\sim y$, then $a|\!\sim x \wedge y$ (AND : conjunction in conclusion).

Notice that each of the last three rules for **P** is a stronger version of the corresponding rules for **O**. However, the OR rule looks rather more like the rule XOR than like WOR. Furthermore, some versions of **P** use the following rule as a basic rule in place of AND.

If $a|\!\sim x$ and $a \wedge x|\!\sim y$, then $a|\!\sim y$ (CT: cumulative transitivity (a.k.a. CUT)).

Given the other rules, CT and AND are derivable from each other.

It turns out that the usual rules for family **P** are stronger than necessary. That is, consider the following family of rules.

Definition 4. The family of rules **P*** consists of the rules of **O** with WAND replaced by AND.

David and I showed that all of the rules of **P** are derivable from those of **P***, and *vice versa* (also see Hawthorne 1996). Thus, the rule AND is the watershed rule that takes one from the *ProbCR* to the *preferential consequence relations*. Furthermore, AND is clearly not sound for *ProbCR*, because it's often the case that $p(x \wedge y|a) < p(x|a)$. So, when either $p(x|a)$ or $p(y|a)$ is very close to the threshold t for a relation $|\!\sim_{p,t}$, it can be the case that $p(x \wedge y|a) < t$.

Not all sound rules for *ProbCR*s are Horn rules. Indeed, as already mentioned in the Introduction, the well-known rule called *negation rationality* is sound for *ProbCR*s.

If $a|\!\sim x$, then either $a \wedge b|\!\sim x$ or $a \wedge \neg b|\!\sim x$ (NR).

Definition 5. The family of rules $\mathbf{Q} : \mathbf{Q} = \mathbf{O} \cup \{\text{NR}\}$.

All rules of **Q** are sound for *ProbCR*s.

Although David and I show that **Q** is sound, what we hadn't realized at that time is the surprising result that the non-Horn rule NR permits the derivation in system **Q** of infinitely many additional Horn rules that are not derivable from **O** alone.

Before moving on to the new results, one additional point is worth making clear. It is easy to specify an infinite-premised Horn rule that is sound for *ProbCR*s. Just consider any infinite set of distinct sentences $\{x_1, x_2, \ldots, x_n, \ldots\}$ such that for any pair of them, $a|\!- \neg(x_i \wedge x_j)$. Then it's not possible to have a probability function p and threshold $t > 0$ such that for each x_i, $p(x_i|a) \geq t$. For, suppose there is such a p and $t > 0$. Then there is an integer $n > 1$ such that $t > 1/\text{n}$; so that $n \times t > 1$. Now notice that $1 = p(x_1 \vee \neg x_1|a) = p(x_1|a) + p(\neg x_1|a) = p(x_1|a) + p(\neg x_1 \wedge (x_2 \vee \neg x_2)|a) = p(x_1 \mid a) + p(\neg x_1 \wedge x_2|a) + p(\neg x_1 \wedge \neg x_2|a) = p(x_1|a) + p(x_2|a) + p(\neg x_1 \wedge \neg x_2|a) = \ldots = p(x_1 \mid a) + p(x_2|a) + \ldots + p(x_n|a) + p(\neg x_1 \wedge \neg x_2 \wedge \ldots \wedge \neg x_n|a) \geq n \times t > 1$. Contradiction! Thus, the only way to have a *ProbCR* for which $a|\!\sim x_i$ for an infinite set $\{x_1, x_2, \ldots, x_n, \ldots\}$ where $a \vdash \neg(x_i \wedge x_j)$ for each pair of them is to have $p(a) = 0$; that's the degenerate case where $a|\!\sim_{p,t}\, y$ for all y (including $\perp$). Thus, the following rule is sound for *ProbCR*:

If for each x_k in $\{x_1, x_2, \ldots, x_n, \ldots\}$, $a|\!\sim x_k$, and for each pair, $a|\!-\neg(x_i \wedge x_j)$, then $a|\!\sim \perp$ (INF).

A number of such infinitary Horn rules are sound for *ProbCR*s. The "Archimedean rule" we presented in our paper is a more complex variation on the same idea. However, such rules are not really that "troubling" for the project of characterizing *ProbCR*s because for any specific threshold *t*, a finite Horn rule will subsume the infinitary rule. That is, let's define *ProbCR(q)* as the set of all *ProbCR*s for which the threshold $t > q$.

Definition 6. Probabilistic Consequence Relations for thresholds above q. Let *ProbCR*(q) be the set of all *ProbCR*s, $|\!\sim_{p,t}$, such that the threshold $t > q > 0$.

In our paper David and I explored sound threshold-sensitive rules for various threshold levels q. However, these rules turn out to be rather complex, so I'll not discuss them in any detail here. The point is that for any given threshold $t \geq \text{q} > 1/n$ for integer $n > 1$, the following finite-premised Horn rule will be sound, and will subsume the INF rule.

If $a|\!\sim x_1, a|\!\sim x_2, \ldots, a|\!\sim x_n$, and for each pair, $a|\!-\neg(x_i \wedge x_j)$, then $a|\!\sim \perp$ (PLAUS(n)).

Given the fact that all known infinite-premised Horn rules that are sound for *ProbCR*s are subsumable by finite-premised rules that are sound for greater than zero bounds on thresholds, perhaps only finite-premised Horn rules should be of any real interest for *ProbCR*s.

In any case, in our 2007 paper David and I conjectured that **O** might suffice for generating all sound finite-premised Horn rules for *ProbCR*s. Within the year after our paper was published, Paris and Simmonds proved otherwise. Although their result did not came out until 2009, they contacted us about their result in February of 2008, and got us thinking.

3 Why O isn't Enough

The Paris and Simmonds (2009) paper is brilliant, and mathematically extremely sophisticated. They provide an algorithm for generating the complete set of sound finite-premised Horn rules for *ProbCR*s, and prove that it does so. But it's difficult to see what these rules look like without simply generating them one at a time. However, their algorithm establishes that there must be an infinite set of sound independent rules (not derivable from any finite subset of the rules). Indeed, for each positive natural number n, there must be such a rule consisting of at least n premises, a rule of the form

$$\text{If } a_1|\!\sim x_1, a_2|\!\sim x_2, \ldots, a_n|\!\sim x_n, \text{ then } b|\!\sim y$$

together with side conditions about logical entailments among the various sentences involved.

Paris and Simmonds (2009) show that all rules of **O** are generated via the first few iterations of their algorithm. All of the additional examples *not* derivable from **O** that they have explicitly generated are very similar to the following example:

$$\text{If } (a \wedge \neg b) \vee (b \wedge \neg a)|\!\sim x, a|\!\sim x, b|\!\sim x, \text{ then } a \vee b|\!\sim x \text{ (PS)}.$$

Let's call this the PS rule for "Paris-Simmonds". Some of their examples have more premises, but all that we are aware of are analogous in a way I'll explain in a moment.

Upon seeing such examples, it occurred to me and David that these examples are somewhat like the XOR rule, which is derivable from **O**. To see the pattern, consider:

$$\text{If } |\!- \neg(a \wedge b) \text{ and } a|\!\sim x \text{ and } b|\!\sim x, \text{ then } (a \vee b)|\!\sim x \quad \text{(XOR)}.$$

Now, put the PS rule into the following form:

$$\text{If } (a \wedge \neg b) \vee (\neg a \wedge b)|\!\sim x, (a \wedge b) \vee (a \wedge \neg b)|\!\sim x, (a \wedge b) \vee (\neg a \wedge b)|\!\sim x,$$
$$\text{then } (a \wedge b) \vee (a \wedge \neg b) \vee (\neg a \wedge b)|\!\sim x \text{(PS)}.$$

Here, the antecedent of each premise is logically equivalent to the antecedent in the original version of the PS rule, above. But this version makes it clear that the antecedent of the conclusion is simply the disjunction of all disjuncts from antecedents of the premises. Furthermore, in this case each of the disjuncts is mutually inconsistent with each of the other disjuncts. This suggests the following rule, which is analogous to XOR but not derivable from it:

$$\text{If } |\!- \neg(a \wedge b), |\!- \neg(a \wedge c), |\!- \neg(b \wedge c), \text{ and, } a \vee b|\!\sim x, a \vee c|\!\sim x, b \vee c|\!\sim x,$$
$$\text{then } (a \vee b \vee c)|\!\sim x \text{ (XOR [3, 2])}.$$

I call this rule XOR [3, 2] because it goes from the support of x by pairs of exclusive disjuncts to the support of x by all three disjuncts. By this labeling scheme the usual XOR rule is XOR [2, 1].

Is this rule sound for *ProbCR*s? It turns out that it is, and the easiest way to prove that is to derive it from the rules of **Q**. In particular, the non-Horn rule *negation rationality*, NR, provides just the boost **O** needs to permit a derivation of this new rule. And since all of the **Q** rules are sound for *ProbCR*, whatever rules we derive from **Q** will also be sound.

To see how NR helps with the derivation of XOR [3, 2], let's first derive an alternative version of NR:

If $\vdash \neg(a \wedge b)$ and $a \vee b |\!\sim x$, then either $a|\!\sim x$ or $b|\!\sim x$ (XNR : eXclusiveNR).

Here's the derivation of XNR from NR:

> Suppose $\vdash \neg(a \wedge b)$ and $a \vee b|\!\sim x$. Then, from NR, either $(a \vee b) \wedge a|\!\sim x$ or $(a \vee b) \wedge \neg a|\!\sim x$. Thus, since $b \vdash \neg a$, we have, either $a|\!\sim x$ or $b|\!\sim x$, via LCE applied to each antecedent.

The implication goes the other way, from XNR to NR, as well. Here is that direction from O ∪ {XNR} (actually we only need LCE, as in the previous proof):

> Suppose $a|\!\sim x$. From LCE we have $(a \wedge b) \vee (a \wedge \neg b)|\!\sim x$. Then, since $\vdash \neg((a \wedge b) \wedge (a \wedge \neg b))$, from (XNR) we get, either $(a \wedge b)|\!\sim x$ or $(a \wedge \neg b)|\!\sim x$.

Now, I brought up XNR is as a means of proving that XOR [3, 2] is sound for *ProbCR*, by deriving XOR[3, 2] from **Q** via XNR. That derivation is pretty straightforward.

Observation 1: XOR[3, 2] is sound for *ProbCR*.

> Proof: Suppose $\vdash \neg(a \wedge b)$, $\vdash \neg(a \wedge c)$, $\vdash \neg(b \wedge c)$, and $a \vee b|\!\sim x$, $a \vee c|\!\sim x$, $b \vee c|\!\sim x$. From $a \vee b |\!\sim x$ we get (by XNR) that either (i) $a|\!\sim x$ or (ii) $b|\!\sim x$. In case (i) we apply XOR together with $b \vee c |\!\sim x$ to get $a \vee b \vee c|\!\sim x$ (since it follows from the side conditions that $\vdash \neg(a \wedge (b \vee c))$). Similarly, in case (ii) we apply XOR together with $a \vee c|\!\sim x$ to get $a \vee b \vee c|\!\sim x$ (since it follows from the side conditions that $\vdash \neg(b \wedge (a \vee c))$). Thus, XOR[3, 2] follows from sound rules for *ProbCR*.

The structure of the XOR[3,2] rule and the XOR[2,1] rule, (a.k.a. the XOR rule) suggests a host of much more general rules of the following sort.

For each pair of integers n, m such that $n > m \geq 1$, define the rule XOR (n, m) as follows:

> XOR (n, m): Consider any list of n pairwise inconsistent sentences. Suppose that for each sentence ***e*** that consists of a disjunction (in the order provided by the list, just to be concrete about it) of exactly m of them we have $e |\!\sim x$. Then, for the sentence d that consists of the disjunction of all n of them (in the order provided by the list) it follows that $d |\!\sim x$.

All of the XOR(n, m) rules (for $n > m \geq 1$) turn out to be sound for *ProbCR*. However, these rules are not independent. It turns out that from the set of sound Horn rules **O** together with only the new rules of form XOR($n+1, n$) for each $n \geq 2$ (since XOR(2, 1) is already part of **O**) we can derive all of the other XOR(n, m) rules.

To establish these claims we'll proceed as follows. I'll first define the set of Horn rules **O+**, which consists of **O** together with the XOR($n+1, n$) rules, for each $n \geq 2$. We then show that all of the new XOR($n+1, n$) rules follow from **Q**. That will establish the soundness of **O+** for *ProbCR*s. Then we show that the remaining XOR(n, m) rules, for each $n > m \geq 1$, are derivable in **O+**.

4 The Soundness of the Horn Rule System O+ and the Derivation of the XOR(n, m) Rules

Definition 7. The family of rules **O+** : **O+** = **O** ∪ {xor($n+1, n$) rules for each $n \geq 2$}.

Observation 2: The family of rules **O+** is sound for *ProbCR*s.

We establish the soundness of **O+** by showing that each XOR($n+1, n$) rules for each $n \geq 1$ follows from the sound set of rules **Q**. The proof is by induction on n.

Proof: basis: XOR(2,1) is just XOR, which has already been established as sound for ProbCRs.

Induction hypothesis: Now, suppose that for all k such that $1 \leq k \leq n-1$, the rules XOR($k+1, k$) hold. (We show that XOR($n+1, n$) must also hold.)

Induction step: Let the members of the list $< a_1, \ldots, a_{n-1}, a_n, a_{n+1} >$ consist of pairwise inconsistent sentences, and suppose that for each disjunction (in order) of any n of them, e, we have $e \mid\sim x$. (We want to show that for the disjunction of all of them, $a_1 \vee \ldots \vee a_{n-1} \vee a_n \vee a_{n+1} \mid\sim x$.)

Notice that $a_1 \vee \ldots \vee a_{n-1} \vee a_n \mid\sim x$ and $a_1 \vee \ldots \vee a_{n-1} \vee a_{n+1} \mid\sim x$. Applying XNR to $a_1 \vee \ldots \vee a_{n-1} \vee a_{n+1} \mid\sim x$ yields that either $a_1 \vee \ldots \vee a_{n-1} \mid\sim x$ or $a_{n+1} \mid\sim x$.

(i) Suppose $a_{n+1} \mid\sim x$. Then, since $\vdash \neg(a_{n+1} \wedge (a_1 \vee \ldots \vee a_{n-1} \vee a_n))$, putting this with $(a_1 \vee \ldots \vee a_{n-1} \vee a_n) \mid\sim x$ via XOR(2, 1) yields $a_1 \vee \ldots \vee a_{n-1} \vee a_n \vee a_{n+1} \mid\sim x$, and we are done.

(ii) Alternatively, if $a_{n+1} \mid\sim x$, then $a_1 \vee \ldots \vee a_{n-1} \mid\sim x$. Now, consider the sequence of n sentences $S = <a_1, \ldots, a_{n-1}, (a_n \vee a_{n+1})>$. The members of S are pairwise inconsistent (since $\vdash \neg(a_j \wedge (a_n \vee a_{n+1}))$ holds for each j such that $1 \leq j \leq n-1$). For each a_j with $1 \leq j \leq n-1$ we already had that $a_1 \vee \ldots \vee a_{j-1} \vee a_{j+1} \vee \ldots \vee a_{n-1} \vee (a_n \vee a_{n+1}) \mid\sim x$ (this is a supposition of the present induction step), and we also have $a_1 \vee \cdots \vee a_{n-1} \mid\sim x$. So, $S = <a_1, a_2, \ldots, a_{n-1}, (a_n \vee a_{n+1})>$ is a list of n pairwise inconsistent sentences , where for each ordered sequence e of $n-1$ of them, we have $e \mid\sim x$. Thus, by the induction hypothesis, for the disjunction of all of them, $a_1 \vee \ldots \vee a_{n-1} \vee (a_n \vee a_{n+1})|\!\sim x$. Then $a_1 \vee \ldots \vee a_{n-1} \vee a_n \vee a_{n+1}|\!\sim x$.

Now let's establish that all of the XOR(n, m) rules (for $n > m \geq 1$) are derivable in **O+**.

Observation 3: From $\mathbf{O}+ = \mathbf{O} \cup \{XOR(n+1, n)$ rules for each n $\geq 2\}$ it follows that all XOR(n, m) rules hold for all $n > m \geq 1$.

Proof: Let m be any natural number such that $m \geq 1$.

(1) Basis: $n = \mathrm{m} + 1$: Let S be any sequences of $n = m + 1$ pairwise inconsistent sentences such that for each disjunction e of m members of S (in the order specified by S) we have $e|\!\sim x$. Then by XOR($m + 1, m$) we have, for the disjunction d of all members of S (in the order specified by S), $d|\!\sim x$.

(k) Induction hypothesis: $n = m+k$: Suppose that for any sequence S of $n = m+k$ pairwise inconsistent sentences, if for each disjunction e of m members of S (in the order specified by S) we have $e\,|\!\sim x$, then also we have, for the disjunction d of all members of S (in the order specified by S), $d|\!\sim x$.

(k+1) Induction step: $n = m + k + 1$: Let S be any sequences of $n = m + k + 1$ pairwise inconsistent sentences such that for each disjunction e of m members of S (in the order specified by S) we have $e|\!\sim x$.
Let S^* be any subsequence of S consisting of $m + k$ members of S. S^* is a sequence of $m + k$ inconsistent sentences, where, for each disjunction e of m members of S^* (in the order specified by S) we have $e\,|\!\sim x$. So for the disjunction d of all $m + k$ members of S^* (in the order specified by S) we have $d|\!\sim x$.
So, by the induction hypothesis, for each disjunction d of $m + k$ members of S (in the order specified by S) we have $d|\!\sim x$. Then by XOR($m + k + 1, m + k$) we have, for the disjunction d^+ of all members of S (in the order specified by S), $d^+|\!\sim x$.

Thus, all of the XOR(n, m) rules (for all $n > m \geq 1$) are sound for the *ProbCR*s.

5 Is O+ Enough? If Not, Then How About Q?

Notice that the Family **O+** provides an infinite list of new Horn rules sound for *ProbCR*s. Paris and Simmonds established that **O** isn't enough by itself, and that only an infinite list of rules can provide a set of axioms that are sufficiently complete to permit the derivation of every finite-premised Horn rules that is sound for *ProbCR*s. Furthermore, **O+** provides derivations for all of the specific rules we know of that have been explicitly calculated via the Paris-Simmonds algorithm. So perhaps **O+** is enough. Is it? That's the central outstanding issue for *ProbCR*s at present.

If the answer to this question turns out to be negative, then the further question remains: Is **Q** enough? For, **O+** is derivable from the *ProbCR* sound rules that make up **Q**, even though the extra power of **Q** comes from the non-Horn rule NR. So, whatever the complete set of finite-premised Horn rules may be, perhaps **Q** suffices to derive them all.

References

Hawthorne, J. (1996). On the logic on non-monotonic conditionals and conditional probabilities. *Journal of Philosophical Logic*, *25*, 185–218.

Hawthorne, J., & Makinson, D. (2007). The qualitative/quantitative watershed for rules of uncertain inference. *Studia Logica*, *86*, 247–297.

Krauss, S., Lehmann, D., & Magidor, M. (1990). Nonmonotonic reasoning preferential models and cumulative logics. *Artificial Intelligence*, *44*, 167–207.

Makinson, D., (1989). General Theory of Cumulative Inference. In M. Reinfrank, J. de Kleer, M. L. Ginsberg, & E. Sandewall (Eds.) *Non-Monotonic Reasoning—Proceedings of the 2nd International Workshop 1988* (pp. 1–18). Berlin: Springer.

Makinson, D. (1994). General Patterns in Nonmonotonic Reasoning. In Dov M. Gabbay, C. J. Hogger, & J. A. Robinson (Eds.), *Handbook of Logic in Artificial Intelligence and Logic Programming, Non-Monotonic and Uncertainty Reasoning* (Vol. 3, pp. 35–110). Oxford: Oxford University Press.

Paris, J., & Simmonds, R. (2009). O is not enough. *Review of Symbolic Logic*, 2, 298–309.

Non-Monotonic Logic: Preferential Versus Algebraic Semantics

Karl Schlechta

Abstract Preferential logic can be seen as the result of manipulation with an abstract notion of size. This intermediate level between relation-based semantics and proof theory is free from the details of the latter two. It gives an unobstructed view of the central mechanism of preferential logic, and opens ways to generalizations and modifications.

Keywords Non-monotonic logic · Preferential semantics · Algebraic semantics · Defaults

1 Introduction

The material of this introduction will be elaborated in the rest of the paper.

We should perhaps emphasize that the present text does not so much repeat well known results, as published, e.g., in Kraus et al. (1990) and Lehmann and Magidor (1992), and also the author's Schlechta (2004), but tries to take a more distanced and abstract view of the area, of underlying - hidden or well visible - ideas, concepts, and assumptions. Thus, the text is destined mostly to the advanced reader. It is a look behind the scene, not at the performance.

It is a great pleasure for me to dedicate this paper to David Makinson, who, with his insights and criticisms, accompanied the present author's career in non-monotonic and related logics from the beginning to the present day. We did not always agree, in methods and concepts, but this is the way fruitful exchange works: enough in common to understand each other, enough difference for stimulation.

K. Schlechta (✉)
CNRS, LIF UMR 7279 , F-13000 Marseille, France
e-mail: schcsg@gmail.com

S. O. Hansson (ed.), *David Makinson on Classical Methods for Non-Classical Problems*, Outstanding Contributions to Logic 3, DOI: 10.1007/978-94-007-7759-0_10,

Lack of space prevents doing both: presenting an overview of present day knowledge and results, and outlining the deeper problems, concepts, and ideas behind the results. To do both thoroughly would probably result in another book.

Consequently, we also chose to put many details into the appendix, so the flow of argumentation is not too much interrupted by material known already—at least roughly—by the reader. The appendices make the text self-contained, however.

1.1 Thinking on the Semantical Side

The author of these lines usually prefers to work on the semantical side. First, semantical arguments are often more robust, as they are insensitive to logically equivalent reformulations: we need not bother about equivalent reformulations, and can concentrate on the main conditions. Second, semantical arguments are often simpler, for instance, drawing a few circles can solve a problem in theory revision, or make it transparent, which would otherwise require considerable familiarity with the mechanism. Likewise, translating problems from one domain to another, say, e.g., from theory revision to update, is often helped by a little drawing of a semantical situation. Moreover, the abstract size semantics discussed below illustrate how small exception sets and "big" sets of normal elements work together to give a principled way of reasoning with information of different degrees of quality. This not only helps intuition, but provides a deeper understanding, too.

Finally, there is also a philosophical justification, which can be summarized as "semantics first": logic should speak about something, and should not be an exercise in thin air. So we should first have a good intuition about our object of reasoning, should then fix this in a formal semantics, and only afterwards look for an adequate language or logic to speak about our object.

Other researchers prefer other approaches. The syntactic approach, see e.g., Gabbay (1989, 1991), and Makinson (1994), starts with rules which seem desirable or natural. Thus, e.g., Cumulativity was discussed before the introduction of smooth models, their semantical counterpart. The present author perhaps just lacks the intuition for this approach—aside from the positive arguments cited above for the semantical approach.

1.2 Preferential Structures and Model Choice Functions, Structural and Abstract Semantics

1.2.1 Preferential Structures and Their Model Choice Functions

A preferential structure is, basically, a set U with a binary relation $\prec$ on U, and a policy about what to do with this relation. Given $X \subseteq U$, we consider elements $x \in X$, which are minimal in X, i.e., there is no $x' \in X$ such that $x' \prec x$. So, if $\mu(X)$ is the set of those minimal elements, we have

$$\mu(X) := \{x \in X : \neg\exists x' \in X . x' \prec x\}.$$

(More precisely, one often works with copies, see Sect. 6, or, equivalently, not necessarily injective labelling functions, but we neglect this in the present section in order to make the concepts clearer.)

Note that preferential structures are then just like Kripke structures, but used differently. In the latter case, we consider $K(X) := \{y \in U : \exists x \in X . x \prec y\}$, where, e.g., $X \subseteq K(X)$ if $\prec$ is reflexive.

The intuitive interpretation is that minimal elements are the most normal or least abnormal elements, they thus represent ideal cases (of normality).

1.2.2 Model Choice Functions as Abstract Semantics

In the intended application, U is the set of models of a given, fixed language, which will mostly be propositional here. We can then define a logical (consequence) relation $\mid\sim$ from U and $\prec$ via μ by $\phi \mid\sim \psi$ iff $\mu(M(\phi)) \models \psi$—where $\models$ is classical satisfaction, and the usual associated semantical consequence relation. We sometimes also use $\vdash$, the classical proof relation, which is, of course, by classical soundness and completeness, equivalent to $\models$.

Conversely, we can define μ from a logic $\mid\sim$ by setting

$$\mu(M(\phi)) := \{m : m \models \psi \text{ for all } \psi \text{ such that } \phi \mid\sim \psi\}.$$

If $\phi \mid\sim \phi$, we then have $\mu(M(\phi)) \subseteq M(\phi)$.

Seen this way, we have a semantics for $\mid\sim$ by a model choice function μ, which may or may not correspond to the model choice function of a preferential structure. It is abstract, as it does not look for an underlying structure, and there may be many different structures which generate the same μ.

Thus, we can go in both directions:

structure $\Leftrightarrow \mu \Leftrightarrow \mid\sim$,

each time considering corresponding properties.

For these reasons, we call it an abstract semantics, whereas the preferential structure will be called a structural semantics. On the other hand, we will also look at the algebraic properties of the model choice function μ, like $\mu(X) \subseteq X$, so we will also call it an algebraic semantics. We use both terms interchangeably, when we think more in terms of properties we use "algebraic", when we think more as independent from an actual structure we use "abstract".

See Sect. 2 for further discussion.

1.2.3 Model Choice Functions as an Object of Research

Thus, μ chooses for $X \subseteq U$ a subset $\mu(X) \subseteq X$, and we can make μ itself an object of our research, examine its properties, as well the relation of those properties, to preferential structures, in particular to conditions on the relation $\prec$ on the one hand side, and to the logic $\mid\!\sim$, and conditions on $\mid\!\sim$ on the other side.

Relation of the model choice function μ to $\prec$ and $\mid\!\sim$
For example, we see that when μ is generated by a relation $\prec$ in the manner described, it will satisfy $\mu(X) \subseteq X$, that $\mu(\mu(X)) = \mu(X)$, etc. We can ask whether imposing transitivity on $\prec$ results in additional properties of μ (it often does not), and, conversely, when we impose conditions on μ, if they are strong enough to guarantee that there is a relation $\prec$ which generates this μ, and perhaps with some additional properties on $\prec$.

On the other hand, we see that the property $\mu(X) \subseteq X$ corresponds to the logical rule $\phi \models \psi \Rightarrow \phi \mid\!\sim \psi$, and that $\models \phi \leftrightarrow \phi'$ implies $\phi \mid\!\sim \psi$ iff $\phi' \mid\!\sim \psi$ as we work with model sets, etc.

Definable model sets and closure conditions
In classical propositional logic, when the language is based on an infinite set of propositional variables, not all model sets X are definable, whether by formulas or even by theories (sets of formulas): there need not be a formula ϕ or a theory T such that $X = M(\phi)$ or $X = M(T)$, see Example 3.1. This is a problem of the expressive strength of the language. In addition, an arbitrary μ, be it generated by a preferential structure and its relation $\prec$ or not, need not preserve definability: it may very well be that X is definable, but $\mu(X)$ is not. This is a closure condition for μ : The domain is closed under μ, or not.

Moreover, in classical logic, the set of definable model sets is closed under finite intersections for formulas, infinite intersections for theories, and closed under finite unions for both formulas and theories. In sequent calculi (see Sect. 3.2), however, we may not have an "or" on the left hand side of $\mid\!\sim$, so the set of model sets for which we can obtain results by the definition $\mu(M(\phi)) := \{m : m \models \psi$ for all ψ such that $\phi \mid\!\sim \psi\}$ need not be closed under finite unions. Thus, μ need not be defined for all $X \subseteq U$, but only on a subset $\mathcal{Y} \subseteq \mathcal{P}(U)$($\mathcal{P}$ the power set operator), and the domain of μ may have only limited closure conditions under the set operators $\cap$ and $\cup$.

Again, we may ask questions about μ defined on such $\mathcal{Y}$, like:

- Are closure conditions needed to obtain certain conditions on $\prec$ of a representing structure? (Yes, for smooth structures, to obtain cumulativity.)
- Do these definability and closure conditions have any incidence on the correspondence between μ and $\mid\!\sim$? (Yes, for definability preservation, see below.)

1.3 Systems of Chosen Subsets

We can also soften the condition that we have one "ideal" subset $\mu(X) \subseteq X$, of "ideal cases" in X, instead we may have a set $\Sigma(X)$ of "good" subsets, without having an ideal one, but which is perhaps approximated, in the sense that we have "better and better" cases, but no ideal ones. This is, e.g., motivated by preferential structures where we may not always have minimal elements (due to infinite descending chains), but they may be "better and better". Note that absence of minimal elements results in inconsistencies of the logic defined by $\phi \mid\!\sim \psi$ iff $\mu(M(\phi)) \subseteq M(\psi)$, as any ψ holds in the empty model set, so this approach is then not useful. See Sect. 4.

It is plausible that such systems are more complicated, also in their correspondence to preferential and logical properties, but one can pose interesting questions, and have strong results, like:

- If, for all $A, A' \in \Sigma(X)$, there is $A'' \in \Sigma(X)$ with $A'' \subseteq A \cap A'$, then AND will hold in the corresponding logic. This property holds for suitably defined $\Sigma(X)$ in preferential structures, if the relation is transitive.
- But this AND property might not always be desired, as the lottery paradox (see Example 5.1) shows.
- One can also show that, sometimes, the resulting logic has the same properties as the logic defined by $\mu(X)$ for a (usually different) preferential structure.

1.4 Model Choice Functions and Size

It is natural to interpret systems of $\mu(X)$ for $X \subseteq U$ by a notion of size, assuming that the normal cases are also the most frequent ones. The reader should note that this interpretation goes beyond the abstract concept of an algebraic semantics. The origins of non-monotonic logics, however, motivate the view that these logics are tools for reasoning about the normal cases, or, about all but a few exotic cases, or the majority of cases. When we look from $\mid\!\sim$ to μ, going from $\phi \mid\!\sim \psi$ to $\mu(M(\phi)) \models \psi$, it is thus natural to interpret $\mu(X)$ as a "big" subset of X, and $X - \mu(X)$ as a small set of exceptions. More precisely, $\mu(X)$ will be the $\subseteq$-smallest big subset of X, and $X - \mu(X)$ the $\subseteq$-biggest small subset of X. We have thus defined a principal filter over X, generated by $\mu(X)$ (or its dual, an ideal, generated by $X - \mu(X)$). See Sect. 5.

It is a natural extension of the language to introduce a new operator ∇ which will be interpreted by μ. $\nabla\phi$ will then denote $\mu(M(\phi))$ in the propositional case, and ∇ will be a new quantifier in first order logic, $\nabla x\phi(x)$ will denote the "normal" ϕ-cases, and can be read as "almost all".

Note that the intersection property of filters corresponds to (an even infinite)
AND: $\phi \mid\!\sim \psi, \phi \mid\!\sim \psi' \Rightarrow \phi \mid\!\sim \psi \wedge \psi'$
But logical laws like
OR: $\phi \mid\!\sim \psi, \phi' \mid\!\sim \psi \Rightarrow \phi \vee \phi' \mid\!\sim \psi$
motivate us to consider not only individual filters over (definable) model sets, but coherent systems of such filters, where, e.g. $\mu(X \cup Y) \subseteq \mu(X) \cup \mu(Y)$ (which results in above OR). This is, of course, conceptually very different from the laws of single filters or ideals.

Conversely, preferential structures motivate us to consider conditions like $X \subseteq Y \Rightarrow \mu(Y) \cap X \subseteq \mu(X)$, so, we have again connections to both sides, to logical rules and to structural semantics. We will thus examine:

1.4.1 Coherence Conditions for Filter Systems

Considering such filter systems and the notion of size as an independent object of research, we can ask for other natural properties of these systems. E.g., if $A \subseteq X$ is small (i.e., $A \subseteq X - \mu(X)$), then it seems natural that, if $X \subseteq Y$, then $A \subseteq Y$ is small too, i.e. $A \subseteq Y - \mu(Y)$. (The property $A' \subseteq A \subseteq X$, $A \subseteq X$ small implies $A' \subseteq X$ is small is already entailed by the filter (or ideal) properties, but those properties say nothing about changing the base set.)

We may ask now, e.g.,

- What are natural coherence properties?
- Which coherence properties correspond to interesting logical properties?
- Which coherence properties are already generated by the basic version of preferential structures?
- Are there corresponding properties in preferential structures with stronger conditions on the relation $\prec$? More precisely, can we find additional conditions on $\prec$ which result in stronger coherence conditions?

1.4.2 Coherence Conditions and Sublanguages

Propositonal models are best seen as functions from the set of propositional variables to $\{0, 1\}$ (or $\{true, false\}$ etc.), so model sets are sets of such functions. Such sets can then sometimes be written as non-trivial products $X = X' \times X''$, like $\{pqr, pq\neg r, \neg p\neg qr, \neg p\neg q\neg r\} = \{pq, \neg p\neg q\} \times \{r, \neg r\}$, in obvious shorthand.

We can now ask whether $\mu(X' \times X'') = \mu(X') \times \mu(X'')$, where $\mu(X' \times X'')$ is formed in the full language, $\mu(X')$ in the language of X', and $\mu(X'')$ in the language of X''. This is then a coherence condition between several languages. We may formulate a similar condition in the big language as $\mu(X' \times X'') = \mu(X' \times 1_{X''}) \cap \mu(1_{X'} \times X'')$, where $1_{X'}$ is the set of all possible values in the sublanguage of X', etc. For instance, $1_{X'}$ is $\{pq, p\neg q, \neg pq, \neg p\neg q\}$ in above example.

Again, we can investigate natural properties for such products, compare them to properties of $\prec$, when μ is the model choice function of a structure defined by $\prec$ (these product properties are modularity properties like $xx' \prec yy'$ iff $x \prec y$ and $y \prec y'$), and to properties for $\mid\!\sim$, if $\mid\!\sim$ is defined by μ (they result, e.g., in interpolation properties for $\mid\!\sim$).

1.5 Advantages of Considering Abstract Semantics

Considering abstract semantics has several advantages:

(1) Proofs of epresentation theorems often become more transparent and easier, as we split them into two parts, first from logic to abstract semantics, and then from abstract semantics to structural semantics. The strategy reveals difficulties in

each step, which are independent from the other step, and proofs and problems are not muddled together.

(2) Usually, showing completeness is the more difficult part when we want to prove adequacy of a logic to a semantics. Translating the logical properties into algebraic ones is often mostly routine work, the hard part is the construction of the representing structure. Seen this way, the result of this translation to algebraic conditions is a natural place to take a deep breath before the final ascent, it is a welcome "base camp".
In addition, there are often many ways that lead in both directions from the algebraic semantics. On the one hand, we often see similar logical conditions, which then have the same translation, and we can just look up former work, and know where they lead us to. In the other direction, when we have the same kind of structure (preferential, distance based, etc.), it does not matter what the logical interpretation of the points (classical models, sequences of developments, etc.) is, the algebraic properties of the choice function stay the same. Thus, the separate steps may be re-used for different proofs, where we cannot re-use both steps together.

(3) The robustness of abstract semantics is very important. It is neither victim to trivial reformulations on the logical side, nor to reformulations on the structural side - it captures the essence of logic and structure. In particular, when we look for operations on such logics or structures, like revising one logic with another logic, this is the right level to work on.

(4) Perhaps most importantly, investigating the algebraic semantics independently results in new questions and results, and gives a new and broader perspective on problems, it gives deeper insights, and reveals new connections. In our context, this will be the concept of abstract size, and its ramifications.

1.6 Organization of the Paper

The reader should note that not all subjects discussed here are equally well investigated so far, so there is still ample opportunity to do original research.

- We first summarize the main definitions in Sect. 1.7.
- We then discuss the relation of μ and its properties with those of $\prec$ and $\mid\!\sim$ in Sect. 2.
- Definability problems are discussed in Sect. 3.
- Systems of sets $\Sigma(X)$ are discussed in Sect. 4.
- Questions related to abstract size are discussed in Sect. 5.

The material discussed here has been published before. The reader will find comprehensive discussions in Schlechta (1997, 2004), Gabbay and Schlechta (2009a, c, 2011).

1.7 Basic Definitions

We collect here for easier reference some frequently used definitions.

Definition 1.1

(1) $\mathcal{P}(X)$ will be the power set of X.
(2) $\vdash$ and $\models$ will denote the classical consequence and satisfaction relations, the latter also the classical semantic consequence relation, as in $T \models \phi$ iff $M(T) \subseteq M(\phi)$.
(3) $\mid\!\sim$ will be the consequence relation of some new logic, fixed by context.
(4) We use the following very intuitive notation for closure under logical consequence:
For a set of formulas T (a single formula ϕ), $\overline{T}$ $(\overline{\phi})$ will be the set of all classical consequences of T (ϕ), i.e. $\overline{T} := \{\psi : T \models \psi\}$, etc.
Given a different logic $\mid\!\sim$, for a set of formulas T (a single formula ϕ), $\overline{\overline{T}}$ $(\overline{\overline{\phi}})$ will be the set of all consequences of T (ϕ) under this logic $\mid\!\sim$, i.e. $\overline{\overline{T}} := \{\psi : T \mid\!\sim \psi\}$, etc.
Confusion between $\overline{\overline{T}}$ and $\overline{(\overline{T})}$ will not occur, as it is useless to close twice under classical logic.

Definition 1.2 Basic definitions for preferential models or structures.

(1) A pair $\mathcal{M} := \langle U, \prec \rangle$ with U an arbitrary set and $\prec$ an arbitrary binary relation on U is called a preferential model or structure.
(2) We say that a structure $\mathcal{M}$ is transitive, irreflexive, etc., iff $\prec$ is.
(3) Let $\mathcal{M} := \langle U, \prec \rangle$, and define
$\mu_{\mathcal{M}}(X) := \{x \in X : x \in U \wedge \neg\exists x' \in X \cap U. x' \prec x\}$.
(We may assume that $X \subseteq U$.)
If the context is clear, we will write μ for $\mu_{\mathcal{M}}$.
$\mu(X)$ is called the set of minimal elements of X (in $\mathcal{M}$).
Thus, $\mu(X)$ is the set of elements in $X \cap U$ such that there is no smaller one in $X \cap U$. Note that $\mu(X)$ might well be empty, even if X is not.
(4) Validity in a preferential structure, or the semantical consequence relation defined by such a structure:
If U is a set of classical models, we define

$$T \models_{\mu} \phi \text{ iff } \mu(M(T)) \models \phi, \text{ i.e., } \mu(M(T)) \subseteq M(\phi).$$

Definition 1.3 Fix a base set X.
A filter (resp. weak filter) on or over X is a set $\mathcal{F} \subseteq \mathcal{P}(X)$ such that $(F1) - (F3)$ (resp. (F1), (F2), $(F3')$) hold:
(F1) $X \in \mathcal{F}$,
(F2) $A \subseteq B \subseteq X$, $A \in \mathcal{F}$ imply $B \in \mathcal{F}$,
(F3) $A, B \in \mathcal{F}$ imply $A \cap B \in \mathcal{F}$,
$(F3')$ $A, B \in \mathcal{F}$ imply $A \cap B \neq \emptyset$.

So a weak filter satisfies $(F3')$ instead of (F3). Note that $\mathcal{P}(X)$ is a filter, but not a weak one, but all proper filters are also weak filters.
A filter is called a principal filter iff there is $X' \subseteq X$ s.t. $\mathcal{F} = \{A : X' \subseteq A \subseteq X\}$.
A (weak) ideal on or over X is a set $\mathcal{I} \subseteq \mathcal{P}(X)$ such that $(I1) - (I3)$ (or (I1), (I2), $(I3')$) hold:
(I1) $\emptyset \in \mathcal{I}$,
(I2) $A \subseteq B \subseteq X,\ B \in \mathcal{I}$ imply $A \in \mathcal{I}$,
(I3) $A, B \in \mathcal{I}$ imply $A \cup B \in \mathcal{I}$,
$(I3')$ $A, B \in \mathcal{I}$ imply $A \cup B \neq X$.
So a weak ideal satisfies $(I3')$ instead of (I3).
A filter is an abstract notion of size; elements of a filter on X are called big subsets of X; their complements are called small, and the rest have medium size. The dual applies to ideals; this is justified by the following trivial fact:

Fact 1.1 If $\mathcal{F}$ is a (weak) filter on X, then $\mathcal{I} := \{X - A : A \in \mathcal{F}\}$ is a (weak) ideal on X; if $\mathcal{I}$ is a (weak) ideal on X, then $\mathcal{F} := \{X - A : A \in \mathcal{F}\}$ is a (weak) filter on X.

All other definitions will be found in the appendix to the respective sections.

2 Algebraic or Abstract Versus Structural Semantics

2.1 Preferential Structures and Model Choice Functions

Recall from Sect. 1 that a preferential structure generates a system of principal filters, built on the $\mu(X)$, where the X are subsets of the universe. They have some very simple properties, first, obviously, the filter properties, and then also a "coherence" property between the different filters: $X \subseteq Y \Rightarrow \mu(Y) \cap X \subseteq \mu(X)$.

Those very simple properties generate already strong logical properties, as we will discuss now. A more complete picture is found in the appendices to the respective sections.

2.2 Correspondence Between Logical Rules and Model Choice Functions

Most of the properties of model choice functions generated by preferential structures have an evident and 1-1 correspondence to logical properties of $\mid\sim$, like AND, which corresponds to the filter property. Some logical properties are also trivially true, those which always hold when we work with sets of classical models. They concern logically equivalent reformulations, etc.

We now discuss some correspondences between logical properties and those of model choice functions. Let f be such a function, and suppose $f(X) \subseteq X$.

(1) First, when T and T' are classically equivalent, then they have the same model sets, $M(T) = M(T')$, so $f(M(T)) = f(M(T'))$, and $T \mid\sim \phi$ iff $T' \mid\sim \phi$.
(2) Second, if $\phi \models \psi$, and $T \mid\sim \phi$, then $T \mid\sim \psi$: By $\phi \models \psi$, $M(\phi) \subseteq M(\psi)$, by prerequisite $f(M(T)) \subseteq M(\phi)$, so $f(M(T)) \subseteq M(\psi)$, and $T \mid\sim \psi$.
(3) The AND rule holds, as said already: If $T \mid\sim \phi$ and $T \mid\sim \phi'$, then $f(M(T)) \subseteq M(\phi)$ and $f(M(T)) \subseteq M(\phi')$, so $f(M(T)) \subseteq M(\phi) \cap M(\phi') = M(\phi \wedge \phi')$, so $T \mid\sim \phi \wedge \phi'$.
(4) It is easy to see that the model choice functions of preferential structures have the property $f(X \cup X') \subseteq f(X) \cup f(X')$. So, if, $\phi \mid\sim \psi$ and $\phi' \mid\sim \psi$, then $f(M(\phi)) \subseteq M(\psi)$ and $f(M(\phi')) \subseteq M(\psi)$, thus $f(M(\phi \vee \phi')) = f(M(\phi) \cup M(\phi')) \subseteq f(M(\phi)) \cup f(M(\phi')) \subseteq M(\psi)$, so $\phi \vee \phi' \mid\sim \psi$.
(5) In smooth preferential structures, (see Definition 7.1), the important property $f(X) \subseteq Y \subseteq X \Rightarrow f(X) = f(Y)$ holds.
Then, if $\phi \mid\sim \psi$ and $\phi \mid\sim \psi'$, then $\phi \wedge \psi \mid\sim \psi'$: $f(M(\phi)) \subseteq M(\phi) \cap M(\psi) = M(\phi \wedge \psi) \subseteq M(\phi)$, so $f(M(\phi \wedge \psi)) = f(M(\phi)) \subseteq M(\psi')$.

Table 1, "Logical rules" in the appendix for this section gives an (incomplete) list of such correspondences.

Coming from logical properties, we can translate them in this way into properties about model sets, and have an algebraic semantics.

2.3 Correspondence Between Preferential Structures and Model Choice Functions

We already mentioned that in all preferential structures, the model choice function has the evident property $\mu(X) \subseteq X$, and the strong property $X \subseteq Y \Rightarrow \mu(Y) \cap X \subseteq \mu(X)$. The validity of the latter is easy to see: Suppose $x \in X - \mu(X)$, then there must be $x' \prec x$, $x' \in X$. But then also $x' \in Y$, so $x \notin \mu(Y)$. It is less evident to see that these two properties suffice to characterize general preferential structures, i.e., given a choice function μ with these properties, we find a preferential structure, which has exactly this μ as its choice function. (We neglect here the question of copies, see Sect. 6.) In addition, the representing structure can be chosen to be transitive.

Adding properties to the relation $\prec$ leads to new properties for μ. The two most important additions are smoothness (called stopperedness by Makinson) and rankedness. Smoothness leads to cumulativity: $\mu(X) \subseteq Y \subseteq X \Rightarrow \mu(X) = \mu(Y)$, and rankedness leads to $X \subseteq Y$, $\mu(Y) \cap X \neq \emptyset \Rightarrow \mu(X) = \mu(Y) \cap X$. This works again (with some caveats about copies and domain closure) in both directions: The properties for μ hold in such structures, and given μ with such properties can be represented by suitable structures with the corresponding property of the relation $\prec$. The appendix gives some more details.

Table 1 Logical rules, definitions and connections

Logical rules, definitions and connections					
Logical rule		Corr.	Model set	Corr.	Size Rules
		Basics			
(SC) Supraclassicality $\alpha \vdash \beta \Rightarrow \alpha \mid\sim \beta$	(SC) $\overline{T} \subseteq \overline{\overline{T}}$	$\Rightarrow$ $\Leftarrow$	$(\mu \subseteq)$ $f(X) \subseteq X$	trivial	(Opt)
(LLE) Left Logical Equivalence $\vdash \alpha \leftrightarrow \alpha', \alpha \mid\sim \beta \Rightarrow$ $\alpha' \mid\sim \beta$	(LLE) $\overline{T} = \overline{T'} \Rightarrow \overline{\overline{T}} = \overline{\overline{T'}}$		(trivally true)		
(RW) Right Weakening $\alpha \mid\sim \beta, \vdash \beta \rightarrow \beta' \Rightarrow$ $\alpha \mid\sim \beta'$	(RW) $T \mid\sim \beta, \vdash \beta \rightarrow \beta' \Rightarrow$ $T \mid\sim \beta'$		(upward closure)	trivial	(iM)
(wOR) $\alpha \mid\sim \beta,\ \alpha' \vdash \beta \Rightarrow$ $\alpha \vee \alpha' \mid\sim \beta$	(wOR) $\overline{\overline{T}} \cap \overline{T'} \subseteq \overline{\overline{T \vee T'}}$	$\Rightarrow$ $\Leftarrow$	(μwOR) $f(X \cup Y) \subseteq f(X) \cup Y$	$\Leftrightarrow$	$(eM\mathcal{I})$
$(disjOR)$ $\alpha \vdash \neg\alpha',\ \alpha \mid\sim \beta,$ $\alpha' \mid\sim \beta \Rightarrow \alpha \vee \alpha' \mid\sim \beta$	$(disjOR)$ $\neg Con(T \cup T') \Rightarrow$ $\overline{\overline{T}} \cap \overline{\overline{T'}} \subseteq \overline{\overline{T \vee T'}}$	$\Rightarrow$ $\Leftarrow$	$(\mu disjOR)$ $X \cap Y = \emptyset \Rightarrow$ $f(X \cup Y) \subseteq f(X) \cup f(Y)$	$\Leftrightarrow$	$(I \cup disj)$
(AND) $\alpha \mid\sim \beta, \alpha \mid\sim \beta' \Rightarrow$ $\alpha \mid\sim \beta \wedge \beta'$	(AND) $T \mid\sim \beta, T \mid\sim \beta' \Rightarrow$ $T \mid\sim \beta \wedge \beta'$		(closure under finite intersection)	trivial	(I_ω)
(OR) $\alpha \mid\sim \beta, \alpha' \mid\sim \beta \Rightarrow$ $\alpha \vee \alpha' \mid\sim \beta$	(OR) $\overline{\overline{T}} \cap \overline{\overline{T'}} \subseteq \overline{\overline{T \vee T'}}$	$\Rightarrow$ $\Leftarrow$	(μOR) $f(X \cup Y) \subseteq f(X) \cup f(Y)$	$\Leftrightarrow$	$(eM\mathcal{I}) + (I_\omega)$
$\overline{\overline{\alpha \wedge \alpha'}} \subseteq \overline{\overline{\alpha}} \cup \{\alpha'\}$	(PR) $\overline{\overline{T \cup T'}} \subseteq \overline{\overline{\overline{T}} \cup T'}$	$\Rightarrow$ $\Leftarrow (dp) + (\mu \subseteq)$ $\not\Leftarrow -(dp)$ $\Leftarrow (\mu \subseteq)$ $T' = \phi$	(μPR) $X \subseteq Y \Rightarrow$ $f(Y) \cap X \subseteq f(X)$	$\Leftrightarrow$	$(eM\mathcal{I}) + (I_\omega)$
		Cumulativity			
(CUM) Cumulativity $\alpha \mid\sim \beta \Rightarrow$ $(\alpha \mid\sim \beta' \Leftrightarrow \alpha \wedge \beta \mid\sim \beta')$	(CUM) $T \subseteq \overline{T'} \subseteq \overline{\overline{T}} \Rightarrow$ $\overline{\overline{T}} = \overline{\overline{T'}}$	$\Rightarrow$ $\Leftarrow$	(μCUM) $f(X) \subseteq Y \subseteq X \Rightarrow$ $f(Y) = f(X)$	$\Leftarrow$ $\not\Rightarrow$	$(eM\mathcal{I}) + (I_\omega) + (\mathcal{M}_\omega^+)(4)$
		Rationality			
$(RatM)$ Rational Monotony $\alpha \mid\sim \beta, \alpha \not\mid\sim \neg\beta' \Rightarrow$ $\alpha \wedge \beta' \mid\sim \beta$	$(RatM)$ $Con(T \cup \overline{\overline{T'}}), T \vdash T' \Rightarrow$ $\overline{\overline{T}} \supseteq \overline{\overline{\overline{T'}} \cup T}$	$\Rightarrow$ $\Leftarrow (dp)$ $\not\Leftarrow -(dp)$ $\Leftarrow T = \phi$	$(\mu RatM)$ $X \subseteq Y, X \cap f(Y) \neq \emptyset \Rightarrow$ $f(X) \subseteq f(Y) \cap X$	$\Leftrightarrow$	$(\mathcal{M}^{++})$

3 Expressivity Problems

3.1 Limited Expressivity of the Language

The fit between logic and algebraic semantics need not be perfect, and this may be incorrigibly so. The division of representation between a translation from logic to algebra, and the construction of a representing structure for the algebraic semantics brings the underlying problems to light. These are connected to the expressive power of the language, which, in the infinite case and in usual languages, is too weak to describe all possible model sets - there are far too many of them. So, there must be model sets which do not correspond exactly to any formula, not even to any theory. A very simple example is the following:

Example 3.1 Take an infinite propositional language, and omit 1 model from the set of all models. There is no theory which defines this set (the model we took away is missing from the models of all tautologies, but we cannot obtain a stronger theory either). More precisely: Let M be the set of all models of this language, $m \in M$ arbitrary, $M' := M - \{m\}$. Suppose there is a formula ϕ which holds in all $m' \in M'$, but not in m, so $\neg\phi$ holds in m, but in no $m' \in M'$. But $\neg\phi$ is finite, so it contains a finite number of propositional variables. Thus, any m' which agrees with m on these variables, will also make $\neg\phi$ true. But there are uncoutably many such m', contradicting $M' \models \phi$.

Thus, given $M(\phi)$, $\mu(M(\phi))$ need not be a definable model set any more — the relation on the structure is independent from logic, and nothing guarantees that $\mu(M(\phi))$ is definable. To preserve 1-1 correspondence between logic and the algebraic semantics, we need an additional, extralogical property of the structure, which we call definability preservation, which does just that, i.e., we require that $\mu(M(\phi))$—or $\mu(M(T))$—is a definable model set. The importance of this condition is shown by the following, simple example, first given in Schlechta (1992). It shows that condition (PR), see Table 1, "Logical rules", may fail in models which are not definability preserving, although it does hold when the model structure is definability preserving.

Example 3.2 Let the language be defined by $\{p_i : i \in \omega\}$, and M the set of its models.
Let $n, n' \in M$ be defined by $n \models \{p_i : i \in \omega\}$, $n' \models \{\neg p_0\} \cup \{p_i : 0 < i < \omega\}$.
Let $\mathcal{M} := \langle M, \prec \rangle$ where only $n \prec n'$, i.e. just two models are comparable. Let $\mu := \mu_{\mathcal{M}}$, and $\mid\!\sim$ be defined as usual by μ.
Set $T := \emptyset$, $T' := \{p_i : 0 < i < \omega\}$. We have $M(T) = M$, $\mu(M(T)) = M - \{n'\}$, $M(T') = \{n, n'\}$, $\mu(M(T')) = \{n\}$. So $\mathcal{M}$ is not definability preserving, and, furthermore, $\overline{\overline{T}} = \overline{T}$, $\overline{\overline{T'}} = \overline{\{p_i : i < \omega\}}$, *so* $p_0 \in \overline{\overline{T \cup T'}}$, *but* $\overline{\overline{\overline{T}} \cup T'} = \overline{\overline{\overline{T} \cup T'}} = \overline{T'}$, *so* $p_0 \notin \overline{\overline{\overline{T}} \cup T'}$, contradicting (PR). □

Thus, the correspondence between the structural and the algebraic semantics may be perfect, but not the one between the logic and the algebraic semantics. The division into two correspondences brings this to light. Of course, once we have seen it, we will recognize the same phenomenon also elsewhere, e.g., in representation theorems for theory revision. In the infinite case, there is no easy remedy on the logical side: no usual axioms allow us to treat structures which do not preserve definability, we have to take other steps. There are characterizations which allow for "small" exception sets, where small is not by size, cardinality can be arbitrarily big, but small means here "undetectable by the language" — see Schlechta (2004).

3.2 Closure Conditions

We have already sometimes used unions, intersections, and differences of model sets. When we consider classical propositional logic, we have finite unions (as the language has an $''OR''$), as well as infinite intersections when we consider theories, i.e., sets of formulas, too, and complements for formula-defined model sets - but not always for theory-defined model sets. Rules in sequent calculi are to be read implicitly as the conjunction of the formulas on the left implies the disjunction of the formulas on the right. Thus, we might have no OR in the formulas on the left of $\mid\!\sim$, and the language of the left may be different from the language on the right. Consequently, we might not be able to formulate some of the usual conditions, and some of the usual proof steps might not be available either.

In completeness proofs, we sometimes need to pass from, say $\phi \mid\!\sim \psi$ and $\phi' \mid\!\sim \psi'$, to $\phi \vee \phi' \mid\!\sim \psi''$, but this might then be impossible in the general case, when we do not have an OR on the left, so $\mu(X)$ and $\mu(X')$ may be defined, but $\mu(X \cup X')$ need not be defined, so we do not know its value, and cannot work with it. In some cases, this not only hinders proofs, but makes usual completeness conditions impossible. For instance, without closure under finite unions of the domain, the condition of cumulativity splits up into an infinity of different conditions, see Gabbay and Schlechta (2009b), Sect. 4.2.2.3. System *P* (see Definition 7.2) without cumulativity is still sufficient to guarantee representation by preferential structures, but the full system *P* is no longer sufficient for representation by smooth preferential structures, we need the full infinite set of conditions to guarantee representation.

This is apparently in manifest contradiction to results in Kraus et al. (1990) or Lehmann and Magidor (1992); of course, this is not really the case. A closer look reveals that Lehmann and co-authors use (implicitly) domains closed under finite unions, as their language contains "OR". The author has shown in Schlechta (1996) by a counterexample that additional conditions are necessary for a completeness result in the absence of "OR" (or absence of closure under finite unions of model sets). Consequently, the absence of suitable domain closure conditions can have important consequences on representation results. And this was our point. We see similar problems with representation results for more complicated structures, e.g., when we speak about threads of developments, which are not necessarily closed under union.

Thus, domain closure conditions are far from innocent, and have, to the author's opinion, not received enough attention so far.

4 Systems of Sets

4.1 Model Sets Versus Systems of Model Sets

As said in Sect. 1.3, the motivation for systems of sets is perhaps best illustrated by the example of preferential structures. In preferential structures, one considers minimal models, and the formulas which hold there. If we have infinite descending chains, there might not be any minimal model of ϕ, so this definition of preferential consequence is vacuously satisfied, giving $\phi \mid\!\sim FALSE$. Still, it might be reasonable to speak about ever-improving situations, and to use them. Then, we have to work with sets of ever better models. A suitable definition is given in Definition 4.1. Thus, we might not just have $M(\phi)$ and some subset $\mu(M(\phi))$, but $M(\phi)$ and a whole system $\mathcal{X}$ of such model sets, approximating the ideal case.

Once we have such a system, there are many things we can do with it. We can, e.g., consider formulas that hold in all $X \in \mathcal{X}$, or which hold in some $X \in \mathcal{X}$. We might also have a relation on $\mathcal{X}$, and give more weight to the different X according to this relation, etc. When we consider — as above — essentially approximations to an ideal situation (minimal elements), which does not exist, then considering what holds in all $X \in \mathcal{X}$ is probably not justified. What interests us, is somehow what holds when "we get better", or "sufficiently good". (Note that this is approximation *within* one logic, and not approximation of one logic by other logics.) So the existentially quantified version seems in our context more justified.

In a more general situation, we might have a neighbourhood semantics, perhaps without approximation, so just a system $\mathcal{X} \subseteq \mathcal{P}(X)$.

Definition 4.1
(This is a definition for the version without copies.)
Let $\mathcal{M} := \langle U, \prec \rangle$. Define
$Y \subseteq X \subseteq U$ is a minimizing initial segment, or MISE, of X iff:

(a) $\forall x \in X \exists y \in Y . y \preceq x$, where $y \preceq x$ stands for $x \prec y$ or $x = y$ (i.e., Y is minimizing), and
(b) $\forall y \in Y, \forall x \in X (x \prec y \Rightarrow x \in Y)$ (i.e., Y is downward closed or an initial part).

In minimal preferential systems, we are not only interested in *one* $\mu(X)$, but in $\mu(X)$, $\mu(X')$, etc. For a fixed X, we are not only interested in one MISE $Y \subseteq X$, but in several of them, $Y, Y' \subseteq X$, etc.; after all, we are interested in approximations, and just one approximating set is a bit poor. Thus, usually, we will have many $MISE$'s, $Y, Y' \subseteq X$, $Z, Z' \subseteq X'$, etc., and we will call the whole system a system of MISE.

4.1.1 Properties and Conditions of this System

Let $\mathcal{X}$ be a system of MISE (over Z, Z', etc., see below, e.g., in (5)).
We assume for simplicity that all $A \in \mathcal{X}$ are definable.
Let us discuss a few desirable properties:

(1) We first want the resulting system to be consistent, so $\bigcap \mathcal{X}$ should not be $\emptyset$.
(2) We want AND to hold, which is often desirable. The condition to impose is that for $A, A' \in \mathcal{X}$, there is $B \in \mathcal{X}$, $B \subseteq A \cap A'$.
(3) In our approach, classical (right) weakening will hold - this is a direct consequence of the definition.
(4) Obviously, if $M(\phi) = M(\phi')$, then ϕ and ϕ' will have the same consequences, as the systems will be the same.
(5) If for $X \in \mathcal{X}(Z)$, $X' \in \mathcal{X}(Z')$ there is $X'' \in \mathcal{X}(Z \cup Z')$, $X'' \subseteq X \cup X'$, then we will have OR.

(6) If systems are defined over Z, Z', $Z \cap Z'$, $Z - Z'$, then we have the following: If $X \in \mathcal{X}(Z \cap Z')$, then by $Z - Z' \in \mathcal{X}(Z - Z')$ and above union condition there is $X'' \subseteq (Z - Z') \cup X \in \mathcal{X}(Z)$, and we can show finite, but not necessarily infinite, conditionalization.

In preferential structures and the MISE as defined above:

- Properties (1), (5), and (6) above will always hold.
- If the relation $\prec$ is transitive, then (2) above will hold, too.

Note that we can also treat the distance semantics of counterfactual conditionals and theory revision by a limit approach, just as in preferential systems.

Abstractly, we can see such systems of MISE as a neighbourhood system, with the ideal case as the limit.

4.2 *Other Neighbourhood Semantics*

We considered above the formulas which hold in $\mu(X)$, or in one (or all) $X \in \mathcal{X}$. This implies closure under right weakening. Sometimes, this is not wanted. An infamous example is the Ross paradox from deontic logic. It shows that obligations should not be closed under right weakening. Watering the flowers might be an obligation, but it is wrong to deduce from this that it is an obligation to water the flowers or steal in the supermarket. (Note that we can formulate the Ross paradox without additional operators by interpreting $\mid\!\sim$ as describing "good" situations.)

We can solve this problem by considering formulas which hold *exactly* in one of the $X \in \mathcal{X}$. Right weakening will not necessarily hold any more (but the approach will still be robust under equivalent reformulations).

This results in a neighbourhood semantics, where a neighbourhood is a (more or less perfect) approximation to the "morally" ideal situation. Our solution here is, basically, to take all situations corresponding to single obligations (water the plants, do not steal, etc.), and consider their closure under unions and intersections - but *not* set difference.

5 The Theory of Abstract Size

So far, we have described the algebraic semantics as a construction which facilitates representation proofs, makes proof parts easy to re-use, and brings to light more or less subtle problems in the infinite case. We treat now the abstract size semantics as a theory in its own right. Note that, without an intuitive meaning, we have no reasonable way to construct a theory of the function μ, except by generalizing from the structural semantics or abstracting from the logical rules. The example of $K(X)$ for Kripke structures illustrates this—$K(X)$ behaves and should behave very differently from $\mu(X)$, see Sect. 1.2.1. An abstract theory of size will formalize this intuition.

5.1 The Meaning of Defaults

Default theories were introduced by R. Reiter (1980). In his spirit, a default : ϕ means roughly: "if it is consistent to assume ϕ, do so". The possibility of writing default theories of the form $\{: \phi, : \neg\phi\}$ without an outcry of the system seems quite unsatisfactory to the present author. We think that a system of defaults should reflect a property of the world, vaguely, how most things are. And then $\{: \phi, : \neg\phi\}$ just does not make sense, i.e., should be contradictory. This leads to the question what a default should mean more precisely, and to the interpretation—admittedly not very close to Reiter's formalism—of : $\phi(x)$ by "in the important or normal cases, ϕ holds", and, in a more quantitative interpretation (or, giving important cases more weight) to "in a big set of cases, ϕ holds".
As we wanted to be able to treat the lottery paradox:

Example 5.1
Suppose we have a lottery with 1000 cases, it is safe to asssume that any specific case n will not win, but it is safe to assume that one case will win, so we cannot blithely take the conjunction of all safe assumptions.

Formalizing "big" by the classical notion of a filter (and its dual notion "small" by the dual concept of an ideal, with the complementary notion of "medium" for the remaining subsets) was thus impossible, and we used "weak filters", where the intersection of two "big" sets need not be big any more, but it should at least be non-empty — see Definition 1.3. (Recall that the definition of a filter involves closure under finite, or even countable, intersection, leading to "AND" on the consequence side.)

This solved the initial problem, the default theory $\{: \phi, : \neg\phi\}$ was inconsistent in this reformulation. The idea was first used and published for a semantics for inheritance networks,see Schlechta (1990), and then, and mainly, used in a first order framework, with sound and complete axiomatization using a new, generalized quantifier, ∇, see Schlechta (1995). Of course, the usual rule of AND requires the filter property, more precisely, that any finite intersections of big sets contains a big set again.

5.2 Additive Laws

Recall that we have systems of "big" subsets - for each (definable) model set X, a system of big subsets, generated by $\mu(X)$. But these systems are not totally independent. Under the small/big set perspective, there are some very natural "coherence conditions". For instance, if $A \subseteq B \subseteq C$, and A is considered big in C, then it should *a fortiori* be considered big in B, as we are improving the proportion of the "good" elements. This is a coherence property, as we change the reference or base set, here from C to B. But there are less obvious, and also more doubtful coherence properties, such as the one corresponding to Cumulativity. We change language slightly, and speak now about "small" sets instead of big ones, this seems more natural here. Suppose $A, B \subseteq C$ are small, and $A \cap B = \emptyset$ (for simplicity). Is A still small in $C - B$? This is not so obvious, as we made the reference set smaller, so A is "less smaller" in $C - B$ than it is in C. Cumulativity gives a positive answer. If we look closer, we see that, by this rule, if we change the base set by a small quantity, proportions stay the same. It is important to note here that the "small" hidden in $\phi \mid\!\sim \psi$ (the set of $\neg\psi$-cases among the ϕ cases is small) is conceptually different from the "small" of base set change. (The similar use hints perhaps at a deeper common trait than discovered so far.)

When we examine these things in detail, we are led to a theory of addition of abstract size (like, for instance, $small + small = small$, corresponding to the basic ideal property), which also allows us, in a very natural way, to discover and examine many different variants of existing rules. In a much more modest way, it is like a periodic table of the elements, where all logical rules find their natural place, ordered by very few ruling principles. We explain some cases of Table 1.

- (Opt) means that we cannot be smaller than $\emptyset$, or bigger than the base set, consequently, $\alpha \vdash \beta$ implies $\alpha \mid\!\sim \beta$, the exception set is $\emptyset$.
- Increasing proportions, i.e., the proportion of positive cases increases: (iM) says that subsets of small sets are small, so, if $\alpha \mid\!\sim \beta$, and $\beta \vdash \beta'$, $\alpha \mid\!\sim \beta'$, as the exception set can only become smaller. (eMI) expresses that, if $A \subseteq X$ is small, and $X \subseteq Y$, then $A \subseteq Y$ is small, too.
- Keeping proportions: $(I \cup disj)$ says that, if $A \subseteq X$ and $B \subseteq Y$ are small, and $X \cap Y = \emptyset$, then $A \cup B \subseteq X \cup Y$ is small. The condition $X \cap Y = \emptyset$ is important, as, otherwise, the "positive part" of both may be the intersection, decreasing the proportion of positive elements.
- Robustness of proportions is expressed by the main filter/ideal property: finite unions of small sets are still small. For instance, in the rules AND, if $\alpha \wedge \neg\beta$ and $\alpha \wedge \neg\beta'$ are small, then so is $\alpha \wedge (\neg\beta \vee \neg\beta')$, so, if $\alpha \mid\!\sim \beta$ and $\alpha \mid\!\sim \beta'$, then $\alpha \mid\!\sim \beta \wedge \beta'$.
- Robustness of $\mathcal{M}^+$ expresses the following: If A has at least medium size in X, and X at least medium size in Y, then A has at least medium size in Y.

5.3 *Multiplicative Laws*

The rule of rational monotony sticks out of this picture. Basically, it says, even if we change the base set by a "medium size" subset, small stays small. (Note that, in this

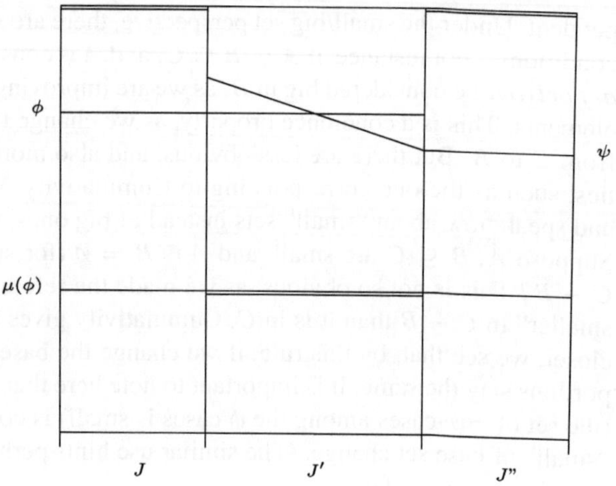

Fig. 1 Non-monotonic interpolation, $\phi \mid\sim \alpha \mid\sim \psi$ Double lines: interpolant $\Pi J \times (\mu(\phi) \upharpoonright J') \times \Pi J''$ Alternative interpolants (in center part): $\phi \upharpoonright J'$ or $(\phi \wedge \psi) \upharpoonright J'$ (Π denotes the Cartesian product)

definition, rational monotony becomes a Horn rule, too, when we have "medium" at our disposal, reflecting the fact that "medium" can be seen as the complement of the union of two positive properties, i.e., as "neither big nor small".)

So, in the above picture, if $A, B \subseteq C$, A is small in C, B is medium size in C, $A \cap B = \emptyset$, then A will still be small in $C - B$. But this property does not fit in well with the other additive rules. It fits in well with multiplicative laws of abstract size. Such multiplicative laws have, however, an even more interesting aspect for interpolation and independence in the following sense: Take two disjoint sublanguages, L and L', let X be a set of models in L, X' in L', and consider now the product $X \times X'$ in the language $L \cup L'$. Suppose $Y \subseteq X$ is big, and $Y' \subseteq X'$ is big, is then $Y \times Y'$ big in $X \times X'$? If this holds, this is a multiplicative law about big sets. We have shown that interpolation for non-monotonic logics is intricately linked to such multiplicative laws, as follows, see Gabbay and Schlechta (2011).

Definition 5.1 $(\mu * 1)$ $\mu(X \times X') = \mu(X) \times \mu(X')$

Proposition 5.1 $(\mu * 1)$ entails interpolation in preferential logic in the following sense:
If $\phi \mid\sim \psi$, then there is α which can be expressed in the common language of ϕ and ψ, such that $\phi \mid\sim \alpha$ and $\alpha \mid\sim \psi$.

The idea of the proof is illustrated in the following diagram:

We argue semantically, whether the model sets can be described also syntactically is another question, and not addressed here. We abbreviate $\mu(M(\phi))$ by $\mu(\phi)$. The language is defined by the disjoint sets of propositional variables $J \cup J' \cup J''$. If X is a model set, we denote the projection of X on J by $X \upharpoonright J$, etc.

The prerequisite is $\phi \mid\sim \psi$, so $\mu(\phi) \subseteq M(\psi)$. Moreover, we assume that ϕ uses only variables in $J \cup J'$, ψ only variables in $J' \cup J''$. As $M(\phi)$ can be written $M(\phi) \upharpoonright (J \cup J') \times TRUE \upharpoonright J''$, and by $(\mu * 1)$, we have $\mu(\phi) \subseteq \mu(\phi) \upharpoonright$

$(J \cup J') \times \mu(TRUE) \restriction J''$, so $\mu(TRUE) \restriction J'' \subseteq \mu(\psi) \restriction J''$. By the same argument, $\mu(\phi) \restriction J' \subseteq M(\psi) \restriction J'$. Consider now $\alpha := TRUE \restriction J \times \mu(\phi) \restriction J' \times TRUE \restriction J''$, then $\mu(TRUE \restriction J \times \mu(\phi) \restriction J' \times TRUE \restriction J'') \subseteq M(\psi)$, as required.

In Fig. 1, the set of models of ϕ is described by the horizontal line labelled "ϕ", continued on J'' arbitrarily (ϕ has no variables in J''). The ψ-models are described by the corresponding line, and continued arbitrarily on J. The common variables are in J'. $\mu(\phi)$ is described by the lower line, labelled $\mu(\phi)$, by prerequisite, it is a subset of $M(\psi)$. The U-shaped set of $M(\alpha)$, with $\alpha := (TRUE \restriction J) \times (\mu(\phi) \restriction J') \times (TRUE \restriction J'')$, described by the double lines, is an interpolant, as we can calculate $\mu(\alpha)$ componentwise.

But such laws also bring to light operations one might take for granted, and which are not innocent. Suppose again $Y \subseteq X$ is big, where X and Y are model sets in a given language L. We now take $X' = Y'$ the set of all models in a new language L'. Is the combined model set $Y \times Y'$ big in $X \times X'$? In other words, is size robust to the change from sublanguage to superlanguage, as long as we do not touch the elements of the new language? The natural reaction is probably, yes, of course — still, it has to be made explicit. These problems are problems of independence, as they were, e.g., discussed by Parikh and his co-authors, see, e.g., Chopra and Parikh (2000). In our opinion, further research is needed here.

5.4 Other Approaches to Abstract Size

B.van Fraasen and D. Makinson, see van Fraasen (1976) and Makinson (2011), have considered a binary operator $p(.,.)$ in the domain of logic. B.van Fraasen uses the reals and their multiplication, and thus goes beyond things we want to commit to here. D. Makinson discusses a "proto-probability function", which assigns to two formulas in a given language a value $p(\alpha, \beta)$ in a set D equipped with a relation $\leq$, which is transitive, complete ($d \leq e$ or $e \leq d$ for all $d, e \in D$), and has distinct greatest and least elements 1 and 0. It is more general, as the set D might have more elements than our $\{small, big, medium\}$ (with perhaps $\emptyset$ and $1 = All$ added), and less general, as we do not only consider formula-defined sets. Still, the idea is the same, and the translation is straightforward, once we put our values into the set D : $p(A, X) = x$ iff $A \subseteq X$ has size x relative to X. In Table 2, the connection to our own approach is pointed out.

We should also mention here the work by Ben-David, Halpern, and others, see Ben-David and Ben-Eliyahu (1994) and Friedman and Halpern (1995).

5.5 Size Operators in the Language

As mentioned, the ∇ operator discussed above in Sect. 1.4 can be used to import size and non-monotonic consequence into the object language, by the definition $\phi \mid\!\sim \psi$ iff $(\nabla\phi) \to \psi$, where $\to$ is classical implication. This has several advantages:

(1) We have contraposition. If $(\nabla\phi) \to \psi$, then $\neg\psi \to \neg\nabla\phi$, this makes sense. If $\phi \mid\!\sim \psi$ holds, and $\neg\psi$ holds, then either ϕ is false, or ϕ is correct, but $\nabla\phi$ fails.
(2) We can nest the operator, e.g., $(\nabla\phi) \to ((\nabla\phi') \to \psi)$ makes sense, it is a formula like any other formula. (This was also the case with our FOL quantifier ∇, see Sect. 7.5.)
(3) We can also formulate and justify a limited form of transitivity: If $(\nabla\phi) \to \nabla\psi$, and $(\nabla\psi) \to \sigma$, then, of course, $(\nabla\phi) \to \sigma$ - and we see exactly why general transitivity may fail, it is the difference between $(\nabla\phi) \to \nabla\psi$ and $(\nabla\phi) \to \psi$.
(4) We can negate $\mid\!\sim$: $\neg((\nabla\phi) \to \psi)$, although this follows trivially from the exact correspondence of models and axioms, it might be important to give the user this possibility, as it might reveal an inconsistency which he is not aware of.

6 The Role of Copies

This section is particularly destined to readers which are interested in the "machinery behind the scene". It discusses how the expressive power of a semantic structure can be increased by allowing copies (or non-injective valuation functions). Readers who are more interested in more tangible, "on the scene" results are referred to, e.g., the work by M. Freund, see Freund (1993).

In the usual treatment of preferential structures, copies or an equivalent construction are used, see Definition 7.5. (Most authors work with valuation functions instead of copies. Both definitions — copies and valuation functions — are equivalent; a copy $\langle x, i\rangle$ can be seen as a state $\langle x, i\rangle$ with valuation x.) This is in the tradition of, e.g., modal logic, where a valuation function need not be injective; or, in other words, states can be logically the same, but not necessarily structurally, too. Tradition is, however, not a sufficient justification.

Copies in preferential structures may code our ignorance, i.e., we know that x is minimized by X, but we do not know by which element of X, they are in this view artificial. But, on the other hand, they have an intuitive justification: they allow minimization by sets of other elements. If we have no copies, and $x \in X - \mu(X)$, then there is $y \in X$ such that $x \notin \mu(Y)$ if $y \in Y$. One single element suffices. We can interpret this as follows: We may consider an element m only abnormal in the presence of several other elements together. E.g., considering penguins, nightingales, woodpeckers, ravens, they all have some exceptional qualities, so we may perhaps not consider a penguin more abnormal than a woodpecker, etc., but when we see all these birds together, the penguin stands out as the most abnormal one. But we cannot code minimization by a set, without minimization by its elements, without the use of copies. Copies will then code the different aspects of abnormality. In our example, a nightingale is abnormal by its songs, a raven is more normal, a woodpecker by its beak, etc., but none of the others is abnormal in so many aspects.

We take a second look at their combinatorial consequences. First, with or without copies, we may have $x \prec y$ and $x' \prec y$. Thus, the presence of x or of x' together with y in a set X assures that y cannot be minimal in X. So, we have an "OR", one of x or x' suffices. With copies, we can also code an AND: If $\langle y, 0\rangle$ and $\langle y, 1\rangle$ are two copies, and we have $x \prec \langle y, 0\rangle$ and $x' \prec \langle y, 1\rangle$ (but no other elements smaller than the y-copies), then we need both x and x' to make y non-minimal. So we also have an AND. This property, presence of OR and AND, is used in the author's basic construction for representation: We consider elements of the form $\langle x, f\rangle$, where $f \in \Pi\{X : x \in X - \mu(X)\}$, with the relation $\langle x, f\rangle \prec \langle x', f'\rangle$ iff $x \in range(f')$. The only remaining property is the coherence condition $X \subseteq Y \Rightarrow \mu(Y) \cap X \subseteq \mu(X)$, expressing the fact that what is not minimal in X cannot be minimal in Y if $X \subseteq Y$. (In reactive structures, see Gabbay and Schlechta (2011), where we can not only attack points, but also arrows (or the $\prec$ between elements), we also have negation, and, as a matter of fact, we can represent *any* choice function as long $\mu(X) \subseteq X$—no coherence conditions need to hold.)

7 Appendix

7.1 Appendix for Section 2

Definition 7.1

(1) A preferential structure $\mathcal{M} = \langle U, \prec\rangle$ is called smooth iff for all X and all $x \in X$ x is either minimal in X or above an element which is minimal in X. More precisely: If $x \in X$, then either $x \in \mu(X)$ or there is $x' \in \mu(X).x' \prec x$.

(2) The relation $\prec$ is called ranked iff, for all a, a' which are incomparable, i.e., $a \not\prec a'$ and $a' \not\prec a$, and all b, we have: $a \prec b$ iff $a' \prec b$, and $b \prec a$ iff $b \prec a'$. This is almost a total order, where the classes of incomparable elements are totally ordered.

(3) $\mathcal{M}$ will be called definability-preserving or dp, iff for all sets of formulas T, $\mu(M(T))$ is again exactly the $M(T')$ for some set of formulas T'. This is an additional property, and indeed quite a strong one.

A remark for the intuition: Smoothness is perhaps best motivated through Gabbay's concept of reactive diagrams; see, e.g., Gabbay (2004) and Gabbay (2008), and also Gabbay and Schlechta (2009a), Gabbay and Schlechta (2009b). In this concept, smaller, or "better", elements attack bigger, or "less good", elements. But when a attacks b, and b attacks c, then one might consider the attack of b against c weakened by the attack of a against b. In a smooth structure, for every attack against some element x, there is also an uncontested attack against x, as it originates in an element y, which is not attacked itself.

7.2 Appendix for Section 2.2

Definition 7.2 The definitions of the rules are given here or in Table 1, "Logical rules, definitions and connections" which also show connections between different versions of rules, the semantics, and rules about size.

Explanation of the table:

(1) The difference between the first two columns is that the first column treats the formula version of the rule, the second the more general theory (i.e., set of formulas) version.
(2) "Corr." stands for "Correspondence".
(3) The third column, "Corr.", is to be understood as follows:
Let $\boldsymbol{D}$ be the set of definable model sets, i.e., the set of all $M(T)$ which are definable by some formula set T. Let a logic $\mid\!\sim$ satisfy (LLE) and (CCL), and define a function $f : \boldsymbol{D} \to \boldsymbol{D}$ by $f(M(T)) := M(\overline{\overline{T}})$. Then f is well defined and satisfies (dp), and $\overline{\overline{T}} = Th(f(M(T)))$.
If $\mid\!\sim$ satisfies a rule on the left-hand side, then — provided the additional properties noted in the middle for $\Rightarrow$ hold, too — f will satisfy the property on the right-hand side.
Conversely:
If $f : \mathcal{Y} \to \mathcal{P}(M_{\mathcal{L}})$ is a function, with $\boldsymbol{D} \subseteq \mathcal{Y}$, and we define a logic $\mid\!\sim$ by $\overline{\overline{T}} := Th(f(M(T)))$, then $\mid\!\sim$ satisfies (LLE) and (CCL). If f is definability-preserving, abbreviated (dp), then $f(M(T)) = M(\overline{\overline{T}})$.
If f satisfies a property on the right-hand side, then — provided the additional properties noted in the middle for $\Leftarrow$ hold, too — $\mid\!\sim$ will satisfy the property on the left-hand side.
(4) We use the following abbreviations for those supplementary conditions in the "Correspondence" columns:
"$T = \phi$" means that if one of the theories (the one named the same way in the present definition) is equivalent to a formula, we do not need (dp).
$-(dp)$ stands for "without (dp)".

Further comments: ((CP) stands for consistency preservation, $T \not\vdash \bot \Rightarrow T \not\mid\!\sim \bot$. (CM) is the more interesting half of (CUM). (CCL) stands for "classically *closed*".)

(1) (*PR*) is also called infinite conditionalization. We choose this name for its central role for preferential structures (*PR*) or (μPR).
(2) The system of rules (AND) (OR) (LLE) (RW) (SC) (CP) (CM) (CUM) is also called system P (for preferential). Adding (*RatM*) gives the system *R* (for rationality or rankedness).
Roughly, smooth preferential structures generate logics satisfying system *P*, while ranked structures generate logics satisfying system R.
(3) (*LLE*) and (*CCL*) will hold automatically, whenever we work with model sets.

(4) (AND) is obviously closely related to filters, and corresponds to closure under finite intersections. (RW) corresponds to upward closure of filters.
More precisely, validity of both depend on the definition, and the direction we consider.
Given f and $(\mu \subseteq)$, $f(X) \subseteq X$ generates a principal filter, $\{X' \subseteq X : f(X) \subseteq X'\}$, with the definition, if $X = M(T)$, then $T \mid\!\sim \phi$ iff $f(X) \subseteq M(\phi)$. Validity of (AND) and (RW) are then trivial.
Conversely, we can define for $X = M(T)$
$\mathcal{X} := \{X' \subseteq X : \exists\phi(X' = X \cap M(\phi)$ and $T \mid\!\sim \phi)\}$.
(AND) then makes $\mathcal{X}$ closed under finite intersections, and (RW) makes $\mathcal{X}$ upward closed. In the infinite case this is usually not yet a filter, as not all subsets of X need to be definable this way. In this case, we complete $\mathcal{X}$ by adding all X'' such that there is $X' \subseteq X'' \subseteq X$, $X' \in \mathcal{X}$.

(5) (SC) corresponds to the choice of a subset.

(6) (CP) is somewhat delicate, as it presupposes that the chosen model set is non-empty. This might fail in the presence of ever better choices, without ideal ones; the problem is addressed by the limit versions.

(7) (PR) is an infinitary version of one half of the deduction theorem: Let T stand for ϕ, T' for ψ, and $\phi \wedge \psi \mid\!\sim \sigma$, so $\phi \mid\!\sim \psi \rightarrow \sigma$, but $(\psi \rightarrow \sigma) \wedge \psi \vdash \sigma$.

(8) (CUM) (whose more interesting half in our context and by our motivation is (CM)) may best be seen as a normal use of lemmas: We have worked hard and found some lemmas. Now we can take a rest, and come back again with our new lemmas. Adding them to the axioms will neither add new theorems, nor prevent old ones from holding.

7.3 Appendix for Section 2.3

Proposition 7.1 We are cheating slightly, as the completeness parts of this proposition are valid only for the version with copies, see below in Sect. 6. Moreover, we neglect conditions on the domain here.

(1) A choice function μ is representable by a preferential structure, i.e., there is a preferential sturcture $\mathcal{M}$ such that $\mu = \mu_{\mathcal{M}}$, iff the following two conditions hold:
$(\mu \subseteq)$ $\mu(X) \subseteq X$
(μPR) $X \subseteq Y \Rightarrow \mu(Y) \cap X \subseteq \mu(X)$
The structure can be chosen transitive.
The structure can be chosen smooth, iff, in addition, the following holds:
(μCUM) $\mu(X) \subseteq Y \subseteq X \Rightarrow \mu(X) = \mu(Y)$
The structure can be chosen ranked, iff, in addition, the following holds:
(μRM) $X \subseteq Y$, $\mu(Y) \cap X \neq \emptyset \Rightarrow \mu(X) = \mu(Y) \cap X$

(2) We give the translation into logic right away:
A logic $\mid\!\sim$ can be represented by a dp pref. structure iff
(LLE) $\overline{T} = \overline{T'} \Rightarrow \overline{\overline{T}} = \overline{\overline{T'}}$

Table 2 Rules on size

		Rules on size	
	"Ideal", $\mathcal{M}^+$	$p(.,.)$ notation	Rules
		Optimal proportion	
(*Opt*)	$\emptyset \in \mathcal{I}(X)$	$p(\emptyset, X) = 0$	(*SC*): $\alpha \vdash \beta \Rightarrow \alpha \mid\!\sim \beta$
		Monotony (Improving proportions). (*iM*): internal monotony, (*eM$\mathcal{I}$*): external monotony for ideals, (*eM2*): external monotony for $p(.,.)$	
(*iM*)	$A \subseteq B \in \mathcal{I}(X) \Rightarrow A \in \mathcal{I}(X)$	$A \subseteq B \subseteq X, p(B,X) = s \Rightarrow p(A,X) = s$	(*RW*): $\alpha \mid\!\sim \beta, \beta \vdash \beta' \Rightarrow \alpha \mid\!\sim \beta'$
(*eM$\mathcal{I}$*)	$X \subseteq Y \Rightarrow \mathcal{I}(X) \subseteq \mathcal{I}(Y)$	$A \subseteq X \subseteq Y, p(A,X) = s \Rightarrow p(A,Y) = s$	(*wOR*): $\alpha \mid\!\sim \beta, \alpha' \vdash \beta \Rightarrow \alpha \vee \alpha' \mid\!\sim \beta$ (*μwOR*): $\mu(X \cup Y) \subseteq \mu(X) \cup Y$
(*eM2*)		$p(X,A) \leq p(X,B), A \cap B = \emptyset \Rightarrow p(X,A) \leq p(X, A \cup B) \leq p(X,B)$	
		Keeping proportions	
(≈)	$(\mathcal{I} \cup disj)$: $A \in \mathcal{I}(X), B \in \mathcal{I}(Y), X \cap Y = \emptyset \Rightarrow A \cup B \in \mathcal{I}(X \cup Y)$	$A, B \subseteq X, p(A,X) = s, p(B,X) = s, A \cap B = \emptyset \Rightarrow p(A \cup B) = s$	(*disjOR*): $\alpha \mid\!\sim \beta, \alpha' \mid\!\sim \beta', \alpha \vdash \neg\alpha', \Rightarrow \alpha \vee \alpha' \mid\!\sim \beta \vee \beta'$ (*μdisjOR*): $X \cap Y = \emptyset \Rightarrow \mu(X \cup Y) \subseteq \mu(X) \cup \mu(Y)$
		Robustness of proportions: $n * small \neq All$	
$(< \omega * s)$	$(\mathcal{I}_\omega)$ $A, B \in \mathcal{I}(X) \Rightarrow A \cup B \in \mathcal{I}(X)$	$A, B \subseteq X, p(A,X) = s, p(B,X) = s \Rightarrow p(A \cup B) = s$	(*AND*): $\alpha \mid\!\sim \beta, \alpha \mid\!\sim \beta' \Rightarrow \alpha \mid\!\sim \beta \wedge \beta'$ (*OR*): $\alpha \mid\!\sim \beta, \alpha' \mid\!\sim \beta \Rightarrow \alpha \vee \alpha' \mid\!\sim \beta$ (*CM*): $\alpha \mid\!\sim \beta, \alpha \mid\!\sim \beta' \Rightarrow \alpha \wedge \beta \mid\!\sim \beta'$ (*μOR*): $\mu(X \cup Y) \subseteq \mu(X) \cup \mu(Y)$ (*μCM*): $\mu(X) \subseteq Y \subseteq X \Rightarrow \mu(Y) \subseteq \mu(X)$
		Robustness of $\mathcal{M}^+$	
$(\mathcal{M}^{++})$	$A \in \mathcal{M}^+(X), X \in \mathcal{M}^+(Y) \Rightarrow A \in \mathcal{M}^+(Y)$	$A \subseteq X \subseteq Y, p(A,X) = m, p(X,Y) = m \Rightarrow p(A,Y) = m$	(*RatM*): $\alpha \mid\!\sim \beta, \alpha \not\mid\!\sim \neg\beta' \Rightarrow \alpha \wedge \beta' \mid\!\sim \beta$ (*μRatM*): $X \subseteq Y, X \cap \mu(Y) \neq \emptyset \Rightarrow \mu(X) \subseteq \mu(Y) \cap X$

(CCL) $\overline{\overline{T}}$ is classically closed
(SC) $T \subseteq \overline{\overline{T}}$
(PR) $\overline{\overline{T \cup T'}} \subseteq \overline{\overline{\overline{\overline{T}} \cup T'}}$.
The structure can be chosen transitive.
The structure can be chosen smooth, iff, in addition,
(CUM) $T \subseteq \overline{\overline{T'}} \subseteq \overline{\overline{T}} \Rightarrow \overline{\overline{T}} = \overline{\overline{T'}}$
holds.
The structure can be chosen ranked, iff, in addition, the following holds:
(RatM) $T \vdash T', Con(\overline{\overline{T'}} \cup T) \Rightarrow \overline{\overline{\overline{\overline{T'}} \cup T}} \subseteq \overline{\overline{T}}$

7.4 Appendix to Section 5.2

The purpose of this table is mainly to connect the author's concept of size to the $p(.,.)$ notation of D. Makinson, see Makinson (2011). Rule $(eM2)$ (called disjunctive interpolation in Makinson (2011)) is given only for the $p(.,.)$ notation, as writing down the different cases for filters and logical rules is (easy, but) cumbersome. Table 1 and Table 2 are best read together, but they are too big to fit on one page.

Recall that $'' = s''$ means "is small", and "m" stands for "medium size".

7.5 Appendix for Section 5.5

Definition 7.3 Let $\mathcal{L}$ be a first order language, and M be a $\mathcal{L}$-structure. Let $\mathcal{N}(M)$ be a weak filter, or $\mathcal{N}$-system - $\mathcal{N}$ for normal - over M. Define $\langle M, \mathcal{N}(M)\rangle \models \phi$ for any $\nabla - \mathcal{L}-$formula inductively as usual, with one additional induction step:
$\langle M, \mathcal{N}(M)\rangle \models \nabla x\phi(x)$ iff there is $A \in \mathcal{N}(M)$ s.t. $\forall a \in A\ (\langle M, \mathcal{N}(M)\rangle \models \phi[a])$.

Definition 7.4 Let any axiomatization of predicate calculus be given. Augment this with the axiom schemata

(1) $\nabla x\phi(x) \wedge \forall x(\phi(x) \rightarrow \psi(x)) \Rightarrow \nabla x\psi(x)$,
(2) $\nabla x\phi(x) \Rightarrow \neg\nabla x\neg\phi(x)$,
(3) $\forall x\phi(x) \Rightarrow \nabla x\phi(x)$ and $\nabla x\phi(x) \Rightarrow \exists x\phi(x)$,
(4) $\nabla x\phi(x) \leftrightarrow \nabla y\phi(y)$ if x does not occur free in $\phi(y)$ and y does not occur free in $\phi(x)$.(for all ϕ, ψ).

7.6 Appendix for Section 6

Definition 7.5

(1) A pair $\mathcal{M} := \langle \mathcal{U}, \prec\rangle$ with $\mathcal{U}$ an arbitrary set of pairs and $\prec$ an arbitrary binary relation on $\mathcal{U}$ is called a preferential model or structure.
If $\langle x, i\rangle \in \mathcal{U}$, then x is intended to be an element of U, and i the index of the copy.
(2) We sometimes also need copies of the relation $\prec$. We will then replace $\prec$ by one or several arrows α attacking non-minimal elements, e.g., $x \prec y$ will be written $\alpha : x \rightarrow y$, $\langle x, i\rangle \prec \langle y, i\rangle$ will be written $\alpha : \langle x, i\rangle \rightarrow \langle y, i\rangle$, and finally we might have $\langle \alpha, k\rangle : x \rightarrow y$ and $\langle \alpha, k\rangle : \langle x, i\rangle \rightarrow \langle y, i\rangle$.
(3) Let $\mathcal{M} := \langle \mathcal{U}, \prec\rangle$ be as above, and let $X \subseteq U$. Define
$\mu_{\mathcal{M}}(X) := \{x \in X : \exists\langle x, i\rangle \in \mathcal{U}.\neg\exists\langle x', i'\rangle \in \mathcal{U}(x' \in X \wedge \langle x', i'\rangle' \prec \langle x, i\rangle)\}$.
Thus, $\mu_{\mathcal{M}}(X)$ is the projection on the first coordinate of the set of elements such that there is no smaller one in X.
Again, by abuse of language, we say that $\mu_{\mathcal{M}}(X)$ is the set of minimal elements of X in the structure. If the context is clear, we write just μ.
We sometimes say that $\langle x, i\rangle$ "kills" or "minimizes" $\langle y, j\rangle$ if $\langle x, i\rangle \prec \langle y, j\rangle$. We also say a set X kills or minimizes a set Y if for all $\langle y, j\rangle \in \mathcal{U}$, $y \in Y$ there is $\langle x, i\rangle \in \mathcal{U}$, $x \in X$ s.t. $\langle x, i\rangle \prec \langle y, j\rangle$.
$\mathcal{M}$ is also called injective or 1-copy iff there is always at most one copy $\langle x, i\rangle$ for each x. Note that the existence of copies corresponds to a non-injective labelling function — as is often used in nonclassical logic, e.g., modal logic.

(4) We define the consequence relation of a preferential structure for a given propositional language $\mathcal{L}$.
If m is a classical model of a language $\mathcal{L}$, we say by abuse of language
$\langle m, i\rangle \models \phi$ iff $m \models \phi$,
and if X is any set of such pairs, that
$X \models \phi$ iff for all $\langle m, i\rangle \in X$ $m \models \phi$.
(5) A structure $\mathcal{M}$ as above will be called smooth iff:
If $x \in X \in \mathcal{Y}$, and $\langle x, i\rangle \in \mathcal{U}$, then either there is no $\langle x', i'\rangle \in \mathcal{U}$, $x' \in X$, $\langle x', i'\rangle \prec \langle x, i\rangle$ or there is a $\langle x', i'\rangle \in \mathcal{U}$, $\langle x', i'\rangle \prec \langle x, i\rangle$, $x' \in X$, s.t. there is no $\langle x'', i''\rangle \in \mathcal{U}$, $x'' \in X$, with $\langle x'', i''\rangle \prec \langle x', i'\rangle$.
(6) We see that smoothness does a bit too much, as it looks at all copies, irrespective of whether any copy will survive or not, we therefore modify the definition, and add $x \notin \mu(X)$ to the prerequisite. We give the full definition, as we think that this is the "right" definition of smoothness.
A structure $\mathcal{M}$ as above will be called essentially smooth iff:
If $x \in X \in \mathcal{Y}$, $x \notin \mu(X)$, and $\langle x, i\rangle \in \mathcal{U}$, then either there is no $\langle x', i'\rangle \in \mathcal{U}$, $x' \in X$, $\langle x', i'\rangle \prec \langle x, i\rangle$ or there is a $\langle x', i'\rangle \in \mathcal{U}$, $\langle x', i'\rangle \prec \langle x, i\rangle$, $x' \in X$, s.t. there is no $\langle x'', i''\rangle \in \mathcal{U}$, $x'' \in X$, with $\langle x'', i''\rangle \prec \langle x', i'\rangle$.

(This is the author's notation, other people use labelling functions, which are then not injective.)

Example 7.1 It is easy to see that we can distinguish plain (total) and essential smoothness, as this example shows:
Let e.g., $a \prec b \prec \langle c, 0\rangle, \langle c, 1\rangle$, without transitivity. Thus, only c has two copies. This structure is essentially smooth, but of course not totally so.
A similar example can be constructed with a transitive relation and infinite descending chains.

Acknowledgments The author would like to thank David Makinson for very valuable advice, and two anonymous referees who helped to make the present article clearer and more readable.

References

Ben-David, S., & Ben-Eliyahu, R. (1994). A modal logic for subjective default reasoning. In S. Abramsky (Ed.), *Proceedings LICS-94* (pp. 477–486), July 1994, Paris, France: IEEE Computer Science Press.

Chopra, S., & Parikh, R. (2000). Relevance sensitive belief structures. *Annals of Mathematics and Artificial Intelligence*, *28*(1–4), 259–285.

Friedman, N., & Halpern, J. (1995). *Plausibility Measures and Default Reasoning*. IBM Almaden Research Center Tech. Rept..

Freund, M. (1993). Injective models and disjunctive relations. *Journal of Logic and Computation*, *3*(3), 231–247.

Gabbay, D., & Schlechta, K. (2009a). Roadmap for preferential logics. *Journal of Applied Non-classical Logic19/1*, 43–95 (see also hal-00311941, arXiv 0808.3073).

Gabbay, D., & Schlechta, K. (2009b). *Logical tools for handling change in agent-based systems*. Berlin: Springer.
Gabbay, D., & Schlechta, K. (2009c). Independence and abstract multiplication. In preparation.
Gabbay, D., & Schlechta, K. (2011). *Conditionals and modularity in general logics*. Heidelberg, August: Springer. ISBN 978-3-642-19067-4, preliminary version in arxiv.org.
Gabbay, D. M. (2004). Reactive Kripke semantics and arc accessibility. In W. Carnielli, F. M. Dionesio, & P. Mateus, (Eds.), *Proceedings CombLog04*, July 28–30 (pp. 7–20). Centre of Logic and Computation, University of Lisbon.
Gabbay, D. M. (2008). Reactive Kripke models and contrary to duty obligations, DEON-2008, Deontic Logic in Computer Science, July 15–18, 2008, Luxembourg. In R. V. D. Meyden, & L. van der Torre, (Eds.), LNAI 5076, pp. 155–173. Berlin: Springer.
Gabbay, D. M. (1989). Theoretical foundations for non-monotonic reasoning in expert systems. In K. R. Apt (Ed.), *Logics and Models of Concurrent Systems*(pp. 439–457). Berlin: Springer.
Gabbay, D. M. (1991). Theoretical foundations for non-monotonic reasoning part 2: Structured non-monotonic theories, in *SCAI*, pp. 19–39.
Kraus, S., Lehmann, D., & Magidor, M. (1990). Nonmonotonic reasoning, preferential models and cumulative logics. *Artificial Intelligence*, *44*(1–2), 167–207.
Lehmann, D., & Magidor, M. (1992). What does a conditional knowledge base entail?. *Artificial Intelligence55*(1), 1–60.
Makinson, D. (2011). Conditional probability in the light of qualitative belief change. *Journal of Philosophical Logic*, *40*, 121–153.
Makinson, D. (1994). General patterns in nonmonotonic reasoning. In D. Gabbay, C. Hogger, & Robinson, (Eds.), *Handbook of Logic in Artificial Intelligence and Logic Programming, Vol. III: Nonmonotonic and Uncertain reasoning*, (pp. 35–110). Oxford University Press.
Reiter, R. (1980). A logic for default reasoning. *Artificial Intelligence*, *13*(1–2), 81–132.
Schlechta, K. (2004). *Coherent systems*. Amsterdam: Elsevier.
Schlechta, K. (1990). Semantics for defeasible inheritance. In L. G. Aiello (Ed.), *Proceedings ECAI 90*, Stockholm, Sweden, 1990, (pp. 594–597). Berlin: Springer.
Schlechta, K. (1992). Some results on classical preferential models. *Journal of Logic and Computation*, 2(6), 675–686.
Schlechta, K. (1995). Defaults as generalized quantifiers. *Journal of Logic and Computation*, *5*(4), 473–494.
Schlechta, K. (1996). Completeness and incompleteness for plausibility logic. *Journal of Logic, Language and Information*, *5*(2), 177–192.
Schlechta, K. (1997). Nonmonotonic logics—basic concepts, results, and techniques. *Springer Lecture Notes series, LNAI*, *1187*, 243.
van Fraasen, B. (1976). Representation of conditional probabilities. *Journal of Philosophical Logic5*, 417–430.

Towards a Bayesian Theory of Second-Order Uncertainty: Lessons from Non-Standard Logics

Hykel Hosni

Abstract Second-order uncertainty, also known as model uncertainty and Knightian uncertainty, arises when decision-makers can (partly) model the parameters of their decision problems. It is widely believed that subjective probability, and more generally Bayesian theory, are ill-suited to represent a number of interesting second-order uncertainty features, especially "ignorance" and "ambiguity". This failure is sometimes taken as an argument for the rejection of the whole Bayesian approach, triggering a Bayes versus anti-Bayes debate which is in many ways analogous to what the *classical versus non-classical* debate used to be in logic. This paper attempts to unfold this analogy and suggests that the development of non-standard logics offers very useful lessons on the contextualisation of justified norms of rationality. By putting those lessons to work I will flesh out an epistemological framework suitable for *extending* the expressive power of standard Bayesian norms of rationality to second-order uncertainty. in a way which is both formally and foundationally conservative.

Keywords Second-order uncertainty · Bayesian epistemology · Admissibility · Imprecise probabilities

---

To appear in Hansson, Sven Ove, (ed.) *David Makinson on classical methods for non-classical problems*. Trends in logic. Springer.

H. Hosni (✉)
Scuola Normale Superiore, Pisa and Centre for Philosophy of Natural and Social Science, London School of Economics, London, UK
e-mail: h.hosni@gmail.com; hykel.hosni@sns.it

S. O. Hansson (ed.), *David Makinson on Classical Methods for Non-Classical Problems*, Outstanding Contributions to Logic 3, DOI: 10.1007/978-94-007-7759-0_11,

1 An Imaginary Opinion Poll

What is the right logic? If logicians worldwide were polled I would imagine the vast majority of answers falling into two categories:

Type 1: "There is no right logic—the question is ill-posed."
Type 2: "Logic *L* is obviously the right logic."

As to the distribution of the answers, I would expect mostly Type 1 responses, a non-negligible proportion of which possibly adding "You should rather be asking *Is logic L adequate for context C?* Those are the questions that logicians today find well-posed and indeed very much worth attacking."

The same opinion poll 30 years ago would have very probably resulted in the opposite outcome, with respondents readily picking their favourite *L*.

Needless to say some (substantial) disagreement is likely to emerge on my projections here. Be it as it may, it is hard to question that over the past few decades the logician's way of thinking about logic has changed. It is too early to say, but it is quite plausible that our way of thinking about logic has changed (and keeps changing) as a consequence of our way of doing logic, especially in connection to computer science, artificial intelligence and cognitive science. Those research areas added a whole new stock of pressing problems which largely contributed to revamping a number of long-standing ones arising from philosophy and linguistics. Those questions, in turn, could be very sharply defined and attacked from the vantage point of an unprecedented mathematical understanding of classical logic, leading to a virtuous circle which provided great momentum for the development of non-standard logics.

Let us now consider a slight variation on our imaginary poll, so that the question now becomes *What is the right measure of uncertainty?* Now I suppose the vast majority of uncertain reasoners would give responses of Type 2, with many adding "why are you asking at all?". In many ways, the current debate on the foundations of uncertain reasoning mirrors what the debate about classical versus non-classical logics used to look like during the first half of the twentieth century, a discussion about who is right and who is hopelessly wrong.

The purpose of this paper is to suggest that uncertain reasoning can greatly benefit from undergoing a change in perspective similar to the one which provided such a favourable context for the development of non-standard logics, some of which today compare in formal depth and philosophical interest to "mathematical logic" as epitomised by Barwise and Keisler (1977). Just as non-standard logics' initial formal development led to a gradual change in foundational perspective, which in turn gave rise to enough formal advance to change our way of thinking about logic, a similar virtuous circle may pave the way for increasingly more expressive formal models of rationality. This will throw new light on the Bayes versus non-Bayes contrast which continues to be the focus of much theoretical work in uncertain reasoning, but rarely provides enough insight to facilitate the much-needed formal advance that the broad area of rational reasoning and decision under uncertainty so very urgently needs.

The paper is organised as follows. Section 2 presents very briefly the Type 2 position in uncertain reasoning. Section 3 focusses on one specific thread in the development of non-standard logics, which I will refer to as the contextualisation of reasoning. This will provide the background against which I will draw an analogy and a contrast between logic and uncertain reasoning. The resulting suggestion will be for uncertain reasoners to abandon positions of Type 2. The second part of the paper is devoted to outlining an epistemological framework which may enable this transition. Section 4 fleshes out the main rationale of such a framework whilst Sect. 5 informally illustrates its applicability to produce conservative extensions of standard Bayesian theory which capture interesting aspects of second-order uncertainty. Section 6 concludes.

2 A Snapshot of the Foundations of Uncertain Reasoning

Bayesian theory abounds with Type 2 positions. Consider for instance de Finetti

> Bayesian standpoint is noways [sic] one among many possible theories but it is an almost self-evident truth, simply and univocally relying on the indisputable coherence rules for probability. (de Finetti 1973, p. 468)

Similar uncautiousness is easily found in connection to Bayesian methods, as opposed to Bayesian theory.[1] Jaynes, for instance, put it as follows:

> The superiority of Bayesian methods is now a thoroughly demonstrated fact in a hundred different areas. One can argue with philosophy; it is not so easy to argue with a computer printout, which says to us: "Independently of all your philosophy, here are the facts of actual performance". (Jaynes 2003, p. xxii)

Interestingly enough, de Finetti and Jaynes endorsed radically different forms of Bayesianism, which prompts a clarification of the use I will make of the adjective Bayesian for the purposes of this work.

Bayesian theory justifies two interdependent norms of individual rationality, namely (i) degrees of belief should be probabilities and (ii) decisions should maximise the subjective expected utility of the outcomes. Anti-Bayesian approaches reject the adequacy of such norms and put forward alternative research programmes. The approach exemplified by Gilboa et al. (2011) shares with Bayesian theory the concern for norms of rationality, but departs from it in that it presupposes the existence of uncertainties which cannot be quantified probabilistically. This line of criticism, which goes back to both Keynes and Knight, albeit in rather distinct forms, is effectively summed up in the opening lines of Schmeidler (1989):

> The probability attached to an uncertain event does not reflect the heuristic amount of information that led to the assignment of that probability. For example, when the information on the occurrence of two events is symmetric they are assigned equal probabilities. If the events are complementary the probabilities will be $1/2$ independent of whether the symmetric information is meager or abundant.

---

[1] For a recent appraisal of the distinction see Gelman (2011).

Gilboa (2009) interprets Schmeidler's observation as expressing a form of "cognitive unease", namely a feeling that subjective probability is unfit as a norm of rational belief. Suppose that some matter is to be decided by the toss of a coin. According to Schmeidler's line of argument, I should prefer tossing my own, rather than some one else's coin on the grounds that, say I have never observed signs of "unfairness" in my coin, whilst I just do not know anything about the stranger's coin. This familiar argument against the completeness of the revealed belief approach[2] goes hand in hand with a radical challenge to probability as a norm of rational belief:

> The main difficulty with [...] the entire Bayesian approach is in my mind the following: for many problems of interest there is no sufficient information based on which one can define probabilities. Referring to probabilities as subjective rather than objective is another symptom of the problem, not a solution thereof. It is a symptom because, were one capable of reasoning one's way to probabilistic assessments, one could have also convinced others of that reasoning and result in a more objective notion of probability. Subjective probabilities are not a solution to the problem: subjectivity [...] does not give us a reason to choose one probability over another. (Gilboa 2009, pp. 130–131)

The Bayesian failure to represent ignorance leads to a plea for rational modesty:

> It is sometimes more rational to admit that one does not have sufficient information for probabilistic beliefs than to pretend that one does. (Gilboa et al. 2011)

Similar lines of argument had been put forward by Shafer (1986). Colyvan (2008) argues against the suitability of probability by suggesting that the so-called *probabilistic excluded middle* misrepresents ignorance to such an extent that it can hardly be taken as a sound principle of uncertain reasoning.[3]

3 An Analogy and a Contrast

This sort of debate is reminiscent, in style and spirit, of the heated disputations[4] over "the right logic", of which the classical versus intuitionistic logic one is certainly the best-known case. The purpose of this section is to put forward a suggestion as to why the success of non-standard logics, and hence of Type 1 mentality, is intimately connected with the progressive loss of appeal of quarrels of that sort.

---

[2] The so-called Ellsberg paradox, of which the coin tossing problem is the simplest example, appears as far back as in Keynes (1921) and Knight (1921). This is often acknowledged in the literature by referring to probabilistically unquantifiable belief as "Knightian uncertainty". The synonym "ambiguity" is due to Ellsberg (1961). Much of the recent revival of the interest in "Knightian decision theory" owes to Bewley (2002).

[3] Roughly speaking, what is objected is that a rational agent must assign probability 1 to any tautology of the form $\theta \vee \neg\theta$, even when the agent knows nothing about θ.

[4] See e.g. Heyting (1956).

3.1 Non-Classical and Non-Standard Logics: A Terminological Remark

In everyday usage the adjective "classical" bears at least two connotations. One refers to something which is traditional. The other refers, in stricter adherence to the Latin etymology, to the class of the "best", the privileged and the important. The expression "classical logic" retains both meanings. Yet I take the current state of research in logic, broadly construed, to support the claim to the effect that "classical logic" is somewhat a misnomer with respect to both connotations.

The history of logic clarifies that "classical logic" is certainly not the invariant to be found in the millennial development of the subject. On the other hand, the mathematical depth of the results and the wealth of applications, from philosophy to computer science to game theory, of whole families of modal, non-monotonic and many-valued logics, stand as obvious evidence to the effect that classical logic is certainly not the only citizen of the "first-class". In light of this, *non-standard logics* appears to be far better a terminology for those logics which arise by making more imaginative uses of the concepts and techniques of classical logic, which is therefore not rejected as ill-founded.

This leads to a general pattern in the construction of non-standard logics. Certain reasoning schemes are not adequately modelled by the classical notion of consequence so that the goal is to define more *realistic* notions of "follows from". In this attempt logicians are often guided by contextually clear modelling needs, which are translated into suitable restrictions on the applicability of classically valid principles of inference. A paramount example of this *extension by restriction* pattern is provided by the logic of defeasible reasoning.

3.2 Context in Action I: Defeasible Reasoning

Consider the naive logical modelling setting in which an agent's reasoning is identified with a consequence relation.[5] The repeated use of a consequence relation is interpreted as determining those formulae (or equivalently, for present purposes, sentences) which the agent is forced to accept—on pain of violating the underlying norms of rationality—given that the agent is accepting a (possibly empty) set of formulae which are interpreted as the premisses of the agent's reasoning.

"Accepting a formula" can be formalised in a number of essentially equivalent ways in classical logic. Say that an agent accepts the formula θ if $v(\theta) = 1$, where $v : \mathcal{L} \to \{0, 1\}$ is the usual notion of valuation which extends uniquely to the set of sentences $S\mathcal{L}$ recursively built from propositional language $\mathcal{L}$. As usual a set

[5] As the content of this section is purely heuristic, I will not burden the reading with otherwise unnecessary definitions. On the general questions of providing rigorous characterisations of logical systems and context of reasoning, Gabbay (1995) is a slightly dated yet still very valuable reference.

$\Gamma \subseteq \mathcal{SL}$ has a model if a valuation v on $\mathcal{L}$ exists such that $v(\gamma) = 1$ for all $\gamma \in \Gamma$. This minimal logical setting leads to the Tarskian analysis of consequence:

> θ is a logical consequence of a set of assumptions Γ, written $\Gamma \models \theta$, if every model of Γ is also a model of θ.

Under the present naive modelling, $\Gamma \models \theta$ can be interpreted as Γ giving the agent reasons to accept θ. Since classical logic is monotonic, an agent whose reasoning is captured by the relation $\models$ must reason monotonically, that is it must satisfy

$$\frac{\Gamma \models \theta}{\Gamma, \psi \models \theta}, \qquad \text{(MON)}$$

where ψ is an arbitrary formula. (MON) can thus be interpreted as saying that ψ does not provide relevant reasons either for or against accepting θ beyond those already available to the agent who accepts Γ. Since ψ is arbitrary[6] we can say something rather stronger, namely that *anything* beyond Γ is actually irrelevant to the acceptance of θ. Thus $\models$ captures a notion of acceptance which is based on having *sufficient reasons*.

There are many situations in which sufficient reasons are never available to the rational agent, no matter how accurate or thoughtful they may turn out to be. Any minimally realistic scenario will feature exogenous, dynamic aspects which make reasoning according to sufficient reasons largely inapplicable. Indeed, with the notable exception of formal deduction, there hardly seems to be a "real-world" context for which *unconstrained* monotonicity can be taken as generally adequate principle of rational reasoning. *This* goes some way towards vindicating the use of the term "classical logic" to denote the logic of formal deduction.

Virtually all the development of defeasible logics, from the early syntactic and modal approaches, to abstract theory of non-monotonic consequence relations, thus focussed on modelling inference based on reasons which are responsive to potentially invalidating refinements of the agent's currently held information. By the end of the 1980s, suitable constraints on the applicability of (MON) were identified—notably *cautious* and *rational* monotonicity[7]—which led to the model-theoretic analysis of defeasible logic as Tarskian consequence nuanced by some suitable minimisation:

> θ is a defeasible logical consequence of a set of assumptions Γ if all minimal models of Γ are also models of θ.

The resulting supraclassical consequence relations extend the expressive power of classical reasoning from "acceptance based on sufficient reasons" to "acceptance based on good reasons" in way which is formally and foundationally conservative.[8]

---

[6] The restriction to single sentences is clearly immaterial here.

[7] The reader who is not familiar with the details may wish to consult Makinson (1994, 2005).

[8] No doubt some meta-mathematical properties are lost in this extension! Supraclassical consequence relations, for instance, need not be closed under substitution (Makinson 2005).

3.3 Context in Action II: Rationality, Coherence and Consistency

The contextual analogy offers also a way of contrasting uncertain reasoning and non-standard logics on the relation between (in)consistency and (ir)rationality. Whilst rational norms of uncertain reasoning appeal to a variety of distinct and often mutually incompatible intuitions about what "rational" actually means,[9] the formal notion of consistency offers much less room for controversy: A set of sentences Γ is *inconsistent* if it logically implies any sentence $\alpha \in \mathcal{SL}$.[10]

According to the naive agent-representation of consequence relations recalled above, the standard relation between logical inconsistency and rationality can be stated as follows: Rational agents cannot accept inconsistent sets of sentences. This view lies at the heart of the belief change approach of which AGM has been a vastly successful example. Inconsistency triggers revision exactly because logical inconsistencies are normatively incompatible with the epistemic state of a rational agent.

Yet, it is easy to imagine situations in which an agent can rationally find themselves accepting an inconsistent set of beliefs. Since the definition of inconsistency is essentially unique, it is again the context of reasoning which must be invoked to justify the rationality—i.e. normative adequacy—of entertaining inconsistent beliefs which however need not trigger a revision.

Interestingly enough one such context was also noted by David Makinson in his *Preface Paradox*. Makinson considers a situation in which the author of a monograph who believes that each statement in the book is true is apologising for the possible persistence of errors. According to Makinson's analysis of the problem, the author

> is being rational even though inconsistent. More than this: he is being rational even though he believes each of a certain collection of statements, which he knows are logically incompatible [...] This appears to present a living and everyday example of a situation [in] which it is [...] rational to hold incompatible beliefs. (Makinson 1965)

The Preface Paradox constitutes an early exploration of the idea that under suitable circumstances it may be rational to violate consistency. This rather subversive intuition paved the way for a number of subsequent investigations which contributed to making inconsistency respectable to logicians.[11] Hence, the relaxation of the tight connection between irrationality and logical inconsistency.

In contrast to this, the wider domain of uncertain reasoning is currently dotted with disagreement on the very definition of consistency, i.e. what should be taken as the formal counterpart of a rational norm of belief. This essentially accounts for the Type

---

[9] Daston (1988) emphasises how the virtually unanimous agreement on the "expectations of reasonable men"—one of the trademark of the Enlightment—played an important role in the construction of the theory of probability. The objective Bayesian approach of Paris (1994) takes for granted that our intuitions about rationality are largely intersubjective, an idea which is further developed in Hosni and Paris (2005).

[10] A set of sentences is consistent if it is not inconsistent.

[11] The expression is borrowed from Gabbay and Hunter (1991). See also Paris 2004. Carnielli et al. (2007) offer a comprehensive survey of the field.

2 majority which I suggest would dominate the imaginary poll of Sect. 1. The reason is quite simply that there is currently no space for a Preface Paradox-like manouvre in Bayesian uncertain reasoning: Since choice behaviour reveals the epistemic state of an agent, there is clearly no distinction between displaying irrational behaviour and entertaining an inconsistent epistemic state.

3.4 Methodological Lessons

It is useful to single out more explicitly the key methodological lessons that emerge from the informal analysis of the development of non-standard logics sketched above. Since reasoning does not occur in a vacuum, logical modelling of rational agents is best relativised to an intended domain of application, or in case our interest lies in "pure" modelling, to some well-specified reasoning *context* which pins down the salient properties of the domain of interest.

I have mentioned some of the motivations for extending Tarskian consequence to capture defeasible reasoning, but similar arguments are easily put forward for a large number of non-standard logics which entered the stage with the dynamic, practical and many-valued turns in logic.[12] Those are just a handful of very familiar examples of how new ways of *doing* logic contributed essentially to changing our way of *thinking about logic*. In turn, this clearly affected our way of doing logic, thus igniting a "doing-thinking-doing" virtuous circle which largely accounts for the intuition that respondents of Type 1 in our imaginary logic poll outnumber those of Type 2. This virtuous circle can in fact be credited for diluting the excitement over the disputes about "the right logic". Take, as an illustration, intuitionistic logic. The so-called Brower-Heyting-Kolmogorov semantics is solidly interpreted within $S4$, a modal logic which conservatively extends classical logic.[13]

Thanks to the virtuous circle, logicians who are sensitive to the pressing need for modelling rational yet defeasible reasoning, may dispense with (unconstrained) monotonicity without plunging into foundational discomfort. Similarly for many other cornerstones of classical logic, should those turn out to be inappropriate for the specific context of interest. Even consistency, as recalled above in connection to the Preface Paradox, might turn out to be undesirable for a logic which aims at modelling rational norms of reasoning. And we can go on restricting and relaxing even more "entrenched" principles of classical logic, such as compositionality, which is clearly

---

[12] See, e.g. Gabbay and Woods (2005, 2007) and van Ditmarsch et al. (2007). See also Moss (2005).

[13] The first representation theorem to this effect goes all the way back to 1933 when Gödel introduced what we now call the Gödel translation and then proved that the algebra of open elements of every modal algebra for S4 is a Heyting algebra and, conversely, every Heyting algebra is isomorphic to the algebra of open elements of a suitable algebra for S4 (see, e.g. Blackburn et al. 2007especially Sect. 7.9).

incompatible with a number of features of interest in (qualitative and quantitative) uncertain reasoning.[14]

This delivers a rather unequivocal message. Classical logic provides the starting point for a number of non-standard logics which aim at capturing more realistic aspects of rational reasoning than those formalised by the classical notion of consequence. Once a particular context of reasoning is sufficiently well defined so as to give rise to sufficiently sharp modelling intuitions, we can set out to construct logics which address what then appear as clear limitations of classical logic.[15] Increasingly more realistic contexts of reasoning demand that we use classical logic in increasingly more imaginative ways, yet none of the above-mentioned restrictions and relaxations of classical logic is motivated by the desire to replace classical logic because it is wrong.

This *contextualisation* strategy lends a precious methodological lesson to uncertain reasoning. Reasoning by analogy naturally suggests the following question: What is the context in which standard Bayesian norms are justified? I will put forward an answer in Sect. 4 (in the limited case of belief norms) and it can hardly be surprising that such a context will be rather narrow. Hence the comprehensible dissatisfaction with standard Bayesian theory to which—contextualisation suggests—we should *not* reply by denying the relevance of Bayesian theory to normative models of rationality.[16] This would be like saying that logic is irrelevant to rational reasoning because classical logic is inadequate for air-traffic control. The success story of non-standard logics recommends a different, more foundationally conservative reaction: Identify the contexts of interest, and then work hard to extend the expressive power of standard Bayesian norms to capture those specific aspects of rational reasoning and decision under uncertainty. If we set ourselves free from the ill-founded quest for the right measure of uncertainty and start looking for what measures are justified in a certain context, we can hope to ignite a doing-thinking-doing virtuous circle similar to the one which proved so useful in the development of non-standard logics.

4 First-Order Uncertainty and Classical Bayesianism

Uncertain reasoning is best understood in terms of a fact and an assumption. The fact, roughly speaking, is that whenever we face a (non-trivial) choice problem we find ourselves in an epistemic state of uncertainty concerning the outcome of our choices. The assumption, on the other hand, is that we can make sense of such an

---

14 Arieli and Zamansky (2009) illustrate the applicability of non-deterministic matrices in modelling non-deterministic structures, where uncertainty is taken to be an intrinsic feature of the world, in addition to being a subjective epistemic state of the reasoning agents. Adams (2005) gives a feel for the *taboo* of abandoning compositionality in connection with Lewis's Triviality results.

15 This process is mostly, but not necessarily, application-driven. As Banach is often reported to have said, "Good mathematicians see analogies between theorems; great mathematicians see analogies between analogies" (as quoted by Jaynes who quotes from Ulam).

16 This is precisely what the title of Gilboa et al. (2011) provocatively recommends.

epistemic state. Jacob Bernoulli's *Ars Conjectandi* (1713) is widely recognised as putting forward the first convincing framework based on the idea that making sense of uncertainty goes hand in hand with *measuring* it. As briefly outlined in Sect. 1 however, the intervening three centuries have brought a rather limited consensus among the wider community of uncertain reasoner as to the precise details of how uncertainty should be measured. Our running logical analogy suggests not only that consensus may, in fact, never be achieved, but also and perhaps more radically, that we should stop looking for it and start articulating in some detail the lack of it instead.

This clearly requires an epistemological framework, to which I now turn.

4.1 Choice Roots for Bayesian Theory

As a starting point I will assume that making sense of our epistemic state of uncertainty—measuring uncertainty—matters to us insofar as we are faced with suitably defined choice problems. Whilst such a "behavioural" perspective might not account for the whole story, it certainly constitutes a very important part of it as illustrated by the recent analysis of *objective* Bayesian epistemology put forward by Williamson (2010).

Williamson singles out three characterising norms of rational belief which jointly pin down the bulk of the objective Bayesian framework. The *Probability Norm* demands that rational agents' degrees of belief should conform to the laws of probability. The *Calibration Norm* requires that the subjective probabilities licensed by the Probability Norm should be further constrained by known frequencies or, if these exist at all, single-case physical probabilities.[17] Finally, the *Equivocation Norm* further refines the choice of subjective probabilities by excluding extreme probability values unless these are being prescribed by the previous norms, and subject to this requirement, it constrains probabilities to be otherwise minimally prejudiced, or equivalently, maximally equivocal.[18] Objective Bayesianism is the epistemological framework in which Probability, Calibration and Equivocation are endorsed as *norms* of rational belief.

Williamson points out that Probability, Calibration and Equivocation are individually justified[19] by appealing to (formal) variations of essentially the same argument which involves the minimisation of a certain loss function. The gist of the argument can be described as follows. Provided that an individual is faced with a suitably defined choice problem, each of the three above norms can be justified by showing that contravening them would increase the agent's expectation of incurring a loss, possibly in the long run. The basic instantiation of this line of argument will be

---

[17] A familiar instantiation of this Norm is Lewis's *Principal Principle*.

[18] The best-known instantiation of the Equivocation norm is the Maximum entropy principle, which has been extensively discussed over the past three decades, often from heterogeneous points of view. For two comprehensive presentations, see Jaynes (2003) and Paris (1994)

[19] See Chap. 3 of Williamson (2010) for a detailed analysis of such justifications.

recalled in some detail in Sect. 4.4 below, where the construction of de Finetti's betting problem is seen to force a choice of betting odds which prevents the bookmaker from facing a sure loss.

This account of objective Bayesian epistemology lends itself to two considerations. First, the framing of the choice problem which motivates the quantification of uncertainty is fundamental in the definition and justification of the candidate norms of rational belief. This makes a strong case for connecting the context of reasoning to the specific description of the choice problem, as discussed above in comparison with non-standard logics. Hence the choice problem should be centrally involved in the definition of the formal counterpart to the intuitive notion of "irrationality". Second, and related to the first point, there seems to a very general and very basic principle which gives rise, through distinct instantiations, to distinct norms of rationality. I propose to render this principle as follows:

Choice Norm: A rational agent must not choose inadmissible alternatives.

The Choice[20] Norm captures a central aspect of Bayesian epistemology by making explicit how the subjective component of individual uncertainty is intertwined with the objective features of the underlying choice problem. This connects choice, belief and decision in a way which is typical of Bayesian theory. De Finetti, for instance, put it as follows:

> A decision must [...] be based on probabilities: i.e. the posterior probabilities as evaluated on the basis of all information so far available. This is the main point to note. In order to make decisions, we first require a statistical theory which provides conclusions in the form of posterior probabilities. The Bayesian approach does this: other approaches explicitly refuse to do this. (de Finetti 1974, p. 252)

This virtuous circle connecting choice, belief and decision accounts for the ubiquitous synergies of the broad concepts of "rationality" and "uncertainty" in statistics, epistemology, economics and related fields. Rukhin (1995), for example, puts it as follows:

> How should one choose a decision rule whose performance depends upon the unknown state of Nature? Since there is no uniquely recognized optimality principle that would provide a complete ordering of all statistical decision rules, this question is probably unanswerable when posed with such generality. However, it seems clear which procedures should not be used—the inadmissible ones which can be improved upon no matter what the unknown state of Nature.[21]

The Choice Norm can thus be seen as a good candidate to address the contrast outlined in Sect. 3.2 above and provide unity in the formalisation of the intuitive notion of "irrationality". To see this note that the Choice Norm is best seen as an attempt

---

20 Alternative denominations might have included "Dominance", "Pareto" or even of course "Admissibility". As they all have rather specific connotations in distinct areas of the uncertain reasoning literature, from statistics to decision and game theory to social choice theory, it appears that "Choice Norm" sits more comfortably at the desired level of generality whilst avoiding potential confusion.

21 See Levi (1986) and Bossert and Suzumura (2012) for an appraisal of similar ideas in epistemology and economics, respectively.

to define rationality in terms of *avoiding blatantly irrational* behaviour which is arguably more readily identified than its positive counterpart.[22] This negative characterisation of rationality is ubiquitous in Bayesian epistemology and is tightly connected to the idea of rationality as maximisation. Wald's seminal "idea of associating a loss with an incorrect decision" (de Finetti 1975, p. 253), for instance, paved the way to the analysis of subjective expected utility as the standard norm of rational decision.

The Choice Norm captures a notion of rationality which is general enough to be applicable even when no uncertainty enters directly the picture. In fact it builds on a very weak, yet non-empty, characterisation of "purposeful behaviour".[23] It is non-empty because the occurrences of "rational" and "inadmissible" appearing in the above phrasing of the Choice Norm have distinct epistemological statuses. Whilst the former is intended intuitively, the latter is taken in a precise, technical sense. So, if "rationality" means anything at all, it cannot be rational to behave in a way which blatantly contradicts the purpose of our own behaviour. It is the ordinary parlance meaning of "irrational" that is being used here, i.e. "stupid", "against commonsense", "illogical", etc. On the other hand the term "inadmissible" which occurs in the Choice Norm aims at capturing what we intuitively regard as being self-defeating—as opposed to purposeful behaviour—in the context of a formally specified choice problem.

What is perhaps the simplest example goes as follows. Let X be a set of feasible alternatives, $R_i \subseteq X^2$ a binary relation such that xR_iy interpreted as "i does not prefer y to x", and $\emptyset \neq C_i(X) \subseteq X$ be i's *choice set* from X (i.e. the non-empty subset of feasible alternatives selected by i). We say that i's choices are *inadmissible* if and only if there exists $y \in X$ such that yR_iz, for all $z \in X$ but $y \notin C_i(X)$.

Going back to the comparison of the previous section, "irrationality" is to "inadmissibility" what "incoherence" is to logical "inconsistency". This analogy suggests that a stupid choice is clearly *always* inadmissible. To see that an inadmissible choice is intuitively irrational, suppose i's preferences are modelled by the binary relation R_i. Under this assumption, an inadmissible choice captures the idea of self-inconsistency—blatant irrationality—so that the Choice Norm specifies the conditions under which an informal problem can be modelled as a *choice problem*, i.e. a formalised situation in which inadmissible alternatives cannot be rationally chosen.[24] Put the other way round, if an agent is normatively justified in selecting inadmissible choices, the situation at hand falls short of being a choice problem. But if there is no well-posed choice problem, we have no reason to worry about making

---

[22] In this respect, the concept of rationality seems to be analogous with that of democracy: there is usually more disagreement on what democracy should be than on what counts as a violation of a democratic society.

[23] I am adapting the terminology from Bossert and Suzumura (2010) who use it in connection to their notion of *Suzumura consistency*, arguably the weakest requirement in the formal theory of rational choice.

[24] This is one way of interpreting the axiom of the Independence of irrelevant alternatives, according to which a dominated alternative should not be chosen from any superset of the original set of feasible alternatives Sen (1970).

irrational choices, a situation in which the problem of quantifying uncertainty need not arise at all.

This suggests that Bayesian theory—in the rendering which I am emphasising here—takes rationality as a function of two arguments: a choice problem and an individual facing it. As a consequence, whether a certain norm of rationality is justified or not, will have to be discussed in relation to the assumptions we make on the choice problem, and those we make on the agent who is confronting it. The remainder of this section is devoted to illustrating how, under this interpretation, standard Bayesianism can be fruitfully seen as the solution to the following problem:

> **First-order uncertainty:** How should a maximally idealised agent behave when facing a maximally abstract choice problem?

Maximum idealisation will be shown to go hand in hand with the requirement that our model be normative. The role of maximal abstraction will be, as in all modelling, essential to set off mathematical formalisation.

4.2 *Idealisation*

Bayesian theory is normative at root, a feature which is abundantly emphasised by its proponents. In a number of early presentations of his ideas, de Finetti likens probability to the "logic of the uncertain", a view which is fully articulated in de Finetti (1972, Chap. 2). Savage (1954) refers to his postulates as "logic-like criteria". More recently, the logical characterisation of "common sense principles" has been the focus of the objective Bayesian approach of Paris and Vencovská (1990) and Paris (1994, 1998).[25]

According to the received view, normative models of rational behaviour should not take into account the potential cognitive limitations of the agents which are being modelled. This is usually motivated by the fact that building cognitive limitations into a normative model deprives it from its role in correcting mistakes, arguably one of the key reasons for developing normative models in the first place.[26] So, whilst the received view is not unchallenged (see, e.g. Gabbay and Woods 2003), I think it offers a very useful starting point and I will therefore largely conform to it. Not completely though, because idealisation is usually thought as a binary, all-or-nothing, property of agents, whereas it is much more natural, in the view I am articulating here, to think of idealisation as coming in degrees.

---

[25] The consensus on granting logic a normative status is far from being unanimous, and indeed the question might turn out to be ill-posed. A discussion of this point would take us too far, but the interested reader might wish to consult, among others (Gabbay and Woods 2003; van Benthem 2008; Wheeler 2008).

[26] As a well-known story goes, Savage was initially tricked into Allais's "paradox", but once Allais pointed that out, Savage acknowledged his mistake and corrected his answer accordingly. A similar reaction to the descriptive failures of Bayesian theory is condensed in the one-page paper (de Finetti 1979).

The idea is as follows. Students recruited by economists for their experiments, certainly qualify as minimally idealised agents, as do all of us individual, real, agents. A Turing machine, with its indefinitely extensible tape, is certainly an example of a maximally idealised agent, one whose memory limitations are of no consequence for the model of computation it defines. Arguably all agents of interest sit somewhere in between the spectrum delimited by those two examples. It is indeed tempting to go on and define an ordering relation on the minimum-maximum abstraction interval, which could be interpreted as "is less subject to cognitive limitations than" and investigate the consequences of this for uncertain reasoning modelling.[27] For present purposes, however, I will restrict the attention to maximally idealised agents only and refer to all other agents as "non-idealised".

4.3 Abstraction

All modelling requires abstraction, a fact that certainly contributes to making all models wrong, in Box's notorious dictum. Abstraction is the inevitable price that we must pay to grant ourselves the privilege of quantitative thinking.[28] The specific aspect of abstraction that will be of direct interest for present purposes concerns the features of the choice problem which, as recalled above, motivates the need for the quantification of the agent's uncertainty, or as I will simply say from now on, *the choice problem.*

De Finetti's betting problem, to be discussed shortly, and Savage's decision matrix are two very familiar examples of maximally abstract choice problems which intuitively can be associated with the formal and complete description of a "real-world" problem. I am stressing *intuitively* here because a precise characterisation of the abstraction of the choice problem is exactly what the framework under construction aims at achieving.

The key idea is that a maximally abstract choice problem includes all and only the information which is relevant to the decision-maker who is facing the choice problem. Put otherwise, when an agent is facing a maximally abstract choice problem it is *that* problem that the agent is facing, and not some other problem, however related. This is a fundamental yet too often overlooked modelling principle which has been given various names in the uncertain reasoning literature, including *the Watts assumption*[29] by Paris (1994), where the idea is presented as follows:

---

[27] Gabbay and Woods (2005) go some way towards developing the idea of a hierarchy of agents based on a similar relation.

[28] This need not pertain only to applied mathematics. In a number of widely known mathematical expositions George Polyá insists that abstraction is intrinsic to mathematical reasoning for the solution to hard mathematical problems is often best achieved by solving simpler problems from which the general idea can be extrapolated.

[29] This assumption is clearly related to Carnap's "Principle of Total Evidence" and Keynes' "Bernoulli's maxim". The fact that experimental subject systematically violate this principle motivates the introduction of the "editing phase" in Prospect Theory.

> The [linear constraints on a probability distribution are] not simply the shadow or description of the expert's knowledge but [...] *(essentially) all the expert's relevant knowledge*. [...] If we make this assumption then our task of giving a value to $Bel(\theta)$ given K is exactly the task that the expert himself carries out. (p. 67)[30]

In this spirit, we can naturally think of an agent facing a maximally abstract choice problem as a *decision-maker with no modelling privileges*. The distinction between decision-makers and decision-modellers can be elusive and this goes some way towards explaining why so much foundational confusion arises on this point, some of which underlies the discussion recapped in Sect. 2. For definiteness, recall Schmeidler's "two coins" example,[31] in which we are told that an agent's preference for tossing their own coin is normatively rational on the grounds that it has never given any sign of being unfair, whereas nothing is known about the stranger's coin. But what is precisely the choice problem here? We might certainly be justified in distrusting the stranger and their coin, but the Watts assumption requires this to be explicitly represented in the choice problem. If it is, no paradoxical situation arises, for the preference for one's own coin is so to speak, tautological. If the extra information concerning one's beliefs about the stranger and their coin are *not* represented in the choice problem, then there is no reason to prefer one's own coin.

More generally, the main source of confusion here lies in the fact that in real life we very often play *both* the role of decision-makers and that of decision-modellers. Take an agent who is about to do their shopping. This real-world problem can be given a maximally abstract representation as a consumer's choice problem (see, e.g. Rubinstein 2006). When doing our shopping, however, we make some choices as modellers, say whether to consider items without the fair-trade certification as feasible alternatives, and some as decision-makers, e.g preferring pomegranate juice to orange juice. Standard Bayesian theory relies, albeit only implicitly, on the rigid distinction between decision-making and decision-modelling. This is in fact central to the very idea of revealing consistent beliefs and preferences through a formally defined elicitation mechanism. Whenever such an elicitation device is assumed to capture all the relevant features of the real-world problem for which uncertainty needs to be quantified, we can think of the agent whose degrees of belief are being elicited as facing a maximally abstract choice problem.

This allows us to reframe de Finetti's Dutch book argument as the result of instantiating the Choice Norm with a maximally abstract betting problem. The purpose of this reframing is to make explicit the context for which standard Bayesian norms are justified. This will then constitute our starting point for the extension of standard Bayesian norms to second-order uncertainty.

---

30 Here $Bel()$ denotes the expert's belief function, θ is a sentence and K is a finite set of expressions of the form $Bel(\theta_i) = \beta_i$, where all the $\beta_i \in [0, 1]$.

31 Which in turn is a variant of a problem known to both Keynes and Knight before it was revamped by Ellsberg.

4.4 The Abstraction of de Finetti's Betting Problem

In a nutshell[32] de Finetti's betting problem is one in which a bookmaker is asked to write their betting odds for a set of events of interest, otherwise known as a *book*. In this context it is natural to take as blatantly irrational a choice of odds that exposes the bookmaker to potential loss independent of the outcome of the events in the book. Admissibility, which de Finetti calls *coherence*[33] can thus be rigorously characterised as "avoiding sure loss" (otherwise known as a *Dutch book*). As Ramsey independently anticipated and de Finetti proved, a Dutch book is avoided exactly if the betting odds conform to the laws of probability.

The formalisation of admissibility as avoiding sure loss rests on a rather convoluted elicitation framework which has caused much discussion over the past eight decades or so and which certainly justifies the present choice of considering de Finetti's betting problem as maximally abstract. Central to this is a complete contract[34] which is introduced to regulate the exchange of money between gamblers and bookmakers. The contract includes the following clauses:

Completeness: The bookmaker's choice is forced for (boolean) combinations of bets and, after the book has been published, the bookmaker is forced to accept all of a potentially infinite number of bets.

Swapping: After reading the published book, the gambler bets by paying to the bookmaker a real-valued stake of her choice. Since the gambler can choose *negative* stakes (betting negative money), she can unilaterally impose a payoff-matrix swap to the bookmaker.

Rigidity: Stakes involved in the betting problem correspond to actual money (in some currency).

Completeness is justified by de Finetti (1931) on the grounds that it provides the following modelling constraints. Were the bookmaker allowed to refuse selling certain bets, the bookmaker's betting odds could not be claimed to reveal his sincere degrees of belief on the relevant events, and as a consequence, the betting problem would fail its fundamental purpose of connecting a rational agent's degrees of belief to their willingness to bet. As he would retrospectively notice, the betting problem is a "device to force the individual to make conscious choices, releasing him from inertia, preserving him from whim" (de Finetti 1974, p. 76).[35]

[32] Since this is one of the most intensely studied aspect of Bayesian epistemology, I will take many details for granted and focus on the specific aspects which are directly relevant to the present discussion. Readers who are not familiar with the argument are urged to consult the original (de Finetti 1931, 1974). Paris (2001) offers a very general proof whilst Williamson (2010) provides ample background.

[33] The term "consistency" is also frequently used in English translations.

[34] In economics and political science a contract is said to be complete if it contemplates all possible contingencies. Despite being blatantly unrealistic, it is a widely used assumption in those areas (see, e.g. Tirole 1999).

[35] The question as to how suitable the betting problem is as an elicitation device is raised by de Finetti in his later work on *proper scoring rules* and especially Brier's (see, especially

In the presence of Completeness, Swapping entails that the bookmaker's degrees of belief should be *fair* betting odds. For suppose the bookmaker were to publish a book with non-zero expectation. Then he could be forced into sure loss by a gambler who put a negative stake on the book. Note that the abstraction leading to fair betting odds is justified only if the agents involved are maximally idealised. This amounts to saying that the bookmaker interacts with gamblers who will exploit any logical possibility of making a Dutch book against him, no matter how computationally demanding this might be. In game theoretic language, publishing fair betting odds constitutes the bookmaker's best response against rational gamblers under the usual common knowledge assumptions. Thus the idealisation of the gamblers is part and parcel of the abstraction of de Finetti's betting problem, which as recalled above, aims at eliciting the *bookmaker*'s degrees of belief.

Rigidity is an immaterial abstraction of de Finetti's betting problem and is motivated by de Finetti's reluctance to appeal to the mathematical theory of utility (see de Finetti 1969 especially Chap. 4). In order to avoid the potential complications arising from the diminishing marginal utility of money, de Finetti assumes that stakes should be *small*, an assumption to which he refers to as the *rigidity hypothesis* (de Finetti 1974, p. 77–78).

The abstraction of de Finetti's betting problem is fundamental to provide an instantiation of the Choice Norm which formalises admissibility as the avoidance of sure loss. De Finetti's (formal) notion of coherence is therefore context-dependent and it is justified only for the specific context defined by the betting problem. Under those restrictions only, the Choice Norm does entail probability as a first-order uncertainty norm of rational belief. De Finetti is explicit about the fact that no second-order uncertainty can be accommodated in his framework:

> Among the answers that do not make sense, and cannot be admitted are the following: "I do not know", "I am ignorant of what the probability is", "in my opinion the probability does not exist". Probability (or prevision) is not something which in itself can be known or not known: *it exists in that it serves to express, in a precise fashion, for each individual, his choice in his given state of ignorance*. To imagine a greater degree of ignorance which would justify the refusal to answer would be rather like thinking that in a statistical survey it makes sense to indicate, in addition to those whose sex is unknown, those for whom one does not even know "whether the sex is unknown or not". (de Finetti 1974, p. 82, my emphasis)

As the quotation clearly suggests, his radically subjectivist position brings de Finetti to deny that there is anything to be modelled outside first-order uncertainty. Whilst de Finetti's radical stance certainly played a role in giving the subjective approach to probability full mathematical citizenship, mostly as a consequence of the much celebrated Representation theorem (see de Finetti 1974), his highly idiosyncratic style contributed to hiding the applicability of the corresponding epistemological framework to modelling interesting and relevant aspects of second-order uncertainty. It is interesting to note that at approximately the same time—de Finetti's

---

(Footnote 35 continued)
de Finetti 1962, 1969, 1972). I will postpone the analysis of admissibility as "miniminum expected loss under Brier's score" to further research.

Fig. 1 A stylised two-dimensional space of models determined by the parameters intervening in the Choice Norm, namely the idealisation of the agent and the abstraction of the choice problem

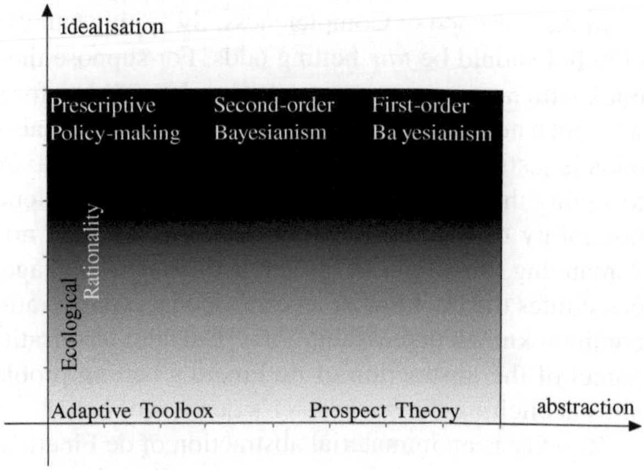

monograph was published in Italian in 1970—Savage added to the second edition of his Savage (1954) the following footnote:

> One tempting representation of the unsure is to replace the person's single probability measure P by a set of such measures, especially a convex set […]. (p. 58)

I. J. Good is one prominent Bayesian who has not resisted the temptation (see Good 1983).

4.5 A Two-Dimensional Framework

Our running logical analogy now triggers the obvious question: If the Probability Norm—the requirement that uncertainty should be measured by probability—is fully justified for first-order uncertainty only, what happens outside its rather narrow borders?

By identifying first-order uncertainty with the epistemic state of a maximally idealised agent who is facing a maximally abstract choice problem, we can tackle the question by either relaxing the abstraction of the choice problem or (inclusively) the idealisation of the agent. Figure 1 illustrates the logical space of modelling which arises in this abstraction-idealisation coordinate system.

Starting from first-order Bayesian theory, we can relax completely the idealisation of the agent and consider abstract choice problems faced by experimental subjects. This identifies the domain of behavioural decision theory for which Prospect Theory offers the best-known framework (Kahneman and Tversky 1979; Wakker 2010). From there we can fully relax also the abstraction of the choice problem, ending up in the domain of the "adaptive toolbox" investigated in (Gigerenzer 1999), where real people use a whole stock of heuristics to make decisions in the real world. Which heuristics should be used for which particular choice problem is a question of clear normative import which is addressed by the ecological approach to rationality (see Gigerenzer 2012 especially Chap. 19). The top-left corner of the rectangle depicted in Fig. 1 finally identifies the context in which idealised agents face real-world problems, of obvious interest in prescriptive policy making.

As recalled above, the present study is only concerned with normative models of rationality so the focus is on the interval delimited by maximal and minimal abstraction, corresponding to the darker area of the rectangle of Fig. 1. The central intuition of the present proposal is that second-order uncertainty can be given a solid foundation by instantiating the Choice Norm on increasingly less abstract choice problems, or equivalently, by granting the decision-maker an increasingly large number of modelling privileges.

The next section illustrates this familiar "extension by restriction" strategy (see Sect. 3.2 above) with an example in which the Choice Norm is instantiated with a second-order uncertainty problem featuring ignorance or ambiguity—as referred to the elicitation of subjective degrees of belief.

Before moving to that, however, it is important to stress that the rectangle in Fig. 1 should not be interpreted as providing a 'grand unified theory of uncertain reasoning and rational decision', as it were. Whilst it accommodates a number of currently popular albeit hardly mutually consistent approaches to the topic, it is not meant to suggest that all models of rationality are or should be commensurable in some particularly precise way. Again in the spirit of our central logical analogy, Fig. 1 is best interpreted as highlighting the variety which is determined by the context-dependence of rational norms of belief.

5 Towards Second-Order Uncertainty

Let us go back to the (Gilboa et al. 2011) plea for rational modesty recalled in Sect. 2 above. In real life, where time and information are scarce, it might well be wiser to hold our judgment and to postpone our decisions until all the facts are in, as it were. However, the completeness of the revealed belief approach normatively requires that subjective degrees be linearly ordered forcing the Bayesian agent to have a probability for all elementary events of interest. As there is no *ought* without a *can*, many anti-Bayesians conclude that the Probability Norm is not necessary for rational belief. This goes hand in hand with a foundational perspective which can be traced back to Keynes (1921) and Knight (1921) who insisted, albeit with rather distinct arguments, that not all (economic) problems admit of a probabilistic quantification of uncertainty. This point of view has been supported over the past four decades by vast yet not uncontroversial experimental evidence.[36] As a consequence, a number of anti-Bayesian proposals currently take issue with probability and the maximisation of expected utility as adequate norms of rationality under uncertainty. The main line

[36] Binmore et al. (2012) report on a recent experiment which casts substantial doubts on the alleged universality of the "ambiguity aversion" phenomenon in Ellsberg-type problems.

of the argument, as recalled in Sect. 2 above, is this: since it fails to reflect the agent's ignorance, the Probability Norm fails as a measure of rational belief.

To fix ideas consider a policy maker who is faced with a problem whose uncertainty is intuitively felt to be "hard to quantify". Take, for example, the event "Greece will exit the eurozone by the end of May 2013", or GREXIT as it is referred to in the financial lingo.[37] There is certainly no obvious state space that captures the relevant scenarios connected to GREXIT. Since nobody can really come up with such a state space, anti-Bayesians suggest that GREXIT relates to the kind of uncertainty which cannot be quantified probabilistically. Putting to one side an analysis of the slippery concept of 'non-probabilistic uncertainty' I would like to focus on how the two-dimensional characterisation of rationality of Fig. 1 helps us to see that whilst intuitively plausible, the anti-Bayesian answer to the issues raised by GREXIT-like problems fails to be normatively persuasive.

Let us begin by noticing that Fig. 1 gives us two main modelling options. The first such option is to take GREXIT as a first-order uncertainty problem, i.e. maximally specified so that no modelling options are available to the decision-maker who is facing it. In this case a maximally idealised policy maker *must* come up with a probability value representing their belief in GREXIT. On a first-order uncertainty reading, it does not matter at all if the process of producing an admissible (i.e. satisfying the Choice Norm) quantification of the policy maker's degree of belief for GREXIT is "hard" or difficult, in any sense. Under these assumptions the idealised policy maker is normatively forced to attach GREXIT a unique point in the real unit interval.

Our second modelling option is to frame GREXIT as a second-order uncertainty problem. In this case, we might want to capture the fact that the scenarios which are relevant to GREXIT are so many and currently so poorly understood that too much information is lost by summarising our uncertainty about GREXIT with one real number. This clearly means allowing some modelling privileges to the decision-maker facing the GREXIT problem. Hence, in a second-order uncertainty framing, the agent might rationally (at the second-order) decide that in the present state of information an interval representation of uncertainty is preferable, or even that it is best not to give a (public) answer in order to prevent the attack of financial speculators and so on, depending on the degree of modelling privileges that we are willing to grant to the decision-maker.

It goes without saying that I will not offer a formal solution to the problem of, say, measuring uncertainty in problems as complex as GREXIT. The purpose of this example is rather that of suggesting a natural framing for the following question: What norms of rationality are adequate if we give decision-makers some modelling privileges, or equivalently, relax the abstraction of the choice problem? The answer must clearly come in two steps. First we need to provide an account of what a principled relaxation of abstraction might be, i.e. the extent to which we are willing to grant modelling privileges to decision-makers. Having done this, we can instanti-

[37] The term appears to have been coined by Willem Buiter of Citigroup, in his 6th February 2012 report.

ate the Choice Norm approach illustrated above and provide a justified notion of admissibility for the specific class of choice problems at hand.

As an example of the applicability of this general framework, I will illustrate how imprecise probabilities can be justified as the rational norm of belief in a for-profit betting problem—a more realistic version of de Finetti's problem which relaxes the Swapping condition.

5.1 Betting for Profit

Recall from Sect. 4.4 above that de Finetti's betting problem is so formulated as to force the (idealised) bookmaker to choose fair betting quotients. Publishing books with zero expectation is necessary and sufficient to protect the bookmaker from being forced into sure loss by rational gamblers, possibly through Swapping.

Fedel et al. (2011) investigate the relaxation of this first-order uncertainty modelling feature by considering a betting scenario in which bookmakers are motivated by making profit in a market-like environment. This clearly imposes the relaxation of Swapping so that gamblers can no longer choose the sign of the stake for their bets. Analogously, bookmakers are allowed to differentiate between buying and selling prices, thus giving rise to the notions of lower and upper probabilities which are well familiar from the theory of imprecise probabilities.[38] How can the Choice Norm be instantiated for this (second-order uncertainty) *for-profit betting problem*? A simultaneous extension of classical Bayesianism to imprecise and fuzzy probabilities is carried out in (Fedel et al. 2011) by constructing the analytic framework of imprecise probabilities on top of a many-valued algebraic semantics. For present purposes I will limit myself to an informal discussion of how the notion of admissibility is arrived at and refer the interested reader to the original paper for precise mathematical details and for further motivation concerning the extension to *fuzzy* events.

The key idea, as anticipated above, is that the bookmaker publishes his book by assigning a pair of real numbers α_i, β_i, intuitively interpreted as the sup of the buying price and the inf of the selling price, respectively, to all events in E_i in the book. So a book (or a system of bets) is defined by a set of pairs $(E_1, [\alpha_1, \beta_1]), \ldots, (E_n, [\alpha_n, \beta_n])$ such that $0 \leq \alpha_i \leq \beta_i \leq 1$ for $i = 1, \ldots, n$. In standard market models, the inequality between buying and selling prices is assumed to be strict, thus defining the so-called *bid-ask spread* (see e.g. Hasbrouck 2007). The assumption is variously motivated: from potential asymmetric information to the bookmaker's need to cover fixed transaction costs. So, in the perspective of reducing the abstraction of the choice problem, it is very natural to consider bookmakers who set a positive spread. Note however that de Finetti's framework is mathematically fully recovered when $\alpha_i = \beta_i$ for all E_i in the book.

The Choice Norm demands that we define an appropriate notion of admissibility that must be satisfied by any justified norm of belief for this refined class of for-profit

---

[38] The standard reference for the field is Walley (1991), which also provides detailed historical background. Miranda (2008) offers an outline of the most recent developments.

betting problems. Let us begin by noting that de Finetti's notion of coherence will not do. For, when publishing his book, the bookmaker knows that the (maximally idealised) gamblers have a choice between:

(1) paying $\beta\lambda$ for the right to receive $\lambda v(E)$, i.e. betting on E.
(2) receiving the payment of $\alpha\lambda$ to pay back $\lambda v(E)$, i.e. betting on E^c.

where $\alpha, \beta \in [0, 1]$, $\lambda > 0$, $v(E)$ maps E to $[0, 1]$ and E^c denotes the complement of E. It is immediate to see that de Finetti's coherence is sufficient to avoid blatant irrationality but is not necessary, in this less abstract choice problem. Thus admissibility cannot be defined as "avoiding sure loss". Consider the simple book $B = \{(E, [0, 1]), (E^c, [.5, 1])\}$. Whilst the bookmaker who published B would certainly protect himself against sure loss, there is a clear sense in which B is blatantly irrational in *for-profit betting problems* as the bookmaker is setting too wide a spread. In the market-like scenario which is motivating the construction of this specific choice problem, by setting too wide a spread between buying and selling prices, the bookmaker is effectively encouraging gamblers to trade with those bookmakers who publish more attractive odds. As this openly violates the notion of purposeful behaviour which the Choice Norm intends to capture, an *inadmissible* choice can be defined as a choice of betting intervals which could be refined without leading to sure loss.[39] In other words, a book is inadmissible if the bookmaker writes odds which are unnecessarily conservative.

One central result of (Fedel et al. 2011) can be stated in the terminology of the present note as follows: inadmissibility is avoided if and only if the bookmaker's odds can be extended to an upper prevision over a suitable algebra of events. Dual results follow from lower previsions and probabilities. Thus, instantiated with for-profit betting problems, the Choice Norm justifies imprecise probability as a norm of rational belief.

There is an increasingly wide consensus on the fact that interval-valued probability captures relevant features of second-order uncertainty, especially ignorance. This was partly acknowledged by Savage in the above-quoted remark (see p. 21) to the effect that the expressive power provided by imprecise probability, and in particular by convex sets of probability measures, would be necessary to model increasingly more interesting classes of decision problems under uncertainty. De Finetti, as one would expect, rejected any such extension.[40] In terms of our opening question, clearly de Finetti thinks of his version of Bayesian theory not as one of the many possible alternatives, but as the only possibility worth considering. Yet, the general framework outlined here suggests, as Savage seems to have clearly anticipated, that a Type 1 attitude can lead to foundationally conservative formal advance just by supplying the Choice Norm with context-dependent formalisations of admissibility.

[39] Inadmissibility is referred to as the *bad bet criterion* in the terminology of (Fedel et al. 2011).
[40] See de Finetti (1975, Appendix 19.3).

5.2 Second-Order Uncertainty Versus Second-Order Probability

As a slightly unfortunate consequence of the terminology which I am using, onc might be led to believe that second-order uncertainty is to be measured by second-order probabilities. The above two-dimensional "agent-problem" characterisation of rationality however, suggests that this is not, in general, the case. Whilst the precise details will have to be addressed separately, the intuitive argument as to why second-order uncertainty, as presently characterised, need not be measured by second-order *subjective* probabilities, goes as follows.

Recall the space of models depicted in Fig. 1 above. The standard Probability Norm is justified, in such a framework, for first-order uncertainty only. By relaxing the abstraction of de Finetti's betting problem we have introduced some second-order uncertainty features in the choice problem, which in turn led to a justification of imprecise probabilities as a second-order norm for rational belief. The relaxation of abstraction can be interpreted as granting the decision-maker some modelling privileges, such as discouraging potential gamblers by setting very wide intervals as a consequence of the bookmakers' "ignorance" about the events in question. Exactly how wide such a spread between buying and selling price should be, is now a choice which the bookmaker makes as modeller, rather than decision-maker. But there is intuitively no reason as to why norms of rational decision-making (i.e. the Probability Norm) should apply "one level up" to rational decision-modelling—say choosing those events E_i of the book on which the bookmaker effectively does not want anyone to bet, a fact which can naturally be expressed by setting the odds for such E_is to [0, 1]. One central feature of the present proposal is that norms for decision-making and norms for decision-modelling, albeit contiguous, are distinct problems.

Note that the general strategy of justifying distinct belief norms for *distinct* choice problems, clearly prevents the resulting second-order norms of rational belief from entering an infinite regress, a concern which typically hovers over second-order (subjective) probability. D. Hume is usually credited with spelling out an early version of this argument, whose contemporary version appears in (Savage 1954, p. 58). Interestingly enough this is the very page in which Savage added the second-edition footnote, recalled above, to the effect that the attitude of being "unsure" could be represented by allowing convex sets of probabilities.

Clearly, no infinite regress can arise when first-order uncertainty is represented by *objective* probabilities as in (Hansson 2009). However, from the point of view of Bayesian theory, which the present framework aims at extending, the assumption that objective probability represents the agent's uncertainty must be made with substantial care. Whether objective probability and possibly second-order probability can be accommodated within the framework outlined in this paper remains an open question.

6 Conclusion

In making the comparison which constitutes the leitmotif of this paper, I have implicitly assumed that non-standard logics and uncertain reasoning are distinct research areas. Whilst this tends to be largely the case if we take the quantitative versus qualitative divide very seriously,[41] it might rightly be objected the opposite is true: Mathematical and philosophical overlaps abound between probability and logic in both the sub-areas of quantitative and qualitative representations of uncertainty.

Whilst logic is in itself central to probability and uncertain reasoning, my present aim was not to discuss the fruitful *formal* interactions between logic and uncertain reasoning. This is the object of a number of thorough investigations including Haenni et al. (2011), Howson (2009), Makinson (2012) and Paris (1994). My present goal was rather to suggest that the way many non-standard logics developed as extensions of the classical (propositional) one offers a potentially very fruitful methodological example of how the limited expressive power of a formal model can be addressed by clarifying its intended domain of application. By comparing uncertain reasoning with very familiar developments in non-standard logics I suggested that the much-needed formal advance of Bayesian theory, especially with regards to second-order uncertainty, may be greatly facilitated if uncertain reasoners take seriously the lessons offered by non-standard logics, especially contextualisation. Since uncertain reasoning does not occur in a vacuum, questions about rational norms are best relativised to an agent and a choice problem.

This simple observation sheds new light on the "big picture" and provides evident prospects for foundational and formal advance which will hopefully give rise to a *doing-thinking-doing* virtuous circle in uncertain reasoning. From the foundational point of view, the two-dimensional contextualisation presented above clearly illustrates how some popular anti-Bayesian criticisms are intuitively appealing, yet normatively inconclusive. For the arguments based on "probability is hard to quantify" to have normative force, one must clearly put forward a classification of choice problems and prove that for some suitable class, probability offers an incomplete formalisation. The analysis of Ellsberg-type problems recalled above does not offer particularly useful insights in this direction.

The strategy of starting with maximally abstract choice problems and then refining them to model increasingly more realistic features of second-order uncertainty appears to be foundationally more transparent than that of postulating the existence of putatively distinct notions of uncertainty, some of which are probabilistically quantifiable, some of which are not. It is hoped that this problem-based approach to measuring uncertainty may prove useful in developing norms of rational belief and decision which are continuous with the real-world applications which eventually motivate much of our interest in this subject. Real-world problems, from climate

[41] Suffice it to mention that much of the popularity enjoyed by non-monotonic logics during the 1980s was more or less directly linked to the idea that logic would better serve the (computational) needs of artificial intellingence than probability.

change to economic uncertainty to biomedical risk, are crying out to be the uncertain reasoner's Tweety.

Acknowledgments I have presented the main ideas of this paper at Kent's Centre for Reasoning and at the LSE Choice Group in London. I would like to thank Jon Williamson and Richard Bradley for inviting me to speak at those seminars and both audiences for their very valuable feedback. I am very grateful to Gregory Wheeler for his comments on an earlier draft and for many stimulating discussions on the topics covered in this paper. Thanks also to two referees, whose thorough reviews helped me to improve the chapter in many ways. Finally, readers familiar with David Makinson's work will certainly have spotted a number of terms and expressions which are easily associated with, and sometimes directly coming from, David's papers and books. I realised this only when the first draft of this chapter was completed. As I started fetching all the originals to give credit where credit was due, it occurred to me that leaving those paraphrases uncredited would probably be the most direct way to express how influential David's way of doing logic is to my way of thinking about logic. So I stopped.

References

Adams, E. W. (2005). What is at stake in the controversy over conditionals. In G. Kern-Isberner, W. Roedder, & F. Kulman (Eds.), *WCII 2002* (pp. 1–11)., Lecture Notes in Artificial Intelligence. New York: Springer.

Arieli, O., & Zamansky, A. (2009). Distance-based non-deterministic semantics for reasoning with uncertainty. *Logic Journal of IGPL*, *17*(4), 325–350.

Barwise, J., & Keisler, H. J. (Eds.). (1977). *Handbook of mathematical logic*. Amsterdam: North-Holland.

Bewley, T. F. (2002). Knightian decision theory, part I (1986). *Decisions in Economics and Finance*, *25*, 79–110.

Binmore, K., Stewart, L., & Voorhoeve, A. (2012). An experiment on the Ellsberg paradox. *Working paper*.

Blackburn, P., van Benthem, J., & Wolter, F. (2007). *Handbook of modal logic*. Amsterdam: Elsevier.

Bossert, W., & Suzumura, K. (2010). *Consistency, choice, and rationality*. Cambridge: Harvard University Press.

Bossert, W., & Suzumura, K. (2012). Revealed preference and choice under uncertainty. *SERIEs*, *3*(1), 247–258.

Carnielli, W., Coniglio, M. E., & Marcos, J. (2007). Logics of formal inconsistency. In D. M. Gabbay & F. Guenthner (Eds.), *Handbook of philosophical logic* (2nd ed., Vol. 14, pp. 1–93). New York: Springer.

Colyvan, M. (June 2008). Is probability the only coherent approach to uncertainty? *Risk Analysis*, *28*(3), 645–652.

Daston, L. (1988). *Classical probability in the enlightenment*. Princeton: Princeton University Press.

de Finetti, B. (1931). Sul significato soggettivo della probabilità. *Fundamenta Mathematicae*, *17*, 289–329.

de Finetti, B. (1962). Does it make sense to speak of good probability appraisers? In I. J. Good (Ed.), *The scientist speculates* (pp. 357–364). Londan: Heinemann.

de Finetti, B. (1969). *Un matematico e L'economia*. Milano: Franco Angeli.

de Finetti, B. (1972). *Probability, induction and statistics*. New York: Wiley.

de Finetti, B. (1973). Bayesianism: Its unifying role for both the foundations and the applications of statistics. *Bulletin of the International Statistical Institute*, *39*, 349–368.

de Finetti, B. (1974). *Theory of probability* (Vol. 1). New York: Wiley.

de Finetti, B. (1975). *Theory of probability* (Vol. 2). New York: Wiley.

de Finetti, B. (1979). A short confirmation of my standpoint. In M. Allais & O. Hagen (Eds.), *Expected utility hypotheses and the allais paradox* (p. 161). Dordrecht: D. Reidel.
de Finetti, B. (2008). *Philosophical lectures on probability*. New York: Springer.
Ellsberg, D. (1961). Risk, ambiguity and the savage axioms. *The Quarterly Journal of Economics*, *75*(4), 643–669.
Fedel, M., Hosni, H., & Montagna, F. (2011). A logical characterization of coherence for imprecise probabilities. *International Journal of Approximate Reasoning*, *52*, 1147–1170.
Gabbay, D. M. (Ed.) (1995). *What is a logical system?*. Oxford: Oxford University Press.
Gabbay, D. M., & Hunter, A. (1991). Making inconsistency respectable 1 : A logical framework for inconsistency in reasoning. In Ph. Jorrand & J. Keleman (Eds.), *Fundamentals of artificial intelligence research*, (pp. 19–32). Lecture notes in Computer Science, Berlin: Springer.
Gabbay, D. M., & Woods, J. (Eds.). (2007). *Handbook of the history of logic: The many valued and non monotonic turn in logic*, (Vol. 8). Amsterdam: North-Holland.
Gabbay, D. M., & Woods, J. (2003). Normative models of rational agency: The theoretical disutility of certain approaches. *Logic Journal of IGPL*, *11*(6), 597–613.
Gabbay, D. M., & Woods, J. (2005). The practical turn in logic. In D. M. Gabbay & F. Guenthner (Eds.), *Handbook of philosophical logic* (2nd ed., Vol. 13, pp. 15–122). Netherland: Springer.
Gelman, A. (2011). Induction and deduction in Bayesian data analysis. *Rationality, Markets and Morals*, *2*, 67–78.
Gigerenzer, G., Todd, P. M., & ABC Research Group. (1999). *Simple heuristics that make us smart*. Oxford: Oxford University Press.
Gigerenzer, G., Todd, P. M., & ABC Research Group. (2012). *Ecological rationality: Intelligence in the world*. Oxford: Oxford University Press.
Gigerenzer, G. (2008). *Rationality for mortals: How people cope with uncertainty*. Oxford: Oxford University Press.
Gilboa, I., Postlewaite, A., & Schmeidler, D. (2011). Rationality of belief or: Why Savage's axioms are neither necessary nor sufficient for rationality. *Synthese*, *187*, 11–31.
Gilboa, I. (2009). *Theory of decision under uncertainty*. New York: Cambridge University Press.
Good, I. J. (1983). *Good thinking: The foundations of probability and its applications*. Minneapolis: The University of Minnesota Press.
Haenni, R., Romeijn, J. W., & Wheeler, G. (2011). *Probabilistic logics and probabilistic networks*. Dordrecht: Springer.
Hansson, S. O. (2009). Measuring uncertainty. *Studia Logica*, *93*(1), 21–40.
Hasbrouck, J. (2007). *Empirical market microstructure: The institutions, economics and econometrics of securities trading*. Oxford: Oxford University Press.
Heyting, A. (1956). *Intuitionism: An introduction*. Amsterdam: North-Holland.
Hosni, H., & Paris, J. B. (2005). Rationality as conformity. *Synthese*, *144*(2), 249–285.
Howson, C. (2009). Can logic be combined with probability? Probably. *Journal of Applied Logic*, *7*(2), 177–187.
Jaynes, E. T. (2003). *Probability theory: The logic of science*. Cambridge: Cambridge University Press.
Kahneman, D., & Tversky, A. (1979). Prospect theory: An analysis of decision under risk. *Econometrica*, *47*(2), 263–291.
Keynes, J. M. (1921). *Treatease on probability*. New York: Harper & Row.
Knight, F. H. (1921). *Risk, uncertainty and profit*. New York: Beard Books Inc.
Levi, I. (1986). The paradoxes of Allais and Ellsberg. *Economics and Philosophy*, *2*, 23–53.
Makinson, D. (1965). The paradox of the preface. *Analysis*, *25*(6), 205.
Makinson, D. (1994). General patterns in nonmonotonic reasoning. In J. Robinson (Ed.), *Handbook of logic in artificial intelligence and logic programming* (Vol. 3, pp. 35–110). Oxford: Oxford University Press.
Makinson, D. (2005). *Bridges from classical to nonmonotonic logic*. London: College Publications.
Makinson, D. (2012). Logical questions behind the lottery and preface paradoxes: Lossy rules for uncertain inference. *Synthese*, *186*, 511–529.

Miranda, E. (2008). A survey of the theory of coherent lower previsions. *International Journal of Approximate Reasoning*, *48*(2), 628–658.
Moss, L. S. (2005). Applied logic: A manifesto. In D. M. Gabbay, S. Gonchargov, & M. Zakharyaschev (Eds.), *Mathematical problems from applied logic I* (pp. 317–343). Netherland: Springer.
Paris, J.B. (2001). A Note on the Dutch Book Method. In T. Seidenfeld, G. De Cooman & T. Fine (Eds.), *ISIPTA '01, Proceedings of the Second International Symposium on Imprecise Probabilities and Their Applications*, Ithaca, NY, USA. Shaler.
Paris, J. B., & Vencovská, A. (1990). A note on the inevitability of maximum entropy. *International Journal of Approximate Reasoning*, *4*(3), 183–223.
Paris, J. B. (1994). *The uncertain reasoner's companion: A mathematical perspective*. Cambridge: Cambridge University Press.
Paris, J. B. (1998). Common sense and maximum entropy. *Synthese*, *117*(1), 75–93.
Paris, J. B. (2004). Deriving information from inconsistent knowledge bases: A completeness theorem. *Logic Journal of IGPL*, *12*, 345–353.
Rubinstein, A. (2006). *Lecture notes in microeconomic theory*. Princeton: Princeton University Press.
Rukhin, A. L. (1995). Admissibility: Survey of a concept in progress. *International Statistical Review*, *63*(1), 95–115.
Savage, L. J. (1972). *The foundations of statistics* (2nd ed.). New York: Dover.
Schmeidler, D. (1989). Subjective probability and expected utility without additivity. *Econometrica*, *57*(3), 571–587.
Sen, A. K. (1970). *Collective choice and, social welfare*. San Francisco: Holden Day.
Shafer, G. (1986). Savage revisited. *Statistical Science*, *1*(4), 463–485.
Tirole, J. (1999). Incomplete contracts: Where do we stand? *Econometrica*, *67*(4), 741–781.
van Benthem, J. (2008). Logic and reasoning: Do the facts matter? *Studia Logica*, *88*, 67–84.
van Ditmarsch, H., van der Hoek, W., & Kooi, B. (2007). *Dynamic epistemic logic*. New York: Springer.
Wakker, P. P. (2010). *Prospect theory for risk and ambiguity*. Cambridge: Cambridge University Press.
Walley, P. (1991). *Statistical reasoning with imprecise probabilities*. New York: Wiley.
Wheeler, G. (2008). Applied logic without psychologism. *Studia Logica*, *88*(1), 137–156.
Williamson, J. (2010). *In defence of objective bayesianism*. Oxford: Oxford University Press.

Part IV
Normative Systems

Abstract Interfaces of Input/Output Logic

Audun Stolpe

Abstract The presents chapter investigates to what extent techniques from belief revision and propositional relevance can be transposed into the context of input/output logic. The study is focused on parallel interpolation, finest splittings, contraction and relevant contraction. We show that basic properties of different input/output logic systems influence the existence of interfaces between the different idioms. E.g. we show that the operator of simple-minded output satisfies parallel interpolation, whilst basic output does not. We use parallel interpolation to prove the existence of finest splittings, and, in analogy with propositional relevance, use finest splittings to define a concept of relevance—one that is attuned to the idiosyncracies of codes of norms. Next, we define an operation of derogation of codes of norms, and temper it by our concept of normative relevance. The chapter ends with an illustration of how the pieces fit together, by giving an analysis of the concept of positive permission.

Keywords Input/Output logic · Finest splittings · Parallell interpolation · Normative systems

1 Introduction

To philosophers and computer scientists David Makinson is undoubtedly best known for his work on non-monotonic reasoning and theory revision, his philosophy of norms and normative systems having gained rather less notoriety. As to the reason why, there is little doubt that the dominant trends and tendencies in artificial intelligence and planning in the beginning of the eighties made the scientific community particularly attentive to the ideas Makinson and his collaborators were putting out there at that time. The AGM-paradigm in particular enters the scene at a moment when artificial intelligence is struggling to reestablish itself on a sound theoretical

A. Stolpe (✉)
Norwegian Defence Research Establishment, Kjeller, Norway

S. O. Hansson (ed.), *David Makinson on Classical Methods for Non-Classical Problems*, Outstanding Contributions to Logic 3, DOI: 10.1007/978-94-007-7759-0_12,

basis. Although the AGM theory was rather more philosophically and epistemologically motivated, it ultimately came to be seen as part of this enterprise.

In contrast, norms and computer systems was never a particularly febrile research area, and although deontic logic continues to receive some attention in philosophical logic and multi-agent systems, it still on the whole tends to be regarded as a somewhat parochial discipline concerned primarily with analysing concepts germane to ethics. In the present author's opinion, a rather more insightful characterisation views deontic reasoning as essentially concerned with adaptiveness, that is, with the potential discrepancy between the actual and the ideal as well as the means to compensate for it. As such, it subsumes intelligent action as a special science, and belongs to artificial intelligence as a legitimate and indispensable part.

Be this as it may, it is an interesting fact of Makinson's authorship that his first forays into the philosophy of norms predate all of his work on non-monotonic reasoning and all of his work on theory revision. Indeed, the AGM paradigm grows out of Makinson's collaboration with Carlos Alchourrón on the logical intricacies of derogating a norm from a legal code (Makinson 1996). Alchourrón and Makinson published a formal treatment of this notion in 1981, entitled "Hierarchies of Regulations and their Logic", which Makinson describes as "groping painfully towards ideas that would later become formulated as maxichoice contraction" (Alchourron and Makinson 1981). Thus, far from being orthogonal to each other, Makinson's earliest works on theory revision and on normative systems are intimately related.

Indeed, this is true of almost all the different threads in Makinson's writings on logic, which display a remarkable unity of approach and orientation. In his retrospective "Ways of Doing Logic" Makinson summarises his approach with reference to a small number of methodological maxims that he sees as integral to his general outlook on the enterprise of philosophical logic as such. Among these we find the maxim *Don't abandon classical logic, use it more creatively*, which urges that whilst there are many forms of reasoning that classical logic as such is not fit to encode—e.g. causal reasoning and default reasoning—the proper way to respond to this situation is sometimes to allocate to classical logic a place in a surrounding body of theory which models the concept in question.

Closely related to this is the maxim *Logic is not just about deduction* which admonishes that not all reasoning systems with a logical component or subsystem are naturally modelled as calculi. For instance, revising one's set of beliefs upon learning of a new and incompatible fact may require a choice of which among several eligible beliefs to give up. Yet, why one would choose to retain one belief over another may have nothing to do with logic—it could be a matter of epistemic priorities or utility rankings, or it could be the outcome of employing a simple tie-breaker such as flipping a coin. The point is that the choice is not deducible within the logic itself, but rather imposed upon it.

Finally, when logic is relegated to a component of a larger theory, Makinson observes, there are often a number of interesting properties that arise before one begins to consider the connectives in the object language. Indeed most of the hard work can usually be carried out on an abstract level in terms of "selection operations, neighbourhoods, qualitative notions of size and of distance, and above all ordering relations on arbitrary sets", as Makinson himself puts it in his foreword to Schlechta (2004). Hence it may be advisable to *Do some logic without logic*.

The reader is referred to Makinson (2003) for the technical implications of Makinson's maxims. Suffice it here to get the general thrust, which is to shift the emphasis from a theory formulated in terms of object language properties to a theory formulated in terms of operations on sets. One may say that Makinson's maxims involve an ascent to the meta-logical level where sets and relations become the primary objects of study. Specifically logical behaviour emerges in cases where (some) sets are sets of formulae and (some) relations are consequence relations. The effect of this ascent is two-fold. First it places philosophical logic in a wider mathematical geography where logic integrates with e.g. order theory, universal algebra and even economics. Secondly it establishes what can be thought of as abstract interfaces between different parts of Makinson's production: many of Makinson's pivotal constructions are entirely agnostic wrt. the precise nature of a closure operator, say, or the underlying logical language. A consequence of this is that there is often an abstract stratum in Makinson's work that allows migration of techniques and results from one formal idiom to another.

The purpose of the present chapter is to study some of these interfaces as they bear on input/output logic. The aim is partly to show that these interfaces are interesting objects of study in their own right, and that the philosophical and technical issues involved can be quite subtle and multi-faceted. However, in so doing we shall also try to advance the state of the art wrt. to certain themes in the theory of norms, specifically those of norm system revision, positive permission and normative relevance—three concepts that turn out to be closely interlinked.

There are several different systems of input/output logic, but we shall focus exclusively on the two systems that according to the nomenclature of Makinson and van der Torre (2000, 2001, 2003) are called *simple-minded output* and *basic output* respectively. All our positive results will be formulated wrt. the former, whereas the latter will be used to provide contrast in the form of negative results. The text thus takes the form of a contrapuntal exposition and development designed to show that although connections exist that can be used to translate results from theory revision and propositional relevance into input/output logic, the connections are fragile and tend to break easily.

The chapter is organised as follows: Section 2 gives a contrasting exposition of the two systems of simple-minded and basic output respectively. It is meant to serve the dual purpose of introducing input/output logic, as well as to pitch one important difference between them. This difference will run as a thread throughout most of the chapter and concerns the property we label input-entailment. Input-entailment is satisfied by simple-minded output but not basic output. In Sect. 3 we show that input-entailment is required in order to have parallel interpolation for norms, a property which in turn facilitates the analysis of normative relevance by way of language-splitting in Sect. 5. In Sect. 4 we show that input-entailment is necessary for correlating maximally non-implying sets of sentences with maximally non-implying sets of norms, thereby directly correlating AGM contraction with derogation of norms. Thus, only in the case of simple-minded output is there a transparent interface between input/output logic and the parts of Makinson's production that concerns respectively propositional relevance and revision. The situation is rather subtle and interesting, though, and it is not always obvious that compatibility on a technical level implies coherence on a philosophical level. We shall comment on these things as we go. Section 6 rounds off the chapter with an illustration of how the pieces we have assembled fit together, by giving a formal treatment of the concept of positive permission based on the notion of derogation from Sect. 4, but tempered by the notion of relevance from Sect. 5.

Several of the results and ideas that will be set out in the following are foreshadowed in earlier publications of the present author (cf. Stolpe 2010a,b,c). However, most of the major theorems of this chapter are new, most importantly the parallel interpolation theorem for simple-minded output in Sect. 3. This theorem enables us to give a more principled and unified account of the relationship between normative relevance and derogation than in previous attempts. It is hoped that the overall picture that emerges will be perceived as proportionally clear.

2 Two Systems of Input/Output Logic

Notation. In the following, we shall use ϵ to denote a finite set of propositional letters, and L to denote the propositional language generated by that set. We include the zero-ary *falsum* f among the connectives in L, and define t as $\neg f$. Lower case letters a, b, c... not including t and f range over formulae of L, whereas sets of formulae are denoted by upper case letters from A to D. When A is a finite set of formulae $\bigwedge A$ denotes the conjunction and $\bigvee A$ the disjunction of its finitely many elements. Upper case letters from F to I range over subsets of $L \times L$—we think of them as sets of norms—and lower case letters r, s, possibly subscripted, range over individual norms, that is, over pairs of formulae (a, b). If H is an n-ary relation, then we shall write H^i for $1 \leq i \leq n$ to denote the projection of H onto its ith coordinate. We extend this notation to elements s of H, writing s^i for the ith coordinate of s. Image-formation will be indicated by supplying an argument in parentheses to a relation, e. g. $(G \cup H)(a)$ denotes the image of a under the relation $G \cup H$. Classical consequence is written with a turnstile $\vdash$ when considered as a relation over $2^L \times L$, and as Cn when viewed as an operation on 2^L onto itself. To make the notation less verbose, we follow the convention of writing $A \cup a$ instead of $A \cup \{a\}$, and similarly for norms.

2.1 Simple-Minded Output

Although input/output logic is usually advertised as a form of specifically deontic reasoning, there is really nothing in the theory itself that forces this interpretation. Rather input/output logic is more aptly considered a logical framework for reasoning about production relations (cf. Bochmann 2005, chp. 8). Here, by a production relation we understand any relation that correlates a state of affairs a with another one b such that a is a sufficient condition for b to obtain, but not necessarily in a logical sense. That is, if (a, b) belongs to the relation in question, then the truth of a is enough to guarantee the truth of b, but the guarantor may not be logic—a need not imply b.

As such, input/output logic ought to have much to offer not only the philosophy of norms, but also artificial intelligence where production relations arise naturally (but are allegedly rarely treated with much formal precision (Bochmann 2005, p. 232)). Examples include production rules from automated planning which provide the mechanism necessary to execute actions in order to achieve some goal, as well as causal links from causal theories which are used to describe the changes ensuant to some event in some environment. These relations are not primarily logical relations; the shattered glass is not implicit in the momentum of the rock (unless physics is logic and logic is metaphysics), rather the behaviour of the rock and the glass is consistent across a large enough sample to justify counting the mass and velocity of the rock as the cause of the shattered glass.

The characteristic feature of all production relations is that they cannot be assumed to be reflexive. A cause is probably never its own effect, nor is the condition under which a goal is pursued necessarily itself a goal. Indeed, no information can in general be assumed to be carried back and forth between the relata, so production relations do not satisfy contraposition either (a butterfly flapping its wings may cause a tornado to alter its path, but if the tornado does not alter its path the butterfly may nevertheless be flapping its wings). In other words, a production relation is logically arbitrary in the sense that the respective coordinates of the relation can from a logical point of view be totally unrelated. It follows that any theory of production relations, if it is to be framed in a logical idiom, faces the fundamental challenge of giving a simple and compelling picture of how a potentially very irregular relation between states of affairs comes to act in consort with the very regularly behaved relation of logical entailment to yield something that is in turn regularly behaved itself. Simple-minded output provides an elegant answer; the two relations align by composition:

Definition 1. *Let* $G \subseteq L \times L$. *Then* $out_1(G, a) = Cn(G(Cn(a)))$.[1]

So defined out_1 is an operation from $2^{L \times L} \times L$ onto 2^L, but input/output operators can also be considered operations from $2^{L \times L}$ into itself, in which case we write $(a, b) \in out(G)$ to mean the same as $b \in out(G, a)$.

To be sure, it *is* true that input/output logic was originally conceived of as a logic of norms (Makinson and van der Torre 2000, 2001, 2003), but that just goes to show that in input/output logic a code of norms is thought of as a production relation in the aforementioned sense: a norm is simply a pair (a, b) correlating an *antecedent, input* or *context* a with a *consequent, output* or *duty* b. The correlation between the

---

[1] Makinson and van der Torre usually allow infinite sets in the right argument of the out operators. This level of generality is not needed in the present chapter, whence we restrict all definitions to single formulae, equivalently finite sets.

antecedent and consequent is perceived as logically arbitrary in the sense that there is nothing to the norm (a, b) over and above the fact that some authority requires that b be done given a.

This reflects a philosophical conviction that runs through all of Makinson's writing on norms since his 1999 paper "On a Fundamental Problem of Deontic Logic", namely that norms are stipulations rather than facts of an independent reality. A norm may be applied or not, recognised as valid or not, and more or less entrenched in our system of values, but norms are not in general true or false (Makinson 1996).

Although Makinson does not discuss the point explicitly, one might be willing to concede that *some* natural classes of norms (if such a thing there be) can be treated as if they bore truth values, for instance moral norms, as is sometimes claimed by philosophers of a realist orientation. They may well be right that it is wrong in a stance-independent, objective sense to persecute people on the basis of religious beliefs, ethnicity or race, and to exploit another's trust solely for personal gain. If so, then calling those norms that have objective support true, and those that don't have it false, seems justified enough. A valid moral argument, then, is one that preserves truth, and the authority of moral norms is proportional to the legitimacy of their truth-claim (Shafer-Landau 2003), p. 248.

However, there are obvious counterexamples in the general case. Consider the regulation from Norwegian law that requires all shops to be closed on Sundays except those that deal mainly in groceries and have a total sales area of less then a hundred square meters. This is not a stance-independent feature of reality, but a logically arbitrary stipulation laid down by the Norwegian authorities for political reasons.

Now, if norms do not in general have truth values, then they are not propositions, and so might not be represented most naturally as conditionals in the object-language. Therefore, input/output logic takes the view that it is the norm-giving authority that is responsible for correlating a state-of affairs with a regulative or duty. Input/output logic therefore substitutes pairs, representing the potentially logically arbitrary decisions of the authority in question, for formulae as the principal blue-print for norms. Yet, since both the antecedent and consequent of a norm remain formulae, norms can still be said to have a logic. Indeed, the operator from Definition 1 can be characterised as follows (Makinson and van der Torre 2000, Observation 1):

Theorem 1. *For any $G \subseteq L \times L$, $out_1(G)$ is the set of norms that are derivable in the system having axioms $(t, t) \cup G$ and the following inference rules:*

$$SI \ \frac{(c, b)}{(a, b)} \ \text{if } a \vdash c \quad AND \ \frac{(a, b), (a, c)}{(a, b \wedge c)} \quad WO \ \frac{(a, b)}{(a, c)} \ \text{if } b \vdash c$$

Here as in Makinson and van der Torre (2003), a *rule* r of arity $n \geq 0$ is an $n + 1$-ary relation over the set $L \times L$ of pairs of formulae in the language L. For any $((a_1, b_1), \ldots, (a_n, b_n), (a_{n+1}, b_{n+1})) \in r$ we call $(a_1, b_1), \ldots, (a_n, b_n)$ the premises of the rule and (a_{n+1}, b_{n+1}) its conclusion. A *derivation* of a pair (a, b) from G, given a set R of rules, is understood to be a tree with (a, b) at the root, each non-leaf node related to its immediate parents by the inverse of a rule in R, and each leaf node either the conclusion of a zero-premise rule in R, or an element of G, or of the form (t, t). We shall henceforth denote *out*-derivability with the subscripted turnstile $\Vdash_n$ where the subscript n will correspond to a particular system. Equivalence relative to a system will similarly be denoted by $\cong_n$ and will informally be referred to as n-equivalence.

Turning now to meta-logical properties, simple-minded output can be viewed as a most natural and immediate generalisation of classical logic. Classical logic becomes the special case where the set of norms G is the diagonal relation over L (cf. Stolpe 2010a, Theorem 2):

Theorem 2. *Let G be the diagonal relation over L. Then $out(G, a) = Cn(a)$.*

Inevitably some properties of classical logic are lost, for instance norms in input/output logic are not reflexive or transitive, nor do they satisfy the principle of disjunctive antecedents (to which we shall return shortly).

Yet, some very interesting, and for certain purposes important, properties also carry over, sometimes in a somewhat diluted form. As an example, take the property of *adjunctiveness* from classical logic, according to which every finite set $A \subseteq L$ is equivalent to the conjunction $\bigwedge A$ of its finitely many formulae. In input/output logic, if we understand adjunction as the operation of taking (a, b) and (c, d) to $(a \wedge c, b \wedge d)$, we have one direction:

Lemma 1. $H \Vdash_1 (\bigwedge H^1, \bigwedge H^2)$.

We shall henceforth denote the adjunction of the finitely many elements of H by $\bigwedge H$, relying on context to disambiguate between sets of formulae and sets of norms. The converse of Lemma 1 does not hold, but we do have a restricted version of it:

Lemma 2. *If $G \Vdash_1 (a, b)$ and $a \vdash s^1$ for each $s \in G$ then $(\bigwedge G^1, \bigwedge G^2) \Vdash_1 (a, b)$.*

Moreover, if $G \Vdash_1 s$ then we can always find a subset of G that satisfies the conditions of Lemma 2.

Lemma 3. *If $G \Vdash_1 (a, b)$ then there is an $H \subseteq G$ such that $H \Vdash_1 (a, b)$ and $a \vdash s^1$ for each $s \in H$.*

As a corollary we have:

Corollary 1. *If $G \Vdash_1 (a, b)$, then there is an $H \subseteq G$ such that $(\bigwedge H^1, \bigwedge H^2) \Vdash_1 (a, b)$.*

Lemma 3 can be generalised to the following property of input-entailment, which says that no non-trivial $\Vdash_1$-derivation diminishes the logical strength of the antecedent of a norm. That is, no $\Vdash_1$-derivation yields a norm with more general applicability than any of its premises:

Lemma 4. (Input entailment). *If $(c, d) \in out_1(F \cup (a, b)) \setminus out_1(F)$ then $c \vdash a$.*

The importance of Lemma 4 is that it allows us to simulate propositional adjunctiveness for each choice of a derivable norm (a, b) from G. Among other things, this is a necessary prerequisite for the existence of parallel interpolants between (a, b) and G—the topic of Sect. 3—which in turn opens up for an analysis of normative relevance along the lines of language splitting. Thus, Lemma 4 will be a pivotal property in subsequent sections.

2.2 Basic Output

Simple-minded output is a generalisation of classical logic that causes some properties known to hold for material implication to fail for norms. Some of these failures may be considered 'benign losses', for instance the failure of reflexivity: it is the purpose of a logic of norms to enable us to reason about the potential discrepancy between the actual and the ideal. If conditional norms are taken to be reflexive, the actual is always ideal so this distinction collapses. On the other hand, one may question whether the failure of disjunctive antecedents is equally desirable. In input/output logic this is the rule of inference that maps premises (a, b) and (c, b) to the conclusion $(a \vee b, c)$ for all choices of a, b and c. Adhering to the nomenclature of Makinson and van der Torre (2000, 2001, 2003) we shall refer to this rule as OR.

By way of intuitive motivation, suppose a penal statute states both that 1) "a person who has shown gross negligence of duty ought to be suspended from office" and 2) "a person who has committed a crime ought to be suspended from office". Is it reasonable to interpret the statute as saying also 3) "a person who has committed a crime *or* shown gross negligence of duty ought to be suspended from office"? Note that 3) does not require it to be known whether it is gross negligence or a crime that the person in question is guilty of, only that it is one or the other. Thus, only if statutory interpretation endorses 3) can the statute be applied to cases for which it is impossible to determine whether the conduct in question carries the necessary elements of a crime: if *mens rea* can be established then perhaps its a crime and not gross negligence, but if not then perhaps its gross negligence and not a crime.

On the semantic level disjunctive antecedents can be added to simple-minded output by varying Definition 1 in the following natural way:

Definition 2. $out_2(G, a) = \bigcap\{Cn(G(V)) : a \in V, V complete\}$.[2]

As observed by Makinson and van der Torre (2000), simple-minded output can be expressed as an intersection $out_1(G, a) := \bigcap\{Cn(G(B)) : a \in B = Cn(B)\}$. Hence, basic output is like simple-minded output except that it restricts the choice of B to complete sets. Observation 2 of Makinson and van der Torre (2000) shows that this operator corresponds to the following system:

---

[2] Here a complete set is one that is maxiconsistent or equal to L.

Theorem 3. *$(a, b) \in out_2(G)$ iff it is derivable by $\Vdash_1$ supplemented with OR (i.e. $\Vdash_2$).*

Interestingly, the principle of disjunctive antecedents pushes input/output logic to the yonder side of the demarcation line drawn by input-entailment (that is, Lemma 4):

Example 1. Put $G := \{(a, b), (c, b)\}$ for distinct and elementary a, b and c, then $G \Vdash_2 (a \vee c, b)$ but since $a \vee c \nvdash a$ and $a \vee c \nvdash c$ Lemma 3 fails.

As announced in the introduction, the property of input-entailment is necessary for parallel interpolation. As such, it is a difference between simple-minded and basic output that is key to the material that follows. The situation is rather subtle and interesting: since disjunctive antecedents is a valid principle of classical logic itself, basic output is in one sense closer to classicality than simple-minded output is. On the other hand, since disjunctive antecedents toggles parallel interpolation off, basic output is in another sense more distant, for as first proved by Kourousias and Makinson (2007), parallel interpolation is a property of classical logic. Parallel interpolation is the topic of the next section, and we postpone a closer examination until then.

2.3 *Basic Output in Terms of Formal Concept Analysis*

Adding OR to the rules of simple-minded output yields a system that is symmetric in the following sense: for every rule that strengthens resp. weakens an input, there is a dual rule that weakens resp. strengthens an output. One of the more interesting and hitherto unnoticed consequences of this duality consists in the fact that any code of norms G can be represented as an equivalent (in a sense to be made precise) formal concept lattice modulo basic output. The present subsection spells out this link in some detail. It is not needed for the main argument of the paper, and may safely be skipped on a first reading.

We recall the central definitions of formal concept analysis (Ganter and Wille 1999; Davey and Priestley 2002): a *formal context* records a set of *objects* and the *attributes* they possess:

Definition 3. **(Context).** *A context is a triple (A, B, I) where A and B are sets and $I \subseteq A \times B$. The elements of A and B are called objects and attributes respectively.*

Obviously, *any* binary relation G can be considered a context, including binary relations over *formulae* as in input/output logic. Next, *polar maps* correlate sets of objects with sets of attributes and vice versa:

Definition 4. **(Polar maps).** *Let (A, B, I) be a context, $C \subseteq A$ and $D \subseteq B$. Then*

$$C^{\rhd} = \{b \in B : (\forall a \in C)(a, b) \in I\}$$
$$D^{\lhd} = \{a \in A : (\forall b \in D)(a, b) \in I\}$$

It is a fundamental result of formal concept analysis that the pair of maps $^{\rhd}$ $^{\lhd}$ forms a Galois connection between 2^A and 2^B in which $^{\rhd}$ is the lower- and $^{\lhd}$ the upper adjoint (Davey and Priestley 2002; Ganter and Wille 1999).

A *concept* is considered to be determined by its *extent* and *intent*: the extent consists of all objects belonging to the concept, while the intent is the collection of attributes that those objects share (Davey and Priestley 2002, p. 64):

Definition 5. (Concepts). *Let (A, B, I) be a context, $C \subseteq A$ and $D \subseteq B$. Then (C, D) is a concept of (A, B, I) iff $C^{\rhd} = D$ and $D^{\lhd} = C$.*

To demand that a concept is determined by its extent ($C^{\rhd} = D$) and by its intent ($D^{\lhd} = C$) means that D should contain just those properties/attributes shared by all objects in C, and similarly that the objects in C should be precisely those that share all the properties in D. The set of concepts of the context (A, B, I) is usually written $\mathfrak{B}(A, B, I)$. In the case where the context in question is (G^1, G^2, G) for a given binary relation G, we simplify and write $\mathfrak{B}(G)$.

Concepts can be ordered by extents or intents by putting $(C_1, D_1) \leq (C_2, D_2)$ either iff $C_1 \subseteq C_2$ or iff $D_2 \subseteq D_1$. The two are the same since upper and lower adjoints of Galois connections are antitone. I.e. $C_1 \subseteq C_2$ implies $C_2^{\rhd} \subseteq C_1^{\rhd}$, whence since $C_2^{\rhd} = D_2$ and $C_1^{\rhd} = D_1$ it follows that $D_2 \subseteq D_1$. The fundamental theorem of concept lattices says that $\langle \mathfrak{B}(A, B, I), \leq \rangle$ (henceforth to be identified with $\mathfrak{B}(A, B, I)$) is a complete lattice in which the context (A, B, I) is encoded in the following manner: put $\gamma(a) = (\{a\}^{\rhd\lhd}, \{a\}^{\rhd})$ for any $a \in A$ and $\mu(b) = (\{b\}^{\lhd}, \{b\}^{\lhd\rhd})$ for any $b \in B$. Then $\gamma(A) := \{\gamma(a) : a \in A\}$ is join-dense and join-irreducible, $\mu(B) := \{\gamma(b) : b \in B\}$ is meet-dense and meet-irreducible and $\gamma(a) \leq \mu(b)$ iff $(a, b) \in I$. In other words, the partial order $\leq$ encodes the relation I in the respective sets of meet- and join-irreducible concepts. Since these concepts are also meet- and join *dense*, the extent and intent of *any* concept $\mathfrak{c}$ in $\mathfrak{B}(A, B, I)$ can be read off like so:

$$(A_{\mathfrak{c}}, B_{\mathfrak{c}}) := (\{a \in A : \gamma(a) \leq \mathfrak{c}\}, \{b \in B : \mathfrak{c} \leq \mu(b)\}) \qquad (1)$$

Thus, if we label each join- and meet-irreducible element of $\mathfrak{B}(A, B, I)$ with the object or property that generates it, then we can read off each concept from the lattice by following the order relation up and down whilst collecting labels as we go.

Example 2. A national labour and welfare service provides social benefits that contribute to the financial security of citizens and other right holders. Being entitled to a social benefit usually depends on certain required pieces of information that are associated by law with the tasks in question. Assume the following lists of tasks and information requirements:

Tasks:	**Information requirements:**
a_1 = declare incapacity to work	b_1 = verify sickness
a_2 = grant sickness pay	b_2 = determine line of work
c_1 = fund further education	d_1 = determine present education
c_2 = calculate insurance benefits	d_2 = obtain family medical history

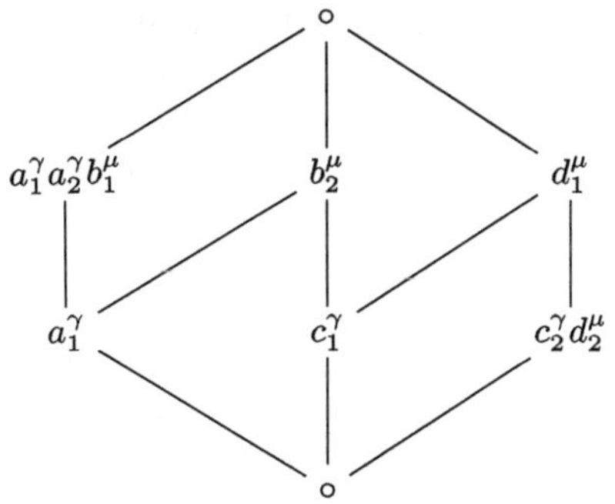

Fig. 1 A concept lattice for (some) social benefits

Suppose the relationship between tasks and information is given by the following relation $G := \{(a_1, b_1), (a_1, b_2), (a_2, b_1), (c_1, b_2), (c_1, d_2), (c_2, d_2), (c_2, d_1)\}$. Then the code corresponds to the concept lattice in Fig. 1.

The labels on the elements correspond to the join- and meet-irreducible elements induced by G^1 and G^2 respectively, i.e. a_1^γ is the join-irreducible concept $\gamma(a_1)$. We can read off the extent and intent of any concept by instantiating equation (1). For instance, the concept labelled a_1^γ is the concept $(\{a_1\}, \{b_1, b_2\})$ which tells us that declaration of incapacity to work requires verification of sickness and line of work. Similarly, the extents/objects below d_1^μ tell us that determining the client's present education is required for funding further education as well as for calculating a client's insurance benefits. The latter also requires a family medical history.

The connection between concept lattices and input/output logic is based on giving concepts a logical interpretation: recall that (C, D) being a concept means that C consists of the set of objects which is such that *each* object in C can be ascribed *all* attributes in D. In other words, each object in C constitutes a sufficient condition for all of D to hold. This suggests that when the context in question is given by a binary relation over *formulae* then the extent of a concept should be understood *disjunctively* and the intent *con*junctively. The concept labelled $a_1^\gamma a_2^\gamma b_1^\mu$ then becomes the norm $(a_1 \vee a_2, b_1)$ whilst $c_2^\gamma d_2^\mu$ becomes $(c_2, d_1 \wedge d_2)$. So understood, $\mathfrak{B}(G)$ derives the same norms as G modulo out_2:

Theorem 4. *Suppose* $(C, D) \in \mathfrak{B}(G)$. *Denote by* $(C, D)^\psi$ *the norm* $(\bigvee C, \bigwedge D)$, *a notation we extend to the set* $\mathfrak{B}(G)^\psi$ *in the obvious way. We have* $out_2(G) = out_2(\mathfrak{B}(G)^\psi)$.

Proof. For the left-to-right inclusion, it suffices by monotony and idempotence for out_2 to show that $G \subseteq out_2(\mathfrak{B}(G)^\psi)$. So suppose $(a, b) \in G$. Then we have $(\{a\}^{\rhd\lhd}, \{a\}^{\rhd}) \in \mathfrak{B}(G)$, $\{a\}^{\rhd} \vdash b$ and $a \vdash \bigvee\{a\}^{\rhd\lhd}$. By the definition of ψ we also have $(\{a\}^{\rhd\lhd}, \{a\}^{\rhd})^\psi \in \mathfrak{B}(G)^\psi$. Therefore, $(a, b) \in out_2(\mathfrak{B}(G)^\psi)$ by WO and SI. For the converse inclusion, it suffices to show that $\mathfrak{B}(G)^\psi \subseteq out_2(G)$. Suppose $(\bigvee A, \bigwedge B) \in \mathfrak{B}(G)^\psi$ and thus *a fortiori* that $(A, B) \in \mathfrak{B}(G)$. Suppose further that $A = \{a_1, \ldots, a_n\}$ and $B = \{b_1, \ldots, b_k\}$. Since $B = A^{\rhd}$ we have $B \subseteq G(a_i)$ for every $1 \leq i \leq n$, whence $(a_i, b_j) \in G$ for every $1 \leq j \leq k$. It follows by AND that $(a_i, \bigwedge_{j=1}^{k} b_k) = (a_i, \bigwedge B) \in out_2(G)$, whence $(\bigvee_{i=1}^{n} a_i, \bigwedge B) = (\bigvee A, \bigwedge B) \in out_2(G)$ by OR.

This simple observation may be expected to be of some significance as a means of analysing a code. Formal concept analysis is a powerful technique for classifying, visualising and analysing complex sets of data, and reveals the inherent hierarchical structure and/or natural clusterings and dependencies among the objects and attributes in a data-set (Ganter and Wille 1999). Theorem 4 thus puts a wide array of well-understood tools at our disposal to help analyse the syntactical form of a code.

It is a well-known fact that many normative notions of theoretical interest supervene on this structure, that is, that they are syntax-dependent. For instance, as we shall see in Sect. 5, different equivalent codes may determine different retraction priorities among individual norms depending on their respective logical form.

Another interesting example is the concept of deontic redundancy: norms come into existence and pass out of existence in the course of the history of a normative system—sometimes as a consequence of acts of legislation, sometimes as commandments of a trusted authority or leader, sometimes as a result of gradually formed societal habits, customs and traditions, explicit agreement or design (von Wright 1998). Norms may become unfulfillable—perhaps they have an expiry date, apply to non-existent subjects, or regulate states of affairs that are no longer a practical possibility. Although such norms are regulatorily inert, they may still clutter the code and make it difficult to process or manage efficiently. Deontic redundancy has therefore been argued to pose an important problem for the theory of norms—though one largely ignored (van der Torre 2010). Consider the following rather abstract definition of redundancy (the reader is referred to van der Torre (2010) for more nuance):

Definition 6. *A non-empty code G contains redundancies (modulo out_2) if there is an $H \subset G$ such that $out_2(H) = out_2(G)$.*

That is, a code contains redundancies if it is equivalent to a proper part of itself. Now, as a simple example of analysing a code of norms using lattice-theoretic means, we take the time to show that $\mathfrak{B}(G)^\psi$ ordered by 2-entailment provides a straightforward sufficient condition for redundancy:

Theorem 5. *Suppose $(A, B), (C, D) \in \mathfrak{B}(G)^\psi$. Then G contains redundancies if $(A, B)^\psi \Vdash_2 (C, D)^\psi$ and $C \times D \neq G$.*

Proof. Suppose the conditions of the theorem hold. For any concept (A, B) we have $(A, B)^\psi \in out_2(A \times B)$, whence $(C, D)^\psi \in out_2(A \times B)$ by supposition. Moreover, for any $(c, d) \in C \times D$ we have $(C, D)^{\psi} \Vdash_2 (c, d)$, so $C \times D \subseteq out_2(A \times B)$. It follows that $out_2(G) \subseteq out_2(G \setminus (C \times D))$ whence $out_2(G \setminus (C \times D)) = out_2(G)$ by monotony for out_2. Since $C \times D \neq G$ we have $(G \setminus (C \times D)) \subset G$, so G contains redundancies.

Stated differently, if the non-trivial concepts of $\mathfrak{B}(G)^{\psi}$ form an anti-chain under $\Vdash_2$, then G is irredundant.[3]

Further exploration of the set of norms $\mathfrak{B}(G)^{\psi}$ would make us stray too far from the main topic of this section. Suffice it for now to say that Theorem 4 may serve as an illustration of how Makinson's methodological approach tends to place philosophical logic in a wider mathematical geography.

3 First Interface: Splitting

A conspicuous difference between input/output logic and most other logics of norms consists in the fact the former does not model the behaviour of a normative system by axiomatising a modal operator and/or an intensional conditional—that is, it is not a modal logic. In input/output logic the system is conceptually prior to the individual norm, and the notion of *a* norm only makes sense in relation to the system of which it forms part. One might say that it is the principal responsibility of a norm to state its own propositional content, whereas it is the system, as a first-class citizen of the theory, that combines, infers or recycles. The content of a norm is therefore simply expressed in propositional logic, whence input/output logic may be said to heed the maxim *don't abandon classical logic, use it creatively*. The idea, in Makinson's own words, is "to work with the simplest possible syntactic apparatus, reserving complex machinery until the exact limits of the more spartan one are clear and only in so far as it is confirmed that its essential ideas are indeed 'on the right track' " (Makinson 1999, p. 3).

This commitment to classical logic is interesting insofar as it allows results and techniques from the extremely well-studied field of propositional logic to be redeployed in the philosophy of norms. In the present section, we shall show that and how this can be done with respect to the technique of *language splitting* aka *propositional relevance*.

Language splitting was originally invented by Parikh (1999) as a way to partition a set of sentences into a set of disjoint modules that hold information about unrelated subject matters. Let $E = \{E_i\}_{i \in I}$ be any partition of the set of all elementary letters of the language L. E is a splitting of A iff there is a family $\{B_i\}_{i \in I}$ with each $\epsilon(B_i) \subseteq E_i$ such that $\bigcup\{B_i\}_{i \in I} \dashv\vdash A$. Parikh's *finest splitting theorem* says that A, no matter how it is selected as long as it is finite, has a finest splitting. Stated differently, the theorem tells us that every finite set of formulae A has a finest representation as a

[3] The converse does not hold.

family of letter-disjoint sets, whence there is a unique way of thinking of a set of formulae as being about a set of *mutually irrelevant* topics.[4]

It is an interesting question whether the same is true of codes of norms, for it is common for a code, if it is of any complexity, to display a modular structure. Take civil law as an example: a civil law system invariably decomposes into different bodies of law which it is natural and convenient to treat as separate social institutions. Property law defines the conditions under which rights and obligations pertaining to the control of physical and intangible objects are transferred between legal persona; criminal law sets out the punishment to be imposed on behaviour that threatens, harms or endangers the safety and welfare of individuals or the society as a whole; administrative law governs eligibility to hold office as well as the activities of administrative agencies of government, and so on and so forth. Intuitively some such modules are semantically independent of each other—for instance motor vehicle codes and stock market regulations. We ought to be able to disentangle them formally.

The finest splitting theorem was generalised to the infinite case by Kourousias and Makinson,[5] who also showed how to use splittings to normalise belief-change operations by making them relevance-sensitive (see also Chopra and Parikh (2000)). In the present section we shall mimic the overall proof procedure devised there (although some steps will require fairly thorough-going adjustment) to obtain an analogue of the finest splitting theorem for sets of norms modulo out_1 (a corresponding concept of normative relevance will be set out in Sect. 5).[6] We shall then contrast this result wrt. the operation of basic output, for which parallel interpolation fails.

3.1 Splitting Modulo Simple-Minded Output

Notation. In the following we shall let ϵ do dual service as a propositional operator, writing $\epsilon(A)$ for the elementary letters of $A \subseteq L$. Note that since we treat the falsum f as a zero-ary connective we have $\epsilon(f) = \epsilon(t) = \emptyset$. Lower-case Greek letters $\alpha, \beta, \gamma, \delta$ will denote subsets of ϵ, $\widehat{\epsilon}$ the set of partitions of ϵ, and $\pi_1, \pi_2, \ldots$ elements of $\widehat{\epsilon}$. Generalising to sets of norms we write Γ and Δ (possibly subscripted) for elements of $2^\epsilon \times 2^\epsilon$. Σ is a function such that $\Sigma(G) := (\epsilon(G^1), \epsilon(G^2))$ for any non-empty $G \subseteq L \times L$, and $\Sigma(\emptyset) = (\emptyset, \emptyset)$. $\widehat{\Sigma}$ denotes the set $\{\pi_i \times \pi_j | (\pi_i, \pi_j) \in \widehat{\epsilon} \times \widehat{\epsilon}\}$, and $\mathscr{E}$, $\mathscr{F}$ (possibly subscripted) are members of $\widehat{\Sigma}$. We also introduce an order relation $\preceq$ over $2^\epsilon \times 2^\epsilon$, and write $\Gamma \preceq \Delta$ whenever $\Gamma^i \subseteq \Delta^i$ for $i \in \{1, 2\}$. If $\Gamma^1 \cap \Delta^1 = \emptyset = \Gamma^2 \cap \Delta^2$ then we shall say that Γ and Δ are coordinatewise disjoint.

---

4 The reader is referred to Parikh (1999); Chopra and Parikh (2000); Makinson (2009); van de Putte (2011) for background and developments.

5 The notion has since migrated from there to description logics (see e.g. Konev et al. (2010); Ponomaryov (2008)) where it is better known as *signature decomposition*.

6 The reader should note that although Kourousias and Makinson's theorem was designed to cover infinite sets of sentences, we nevertheless apply it only to *finite* codes.

There are several ways to define the splitting of a code of norms (modulo an input/output operator); one may split only on the input side, or one may split only on the output side (as in Stolpe (2010b)). However, in the absence of contrary evidence, the most principled approach is to count both antecedents and consequents of norms as relevant for the 'subject' or 'topic' of a norm, and therefore to split on both sides:

Definition 7. (Splitting). *Put* $\mathscr{E} := \{\Gamma_k\}_{k\in I}$ *for* $\mathscr{E} \in \widehat{\Sigma}$. *We say that* $\mathscr{E}$ *splits G iff there is a family* $\{G_k\}_{k\in I}$ *such that each* $\Sigma(G_k) \preceq \Gamma_k$ *and* $G \cong_1 \bigcup\{G_k\}_{k\in I}$.

We shall say of $\{G_k\}_{k\in I}$ that it is *partitioned according to* $\mathscr{E}$ and that it *splits* G. If the reference to $\mathscr{E}$ and $\{G_k\}_{k\in I}$ are unimportant or clear from context, we shall simply say that G is *in finest splitting form*.

Note that Definition 7 partitions all the left letters of a code of norms, and quite independently partitions all the right letters. In other words $\Gamma_i^1 \cap \Gamma_j^2 \neq \emptyset$ is not ruled out. That is, a cell on the left may very well overlap a cell on the right. What matters is that a splitting partitions each side taken separately, i.e. $\Gamma_i^n \cap \Gamma_j^n = \emptyset$ for $n \in \{1, 2\}$, so a cell on the left may never overlap another cell on the left and similarly for the right side.

It is probably not obvious why we do not simply partition the letters occurring in the code, whether on the left or on the right. Peeking ahead to Sect. 5 there is this to say: we shall use the splitting to define a criterion of relevance that applies to norms (in analogy to the classical propositional case). In the general case, a splitting that lumps together antecedents and consequents is too coarse to yield a criterion that is sensitive to the input/output operator in question. To see this, suppose the input/output operator is out_1 and consider the code $G := \{(a \vee b, c), (a, d)\}$ where all letters are elementary and distinct. If each norm is to be expressible in exactly one cell of the splitting, and the splitting treats antecedents and consequents collectively, then it will contain the cell $\{a, b, c, d\}$. Yet, there are good reasons why (a, d) ought not to be deemed relevant to $(a \vee b, c)$: the consequents are independent, and, due to the input-entailment property of out_1, nothing that pertains to the context a will ever affect the weaker context $a \vee b$.

Besides, treating antecedents and consequents independently is really the only strategy that is coherent with the essential idea behind input/output logic, which is that of isolating the condition of application of a norm from its normative consequences so that information about one cannot be carried forwards or backwards to the other. This is the essential feature that separates norms from conditionals and makes them non-reflexive. It needs to be mirrored, therefore, by the definition of a splitting.

It is a key fact of the set of splittings as so defined that it forms a complete lattice under the following fineness relation:

Definition 8. (Fineness). *We use the notation* $\pi_1 \leqslant \pi_2$ *to signify that* π_1 *is* at least as fine as π_2 *according to the usual definition of fineness for partitions. We extend this notion to elements of* $\widehat{\Sigma}$ *by deeming* $\mathscr{E} := \pi_1 \times \pi_2$ *at least as fine as* $\mathscr{F} := \pi_3 \times \pi_4$ *iff* $\pi_1 \leqslant \pi_3$ *and* $\pi_2 \leqslant \pi_4$. *We overload notation and use* $\leqslant$ *for both relations.*

We record this as a separate lemma:

Lemma 5. *$\langle \widehat{\Sigma}, \lessdot \rangle$ is a complete lattice.*

Proof. Let $\mathscr{E}, \mathscr{F} \in \widehat{\Sigma}$. Put $\mathscr{E} := \pi_1 \times \pi_2$ and $\mathscr{F} := \pi_3 \times \pi_4$ and consider the relation $\mathscr{E}_2 := \pi_1 \wedge \pi_3 \times \pi_2 \wedge \pi_4$ where $\wedge$ is *infimum*. It is a standard result of general lattice theory that $\widehat{\epsilon}$ is itself a complete lattice, so $\pi_1 \wedge \pi_3, \pi_2 \wedge \pi_4 \in \widehat{\epsilon}$ whence $\mathscr{E}_2 \in \widehat{\Sigma}$. Now, clearly $\mathscr{E}_2 \lessdot \mathscr{E}$ and $\mathscr{E}_2 \lessdot \mathscr{F}$ since $\pi_1 \wedge \pi_3$ is at least as fine as each of π_1, π_3 and similarly for π_2 and π_4. It remains to show that $\mathscr{E}_2$ is indeed the greatest lower bound: let $\pi_a \times \pi_b \lessdot \pi_1 \times \pi_2, \pi_3 \times \pi_4$. Then $\pi_a \lessdot \pi_1, \pi_3$ and $\pi_b \lessdot \pi_2, \pi_4$, whence $\pi_a \lessdot \pi_1 \wedge \pi_3$ and $\pi_b \lessdot \pi_2 \wedge \pi_4$. Therefore $\pi_a \times \pi_b \lessdot \mathscr{E}_2$, i.e. $\mathscr{E}_2$ is the infimum of $\mathscr{E}$ and $\mathscr{F}$. A similar argument establishes suprema, therefore $\widehat{\Sigma}$ is a lattice. That it is complete follows from the finiteness of ϵ.

Following the strategy of Kourousias and Makinson (2007), we shall eventually prove the finest splitting theorem for simple-minded output by selecting the greatest lower bound $\mathscr{E}$ on a relevant subset $\mathcal{X}$ of $\widehat{\Sigma}$, and then use parallel interpolation to show that $\mathscr{E} \in \mathcal{X}$.

In the classical case, parallel interpolation has an appealingly simple proof. Unfortunately it cannot be rerun for simple-minded output. We give a proof that instead of mimicking the proof of the classical theorem applies that theorem. Note that this proof and the proof of the finest splitting theorem below both require the aforementioned input-entailment property:

Theorem 6. (Parallel 1-interpolation). *Let $G = \bigcup \{G_i\}_{i \in I}$ where the $\Sigma(G_i)$ are coordinatewise disjoint, and suppose $G \Vdash_1 s$. Then there is a set of norms $\{s_i\}_{i \in I}$ such that $\epsilon(s_i^j) \subseteq \epsilon(G_i^j) \cap \epsilon(s^j)$ for $j \in \{1, 2\}$, $G_i \Vdash_1 s_i$ and $\{s_i\}_{i \in I} \Vdash_1 s$.*

Proof. In the limiting case that $G = \emptyset$ we have $s \cong_1 (t, t)$ and $\Sigma(G_i) = \Sigma(\emptyset) = (\emptyset, \emptyset)$ for the $i \in I$. Therefore, since $\epsilon(t) = \emptyset$, we may choose $s_i = (t, t)$, whence the property holds trivially. For the principal case that G is non-empty put $s := (a, b)$. Since $(a, b) \in out_1(G)$, then by Lemma 3 we may choose a set of elements $S := \{s_i\}_{i \in I}$ of G such that

1. $a \vdash s_i^1$ for every $i \in I$, and
2. $\{s_i^2\}_{i \in I} \vdash b$

The proof proceeds from this observation in two steps: for the first step, let J be the least subset of I such that $S \subseteq \bigcup \{G_j\}_{j \in J}$. We show that parallel interpolants exist between $\bigcup \{G_j\}_{j \in J}$ and (a, b). For the second step, we select a suitable norm from each G_k where $k \in I \setminus J$ to add to the stock of interpolants from step one. We thus obtain a full generalisation.

So, sort the norms in S to keep track of their provenance, specifically let f be a function from J to 2^S such that $f(k) := \{s \in S : s \in G_k\}$ and put $r_k := (\bigwedge f(k)^1, \bigwedge f(k)^2)$. Note that $f(k)$ is non-empty since G is non-empty and J is minimal, whence r_k is well-defined. We have $a \vdash r_k^1$ from item 1 since the range of f is S, moreover

3. by Craig interpolation there is a c_k with $a \vdash c_k \vdash r_k^1$ s.t. $\epsilon(c_k) \subseteq \epsilon(r_k^1) \cap \epsilon(a)$

4. $\{r_k^2\}_{k\in J} \vdash b$ from item 2 since the range of f is S
5. $G_k \Vdash_1 r_k$ by Lemma 1
6. $\{\epsilon(r_k^2)\}_{k\in J}$ is a partition

From item 4 and item 6 it follows that we may apply standard parallel interpolation to obtain for every r_k a d_k such that

7. $r_k^2 \vdash d_k$, $\epsilon(d_k) \subseteq \epsilon(r_k^2) \cap \epsilon(b)$ and $\{d_k\}_{k\in J} \vdash b$

By renumbering J from 1 to m inclusive we thus have the following derivation from G:

$$\cfrac{\cfrac{\cfrac{\cfrac{\cfrac{\cfrac{}{(r_1^1, r_1^2), \ldots, (r_m^1, r_m^2)}\textit{ from 5}}{(c_1, r_1^2), \ldots, (c_m, r_m^2)}\;SI\textit{ from item 3}}{(c_1, d_1), \ldots, (c_m, d_m)}\;WO\textit{ from item 7}}{(a, d_1), \ldots (a, d_m)}\;SI\textit{ from 3}}{(a, \bigwedge_{j\in J} d_j)}\;AND}{(a, b)}\;WO\textit{ from item 7}$$

Since $\epsilon(r_k^1) \subseteq \epsilon(G_k^1)$ and $\epsilon(r_k^2) \subseteq \epsilon(G_k^2)$ it follows from item 3 and 7 that the norms $T := (c_1, d_1), \ldots, (c_k, d_k)$ in the third line of the derivation above are parallel interpolants between $\bigcup\{G_j\}_{j\in J}$ and (a, b). This completes the first step of the proof. For the second step it suffices to find for each $m \in I \setminus J$ one norm (c_m, d_m) such that $G_m \Vdash_1 (c_m, d_m)$, $T \cup \{(c_m, d_m)\}_{m\in I\setminus J} \cong_1 T$ and such that the conditions of the theorem hold. Put $(c_m, d_m) = (t, t)$ for each $m \in I \setminus J$ and the proof is complete.

The next lemma generalises Lemma 2.3 from Kourousias and Makinson (2007). In analogy to its classical counterpart it permits us to focus more on individual norms in G and less on G itself when attempting to prove that a given $\mathscr{E} \in \widehat{\Sigma}$ is a splitting of G (Kourousias and Makinson 2007, p. 3). The proof is rather different, though, in that it makes direct use of parallel interpolation:

Lemma 6. *Put $\mathscr{E} := \{\Gamma_k\}_{k\in I}$. Then $\mathscr{E}$ splits G iff for every $s \in G$ there are norms $s_1, \ldots, s_n$ and pairs $\Gamma_1, \ldots, \Gamma_n \in \mathscr{E}$ such that*

1. $\Sigma(s_k) \preceq \Gamma_k$ *for each* $k \leq n$
2. $G \Vdash_1 s_k$ *for each* $k \leq n$
3. $\{s_k\}_{k\leq n} \Vdash_1 s$.

Proof. Suppose $\mathscr{E} = \{\Gamma_k\}_{k\in I}$ splits G. Then there is a family $\{G_k\}_{k\in I}$ such that $\Sigma(G_i) \preceq \Gamma_i$ and $G \cong_1 \bigcup\{G_k\}_{k\in I}$. Take any $s \in G$, then $\bigcup\{G_k\}_{k\in I} \Vdash_1 s$. Since $\Sigma(G_i) \preceq \Gamma_i$ we have that the $\Sigma(G_i)$ are coordinatewise disjoint. We may therefore apply parallel interpolation to obtain a set $\{s_1, \ldots s_n\} \subseteq G$ such that (i) $\{s_k\}_{k\leq n} \Vdash_1 s$ (ii) $G_i \Vdash_1 s_i$ and (iii) $\epsilon(s_i^1) \subseteq \epsilon(G_i^1) \cap \epsilon(s^1)$ and $\epsilon(s_i^2) \subseteq \epsilon(G_i^2) \cap \epsilon(s^2)$ for $i \leq n$. Condition 3 is (i) whereas 2 follows from (ii) by monotony for out_1. It only remains to show that $\Sigma(s_i) \preceq \Gamma_i$. Since $\Sigma(G_i) \preceq \Gamma_i$ we have $\epsilon(G_i^1) \subseteq \Gamma_i^1$, so (iii) yields

$\epsilon(s_i^1) \subseteq \Gamma_i^1$. The argument for s_i^2 is similar, and the converse directions is essentially similar to (Kourousias and Makinson 2007, Lemma 2.3).

We are ready to prove the finest splitting theorem for simple-minded output:

Theorem 7. *Every set G of norms has a unique finest splitting modulo out*$_1$.

Proof. Let $\mathscr{E}$ be the infimum in $\langle \widehat{\Sigma}, \preccurlyeq \rangle$ of the splittings of G. The infimum exists by Lemma 5. It suffices to show that $\mathscr{E}$ also splits G. Let $s \in G$. By Lemma 6 it suffices to find norms $s_1, \ldots, s_n$ and pairs $\Gamma_1, \ldots, \Gamma_n \in \mathscr{E}$ such that conditions (1), (2) and (3) of that lemma hold. Since G is a relation over *formulae* of L, i.e. over finite objects, there is a splitting $\mathscr{F} = \{\Delta_k\}_{k \in I}$ of G that is as fine as can be with respect to s, in the sense that no splitting of G that is finer than $\mathscr{F}$ separates any two letters in $\epsilon(s^1)$ or in $\epsilon(s^2)$ that are not already separated by $\mathscr{F}$. Since $\mathscr{F}$ is a splitting of G, we know by Lemma 6 that there are formulae $r_1, \ldots, r_m$ and sets $\Delta_1, \ldots \Delta_m \in \mathscr{F}$ such that

(a) $\Sigma(r_i) \leq \Delta_i$, for each $i \leq m$
(b) $G \Vdash_1 r_i$ for each $i \leq m$
(c) $\{r_i\}_{i \leq m} \Vdash_1 s$

By Lemma 3 there is a $J \subseteq [1, m]$ such that $\{r_k\}_{k \in J} \Vdash_1 s$ with $s^1 \vdash r_k^1$.

Now, if H is a subset of $\{r_k\}_{k \in J}$ such that $\Sigma(H) \leq \Delta_p$, then let r_p stand for the adjunction of all the norms in H and let I_2 be the index set containing all and only such p. Then

(d) $\bigwedge\{r_k\}_{k \in J} \cong_1 \bigwedge\{r_p\}_{p \in I_2}$ and
(e) the $\Sigma(r_p)$ for $p \in I_2$ are coordinatewise disjoint

Also, since $s^1 \vdash r_k^1$ Lemma 2 together with (d) above yields $\bigwedge\{r_p\}_{p \in I_2} \Vdash_1 s$, whence

(f) $\{r_p\}_{p \in I_2} \Vdash_1 \bigwedge\{r_p\}_{p \in I_2} \Vdash_1 s$

by Lemma 1, so $\{r_p\}_{p \in I_2} \Vdash_1 s$ by the transitivity of $\Vdash_1$. Now, the $\Sigma(r_p)$ are pairwise disjoint, by (e), so we may apply parallel interpolation to obtain a set of norms $\{s_p\}_{p \in I_2}$ such that

(g) $\epsilon(s_p^1) \subseteq \epsilon(r_p^1) \cap \epsilon(s^1) \subseteq \epsilon(\Delta_p^1) \cap \epsilon(s^1)$ (resp. Δ_p^2 and s^2), using also (a)
(h) $r_p \Vdash_1 s_p$ for $p \in I_2$
(i) $\{s_p\}_{p \in I_2} \Vdash_1 s$

We show that $\{s_p\}_{p \in I_2}$ satisfies (1)–(3) of Lemma 6 wrt. $\mathscr{E}$. Condition (3) is just (i). For condition (2) we need to show that $G \Vdash_1 s_p$ for each $p \in I$. By construction, r_p is equal to the adjunction of $\{r_m\}_{m \in J'} \subseteq G$ for some $J' \subseteq J$. We have $G \Vdash_1 r_m$ for $m \in J'$ by b) whence $G \Vdash_1 \bigwedge\{r_m\}_{m \in J'}$ by repeated applications of SI and AND. Thus, since $\bigwedge\{r_m\}_{m \in J'} = r_p$ condition (2) follows from (h) by the transitivity of $\Vdash_1$. To complete the proof it only remains to check condition (1). That is, we need to check, for every s_p with $p \in I$, the existence of a $\Gamma_p \in \mathscr{E}$ such that $\Sigma(s_p) \preceq \Gamma_p$. By g) we have $\epsilon(s_p^1) \subseteq \epsilon(\Delta_p^1) \cap \epsilon(s^1)$. Since $\epsilon(s_p^1) \subseteq \epsilon(s^1)$ then the infimum $\mathscr{E}$ under $\leq$ of the splittings of G does not separate any two elementary letters in $\epsilon(s_p^1)$ that are not already separated by $\mathscr{F}$. But since also $\epsilon(s_p^1) \subseteq \Delta_p^1$, no distinct letters in $\epsilon(s_p^1)$ are separated by $\mathscr{F}$. This argument can be repeated for s_p^2 and Δ_p^2, whence we conclude that there is a $\Gamma_i \in \mathscr{E}$ with $\Sigma(s_p) \leq \Gamma_i$ as desired.

Example 3. Let $G := \{(a, b \to \neg c), (a, \neg c \to d), (a, b \vee e), (a, \neg e), (a, (d \to g) \vee (g \to d))\}$, where we assume that all letters are elementary. Then $G(a)$ is classically equivalent to $b \wedge \neg c \wedge d \wedge \neg e$. By AND and WO, therefore, G is equivalent to $G' := \{(a, b), (a, \neg c), (a, d), (a, \neg e)\}$. Each element of G' is written in one of the respective members of $S := \{(\{a\}, \{b\}), (\{a\}, \{c\}), (\{a\}, \{d\}), (\{a\}, \{e\})\}$. Now, the finest partition of ϵ that contains $\{a\}$ is $\mathscr{L} := \{\{g\} | \, g \in L\}$ which is also the finest partition that contains each of $\{b\}, \{c\}, \{d\}$ and $\{e\}$. Since $S \subseteq \mathscr{L} \times \mathscr{L}$ it follows that $\mathscr{L} \times \mathscr{L}$ is the finest splitting of G.[7]

Example 3 is rather special insofar as all the norms in G have the same condition of application. Note too that G^2 is in finest splitting form in the classical sense.

In the general case, though, splitting a set of norms does not reduce to splitting its respective input and output sets:

Example 4. Put $G := \{(a, b \vee c), (d, b \wedge e)\}$ where all letters are elementary and distinct. The finest splitting of G will separate b from e in the output, since G is equivalent modulo out_1 to $\{(a, b \vee c), (d, b), (d, e)\}$. However there is no way to separate b from c since this would result in a strictly stronger output. Hence the finest splitting of G is $\mathscr{L} \times (\{\{b, c\}\} \cup (\mathscr{L} \setminus \{\{b\}, \{c\}\}))$, where $\mathscr{L}$ is defined as in the preceding example. Yet, since $\{b \vee c, b \wedge e\}$ is equivalent to $\{b \wedge e\}$ the finest splitting of G^2 is simply $\mathscr{L}$.

As the example shows, the splitting of a code G is essentially determined by the properties of the entailment relation induced by the underlying input/output operator. In the case of simple-minded output, one such determining factor consists in the fact that $\Vdash_1$ has no rule that weakens inputs:

Example 5. Put $G := \{(a \wedge b, c)\}$. Then G is already in finest splitting modulo out_1 since G is not 1-equivalent to $\{(a, c), (b, c)\}$. Hence the finest splitting of G is $(\{\{a, b\}\} \cup (\mathscr{L} \setminus \{\{a\}, \{b\}\})) \times \mathscr{L}$.

The example illustrates that no norm can be 1-equivalently rewritten as set of norms with logically weaker antecedents. The splitting of G may require simplification of inputs in some case—e.g. the norm $(a \wedge (c \vee \neg c), b)$ should be replaced by (a, b)[8]—but the finest splitting form of G will never strictly reduce the logical strength of the inputs. We state this without proof:

Lemma 7. *Suppose that $G_1 \cong G_2$ and that G_1 is in finest splitting form. Put $f(a) = b$ iff $a \dashv\vdash b$. Then f is a bijection between G_1^1 and G_2^1.*

---

[7] This is an adaptation of (Makinson 2009, Example 3.1.)

[8] I am grateful to one of the reviewers for pointing this out to me.

Thus, as seen from the perspective of simple-minded output, inputs are always logically singular.

3.2 A Comparison with Basic Output

Does the finest splitting theorem 'scale up' to basic output, that is, does every code of norms have a unique finest splitting modulo out_2? Whilst the jury is still out on this question, we already have enough information to say that a proof cannot in any case proceed in the same manner as for out_1. Recall from Example 1 that basic output fails Lemma 4. We claimed that without it parallel interpolants don't necessarily exist. It is time to substantiate this claim:

Lemma 8. *Let $\{s_i\}_{i\in I}$ be a set of norms such that $s_j^2 = s_k^2$ for all $j, k \in I$. Then if $\{s_i\}_{i\in I} \Vdash_2 t$ then $s_j^2 \vdash t^2$ for any $j \in I$.*

Theorem 8. *out_2 does not have parallel interpolation*

Proof. Let G be partitioned into $G_1 := \{(a, b), (c, d)\}$, $G_2 := \{(g, b)\}$, and assume that all formulae are distinct elementary letters. Note that $\Sigma(G_2)$ and $\Sigma(G_1)$ are coordinatewise disjoint, so the example is legitimate. We have:

$$\dfrac{\dfrac{\dfrac{(a,b) \qquad (g,b)}{(a\vee g, b)}\,OR}{((a\vee g)\wedge c, b)}\,SI \qquad \dfrac{(c,d)}{((a\vee g)\wedge c, d)}\,SI}{((a\vee g)\wedge c, b\wedge d)}\,AND$$

However, it is impossible to select a single rule s such that $G_1 \Vdash_2 s$ and such that $s \cup r \Vdash_2 ((a \vee g) \wedge c, b \wedge d)$ for an r such that $G_2 \Vdash r$. In other words, there is no (parallel) interpolant between G_1 and $((a \vee g) \wedge c, b \wedge d)$. To prove this we devise a two-step argument: First we show that everything in $G_1 \cup G_2$ is essential to the derivation of $((a \vee g) \wedge c, b \wedge d)$. Next, we show that the strongest single norm that is derivable from G_1 when both elements of G_1 are leaves in the derivation, is too weak to act as an interpolant, whence no interpolant exists.

For the first step, note that (c, d) and at least one of (a, b), (g, b) is essential to the derivation, by Lemma 8. Put $H := \{(c, d), (a, b)\}$ and consider the set $A := \{\neg a \wedge g \wedge c\}$. Since g, a and c are all distinct elementary letters, A is consistent. It is a standard result of classical logic that any such set can be extended to a maximal consistent set V. Since $\neg a \wedge g \wedge c \vdash g \wedge c \vdash (a \vee g) \wedge c$ we have $(a \vee g) \wedge c \in V$ whence V is one of the complete sets that define $out_2(H, (a \vee g) \wedge c)$. Note however that $a \notin V$ since $\neg a \in V$ and V is consistent. Therefore $Cn(G(V)) = Cn(d)$, whence $b \wedge d \notin out_2(G, (a \vee g) \wedge c)$. We may therefore exploit the completeness of out_2 wrt. $\Vdash_2$ (Theorem 3) to conclude that $((a \vee g) \wedge c, b \wedge d)$ is not derivable from H. Running the exact same argument once more with (g, b) in place of (a, b) shows that (c, d), (a, b) and (g, b) are all essential to the derivation.

For the second step, it is easy to prove that $(a \wedge c, b \wedge d)$ is the strongest single norm that is derivable from G_1 by a derivation in which both of (a, b) and (c, d) figure as leaves. Put $H := \{(a \wedge c, b \wedge d), (g, b)\}$. If $(a \wedge c, b \wedge d)$ were an interpolant we should be able to derive $((a \vee g) \wedge c, b \wedge d)$ from H. To see that this is not so, put $B := \{(\neg a \vee \neg c) \wedge g \wedge c\}$, B is consistent and has a maximal consistent extension V. Again $(a \vee g) \wedge c \in V$ so V is one of the complete sets defining $out_2(H, (a \vee g) \wedge c)$. However $a \notin V$ since $(\neg a \vee \neg c) \wedge g \wedge c \vdash \neg a$ and since V is logically closed. By the same token $a \wedge c \notin V$, therefore $Cn(G(V)) = Cn(G(b))$ so $b \wedge d \notin out_2(G, (a \vee g) \wedge c)$. It follows that $((a \vee g) \wedge c, b \wedge d)$ is not derivable from H, which completes the proof.

Parallelism is essential to this argument, since basic output satisfies plain interpolation:

Theorem 9. *If $G \Vdash_2 s$ then there is a norm t such that $G \Vdash_2 s \Vdash_2 t$ with $\epsilon(t^j) \subseteq \epsilon(G^j) \cap \epsilon(s^j)$ for $j \in \{1, 2\}$.*

Proof. Proof proceeds by induction on the length of the derivation of a norm s from G. The base case where $s \in G$ is trivial. Moreover, we know from the parallel interpolation theorem for simple-minded output that SI, WO and AND preserve the property. It only remains to show, therefore, that the property is preserved by OR. Put $s := (a \vee c, b)$ and suppose s is derived from (a, b) and (c, b) by OR. By the induction hypothesis the property holds for shorter derivations, so there are interpolants t_1 and t_2 such that (a, b) is derivable from t_1 and (c, b) from t_2. Since these are both single-premise derivations, it follows that $a \vdash t_1^1, c \vdash t_2^1$ and $t_1^2 \vdash b, t_2^2 \vdash b$. Therefore $a \vee c \vdash t_1^1 \vee t_2^1$ and $t_1^2 \vee t_2^2 \vdash b$, so we have the following derivation.

$$
WO\ \dfrac{SI\ \dfrac{OR\ \dfrac{WO\ \dfrac{(t_1^1, t_1^2)}{(t_1^1, t_1^2 \vee t_2^2)} \qquad WO\ \dfrac{(t_2^1, t_2^2)}{(t_2^1, t_2^2 \vee t_1^2)}}{(t_1^1 \vee t_2^1, t_1^2 \vee t_2^2)}}{(a \vee c, t_1^2 \vee t_2^2)}}{(a \vee c, b)}
$$

Since t_1 and t_2 are interpolants, $\epsilon(t_1^j \vee t_2^j) \subseteq \epsilon(G^j) \cap (s^j)$ for $j \in \{1, 2\}$, so $(t_1^1 \vee t_2^1, t_1^2 \vee t_2^2)$ is an interpolant between G and $s = (a \vee c, b)$.

To be sure Theorem 8 does not entail that the finest 2-splitting of an arbitrarily chosen code G does not necessarily exist, only that its existence cannot be proved using parallel interpolation in the form of Definition 7. However, as suggested by one reviewer, the existence of finest splittings may be derivable from a weaker version of parallel interpolation, a version not contradicted by Theorem 8: Suppose $G \Vdash_1 s$ and $G = \bigcup\{G_i\}_{i \in I}$ where the G_i partition G according to letter sets. Then there are finite sets of norms Δ_i $(i \in I)$ such that $\bigcup_i \Delta_i \Vdash s$ and $\epsilon(\Delta_i^j) \subseteq \epsilon(G_i^j) \cap \epsilon(s^j)$

for $j \in \{1, 2\}$. We leave it an open problem to verify whether this property does indeed hold for basic output, and if so to verify whether it suffices to obtain the finest splitting theorem.

What we do know, though, is that since $\Vdash_1$ is included in $\Vdash_2$ any G has a 2-splitting which is at least as fine as the finest 1-splitting:

Theorem 10. *Let $\mathcal{E} \in \widehat{\Sigma}$ be the finest splitting of G. Then there is a $\mathcal{F} \in \widehat{\Sigma}$ with $\mathcal{F} \preceq \mathcal{E}$ such that $\mathcal{F}$ is a splitting of G modulo out_2.*

The refinement relation in this theorem may be proper, for if G is the singleton code $\{(a \vee b, c)\}$—where a, b and c are distinct elementary letters—then it is already on finest splitting form modulo out_1, but not so modulo out_2 since $\{(a \vee b, c)\} \cong_2 \{(a, c), (b, c)\}$. Moreover, $\{(a, c), (b, c)\}$ is readily seen to be a finest splitting form of G modulo out_2. We conjecture that such a form always exists.

Conjecture 1. Every code G has a unique finest splitting modulo out_2

Note also that for $\{(a, c), (b, c)\}$ the input set $\{a, b\}$ is now in finest splitting form in the propositional sense. Yet, this does not hold in general: add $(a \wedge b, c)$, as in Example 5, then the code is still as finely split as it can be modulo out_2, but the set of inputs is no longer in finest splitting form modulo propositional logic.

4 Second Interface: From Contraction to Derogation

The second interface we'd like to draw attention to is that which exists between input/output logic and the classical AGM model of theory revision. This interface may be seen to spring from the way the AGM approach heeds the maxim 'logic is not just about deduction'. Nothing in the AGM model of theory change relies on the properties of a particular deductive system or logic. Rather, the evolution of a theory, considered *in abstracto*, is reduced to addition and subtraction of formulae modulo the intersection of a set of preferred maximally non-implying subsets. Therefore, the AGM model is applicable to any theory that is the value of a closure operator for which the notion of a maximally non-implying subset is well-defined. Of course, the behaviour of such a model will depend on the choice of closure operator.

Simple-minded output is one such operator (as indeed are all the input/output operators defined by Makinson and van der Torre), and it turns out to combine beautifully with the AGM paradigm. Here as always, though, the devil is in the detail, and low-level differences between input/output logic and propositional logic surface in the list of postulates (as tradition has come to call them) that characterise the respective removal operations.

We shall confine our discussion in this section to the operation of *derogation*, which we think of as the norm-theoretic analogue of AGM *contraction*, leaving revision (or in the case of norms; 'amendment') to one side for now.[9] In so doing, theory

[9] For a discussion of this choice of terminology, see Stolpe (2010a).

revision may be said to have come full circle, for as mentioned in the introduction, the logic of theory change grew out of the collaboration between Makinson and Alchourrón in the late 1970s on the fine logical structure of derogation in legal codes (Makinson 1996). Of course, input/output logic was not available at the time, so sets of formulae were pushed into service to act as codes of norms—with the ensuing norm-theoretic shortcomings packaged therein. As in the preceding section, we shall contrast simple-minded with basic output.

4.1 Partial Meet Derogation Modulo Simple-Minded Output

An operation of derogation based on input/output logic was defined in Stolpe (2010a), in complete analogy with AGM-contraction:

Definition 9. (Partial-meet derogation). *Let G be any code of norms. Then $-$ is a partial meet derogation operator if $G - (a, b) = \bigcap \delta(G \perp (a, b))$, for some selection function δ for G.*

Compared to its classical counterpart, Definition 9 brings little new: the operation $\perp$ returns the family of subsets of its left argument (the *derogantes*) that are such that they maximally do not entail the right argument (the *derogandum*) modulo out_1.[10] The operation δ is a so-called selection function which returns a subset of $G \perp (a, b)$ if the latter is non-empty and $\{G\}$ otherwise. Intuitively $G \perp (a, b)$ represents different ways to modify G, with a minimal loss of information, so as not to entail (a, b). The function δ, on the other hand, chooses the most preferred ones among the remainders/derogantes, according to some unspecified criterion of utility. All this is standard.

Partial-meet derogation as so defined is characterised by a list of syntactic conditions the majority of which resemble AGM postulates closely. However, there are some interesting new-comers as well. The following theorem is proved in Stolpe (2010a):

Theorem 11. *Suppose $G = out_1(G)$. Then $-$ is a partial-meet derogation operator iff it satisfies the following conditions:*

D-Closure: $G - (a, b) = out_1(G - (a, b))$
D-Inclusion: $G - (a, b) \subseteq G$
D-Success: *If* $\nvdash b$ *then* $(a, b) \notin G - (a, b)$
D-Extensionality: *If* $(a, b) \cong_1 (c, d)$ *then* $G - (a, b) = G - (c, d)$
Input Dependence: *If* $(c, d) \in G \setminus G - (a, b)$ *then* $a \vdash c$
Local Recovery: *If* $(c, d) \in G$ *&* $a \vdash c$ *then* $(c, d) \in out_1(G - (a, b)) \cup (c, b))$.

[10] The reader should bear in mind $\perp$ and $-$ are both relative to the underlying input/output logic. Had there been any room for confusion, they should therefore have been written $\perp_1$ and $-_1$.

The first four of these conditions are not really idiosyncratic for norms. Of, course closure is now relative to $\Vdash_1$-implicature, as stated by *D-Closure*, and there is a proviso in the condition of *D-Success* that reflects the fact that (a, t) behaves as a tautology wrt. to simple-minded output for any a. Except for these rather obvious adjustments, each of the conditions in question emulates a corresponding postulate for AGM contraction. What these conditions really reflect is the underlying concept of a closure operator, and they will hold, minor variations aside, for all such.

More interesting are the principles of *Input Dependence* and *Local Recovery*. They do not have direct counterparts in the AGM theory, and may therefore be expected to tell us something that is specific to the dynamic behaviour of norms (under $\Vdash_1$-entailment). As the name indicates, *Local Recovery* is a restricted version of the debated recovery postulate from AGM-contraction, which translated into input/output logic reads:

Global Recovery: $G \subseteq (G - (a, b)) \cup (a, b)$.

This principle is not valid for out_1-derogation,[11] not even when $G = out_1(G)$, and it will aid the interpretation of both of *Input dependence* and *Local Recovery* to understand why.

Example 6. (*Counterexample to Global Recovery*). Put $G := \{(a, b), (a \wedge c, b)\}$. Suppose that a and c are logically distinct and that $\nvdash b$. From the latter assumption, it follows by *D-Success* that $(a \wedge c, b) \notin out_1(G) - (a \wedge c, b)$, whence *D-Closure* and *SI* yield $(a, b) \notin out_1(G) - (a \wedge c, b)$. Now, suppose for reductio that $(a, b) \in out_1((out_1(G) - (a \wedge c, b)) \cup (a \wedge c, b))$. Then it follows from Lemma 4 that $a \vdash c$ contradicting the assumption that a and c are logically distinct.

Note that the input-entailment property from Lemma 4 is needed in the verification. Due to input-entailment removal of norms from a system only require incisions into contexts that are weaker than the one described by the antecedent of the derogating norm. Conversely additions only affect stronger contexts. This is essentially the reason why global recovery fails, for when (a, b) is removed, then a norm with a weaker condition of application $(a \vee c, b)$ will go too, whereas if (a, b) is added back in, then only stronger contexts are affected, so $(a \vee c, b)$ is not restored. *Input Dependence* and *Local Recovery* each express one side of this behaviour, which we may refer to as the *locality* of simple-minded output. When $G = out_1(G)$, the locality of simple-minded output can be expressed with the following equivalent principle Stolpe (2010a), Theorem 6:

Locality: If $(c, d) \in G \setminus G - (a, b)$ then $(c, b \to d) \in G - (a, b)$.

That is, *Locality* is equivalent to the conjunction of *Input Dependence* and *Local Recovery*.

As regards the relationship between AGM-contraction and partial-meet derogation, it turns out that for any subset H of G which is maximally such that $H \nVdash_1 (a, b)$, the set $H(Cn(a))$ is a subset of $G(Cn(a))$ and it is maximally such that $H(Cn(a)) \nvdash b$. Moreover the function $\mu : G \perp (a, b) \to G(Cn(a)) \perp b$ defined by

[11] This was first noticed in Boella and van der Torre (2006).

putting $\mu(H) = H(Cn(a))$ is a bijection between $G \perp (a, b)$ and $G(Cn(a)) \perp b$. As a corollary to this we have (Stolpe 2010a, Corollary 2):

Corollary 2. *For every partial-meet derogation operation $-$ and formula a, there is a partial-meet contraction operation $\backsim$ such that $G(Cn(a)) \backsim b \vdash d$ iff $G-(a, b) \Vdash_1 (a, d)$.*

4.2 A Comparison with Basic Output

A quick comparison with the idiom of basic output reveals that locality is not characteristic of input/output logic *per se*, only of simple-minded output. As Example 1 shows, basic output fails input-entailment (i.e. Lemma 4), which in turn causes *Input Dependence* to fail:

Example 7. Let G be the code $\{(a, b), (c, b)\}$ where a, b and c are distinct elementary letters. Then we have that either $(a, b) \notin F$ or $(c, b) \notin F$ for every $F \in out_2(G) \perp (a \vee c, b)$ by OR. Hence, no matter how the selection function δ is defined, it follows that $(a, b) \in out_2(G) \setminus (out_2(G) - (a \vee c, b))$ or $(c, b) \in out_2(G) \setminus (out_2(G) - (a \vee c, b))$, but $a \vee c \nvdash a$ and $a \vee c \nvdash c$ so *Input Dependence* fails.

On the other hand, since the derivable antecedents under $\Vdash_2$ comprise the local ones (in the sense defined above), *Local Recovery* continues to hold. The proof of this requires the following lemma, which is akin to one half of the deduction theorem from classical logic:

Lemma 9. *If $(c, d) \in out_2(G \cup (a, b))$ then $(a \wedge c, b \rightarrow d) \in out_2(G)$.*

Proof. Argument proceeds by induction on the length of a $\Vdash_2$-derivation. In Stolpe (2010a), the property is proved to hold for $\Vdash_1$ so it only remains to show that it is preserved by OR: Suppose (c, d) is derived from (g, h) and (g', h') by OR. Then $c = g \vee g'$ and $h = h' = d$. By the induction hypothesis we have that $(a \wedge g, b \rightarrow h), (a \wedge g', b \rightarrow h')$ are $\Vdash_2$-derivable. Since $h = h'$ it follows that $b \rightarrow h = b \rightarrow h'$ whence $((a \wedge g) \vee (a \wedge g'), b \rightarrow h) = (a \vee (g \vee g'), b \rightarrow h) = (a \vee c, b \rightarrow d) \in out_2(G)$. Therefore, $(a \wedge c, b \rightarrow d) \in out_2(G)$ by SI as desired.

Now, for *Local Recovery* itself:

Theorem 12. *out_2-derogation satisfies Local Recovery.*

Proof. Suppose that $(c, d) \in out_2(G)$ and that $a \vdash c$. We want to show that $(c, d) \in out_2((out_2(G) - (a, b)) \cup (c, b))$. By AND it suffices to show that $(c, b \rightarrow d) \in out_2(G) - (a, b)$. Note that, $(c, b \rightarrow d) \in out_2(G)$, by WO, since $(c, d) \in out_2(G)$. Suppose for reductio ad absurdum that $(c, b \rightarrow d) \notin H$ for some $H \in out_2(G) \perp (a, b)$. Then, by the maximality of H it follows that $(a, b) \in out_2(H \cup (c, b \rightarrow d))$, so Lemma 9 and the fact that $H = out_2(H)$ (easily shown) yield $(c \wedge a, (b \rightarrow d) \rightarrow b) \in H$. Now, $(b \rightarrow d) \rightarrow b \dashv\vdash \neg(b \rightarrow d) \vee b \dashv\vdash \neg(\neg b \vee d) \vee b \dashv\vdash$

$(b \wedge \neg d) \vee b \dashv\vdash b$. Hence $(c \wedge a, b) \in H$, by one application of WO. Since $a \vdash c$ it follows that $(a, b) \in H$, contradicting $H \in out(G) \perp (a, b)$.

Whether a stronger recovery property can be shown to hold now that reasoning isn't local remains a topic for future research, and so does the proper characterisation of derogation modulo basic output. However, we note that the absence of the input-entailment property yet again turns out to have some bearing on the question:

Example 8. Put $G := \{(a, b), (c, b)\}$ for distinct elementary letters a, b and c. Since neither of the singletons $\{(a, b)\}$ or $\{(c, b)\}$ entail $(a \vee c, b)$ modulo $\Vdash_2$, they can be extended to sets $F, G \in out_2(G) \perp (a \vee c, b)$. Define the function μ in the same manner as in Subsect. 4.1, i.e. by putting $\mu(J) = J(Cn(a \vee c))$ for every $J \in out_2(G) \perp (a \vee c, b)$. Then $\mu(H) = \mu(G) = Cn(\emptyset)$ so μ is not injective.

In other words, in the absence of input-entailment, the function μ does not put remainders of sets of formulae and remainders of sets of norms in one-to-one correspondence. Therefore Corollary 2 cannot without further ado be generalised to basic output. Thus, this section ends on the same note as the preceding one: A simple and transparent relationship exists between simple-minded output and, in this case, AGM contraction. However, adding OR to the system causes input-entailment to fail, and the connection is lost.

5 Inheriting from Multiple Interfaces: Relevant Change

In the classical propositional case, there is an interesting relationship between AGM-revision on the one hand and finest splittings on the other. Of course, this is not just a happy accident, but stems from the fact that language splitting was invented by Parikh as a response to a perceived shortcoming in the AGM paradigm. Recall that the existence of finest splittings shows that, as far as classical propositional logic goes, there is a unique way of thinking about any corpus of knowledge as composed of a set of mutually irrelevant subject matters. Parikh noticed the heavy-handedness of AGM-revision in this respect; that the incisions required by AGM-contraction may spill over onto (in this sense) unrelated subjects.

In light of Corollary 2 there is no reason to expect derogation to fare any better (if better it is (cf. Makinson 2009, Sect. 7)). Indeed it can easily be shown that some derogation operations remove material that is intuitively unrelated to that which the derogandum itself addresses:

Example 9. Let b and d be distinct elementary letters and let G be the code $out_1(\{(c, b), (c, d)\}$. Assume that $a \vdash c$ and put $out_1(G) - (a, b) = out_1((c, b \leftrightarrow d))$. Then all the conditions of Theorem 11 including *Local Recovery* are satisfied, so $-$ is a partial meet derogation operation. However, the derogation removes not only (c, b) but also (c, d) which is intuitively irrelevant to (a, b), since b and d are distinct letters.

Letting this example clue us in on how to distinguish relevant from irrelevant incisions into a code, we may as a tentative characterisation deem unrelatedness in *one* coordinate as a sufficient criterion for irrelevance. That is, if G is a code in finest splitting form and $r \in G$ we may say that r is irrelevant to the derogation of s from G if either the consequents or the antecedents of r and s are respectively irrelevant to each other modulo G. Yet, in the case of simple-minded output this is an unnecessarily strong requirement: recall from Lemma 7 that the finest splitting of G will not reduce the logical strength of inputs. Combine this with the easily verifiable fact that if $r \in G$ but $r \notin G - s$ then $s^1 \vdash r^1$ and we have that irrelevance on the left ought to mean simply $s^1 \nvdash r^1$. This gives the following definition:

Definition 10. (Irrelevant norms). *Let G be any set of norms and s a derogandum. Let $\mathscr{E} = \{\Gamma_i\}_{i \in I}$ be the unique finest splitting of G. We say that a norm r is irrelevant to s modulo G iff either $s^1 \nvdash r^1$ or there is no $i \in I$ such that both $\Gamma_i^2 \cap \epsilon(s^2) \neq \emptyset$ and $\Gamma_i^2 \cap \epsilon(r^2) \neq \emptyset$.*[12]

When this definition snaps into place, there are few surprises in the central theorem of this section: the theorem itself says that derogation performed on a code in finest splitting form leaves irrelevant norms untouched.[13] Note though, that there is a proviso: the antecedent of the derogandum must not generate contradictory obligations. We shall have more to say about this proviso shortly.

Theorem 13. *Let $\mathscr{E} = \{\Gamma_i\}_{i \in I}$ be the unique finest splitting of G with $G' := \{G_i\}_{i \in I}$ a family such that $\bigcup\{G_i\}_{i \in I} \cong_1 G$ and $\Sigma(G_i) \preceq \Gamma_i$ for each $i \in I$. Let s be any norm, and $G' - s$ a partial meet derogation of s from G'. Then, whenever $r \in G'$ is irrelevant to s modulo G', and $G(s^1)$ is consistent, we have that $r \in G' - s$.*

Proof. Suppose $r \in G'$ but $r \notin G' - s$ while r is irrelevant to s; we derive a contradiction. It follows from $r \in G' \setminus G' - s$ that $s^1 \vdash r^1$, whence by the irrelevance there is no $i \in I$ such that $\Gamma_i^2 \cap \epsilon(s^2) \neq \emptyset$ and $\Gamma_i^2 \cap \epsilon(r^2) \neq \emptyset$. Let $\{\Gamma_j^2\}_{j \in J}$, where $J \subseteq I$ is the subfamily of elements that overlaps $\epsilon(s^2)$. By the irrelevance, $\bigcup\{\Gamma_j^2\}_{j \in J}$ and $\epsilon(r^2)$ are disjoint. Since $r \notin G' - s$, we have by the definition of partial meet derogation that $r \notin H$ for some derogans H. *Qua* derogans, H is such that $G' - s \subseteq H \subseteq G'$ and maximally $H \nVdash_1 s$. Since $r \in G' \setminus H$, therefore, it follows that $H \cup r \Vdash_1 s$. Put $G_1^* = \{G_j\}_{j \in J}$ and $G_2^* = \{G_k\}_{k \in I \setminus J}$. Then since $H \subseteq G' = \{G_i\}_{i \in I} = G_1^* \cup G_2^*$ we have $(H \cap G_1^*) \cup (H \cap G_2^*) \cup r = H \cap (G_1^* \cup G_2^*) \cup r = H \cup r \Vdash_1 s$. Applying Lemma 3 there are norms $H^- := \{u_1, \ldots, u_m\} \cup \{v_1, \ldots, v_n\} \cup r \Vdash_1 s$ where $u_1, \ldots, u_m$ are elements of

---

12 As observed by David Makinson (personal communication), the presence of the clause $s^1 \nvdash r^1$ prevents the symmetry of irrelevance. Referring back to the example immediately following Definition 7 in Sect. 3, it is tempting to say that that is how it has to be as long as the operator in question is out_1. The norm $(a \vee b, c)$ is relevant to (a, b) since a will trigger both b and c. However, (a, b) is not relevant to $(a \vee b, c)$ since stronger contexts never affect weaker ones modulo out_1. Notwithstanding this, prudence dictates that we suspend a final judgement until we have had time to investigate the matter more closely.

13 The proof of it is an adaptation of (Kourousias and Makinson 2007, Theorem 4.1).

$H \cap G_1^*$, $v_1, \ldots, v_n$ are elements of $H \cap G_2^*$, and such that $H^- \cup r \Vdash_1 s$ and $s^1 \vdash w^1$ for every $w \in H^-$. It follows that $\{u_1^2, \ldots, u_m^2\} \cup \{v_1^2, \ldots, v_n^2\} \cup r^2 \vdash s^2$, whence $\{u_1^2, \ldots, u_m^2\} \cup \neg s^2 \vdash \neg(v_1^2 \wedge \ldots \wedge v_n^2 \wedge r_k^2)$ by contraposition. The left and right sides have no letters in common. For by construction all letters occurring in formulae on the left are in the language of $\bigcup\{\Gamma_j^2\}_{j \in J}$ whereas all letters occurring in formulae on the right are in the language of $\bigcup\{\Gamma_k^2\}_{k \in I \setminus J}$. Since left and right have no letters in common, either the left side is inconsistent or the right side is a tautology. If the left side is inconsistent then $\{u_1^2, \ldots, u_m^2\} \vdash s^2$ so since $u_1, \ldots, u_m \in H$ and $s^1 \vdash u_1^1 \wedge \ldots \wedge u_m^1$, by the construction of H^-, we have $H \Vdash_1 s$ contrary to assumption. If the right side is a tautology, on the other hand, then $\{v_1^2, \ldots, v_n^2, r^2\}$ is inconsistent, which since $s^1 \vdash w^1$ for every $w \in H^-$, $s^1 \vdash r^1$ and $r \in G'$ entails that $G'(s^1)$ is inconsistent. However, $G' \cong_1 G$ so this contradicts the assumption that $G(s^1)$ is consistent.

5.1 Adjusting to an Anomaly

Despite the pleasantness of having Theorem 13 go through, Definition 10 does not always perform well.[14] In particular, it does not handle disjunctive norms with much grace. Consider the following example:

Example 10. Put $G := \{(a, b \vee c), (e, b \wedge c)\}$ where all letters are elementary and distinct. The finest splitting of G still contains $\{b, c\}$, so G is still in finest splitting form. Yet, $(e, c) \notin out_1(G - (e, b))$, even though $(a, b \vee c)$ and $(e, b \wedge c)$ are not relevant to each other according to Definition 10.

The problem, stated somewhat loosely, is that the global nature of the notion of a finest splitting modulo 1-entailment, prevents the refinement process from propagating into a context (e in the example above) the output of which $(b \wedge c)$ is covered by the output $(b \vee c)$ of another context (a) that corresponds to a lower bound $(\{a\}, \{b, c\})$ in the splitting. This allows one context to affect the dynamic behaviour of another for which it is not even relevant. In the example above $(\{a\}, \{b, c\})$ belongs to the finest splitting, whence neither $(\{e\}, \{b\})$ nor $(\{e\}, \{c\})$ does. In a sense, therefore, the norm $(a, b \vee c)$ protects $(e, b \wedge c)$ from the refinement process.

One possible solution to this problem begins by noting that Definition 7 does not in fact distinguish between $G := \{(a, b \vee c), (e, b \wedge c)\}$ and $H := \{(a, b \vee c), (e, b), (e, c)\}$—they are both finest splitting forms. The finest splitting of G contains $\Gamma_1 = (\{a\}, \{b, c\})$ and $\Gamma_2 = (\{e\}, \{b, c\})$, and any norm s in either of G and H is such that $\Sigma(s) \preceq \Gamma_i$ for some $i \in \{1, 2\}$. There is no such slack in the classical case: the only finest splitting form of the set $\{a \vee b, a \wedge b\}$ modulo classical entailment is $\{a, b\}$, which is as fine-grained as can be. Thus, language splitting is less demanding and allows a greater variety of forms in input/output logic than in classical logic.

[14] The material in this subsection is largely a response to one of the anonymous reviewers, to whom the credit goes for having spotted the problem.

One natural adjustment, in light of Example 10, is to make it count how the finest splitting form is selected. A straightforward way to do that is to utilise the all-some lifting of the $\preceq$ relation introduced in Sect. 3.1 to define an ordering relation on the set of finest splitting forms of a code:

Definition 11. *Put $G \trianglelefteq H$ iff for every $\Gamma \in \{(\epsilon(s^1), \epsilon(s^2)) : s \in G\}$ there is a $\Delta \in \{(\epsilon(t^1), \epsilon(t^2)) : t \in H\}$ such that $\Gamma \preceq \Delta$.*

We may then define the derogation operation in such a manner that it is always performed on $\trianglelefteq$-minimal finest splitting forms (which trivially exist).

It is easy to see that this suffices to iron out the kinks in Example 10, because $\{(a, b \vee c), (e, b), (e, c)\} \trianglelefteq \{(a, b \vee c), (e, b \wedge c)\}$ according to Example 11 (but not the other way around), and when (e, b) is removed from the latter then (e, c) remains. Moreover, since the $\trianglelefteq$-minimal finest splitting form *is* a finest splitting form, this solution requires no alterations in the concept of normative relevance, wherefore Theorem 13 remains valid. It ought to be emphasised, though, that this solution has not been fully researched yet, and that we cannot rule out other forms of 'logical interference' similar to that of Example 10.

5.2 A Note on Consistency

A word of caution is in order at this point, though: while Theorem 13 does underline the general theme of this chapter, namely that Makinson's non-classical uses of classical logic establishes something akin to abstract interfaces between different parts of his production, it is important to stress that the philosophical plausibility of the concepts involved may not translate a hundred percent. With respect to the concept of relevance, an important philosophical difference may be seen to reside in the consistency proviso that in both cases must be added in order to ensure the relevance-sensitivity of the revision operation in question. As an assumption about belief sets or knowledge corpora consistency is a natural idealisation in the sense that if a person believes that a is the case, then it is rational to assume that he or she does not also believe $\neg a$. Even though it is well-known that everything in between scientific theories and everyday reasoning is prone to contain or classically entail contradictions,[15] consistency is nevertheless a regulative ideal. This applies to the interpretation of other people as much as it does to our own reasoning. As famously argued by Donald Davidson, if it is to be possible at all to make sense of the behaviour and/or cognitive processes of another person, then it is necessary to interpret him or her charitably and assume by default that that person's beliefs are by and large true and hence consistent. That is not to say that one needs to assume the global consistency of all the beliefs of the other person. It does mean, though, that whatever one infers about his or her actions, one does so by entertaining the possibility of holding true the particular subset of beliefs that explain the behaviour in question.

[15] I owe this phrase to one of the anonymous referees.

The concept of classical consistency sits much less comfortably with the notion of a normative system, understood as a set of norms that is closed under some relation of entailment. There are good reasons why a derived norm (a, f) does not need to be considered inconsistent or even incoherent. On the contrary, it is common for a code of norms—probably unavoidable if the code is of any complexity—to regulate two distinct states of affairs separately, which even though they are associated with contrary commitments may nevertheless happen to occur simultaneously. Whether or not two events can occur at the same time, is usually a matter of contingent fact, and not a possibility that can be ruled out by stipulation. Therefore, that an agent cannot under unhappy circumstances live up to all his commitments, is something a code should give us the resources to anticipate, it is not something the code should seek to prevent. That is, one of the things we should expect a formal model of normative systems to do is precisely to alert us to such conflicts, and the circumstances under which they arise. A derived norm (a, f) ought to be interpreted this way. It tells us that the state of affairs described by a triggers contrary commitments and is therefore one to avoid.

Since the concept of relevance from Definition 10 only works when the antecedent of the derogandum does not trigger conflicting obligations, Theorem 13 does not apply to contexts such as a. It remains for future work to assess the severity of this restriction.

6 An Application to Conceptual Analysis

We shall round off this chapter by revisiting the analysis of the concept of positive permission, as developed in Stolpe (2010c) and Stolpe (2010b), as an illustration of how the pieces we have assembled may be combined and applied. The question about the proper characterisation of positive permission aka explicitly permissive norms is a long standing one in legal theory and the philosophy of norms, and one that has an important bearing on e.g. how the evolution of a normative system is to be conceived.

In Stolpe (2010c) it is argued that a positive permission is a liberty that is implied by an explicit rule whereby some institutionally recognised number of agents are released from an obligation or shielded from the imposition of one—the keywords here being 'release' and 'shield'. That is, a permissive norm is conceived of as a rule whose primary function is to *restrict* the scope and influence of an already existing obligation, or to *preempt* one that could possibly be passed. Therefore Stolpe (2010c) approaches the concept of positive permission in terms of the elimination or preemption (temporary or not) of a contrary prohibition from a code, which in turn is analysed as an operation of derogation on the given code or on one of its possible extensions. Unfortunately, it was shown in Stolpe (2010b) that if the derogation operation in question is cast as an operation on closed sets of norms (modulo $\Vdash_1$)—as it is in Stolpe (2010c)—then, due to certain notorious quirks of material implication, problems ensue that threaten to trivialise the account. Nevertheless, the central result of Stolpe (2010b) shows that one can escape this predicament by substituting a *base contraction* (read; derogation) paradigm for a *theory contraction* paradigm in the fundamental definition of positive permission, and it shows that a concrete way to implement this strategy is to make sure that only those prohibitions that are relevantly related to the permissive rule in question will be overridden by it.

The material in the present section improves upon this analysis in the following respects: First, we give a pair of representation theorems that completely characterise the concepts of *exemption* and *antithetic permission* as defined in Stolpe (2010c), which we continue to regard as the two principal concepts of positive permission. This involves correcting an error from Stolpe (2010c) and Stolpe (2010b) related to a limiting case. More fundamentally, however, we substitute the analysis of normative relevance given in the preceding section for that given in Stolpe (2010b). In Stolpe (2010b) normative relevance was defined rather tentatively in terms of the so-called right-splitting of a code of norms—an equivalent form of a code such that each antecedent of a norm generates an output set in finest splitting form in the classical propositional sense. This way of going about it is, however, rather ad hoc from the point of view of norm-theory, as it is lopsided towards the propositional concept of relevance. In the following we shall instead plug in Theorem 13, which is much more transparent wrt. to the relationship between the concept of relevance and that of normative implication, insofar as it utilises the concept of a splitting as it applies to a code of norms rather than to a set of sentences.

The unit of analysis in the remainder of this section is a *code* $\langle G, P\rangle$ consisting of a set of *explicitly stated mandatory norms* G and a set of *explicitly stated permissive norms* P. For brevity, P will sometimes simply be called explicit permissions, although, strictly speaking, it is a set of *permissive norms*. It will be assumed that P does not contain trivial norms of the form (a, t) nor unsatisfiable norms of the form (a, f). Where (a, b) is an implied norm, we shall say that it is a mandatory or a permissive norm (as the case may be) *according to* or *in* such a code, in which case it means that b is required or permitted by that code whenever a is *true*.

Now, as mentioned already, Stolpe (2010c) takes the view that the function of a permissive norm is to restrict the scope and influence of an already existing obligation, or to preempt one that could possibly be passed. The first case is taken to correspond to the concept of an exemption, that is, to an an institutionally recognised case for which an existing covering prohibition is declared to have no force. The latter is taken to correspond to the concept of an antithetic permission, that is, to a norm (a, b) such that given the obligations already present in the code in question we can't forbid b under the condition a without thereby contradicting something d that is implicit in that code's explicit permissions Makinson and van der Torre (2003). A paradigm example of an exemption is the permission to process personal and otherwise not legally accessible information about a person given that person's express consent. A paradigm example of an antithetic permission is an action which is protected by constitutional law, e.g. by freedom of speech, and which cannot therefore be forbidden on pain of contradicting the constitution. The former of these concepts was defined in Stolpe (2010c) as follows:

Definition 12. (Exemptions). *(a, b) is an exemption according to the code $\langle G, P\rangle$ iff $(a, \neg b) \in out(G) \setminus out(G) - (c, \neg d)$ for some $(c, d) \in P$ such that $c \dashv\vdash a$.*

Thus, (a, b) is an exemption if a prohibition $(a, \neg b)$ can be derived from the code, but this prohibition is overriden by an explicit permissive norm (c, d). Exemptions are thus cast as cut-backs on the code required to respect the explicit permissions in P. More precisely (a, b) is an exemption if the norms in G entail a prohibition that regulates the state of affairs a by prohibiting b, and $(a, \neg b)$ is such that, unless it is removed, the code will contradict an explicit permission in P. We shall say that (a, b) is an exemption *by* the explicit permission (c, d).

As regards, antithetic permissions, their function consists, one might say, in preventing a set of mandatory norms from growing in such a way as to render explicitly permitted actions forbidden. This checked-growth perspective is expressed in Stolpe (2010c) as follows:

Definition 13. (Antithetic permission). *(a, b) is antithetically permitted in $\langle G, P\rangle$ iff $(a, \neg b) \notin out_1(G \cup (a, \neg b)) - (c, \neg d)$ for some $(c, d) \in P$ with $a \dashv\vdash c$.*

That is, (a, b) is antithetically permitted in $\langle G, P\rangle$ if $(a, \neg b)$ would be annulled by an explicit permission (c, d) in P were it to be added to the code G. As with exemptions, we shall say that (a, b) is antithetically permitted *by* (c, d).

Exemptions and antithetic permission are closely related, as witnessed by the following results from Stolpe (2010c):

Theorem 14. *If (a, b) is an exemption in $\langle G, P\rangle$ then (a, b) is antithetically permitted in the same code.*

Theorem 15. *If (a, b) is antithetically permitted in $\langle G, P\rangle$, then it is an exemption in $\langle G \cup (a, \neg b), P\rangle$.*

Putting Theorem 14 and Theorem 15 together, we immediately get the following strengthened forms:

Theorem 16. *(a, b) is an exemption in $\langle G, P\rangle$ iff (a, b) is antithetically permitted in $\langle G, P\rangle$ and $(a, \neg b) \in out_1(G)$.*

Theorem 17. *(a, b) is antithetically permitted in $\langle G, P\rangle$ iff it is an exemption in $\langle G \cup (a, \neg b), P\rangle$.*

These theorems show that the difference between exemption and antithetic permission is slight. Basically, an exemption is just an antithetic permission (a, b) for which there is a corresponding mandatory norm $(a, \neg b)$. This is a somewhat surprising result, considering the fact that the intuitive ideas we started out with were rather different. An exemption was initially construed as a permissive norm (a, b) which is implicit in something that has been explicitly promulgated, whilst an antithetic permission was construed as a norm (a, b) which is such that, given the obligations already present in the code G in question, we can't forbid b under the condition a without thereby contradicting something d that is implicit in what has been explicitly permitted. Our analysis indicates that these ideas are strongly convergent, albeit not equivalent.

6.1 Syntactic Characterisation

The respective sets of antithetic permissions and exemptions both admit a syntactic representation that displays in a very clear form what it means to reason with them, and how they interact with obligations. We take the time to record these results, as they contribute, we believe, to the overall transparency of our analysis. We shall need a few lemmata:

Lemma 10. *If* $(a, b) \in P$ *then* (a, b) *is antithetically permitted in* $\langle G, P\rangle$.

Lemma 11. *If* $(a, \neg b \rightarrow \neg d) \in out_1(G)$ *for some* $(a, d) \in P$ *such that* $b \wedge d \nvdash f$ *then* (a, b) *is antithetically permitted in* $\langle G, P\rangle$.

Proof. Assume the conditions of the lemma hold. We show that $(a, \neg b) \notin out_1(G \cup (a, \neg b)) - (a, \neg d)$. It suffices to find a derogans $F \in out_1(G \cup (a, \neg b)) \perp (a, \neg d)$ such that $(a, \neg b) \notin F$. Consider the set $F^- := \{(a, \neg b \rightarrow \neg d)\}$. If $(a, \neg d) \in out_1(F^-)$ then $\neg b \rightarrow \neg d \vdash \neg d$, so $(\neg b \rightarrow \neg d) \rightarrow \neg d$ is a tautology. Now,

$$
\begin{aligned}
(\neg b \rightarrow \neg d) \rightarrow \neg d &\dashv\vdash \neg(\neg b \rightarrow \neg d) \vee \neg d \\
&\dashv\vdash (\neg b \wedge d) \vee \neg d \\
&\dashv\vdash b \rightarrow \neg d
\end{aligned}
$$

whence $b \vdash \neg d$ contrary to the assumption that $b \wedge d \nvdash f$. We may assume therefore that $(a, \neg d) \notin out_1(F^-)$. Thus, since $F^- \subseteq out_1(G) \subseteq out_1(G \cup (a, \neg b))$, F^- can be expanded to a derogans $F \in out_1(G \cup (a, \neg b)) \perp (a, \neg d)$. Recall that F contains $(a, \neg b \rightarrow \neg d)$ by construction, so we have $(a, \neg b) \notin F$ which completes the proof.

Lemma 12. *If* (a, b) *is antithetically permitted in* $\langle G, P\rangle$ *then* $(a, \neg b \rightarrow \neg d) \in out_1(G)$ *for some* $(c, d) \in P$ *with* $a \dashv\vdash c$.

Taken together, these lemmata yield the following corollary:

Corollary 3. (a, b) *is antithetically permitted in* $\langle G, P\rangle$ *iff* $(a, \neg b \rightarrow \neg d) \in out_1(G)$ *for some* $(a, d) \in P$ *such that* $a \dashv\vdash c$ *and* $b \wedge d \nvdash f$.

The restriction to logically compatible items was overlooked in Stolpe (2010c) and Stolpe (2010b). It expresses the fact that an antithetic permission cannot be permitted by an explicit permission with which it is not consistent.

Corollary 3 now induces the following syntactical representation of antithetic permission:

Theorem 18. *Let $\langle G, P\rangle$ be a code. Label a norm $(a, b)^o$ if $(a, b) \in out_1(G)$, $(a, b)^p$ if $(a, b) \in P$, and label it $(a, b)^a$ if (a, b) is antithetically permitted in $\langle G, P\rangle$ according to Definition 13. Then the antithetic permissions in $\langle G, P\rangle$ is precisely the set of norms of the form $(a, b)^p$ or $(a, b)^a$ that are derivable from axioms $(a, b)^o$ where $(a, b) \in out_1(G)$ using the following rules of inference.*

$$EQL^a \ \frac{(a,b)^p}{(a',b)^p} \ if\, a \dashv\vdash a' \qquad OP^a \ \frac{(a, \neg b \rightarrow \neg d)^o, (a,d)^p}{(a,b)^a} \ b \wedge d \nvdash f$$

$$AX^a \ \frac{}{(a,b)^p} \ (a,b) \in P$$

Proof. The soundness of EQL^a is immediate from definition Definition 13 of antithetic permission together with Lemma 10. OP^a is one direction of Corollary 3 and AX^a is Lemma 10 . For the converse direction suppose (a, b) is antithetically permitted in $\langle G, P\rangle$. By Lemma 12 we have that $(a, \neg b \rightarrow \neg d) \in out_1(G)$ for some $(c, d) \in P$ such that $a \dashv\vdash c$ and $b \wedge d \nvdash f$, hence

$$OP^a \ \dfrac{(a, \neg b \rightarrow \neg d)^o \qquad EQL^a \ \dfrac{Ax^a \ \dfrac{}{(c,d)^p}}{(a,d)^p}}{(a,b)^a}$$

Thus any antithetic permission (a, b) can be generated from $out_1(G)$ using only these rules, which completes the proof.

Turning now to to the concept of exemption, only minor modifications to Theorem 18 are required:

Theorem 19. *Let $\langle G, P\rangle$ be a code. Label a norm $(a, b)^o$ or $(a, b)^p$ under the same conditions as in Theorem 18, and label it $(a, b)^e$ if (a, b) is an exemption in $\langle G, P\rangle$ according to Definition 13. Then the exemptions in $\langle G, P\rangle$ are precisely the set of norms of the form $(a, b)^p$ or $(a, b)^e$ that are derivable from axioms $(a, b)^o$ where $(a, b) \in out(G)$ by the following rules of inference:*

$$EQL^e \ \frac{(a,b)^p}{(a',b)^p} \ if\, a \dashv\vdash a' \qquad OP^e \ \frac{(a, \neg b \rightarrow \neg d)^o, (a,d)^p, (a, \neg b)^o}{(a,b)^e} \ b \wedge d \nvdash f$$

$$AX^e \ \frac{(a, \neg b)^o}{(a,b)^e} \ (a,b) \in P$$

Proof. Similar in all essentials to that of Theorem 18.

6.2 Avoiding Triviality

The overall approach of the foregoing analysis is a sound one, we believe, but as it turns out the definitions are not quite apt as they stand, for it is possible to prove the following triviality property:

Theorem 20. *If* (a, b) *is an exemption by* (c, d) *in* $\langle G, P\rangle$, *then* (a, e) *is antithetically permitted in* $\langle G, P\rangle$ *for any* e *such that* $e \wedge d \nvdash f$.

Proof. If (a, b) is an exemption in $\langle G, P\rangle$, then $(a, \neg b) \in out_1(G) \setminus out_1(G) - (c, \neg d)$ for $(c, d) \in P$ with $a \dashv\vdash c$. There is thus an $F \in out_1(G) \perp (c, \neg d)$ with $(a, \neg b) \notin F$, which, since $(a, \neg b) \in out_1(G)$ means that $(c, \neg d) \in out_1(F \cup (a, \neg b))$. Hence, it follows by Lemma 9 that $(a, \neg b \rightarrow \neg d) \in out_1(G)$. Now, $(a, \neg b) \in out_1(G)$ by assumption, and we have $(a, \neg b \rightarrow (\neg e \rightarrow \neg b))$ from (t, t) by SI, whence $(a, \neg e \rightarrow \neg b)$ by AND. Taking stock we have $(a, \neg e \rightarrow \neg b)$, $(a, \neg b \rightarrow \neg d) \in out_1(G)$, so $(a, \neg e \rightarrow \neg d) \in out_1(G)$ by AND and WO. Since by assumption $d \wedge e \nvdash f$ it thus follows from Lemma 11 that (a, e) is antithetically permitted in $\langle G, P\rangle$.

In other words, the existence of an exemption, *any* exemption, makes everything that is logically compatible with the thus permitted action antithetically permitted under the same circumstances. That is, if anything at all constitutes an exemption under the state of affairs described by a, then everything is permitted (by antithesis) in those circumstances. This is clearly an intolerable result.

Nevertheless, we do not regard Theorem 20 as a refutation of the general idea of analysing permission in terms of derogation. Instead we put the blame on the underlying derogation operator which, according to a common classification scheme, is an operator of the full meet theory-variant. Such an operator can be generalised along two different lines: we may either select only a subset of the *derogantes* to intersect, thereby moving to a partial meet framework, or we may redefine derogation so that it operates on norm bases, that is, on sets of norms that are not assumed to be closed under $\Vdash_1$. Several things indicate that the latter of these is the more promising option. For one, unless the selection function is concretised somehow to exclude it, an abstract partial-meet derogation operation incurs losses in the theory of permission itself. It is well known that an AGM-style selection function induces a one-off revision operator that can only be applied to the set of norms for which it is defined. We would therefore not without further ado be in position to compare and juxtapose different derogation operations, as necessary e.g. for Theorem 15. Secondly, the proof of Theorem 20 indicates that the real culprit behind the triviality property is the propositional schema $a \rightarrow (b \rightarrow a)$, which is present in the output of any derogated code for every choice of a and b (since by *D-Closure* derogation performed on a closed set of norms yields a closed set of norms, and since closed sets of norms have logically closed output sets). All in all, this tilts the trade-offs heavily towards performing derogation on norm bases as the preferable way of avoiding triviality in the concepts of permission.

As hinted at several times already, we are now in position to do this in a manner that is more informative than the abstract quick-fix of simply dropping the closure requirement, insofar as we now have the finest splitting of any code available. Moreover, Theorem 13 tells us that this form is relatively regularly behaved, so we are not just randomly choosing a norm base, but one whose dynamic behaviour we know something about. Indeed, when we perform derogation on a finest splitting form of G rather than on G itself,[16] the positive permission concepts become tempered by relevance completely free of charge.

Theorem 21. *Suppose that derogation is always performed on a finest splitting form of a set of norms, and suppose that* $(G \cup (a, \neg b))(c)$ *is consistent. Then* (a, b) *is antithetically permitted in* $\langle G, P\rangle$ *by* (c, d) *only if* $(a, \neg b)$ *is relevant to the derogation of* $(c, \neg d)$ *from* $G \cup (a, \neg b)$.

We have stated this theorem without proof for the simple reason that it has almost no formal content. The pieces simply fit together without any mathematical welding.

It remains to show that this does indeed suffice to dodge the triviality result from Theorem 20. The following example will do:

Example 11. Put $G := \{(a, \neg b)\}$ and $P := \{(a, b)\}$, where b is an elementary letter. Then $G - (a, \neg b) = \emptyset$, so (a, b) is an exemption in $\langle G, P\rangle$. Let e be an elementary letter distinct from b. Then $\{b, e\}$ is consistent. Therefore, since $G \cup (a, \neg e)$ is in finest splitting form, $(a, \neg e)$ is irrelevant to the derogation of $(a, \neg b)$ from $G \cup (a, \neg e)$, by Definition 10, whence $(a, \neg e)$ is not antithetically permitted in $\langle G, P\rangle$ by Theorem 21.

Thus, implementing relevance in the underlying derogation operation represents one way to counterbalance the negative impact of the more notorious quirks of material implication, and thereby to restore the concepts of positive permission to the status of genuinely informative concepts.

7 Conclusion

We have argued that Makinson's methodological stance involves an ascent to a meta-logical level where abstractions over the underlying logical language establish a shared mathematical stratum that allows general techniques and results to be migrated from one idiom to another. We have showed in particular, that Makinson's work on revision theory and propositional relevance can be ported to the idiom of input/output logic, where it can fruitfully be brought to bear on norm-theoretic questions such as that concerning the nature of positive permission or the proper definition of normative relevance.

However, our investigations have revealed that there are definite limits to how much traffic these interfaces can carry. In particular, the property we have called

[16] There are in general more than one such form (cf. van de Putte (2011)).

input-entailment turns out to be an important one insofar as it is necessary for parallel interpolation as well as for establishing a correlation between maximally non-implying sets of sentences and maximally non-implying sets of norms. Without parallel interpolation, it is not clear how to prove the existence of finest splittings for codes of norms—if indeed a proof exists—and therefore also unclear how to define the notion of normative relevance. Without the correspondence, on the other hand, revision theory for sets of norms will have to be developed largely from first principles, since we cannot directly utilise the known properties of maximally non-implying sets.

On a philosophical level, our investigations have revealed that propositional consistency may constitute an axis along which normative and propositional reasoning diverge. Whereas consistency is a reasonable assumption to make when interpreting another agent, it is not equally reasonable, nor desirable, to demand consistency of commanded contents since one of the things one would want a formal model of a normative system to do is to anticipate and predict normative conflict. Yet, Theorem 13 does not apply to contexts that generate inconsistent output, so relevance does not, as the matter stands, apply to the resolution of normative conflict.

All in all, this chapter leaves many loose ends to be explored. Some of these can be sorted according to the interfaces we have explored. When it comes to the notion of derogation, for instance, an important question is to what extent we can migrate results from classical AGM theory over into input/output logic:

- Is there a natural preference relation between sets of norms that could be used to define a relational partial meet derogation operator?
- Is derogation by way of entrenchment a strategy worth exploring?
- What is the proper characterisation of derogation modulo basic output?

As regards splittings and the concept of relevance, we have at least the following:

- Are there finest splittings beyond the input-entailment border?
- Is the weaker principle of multiple parallel interpolation (cf. 3.2) valid for basic output, and if so
- can it be used to prove the existence of finest splittings?

In addition to these, interesting questions arise in relation to formal concept analysis as well. For instance:

- Is it possible to exploit the Duquenne-Guiges-basis of a concept lattice as a means to calculate implications between obligations?
- Could this give rise to an efficient algorithmic procedure?
- Is it possible to give a concept-theoretic semantics to the different input/output operators?

Finally, there are unresolved question of a more philosophical nature:

- Can we have normative relevance without presupposing output consistency?
- Is the asymmetry of our concept of relevance a bug or a feature?
- Are there other ways to split a code of norms that make intuitive sense?

References

Alchourron, C., & Makinson, D. (1981). Hierarchies of regulations and their logic. In R. Hilpinen (Ed.), *New studies in deontic logic* (pp. 125–148). Dodrecht: Reidel.

Bochmann, A. (2005). Explanatory nonmonotonic reasoning. Volume 4 of avances in logic. Singapore: World Scientific Publishong co.

Boella, G., & van der Torre, L. (2006). Introduction to normative multiagent systems. *Computational and Mathematical Organization Theory*, *12*, 71–79.

Chopra, S., & Parikh, R. (2000). Relevance sensitive belief structures. *Annals of Mathematics and Artificial Intelligence*, *28*, 259–285.

Davey, B. A., & Pricstley, H. A. (2002). *Introduction to lattices and order*. Cambridge: Cambridge University Press.

Ganter, B., & Wille, R. (1999). *Formal concept analysis*. Mathematical foundations: Springer.

Konev, B., Lutz, C., Ponomaryov, D., & Wolter, F. (2010). Decomposing description logic ontologies. In F. Lin & U. Sattler (Eds.), *Proceedings of the 12th International Conference on Principles of Knowledge Representation and Reasoning (KR2010)*, AAAI Press.

Kourousias, G., & Makinson, D. C. (2007). Parallel interpolation, splitting, and relevance in belief change. *Journal of Symbolic Logic*, *7*, 994–1002.

Makinson, D. (1999). On a fundamental problem of deontic logic. In P. MacNamara & H. Prakken (Eds.), *Norms, logics and information systems. New studies in deontic logic and computer science*. Volume 49 of frontiers in artificial intelligence and applications (pp. 29–53). Amsterdam: IOS Press.

Makinson, D. (2003). Ways of doing logic: What was different about AGM 1985? *Journal of Logic and Computation*, *13*, 3–13.

Makinson, D. (2009). Propositional relevance through letter-sharing. *Journal of Applied Logic*, *7*(4), 377–387. Special Issue: Formal models of belief change in rational agents.

Makinson, D. C. (1996). In memoriam carlos eduardo alchourron. *Nordic Journal of Philsophical Logic*, *1*, 3–10.

Makinson, D., & van der Torre, L. (2000). Input/output logics. *Journal of Philosophical Logic*, *29*, 383–408.

Makinson, D., & van der Torre, L. (2001). Constraints for input/output logics. *Journal of Philosophical Logic*, *30*, 155–185.

Makinson, D., & van der Torre, L. (2003a). Permission from an input/output perspective. *Journal of Philosophical Logic*, *32*(4), 391–416.

Makinson, D., & van der Torre, L. (2003b). What is input/output logic? *Foundations of the Formal Sciences II: Applications of Mathematical Logic in Philosophy and Linguistics*, *17*, 163–174.

Parikh, R. (1999). *Beliefs, belief revision, and splitting languages*. Stanford: Center for the Study of Language and Information.

Ponomaryov, D. (2008). On decomposibility in logical calculi. *Bulletin of the Novosibirsk Computing Center*, *28*, 111–120.

Schlechta, K. (2004). Coherent systems. Amsterdam: Elsevier Science.

Shafer-Landau, R. (2003). *Moral realism: A defence*. Oxford: Clarendon Press.

Stolpe, A. (2010a). Norm-system revision: Theory and application. *Artificial Intelligence and Law*, *18*, 247–283.

Stolpe, A. (2010b). *Relevance, derogation and permission*. In DEON (pp. 98–115).

Stolpe, A. (2010c). A theory of permission based on the notion of derogation. *Journal of Applied Logic*, *8*(1), 97–113.

van de Putte, F. (2011). Prime implicates and relevant belief revision. *Journal of Logic and Computation7*, 1–11.

van der Torre, L. (2010). Deontic redundancy: A fundamental challenge for deontic logic. *Proceedings of the 10th International Conference on Deontic Logic in Computer Science* (pp. 11–32). DEON'10, Heidelberg, Springer-Verlag.

von Wright, G.H. (1998). *Is and ought. In normativity and norms*. Oxford: Clarendon Press.

Intuitionistic Basis for Input/Output Logic

Xavier Parent, Dov Gabbay and Leendert van der Torre

Abstract It is shown that I/O logic can be based on intuitionistic logic instead of classical logic. More specifically, it is established that, when going intuitionistic, a representation theorem is still available for three of the four (unconstrained) original I/O operations. The trick is to see a maximal consistent set as a saturated one. The axiomatic characterization is as in the classical case. Therefore, the choice between the two logics does not make any difference for the resulting framework.

Keywords Deontic logic · I/O Logic · Intuitionistic logic · Completeness · Saturated set

1 Introduction

So-called input/output logic (I/O logic, for short) is a general framework devised by Makinson and van der Torre (2000, 2001, 2003) in order to reason about conditional norms. A frequent belief about I/O logic is that it presupposes classical logic. The aim of this chapter is to show that this is a misunderstanding.

From the outset Makinson and van der Torre made it clear that it would be quite misleading to refer to I/O logic as a kind of 'non-classical logic.' Its role, they argue, is to study not some kind of non-classical logic, but "a way of using the classical

X. Parent (✉) · D. Gabbay · L. van der Torre
University of Luxembourg, Luxembourg, Luxembourg
e-mail: xavier.parent@uni.lu

L. van der Torre
e-mail: leon.vandertorre@uni.lu

D. Gabbay
Bar-Ilan University, Ramat Gan, Israel
King's College London, Strand, London, UK
e-mail: dov.gabbay@kcl.ac.uk

S. O. Hansson (ed.), *David Makinson on Classical Methods for Non-Classical Problems*, Outstanding Contributions to Logic 3, DOI: 10.1007/978-94-007-7759-0_13,

one" (Makinson and van der Torre, 2000, p. 384). The basic intuition motivating the I/O framework is explained further thus, by contrasting an old and a new way to think about logic:

> From a very general perspective, logic is often seen as an 'inference motor', with premises as inputs and conclusions as outputs [...]. But it may also be seen in another role, as 'secretarial assistant' to some other, perhaps non-logical, transformation engine [...]. From this point of view, the task of logic is one of preparing inputs before they go into the machine, unpacking outputs as they emerge and, less obviously, co-ordinating the two (Makinson and van der Torre 2000, p. 384).

In input/output logic, the meaning of the deontic concepts is given in terms of a set of procedures yielding outputs for inputs. Thus, the semantics may be called "operational" rather than truth-functional. This is where the black box analogy comes in. To some extent, the system can be viewed solely in terms of its input, output and transfer characteristics without any knowledge of its internal workings, which remain "opaque" (black). Logic is here reduced to an ancillary role in relation to it.

The picture of logic assisting a transformation engine appears to be very general: *prima facie* any base logic may act as a secretarial assistant. Still, the only input-output logics investigated so far in the literature are built on top of classical propositional logic. In particular, one of the key building blocks is the notion of maximal consistent set. In the context of logics other than classical logic, this notion will not do the required job. This raises the question of whether the framework is as general as one might think at first sight, and whether it can be instantiated using other logics. If the answer is positive, then one would like to know what properties of classical logic are essential for the completeness result to obtain.

As a first step towards clarifying these issues, we here consider the case of intuitionistic logic. The main observation of this chapter is that, if I/O logic is built on top of the latter, then a representation theorem for three of the four standard I/O operations is still available. These are the so-called simple-minded, basic, and simple-minded reusable operations. What is more, the axiomatic characterization of the I/O operations remains the same. We will show that the job done by the notion of maximal consistent set in I/O logic can equally be done by its analog within intuitionistic logic, the notion of saturated set. The result is given for the unconstrained version of I/O logics dealing with obligation. It is the only one that comes with a known complete axiomatic characterization. The unconstrained version of I/O logics dealing with permission (see Makinson and van der Torre 2003), and the constrained version of I/O logics dealing with contrary-to-duties (see Makinson and van der Torre 2001) both come with a syntactical characterization or proof-theory. But strictly speaking this one is not an axiomatization. There are no soundness and completeness theorems relating operations to proof systems.

We also complement our positive results by a negative one concerning the fourth and last standard operation, called basic reusable. It is pointed out that the so-called rule of "ghost contraposition", which holds for the classically based operation, fails for its intuitionistic counterpart. The axiomatization problem for the latter I/O operation remains an open one.

This chapter is organized as follows. Section 2 provides the required background. Section 3 gives the completeness results. Section 4 concludes, and provides some suggestions as to useful ways forward.

2 Background

Section 2.1 presents the framework as initially defined by Makinson and van der Torre (2000). We shall refer to it as the initial framework. Section 2.2 explains how to change the base logic from classical to intuitionistic logic.

2.1 I/O Logic

In input/output logic, a normative code is a (non-empty) set G of conditional norms, which is a (non-empty) set of ordered pairs (a, x). Here a and x are two well-formed formulae (wff's) of propositional logic. Each such pair may be referred to as a generator. The body a is thought of as an input, representing some condition or situation, and the head x is thought of as an output, representing what the norm tells us to be obligatory in that situation.

Some further notation. L is the set of all formulae of propositional logic. Given an input $A \subset L$, and a set of generators G, $G(A)$ denotes the image of G under A, i.e., $G(A) = \{x : (a, x) \in G \text{ for some } a \in A\}$. $Cn(A)$ denotes the set $\{x : A \vdash x\}$, where $\vdash$ is the consequence relation used in classical logic. The notation $x \dashv\vdash y$ is short for $x \vdash y$ and $y \vdash x$.

2.1.1 Semantics

As mentioned, I/O logic comes with an operational rather than truth-functional semantics. The meaning of the deontic concepts is given in terms of a set of procedures yielding outputs for inputs. The following I/O operations can be defined.

Definition 1 (I/O operations). Let A be an input set, and let G be a set of generators. The following input/output operations can be defined, where a complete set is one that is either maximal consistent[1] or equal to L:

[1] The set is consistent, and none of its proper extensions is consistent.

$$out_1(G, A) = Cn(G(Cn(A)))$$
$$out_2(G, A) = \cap\{Cn(G(V)) : A \subseteq V, V\ complete\}$$
$$out_3(G, A) = \cap\{Cn(G(B)) : A \subseteq B \supseteq Cn(B) \supseteq G(B)\}$$
$$out_4(G, A) = \cap\{Cn(G(V)) : A \subseteq V \supseteq G(V), V\ complete\}$$

$out_1(G, A)$, $out_2(G, A)$, $out_3(G, A)$ and $out_4(G, A)$ are called "simple-minded" output, "basic" output, "simple-minded reusable" output, and "basic reusable" output, respectively.

These four operations have four counterparts that also allow throughput. Intuitively, this amounts to requiring $A \subseteq out_i(G, A)$ for $i \in \{1, 2, 3, 4\}$. In terms of the definitions, it is to require that G is expanded to contain the diagonal, i.e., all pairs (a, a). The throughput version of the I/O operations will be put to one side in this preliminary study.

2.1.2 Proof-Theory

Put $out_i(G) = \{(A, x) : x \in out_i(G, A)\}$ for $i \in \{1, 2, 3, 4\}$. This definition leads to an axiomatic characterization that is much alike those used for conditional logic.

The specific rules of interest here are described below. They are formulated for a singleton input set A (for such an input set, curly brackets will be omitted). The move to an input set A of arbitrary cardinality will be explained in a moment.

$$(\text{SI})\quad \frac{(a, x) \qquad b \vdash a}{(b, x)}$$

$$(\text{WO})\quad \frac{(a, x) \qquad x \vdash y}{(a, y)}$$

$$(\text{AND})\quad \frac{(a, x) \qquad (a, y)}{(a, x \wedge y)}$$

$$(\text{OR})\quad \frac{(a, x) \qquad (b, x)}{(a \vee b, x)}$$

$$(\text{CT})\quad \frac{(a, x) \qquad (a \wedge x, y)}{(a, y)}$$

SI, WO, and CT abbreviate "strengthening of the input", "weakening of the output", and "cumulative transitivity" respectively.

Given a set R of rules, a derivation from a set G of pairs (a, x) is a sequence α_1,.., α_n of pairs of formulae such that for each index $1 \leq i \leq n$ one of the following holds:

Table 1 I/O systems

Output operation	Rules
Simple-minded (out_1)	{SI, WO, AND}
Basic (out_2)	{SI, WO, AND} + {OR}
Simple-minded reusable (out_3)	{SI, WO, AND} + {CT}
Basic reusable (out_4)	{SI, WO, AND} + {OR,CT}

- α_i is an hypothesis, i.e., $\alpha_i \in G$;
- α_i is $(\top, \top)$, where $\top$ is a zero-place connective standing for 'tautology';
- α_i is obtained from preceding element(s) in the sequence using a rule from R.

The elements in the sequence are all pairs of the form (a, x). Derivation steps done in the base logic are not part of it.

A pair (a, x) of formulae is said to be derivable from G if there is a derivation from G whose final term is (a, x). This may be written as $(a, x) \in deriv(G)$, or equivalently $x \in deriv(G, a)$. A subscript will be used to indicate the set of rules employed. The specific systems of interest here will be referred to as $deriv_1$, $deriv_2$, $deriv_3$, and $deriv_4$. They are defined by the rules {SI, WO, AND}, {SI, WO, AND, OR}, {SI, WO, AND, CT}, and {SI, WO, AND, OR, CT}, respectively. Note that, given (SI), (CT) can be strengthened into plain transitivity ("From (a, x) and (x, y), infer (a, y)").

When A is a set of formulae, derivability of (A, x) from G is defined as derivability of (a, x) from G for some conjunction $a = a_1 \wedge ... \wedge a_n$ of elements of A. We understand the conjunction of zero formulae to be a tautology, so that $(\emptyset, a)$ is derivable from G iff $(\top, a)$ is.

The following applies:

Theorem 1 (Soundness and completeness).

$$out_1(G, A) = deriv_1(G, A)$$
$$out_2(G, A) = deriv_2(G, A)$$
$$out_3(G, A) = deriv_3(G, A)$$
$$out_4(G, A) = deriv_4(G, A)$$

Proof. See Makinson and van der Torre (2000). □

Table 1 shows the output operations, and the rules corresponding to them.

2.2 *Intuitionistic I/O Logic*

This subsection describes the changes that must be made to I/O logic for it to be based on intuitionistic logic. We start with a few highlights on the latter. Those readers already familiar with intuitionistic logic and its Kripke semantics should pass directly to Sect. 2.2.2, returning to Sect. 2.2.1 if needed for specific points.

2.2.1 Intuitionistic Logic

Intuitionistic logic has its roots in the philosophy of mathematics propounded by Brouwer in the early twentieth century. According to Brouwer, truth in mathematics means the existence of a proof (cf Troelstra and van Dalen 1988). Intuitionistic logic can, thus, be described as departing from classical logic in its definition of the meaning of a statement being true. In classical logic, a statement is 'true' or 'false', independently of our knowledge concerning the statement in question. In intuitionistic logic, a statement is 'true' if it has been proved, and 'false' if it has been disproved (in the sense that there is a proof that the statement is false). As a result, intuitionistic logicians reject two laws of classical logic, among others. One is the law of excluded middle, $a \vee \neg a$, and the other is the law of double negation elimination, $\neg\neg a \rightarrow a$. Under the intuitionistic reading, an assertion of the form $a \vee \neg a$ implies the ability to either prove or refute a. And a statement like $\neg\neg a$ says that a refutation of a has been disproved. You may have an opinion that this is not the same as a proof (or reinstatement) of a; the truth of a may still be uncertain. Of course, even though intuitionistic logic was initially conceived as the correct logic to apply in mathematical reasoning, it would be a mistake to restrict the latter logic to the mathematical domain. The intuitionistic understanding of mathematical language may be generalized to all language, if one takes the notion of proof in a very broad sense. Most notably, Dummett (see e.g. Dummett 1973) championed such a generalization. The key idea is to assume that the meaning of a statement is always given by its justification conditions, i.e., the conditions under which one would be justified in accepting the statement. This is known as the verificationist theory of meaning.

In this chapter, intuitionistic logic will be described as the minimal logic that both contains the *ex falso* rule

(Ex falso) $$a, \neg a \vdash b$$

and allows for the deduction theorem:

(DT) $$\Gamma \vdash a \rightarrow b \text{ iff } \Gamma, a \vdash b$$

Classical logic allows for a stronger equivalence, namely:

(SDT) $$\Gamma \vdash (a \to b) \vee c \text{ iff } \Gamma, a \vdash b \vee c$$

Below we list the main properties of $\vdash$ to which we shall appeal later. They are taken from the discussion by Thomason (1968). From now onwards, $\vdash$ and Cn will refer to intuitionistic rather than classical logic.

Group I

$$\textbf{(Ref)} \text{ If } a \in \Gamma, \text{ then } \Gamma \vdash a$$

$$\textbf{(Mon)}\ \frac{\Gamma \vdash a}{\Gamma \cup \Delta \vdash a}$$

$$\textbf{(Cut)}\ \frac{\Gamma \vdash a \qquad \Gamma \cup \{a\} \vdash b}{\Gamma \vdash b}$$

$$\textbf{(C)} \text{ If } \Gamma \vdash a, \text{ then } \Gamma' \vdash a \text{ for some finite } \Gamma' \subseteq \Gamma$$

The labels (**Ref**), (**Mon**), and (**C**) are mnemonic for "Reflexivity", "Monotony", and "Compactness", respectively. (**Cut**) can be equivalently expressed as

$$\textbf{(Cut')}\ \frac{\Gamma \vdash a \text{ for all } a \in A \qquad \Gamma \cup A \vdash b}{\Gamma \vdash b}$$

Group II

$$(\to\textbf{:I})\ \frac{\Gamma \cup \{a\} \vdash b}{\Gamma \vdash a \to b} \qquad (\to\textbf{:E})\ \frac{\Gamma \vdash a \qquad \Gamma \vdash a \to b}{\Gamma \vdash b}$$

$$(\vee\textbf{:I})\ \frac{\Gamma \vdash a}{\Gamma \vdash a \vee b} \quad \frac{\Gamma \vdash b}{\Gamma \vdash a \vee b} \qquad (\vee\textbf{:E})\ \frac{\Gamma \cup \{a\} \vdash c \qquad \Gamma \cup \{b\} \vdash c \qquad \Gamma \vdash a \vee b}{\Gamma \vdash c}$$

$$(\wedge\textbf{:I})\ \frac{\Gamma \vdash a \qquad \Gamma \vdash b}{\Gamma \vdash a \wedge b} \qquad (\wedge\textbf{:E})\ \frac{\Gamma \vdash a \wedge b}{\Gamma \vdash a} \quad \frac{\Gamma \vdash a \wedge b}{\Gamma \vdash b}$$

$$(\neg\textbf{:I})\ \frac{\Gamma \cup \{a\} \vdash b \qquad \Gamma \cup \{a\} \vdash \neg b}{\Gamma \vdash \neg a} \qquad (\neg\textbf{:E})\ \frac{\Gamma \vdash a \qquad \Gamma \vdash \neg a}{\Gamma \vdash b}$$

$$(\top\textbf{:I})\ \frac{-}{\Gamma \vdash \top}$$

The rules in group I may be called "structural". They are also often referred to as the "Tarski conditions" in honor of Alfred Tarski who first saw their importance. Indeed these can be shown to express conditions that are jointly necessary and sufficient for propositions to be chained together in a derivation (see Makinson 2012, Sect. 10.2).

The rules in group II may be called "elementary". They can be classified as introduction or elimination rules depending on whether they allow us to introduce or eliminate a connective. $\top$ has no elimination rule.

The principle (Ex falso) follows from (**Ref**) and ($\neg$:**E**). The reader may also easily verify that (DT) is derivable. One direction is just ($\rightarrow$:**I**). For the other, assume $\Gamma \vdash a \rightarrow b$. By (**Mon**), $\Gamma, a \vdash a \rightarrow b$. By (**Ref**) and (**Mon**), $\Gamma, a \vdash a$. By ($\rightarrow$:**E**), $\Gamma, a \vdash b$, as required.

Negation in classical logic (so-called classical negation) is defined by ($\neg$:**I**) and the rule

$$(\neg{:}\mathbf{E'})\ \frac{\Gamma, \neg a \vdash b \qquad \Gamma, \neg a \vdash \neg b}{\Gamma \vdash a}$$

In the presence of ($\neg$:**E'**), the laws $\vdash a \vee \neg a$ and $\vdash \neg\neg a \rightarrow a$ become derivable. It is noteworthy that (SDT) becomes derivable too. The proofs are omitted.

We write $x[y/a]$ for the formula obtained by replacing, in x, all occurrences of atom a with y. Where Γ is a set of formulae, we write $\Gamma[y/a]$ for $\{x[y/a] : x \in \Gamma\}$.

Theorem 2 (Substitution theorem for derivability). *If $\Gamma \vdash x$, then $\Gamma[y/a] \vdash x[y/a]$.*

Proof. This is van Dalen (1994, Theorem 5.2.4). □

If some formula y occurs in a formula x, the occurrence itself can be described by a sequence of natural numbers. It gives the so-called position of y in (the construction tree of) x. For instance, the second occurrence of b has position 22 in $(a \wedge b) \rightarrow (a \vee b)$. We write $x[y]_p$ for the result of replacing the subformula at position p in x by y. The formal definition of the notion of position is as usual, and so is that of the '$x[y]_p$' construct.[2] We have:

Theorem 3 (Replacement of equivalents). *If $x_1 \dashv\vdash x_2$, then $x[x_1]_p \dashv\vdash x[x_2]_p$.*

Proof. The proof is standard, and is omitted. □

For future reference, we note the following facts.

Lemma 1

(1) $a \vdash a \vee b$ *and* $b \vdash a \vee b$

(2) $a \wedge b \vdash a$ *and* $a \wedge b \vdash b$

(3) $a \vdash b \rightarrow a$

(4) *If* $a \vdash c$ *and* $b \vdash c$ *then* $a \vee b \vdash c$

(5) *If* $a \vdash c$ *then* $a \wedge b \vdash c$

---

[2] See, for example, Nonnengart and Weidenbach (2001, Sect. 2).

(6) $a, b \vdash c$ *iff* $a \wedge b \vdash c$

(7) $a \dashv\vdash a \wedge \top$

(8) $\top \vdash x$ *whenever* $x \in Cn(\emptyset)$

(9) $a \wedge (a \rightarrow b) \vdash b$

(10) $a \wedge b \wedge (c \vee d) \vdash (a \wedge c) \vee (b \wedge d)$

(11) $Cn(\Gamma) = Cn\,Cn(\Gamma)$

Proof. For (1), we have $a \vdash a$ by (**Ref**). Using ($\vee$:**I**) it follows that $a \vdash a \vee b$. The argument for $b \vdash a \vee b$ is similar.

For (2), by (**Ref**), $a \wedge b \vdash a \wedge b$. By ($\wedge$:**E**), $a \wedge b \vdash a$. The argument for $a \wedge b \vdash b$ is the same.

For (3), by (**Ref**), $a \vdash a$. By (**Mon**), $a, b \vdash a$. By ($\rightarrow$:**I**), $a \vdash b \rightarrow a$.

For (4), assume $a \vdash c$ and $b \vdash c$. By (**Mon**), $a \vee b, a \vdash c$ and $a \vee b, b \vdash c$. By (**Ref**), $a \vee b \vdash a \vee b$. By ($\vee$:**E**), $a \vee b \vdash c$ as required.

For (5), assume $a \vdash c$. By (**Mon**), $a \wedge b, a \vdash c$. By (2) in this Lemma, $a \wedge b \vdash a$. By (**Cut**), $a \wedge b \vdash c$.

For (6). From left-to-right, assume $a, b \vdash c$. By (**Mon**), $a \wedge b, a, b \vdash c$. By (2) in this Lemma, $a \wedge b \vdash a$, and $a \wedge b \vdash b$. So, by (**Cut'**), $a \wedge b \vdash c$ as required. From right-to-left, assume $a \wedge b \vdash c$. By (**Ref**) and (**Mon**), $a, b \vdash a$ and $a, b \vdash b$. By ($\wedge$:**I**), $a, b \vdash a \wedge b$. By (**Mon**), $a, b, a \wedge b \vdash c$. By (**Cut**), $a, b \vdash c$.

For (7), $a \wedge \top \vdash a$ is just a special case of (2). By (**Ref**), $a \vdash a$. By ($\top$:**I**), $a \vdash \top$. By ($\wedge$:**I**), $a \vdash a \wedge \top$.

For (8), assume $x \in Cn(\emptyset)$. So $\emptyset \vdash x$. By (**Mon**), $\top \vdash x$ as required.

For (9). By (2) in this Lemma, $a \wedge (a \rightarrow b) \vdash a$ and $a \wedge (a \rightarrow b) \vdash a \rightarrow b$. By ($\rightarrow$:**E**), $a \wedge (a \rightarrow b) \vdash b$ as required.

For (10), using (2) in this Lemma and (**Mon**), we get $c, a \wedge b \vdash a$. By (**Ref**) and (**Mon**), $c, a \wedge b \vdash c$. So $c, a \wedge b \vdash a \wedge c$ by ($\wedge$:**I**). And, by (1) in this Lemma, $a \wedge c \vdash (a \wedge c) \vee (b \wedge d)$. By (**Mon**), $c, a \wedge b, a \wedge c \vdash (a \wedge c) \vee (b \wedge d)$. By (**Cut**), $c, a \wedge b \vdash (a \wedge c) \vee (b \wedge d)$. By ($\rightarrow$:**I**), $c \vdash (a \wedge b) \rightarrow ((a \wedge c) \vee (b \wedge d))$. A similar argument yields $d \vdash (a \wedge b) \rightarrow ((a \wedge c) \vee (b \wedge d))$. By (4) in this Lemma, $c \vee d \vdash (a \wedge b) \rightarrow ((a \wedge c) \vee (b \wedge d))$. Using the left-to-right direction of (DT), we get $c \vee d, a \wedge b \vdash (a \wedge c) \vee (b \wedge d)$. After rearranging the premisses, the conclusion $(a \wedge b) \wedge (c \vee d) \vdash (a \wedge c) \vee (b \wedge d)$ follows, by a straightforward application of (6) in this Lemma.

(11) is the property of idempotence for Cn. The left-in-right inclusion follows from (**Ref**). For the right-in-left inclusion, let $x \in Cn\,Cn(\Gamma)$. By (**C**), there are $x_1, \ldots, x_n (n \geq 0)$ in $Cn(\Gamma)$ such that $x_1, \ldots, x_n \vdash x$. On the one hand, $\Gamma \vdash x_1, \ldots,$ and $\Gamma \vdash x_n$. On the other hand, $\Gamma \cup \{x_1, \ldots, x_n\} \vdash x$ by (**Mon**). So $\Gamma \vdash x$ by (**Cut'**). Hence $x \in Cn(\Gamma)$, as required. □

It is a well-known fact that intuitionistic logic satisfies the so-called disjunction property. Expressed in terms of theoremhood, this is the property that, if $\vdash a \vee b$, then $\vdash a$ or $\vdash b$. This property also holds for the notion of deducibility under a set of assumptions, but in a qualified form. A restriction must be placed on the occurrences of $\vee$ in the premisses set Γ. The proviso in question may be formulated in terms of the notion of Harrop formula. The Harrop formulae, named after Ronald Harrop (1956), are the class of formulae defined inductively as follows:

- $a \in H$ for atomic a;
- $\neg a \in H$ for any wff a;
- $a, b \in H \Rightarrow a \wedge b \in H$;
- $b \in H \Rightarrow a \rightarrow b \in H$.

Intuitively, a Harrop formula is a formula in which all the disjunctions are 'hidden' inside the left-hand scope of an implication, or inside the scope of a negation.

We say that a set Γ is Harrop if it is made up of Harrop formulae only.

Theorem 4 (Disjunction property under hypothesis). *If Γ is Harrop, then, if $\Gamma \vdash a \vee b$, it follows that $\Gamma \vdash a$ or $\Gamma \vdash b$.*

Proof. This is Theorem 23 in van Dalen (2002). For a proof, see Gabbay (1981, Chap. 2, Sect. 3). □

In intuitionistic logic, a set Γ of wff's is said to be consistent – written Con Γ – just in case $\Gamma \nvdash a$ for some wff a. Thus, Γ is inconsistent – Incon Γ – just when $\Gamma \vdash a$ for all wff a. For future reference, Theorem 5 records some properties of the notion of consistency thus conceived.

Theorem 5.

(1) $\quad$ Con$\{a \vee b\}$ *iff:* Con$\{a\}$ *or* Con$\{b\}$

(2) $\quad$ Con$\{b\}$ *implies* Con$\{a\}$ *whenever* $b \vdash a$

(3) $\quad$ Con Γ *iff* Con Γ' *for all finite* $\Gamma' \subseteq \Gamma$

(4) $\quad$ Con Γ *implies* Con $Cn(\Gamma)$

Proof. For (1):

- For the left-to-right direction, assume Con$\{a \vee b\}$. This means that $\{a \vee b\} \nvdash c$ for some c. By Lemma 1 (4), either $\{a\} \nvdash c$ or $\{b\} \nvdash c$, which suffices for Con$\{a\}$ or Con$\{b\}$.
- For the right-to-left direction, we show the contrapositive. Let Incon$\{a \vee b\}$. So $a \vee b \vdash c$ for all c. Consider an arbitrary c. By Lemma 1 (1), $a \vdash a \vee b$. By (**Mon**), $a, a \vee b \vdash c$. By (**Cut**), $a \vdash c$. So, $a \vdash c$ for all c, and thus Incon$\{a\}$. The argument for Incon$\{b\}$ is similar.

For (2), assume $b \vdash a$ and Con$\{b\}$. The latter means that $b \nvdash c$ for some c. From this and $b \vdash a$ using (**Cut**) it follows that $a, b \nvdash c$. By (**Mon**), $a \nvdash c$, which suffices for Con$\{a\}$.

For (3). This uses consistency to express compactness.

- For the left-to-right direction, assume Con Γ. So, $\Gamma \nvdash c$ for some c. Let Γ' be a finite subset of Γ. By (**Mon**), $\Gamma' \nvdash c$, and thus Con Γ' as required.

- For the reverse, assume Con Γ' for any finite $\Gamma' \subseteq \Gamma$, but Incon Γ. From the latter, $\Gamma \vdash c$ and $\Gamma \vdash \neg c$ for an arbitrarily chosen c. By (**C**), $\Gamma_1 \vdash c$ and $\Gamma_2 \vdash \neg c$ for some finite $\Gamma_1, \Gamma_2 \subseteq \Gamma$. Since Γ_1 and Γ_2 are both finite, so is $\Gamma_1 \cup \Gamma_2$. Furthermore, by (**Mon**), $\Gamma_1 \cup \Gamma_2 \vdash c$ and $\Gamma_1 \cup \Gamma_2 \vdash \neg c$. By ($\neg$:**E**), $\Gamma_1 \cup \Gamma_2 \vdash b$ for all b. So Incon $(\Gamma_1 \cup \Gamma_2)$, contradicting the opening assumption.

For (4), assume Con Γ. So $c \notin Cn(\Gamma)$ for some wff c. By Lemma 1 (11), $c \notin Cn\, Cn(\Gamma)$, and thus Con $Cn(\Gamma)$ as required. □

We end this section by describing a Kripke-type semantics commonly used for intuitionistic logic. A Kripke model M for intuitionistic logic is a triplet $(W, \geq, V)$, where

- W is a set of possible worlds, $t, s, ...$;
- $\geq$ is a reflexive and transitive relation on W;
- V is an evaluation function assigning to each propositional letter a the set of worlds at which a is true.

Each world t forces the truth of formulae, and this relation is indicated by $\models$. Following Kripke (1965), one might think of the worlds as representing points in time (or "evidential situations"), at which we may have various pieces of information. If, at a given time point t, we have enough information to prove x, then we say that x has been verified at t, or that t forces x. If we lack such information, we say that x has not been verified at t, or that t does not force x. Persistence over time is required for all atomic a, in the sense that

$$t \geq s \text{ and } s \in V(a) \text{ imply } t \in V(a)$$

The forcing relation $\models$ satisfies the usual conditions except for

$$w \models x \rightarrow y \text{ iff } \forall w'((w' \geq w \,\&\, w' \models x) \Rightarrow w' \models y)$$
$$w \models \neg x \text{ iff } \neg\exists w'(w' \geq w \,\&\, w' \models x)$$

The notions of semantic consequence and satisfiability are defined in the usual way. For a formula x to be a semantic consequence of Γ (notation: $\Gamma \models x$), it must be the case that, whenever all the formulae in Γ are forced at some point in a model, then x is forced in that same model at the same point. A set Γ of formulae is said to be satisfiable if there is some model in which all its component formulae are forced at some point in the model. We have:

Theorem 6 (Soundness and completeness). *$\Gamma \vdash x$ iff $\Gamma \models x$*

Proof. See e.g. Thomason (1968). □

For more on intuitionistic logic, the reader is referred to Gabbay (1981, 2007).

2.2.2 Redefining the I/O Operations

For out_1 and out_3, it is natural to keep the same definitions as in the original framework, but assume that the underlying consequence relation *Cn* is defined as in intuitionistic logic.

For out_2 and out_4, a little more care is needed. The original account uses the notion of maximal consistent set. If the base logic is intuitionistic logic, the latter notion will not be suitable for the problem at hand. We use the notion of saturated set instead. It can be defined thus.

Definition 2 (Saturated set, Thomason 1968**).** Let S be a (non-empty) set of wff's. S is said to be saturated if the following three conditions hold:

(1) Con S (S is consistent)

(2) $a \vee b \in S \Rightarrow a \in S$ or $b \in S$ (S is join-prime)

(3) $S \vdash a \Rightarrow a \in S$ (S is closed under $\vdash$)

Theorem 7. *If S is consistent and Harrop, then Cn(S) is saturated.*

Proof. Let S be consistent and Harrop. By Theorem 5 (4), $Cn(S)$ is consistent. For join-primeness, let $a \vee b \in Cn(S)$. By Theorem 4, either $a \in Cn(S)$ or $b \in Cn(S)$, as required. Closure under $\vdash$ follows from Lemma 1 (11). □

The relationship between maximal consistent set (in the classical sense) and saturated set (in the intuitionistic sense) may be described as follows. A set of wff's is said to be maximal consistent if it is consistent and none of its proper extensions is consistent. In classical logic, this definition can be rephrased (equivalently) using the notion of "$\neg$-completeness". A set S of wff's is said to be $\neg$-complete, whenever $a \in S$ or $\neg a \in S$ for all wff a. Call S maximal consistent⋆ if it is consistent and $\neg$-complete.[3]

Example 1 Suppose the language has two propositional letters a and b only. By Theorem 7, $Cn(a)$ is a saturated set. However, $Cn(a)$ is not $\neg$-complete, since neither $b \in Cn(a)$ nor $\neg b \in Cn(a)$. Therefore, not all saturated sets are maximal consistent⋆.

Theorem 8 shows that, for classical logic, the notion of saturated set coincides with that of a maximal consistent one.

Theorem 8. *The following applies:*

1. *Any maximal consistent⋆ set S is saturated;*
2. *Suppose the law of excluded middle holds, i.e., $\vdash a \vee \neg a$. Any saturated set S is $\neg$-complete, and thus maximal consistent⋆.*

Proof. For the first claim, let S be maximal consistent⋆. For join-primeness, assume $a \vee b \in S$, but $a \notin S$ and $b \notin S$. By $\neg$-completeness, $\neg a \in S$ and $\neg b \in S$. The

[3] We use the superscript ⋆ to emphasize that this version of maximal consistency is, to some extent, peculiar to classical logic.

reader may easily verify that the set $\{a \vee b, \neg a, \neg b\}$ is not satisfiable in any model, and so (by Theorem 6) it is inconsistent. So, by Theorem 5 (3), S is inconsistent, contrary to the opening assumption. For closure under $\vdash$, assume $S \vdash a$ but $a \notin S$. By $\neg$-completeness, $\neg a \in S$. By (**Ref**), $S \vdash \neg a$. This is enough to make S inconsistent, given ($\neg$:**E**).

The second claim is Restall (2000, Lemma 5.6). By (**Mon**), $S \vdash a \vee \neg a$. By Definition 2 (3), $a \vee \neg a \in S$. By Definition 2 (2), either $a \in S$ or $\neg a \in S$. □

The following will come in handy.

Theorem 9. *For any saturated set S,*

(1) $S = Cn(S)$
(2) *If* $a, b \in S$, *then* $a \wedge b \in S$

Proof. For (1). The right-in-left inclusion is Definition 2 (3). The left-in-right inclusion is (**Ref**).

For (2), let $a, b \in S$. By (**Ref**), $S \vdash a$ and $S \vdash b$. By ($\wedge$:**I**), $S \vdash a \wedge b$. By Definition 2 (3), $a \wedge b \in S$. □

An analog of so-called Lindenbaum's lemma is available.

Theorem 10. *If* $\Gamma \nvdash a$, *then there is a saturated set S such that* $\Gamma \subseteq S$ *and* $a \notin S$

Proof. This is Thomason (1968, Lemma 1). □

out_2 can be reformulated as follows. To avoid proliferation of subscripts, we use the exact same name as in the original framework.

Definition 3 (out_2, **intuitionistic basic output).**

$$out_2(G, A) = \begin{cases} \cap\{Cn(G(S)) : A \subseteq S, S \text{ saturated}\}, & \text{if Con } A \\ Cn(h(G)), & \text{otherwise} \end{cases}$$

where $h(G)$ is the set of all heads of elements of G, viz. $h(G) = \{x : (a, x) \in G \text{ for some } a\}$.

The definition is well-behaved, because of Theorem 10. It guarantees that there is at least one saturated set extending A, when A is consistent.

A similar remark applies to out_4, which may be redefined thus:

Definition 4 (out_4, **intuitionistic basic reusable output).**

$$out_4(G, A) = \begin{cases} \cap\{Cn(G(S)) : A \subseteq S \supseteq G(S), S \text{ saturated}\}, & \text{if Con } A \\ Cn(h(G)), & \text{otherwise} \end{cases}$$

3 Completeness Results

This section gives completeness results for the first three intuitionistic output operations described in the previous section. To help with cross-reference, we shall refer to the completeness proof given in Makinson and van der Torre (2000) as the original proof or the classical case. From now onwards it is understood that the output operations are defined on top of intuitionistic logic. We give the full details even when the argument is a re-run of the original one. This, in order to pinpoint what elementary rules are needed, and where.

The original completeness proofs for out_2 and out_3 both make essential use of the completeness result for out_1. We start by noticing that the latter one carries over to the intuitionistic setting. The proof for the classical case is outlined in Makinson and van der Torre (2000); we give it in full detail to make clear that it goes through in the intuitionistic case.

Theorem 11. *out_1 validates the rules of $deriv_1$.*

Proof. The verification is easy, and is omitted. Note that the argument for (AND) uses ($\wedge$:**I**). □

Theorem 12 (Soundness, simple-minded). $deriv_1(G, A) \subseteq out_1(G, A)$

Proof. Assume $x \in deriv_1(G, A)$. By definition, $x \in deriv_1(G, a)$ for some conjunction $a = a_1 \wedge ... \wedge a_n$ of elements of A. We need to show $x \in out_1(G, a)$. (By Lemma 1 (6), this is equivalent to $x \in out_1(G, \{a_1, ..., a_n\})$, from which $x \in out_1(G, A)$ follows by monotony in A.) The proof is by induction on the length n of the derivation.

We give the base case $n = 1$ in full detail in order to highlight what elementary rules are needed. In this case, either $(a, x) \in G$ or (a, x) is the pair $(\top, \top)$. Suppose $(a, x) \in G$. By (**Ref**) $a \vdash a$, and thus $x \in G(Cn(a))$. By (**Ref**) again $G(Cn(a)) \vdash x$, which suffices for $x \in out_1(G, a)$. Suppose (a, x) is the pair $(\top, \top)$. By ($\top$:**I**), $G(Cn(\top)) \vdash \top$, which suffices for $\top \in out_1(G, \top)$.

The inductive case is straightforward, using Theorem 11. Details are omitted. □

Theorem 13 (Completeness, simple-minded). $out_1(G, A) \subseteq deriv_1(G, A)$

Proof. Let $x \in Cn(G(Cn(A)))$. We break the argument into cases depending on whether some elements of G are "triggered" or not.

Suppose $G(Cn(A)) = \emptyset$. By Lemma 1 (8), $\top \vdash x$. And by ($\top$:**I**) $a \vdash \top$ for some (arbitrarily chosen) conjunction $a = a_1 \wedge ... \wedge a_n$ of elements of A. The required derivation of (A, x) is shown below.

$$\cfrac{\cfrac{(\top, \top)}{(\top, x)}\ \text{WO}}{(a, x)}\ \text{SI}$$

Suppose $G(Cn(A) \neq \emptyset$. By (**C**) and Lemma 1(6), there are $x_1, ..., x_n (n > 0)$ in $G(Cn(A))$ such that $x_1 \wedge ... \wedge x_n \vdash x$. So G contains the pairs $(a_1, x_1), ..., (a_n, x_n)$

with $A \vdash a_1$, ..., and $A \vdash a_n$. By ($\wedge$:**I**), $A \vdash a_1 \wedge ... \wedge a_n$. By (**C**) and Lemma 1 (6), $a \vdash a_1 \wedge ... \wedge a_n$ for some conjunction $a = b_1 \wedge ... \wedge b_m$ of elements of A. By definition, each (a_i, x_i) is derivable from G. Furthermore, by ($\wedge$:**E**) $a \vdash a_i$ for all $1 \leq i \leq n$. Based on this one might get a derivation of (A, x) from G as shown below.

$$\text{WO}\;\dfrac{\dfrac{\dfrac{(a_1, x_1)}{(a, x_1)}\text{SI} \quad \cdots\cdots \quad \dfrac{(a_n, x_n)}{(a, x_n)}\text{SI}}{(a, x_1 \wedge \ldots \wedge x_n)}\text{AND}}{(a, x)}$$

This completes the proof. □

The argument for out_2 is more involved, and needs to be adapted.

Theorem 14. *out_2 validates all the rules of $deriv_2$.*

Proof. For (WO) and (AND), the verification is straightforward, and is omitted.

For (SI), assume $x \in out_2(G, a)$ and $b \vdash a$. Assume Incon$\{a\}$. By Theorem 5 (2), Incon$\{b\}$, and thus we are done. Assume Con$\{a\}$ but Incon$\{b\}$. In this case, we are done too, because

$$x \in \cap\{Cn(G(S)) : \{a\} \subseteq S, S \text{ saturated}\} \subseteq Cn(G(h(G)))$$

Assume Con$\{a\}$ and Con$\{b\}$. Consider a saturated set S such that $b \in S$. By (**Ref**), $S \vdash b$. By (**Mon**), $S, b \vdash a$, and thus, by (**Cut**), $S \vdash a$. By Definition 2 (3), $a \in S$. Thus, $x \in Cn(G(S))$, which suffices for $x \in out_2(G, b)$.

For (OR), assume $x \in out_2(G, a)$ and $x \in out_2(G, b)$. Assume Incon$\{a \vee b\}$. In this case, by Theorem 5 (1), Incon$\{a\}$ and Incon$\{b\}$, and so we are done. Suppose Con$\{a \vee b\}$. In this case, by Theorem 5 (1) again, either Con$\{a\}$ or Con$\{b\}$. Suppose the first, but not the second, applies (the argument is similar the other way around, and it still goes through with a minor adjustment if both apply). Consider a saturated set S such that $a \vee b \in S$. By Definition 2 (2) either $a \in S$ or $b \in S$. The latter is ruled out by the consistency of S and Theorem 5 (3). So $a \in S$. From Con$\{a\}$ and $x \in out_2(G, a)$, it follows that $x \in Cn(G(S))$. Thus, $x \in out_2(G, a \vee b)$, as required. □

Corollary 1 (Soundness, basic output). *$deriv_2(G, A) \subseteq out_2(G, A)$*

Proof. By induction on the length of the derivation using Theorem 14. □

Theorem 15 (Completeness, basic output). *$out_2(G, A) \subseteq deriv_2(G, A)$*

Proof. Like in the original proof, we break the argument into cases. The first is a borderline case, and the second is the principal case.

For ease of exposition, we write (SI, AND) to indicate an application of SI followed by that of AND, and similarly for other rules. Thus, (SI, AND) abbreviates the following derived rule

$$\text{(SI, AND)}\ \frac{(a_1, x_1) \qquad \cdots \qquad (a_n, x_n)}{(\wedge_{i=1}^{n} a_i, \wedge_{i=1}^{n} x_i)}$$

Case 1: Incon A. In this case, $out_2(G, A) = Cn(h(G))$. Let $x \in Cn(h(G))$. By (**C**), there are finitely many x_i's ($i \leq n$) in $h(G)$ such that $x \in Cn(x_1, ..., x_n)$. A derivation of (a, x) from G may, then, be obtained as shown below, where a is the conjunction of a finite number of elements in A, and all the pairs (b_i, x_i) are in G.

$$\dfrac{\dfrac{\dfrac{(b_1,x_1) \quad \ldots \quad \ldots \quad (b_n,x_n)}{(\wedge_{i=1}^{n} b_i, \wedge_{i=1}^{n} x_i)}\text{(SI,AND)} \qquad \dfrac{x_1,\ldots,x_n \vdash x}{\wedge_{i=1}^{n} x_i \vdash x}\text{Lem 1 (6)}}{(\wedge_{i=1}^{n} b_i, x)}\text{WO} \qquad \dfrac{\dfrac{\text{Incon}\,A}{A \vdash \wedge_{i=1}^{n} b_i}}{a \vdash \wedge_{i=1}^{n} b_i}\text{C + Lem 1 (6)}}{(a,x)}\text{SI}$$

Case 2: Con A. Assume (for reductio) that $x \in out_2(G, A)$ but $x \notin deriv_2(G, A)$. From the former, we get that $x \in Cn(h(G))$. We use the same kind of maximality argument as in the original proof to derive the contradiction that $x \notin out_2(G, A)$.

We start by showing that A can be extended to some "maximal" $S \supseteq A$ such that $x \notin deriv_2(G, S)$. By maximal, we mean that $\forall S' \supset S$, $x \in deriv_2(G, S')$. Thus, S is amongst the "biggest" input sets S containing A and not making x derivable.

S is built from a sequence of sets $S_0, S_1, S_2, \ldots$ as follows. Consider an enumeration $x_1, x_2, x_3, \ldots$ of all the formulae. Put $S_0 = A$, and

$$S_n = \begin{cases} S_{n-1} \cup \{x_n\}, & \text{if } x \notin deriv_2(G, S_{n-1} \cup \{x_n\}) \\ S_{n-1}, & \text{otherwise} \end{cases}$$

It is straightforward to show by induction that

Claim 1. $x \notin deriv_2(G, S_n)$, for all $n \geq 0$.

Now we define S to be the infinite collection of all the wff's in any of the sets in the sequence:

$$S = \cup\{S_n : n \geq 0\}$$

Note that S includes each of the sets in the sequence:

Claim 2. $S_n \subseteq S$, for all $n \geq 0$.

So in particular S includes A ($= S_0$).

Claim 3. $A \subseteq S$.

Thus, S is an extension of A. To show that $x \notin deriv_2(G, S)$, three more results are needed; their proof is omitted.

Claim 4. $S_k \subseteq S_n$, for $k \leq n \geq 0$.

Claim 5. $x_k \in S_k$, whenever $x_k \in S$, for $k > 0$.

Claim 6. For every finite subset S' of S, $S' \subseteq S_n$, for some $n \geq 0$.

With these results in hand, the argument for $x \notin deriv_2(G, S)$ may run as follows. Assume, to reach a contradiction, that $x \in deriv_2(G, S)$. By compactness for $deriv_2$, $x \in deriv_2(G, S')$ for some finite $S' \subseteq S$. By Claim 6, $S' \subseteq S_n$ for some $n \geq 0$. By monotony in the right argument, $x \in deriv_2(G, S_n)$. This contradicts Claim 1.

Next, we show that S is maximal. Let $y \notin S$. Any such y is such that $y = x_n$, for some $n \geq 1$. By Claim 2 $S_n \subseteq S$, and thus $y \notin S_n$. By construction, $S_{n-1} = S_n$, and $x \in deriv_2(G, S_{n-1} \cup \{y\}) = deriv_2(G, S_n \cup \{y\})$. By Claim 2, $S_n \cup \{y\} \subseteq S \cup \{y\}$. By monotony in the right argument for $deriv_2$, we get that $x \in deriv_2(G, S \cup \{y\})$, as required.

Now, we show that S is a saturated set. This amounts to showing that S is (i) consistent (ii) closed under consequence, and (iii) join-prime.

For (i), assume $\text{Incon}\, S$. We have $x \in Cn(h(G))$. Note that $x \notin Cn(\emptyset)$. Otherwise, for the reason explained in the proof of Theorem 13, (A, x) would be derivable from G:

$$\cfrac{\cfrac{(\top, \top)}{(\top, x)}\ \text{WO}}{(a, x)}\ \text{SI}$$

Here a denotes the conjunction of finitely many elements of A (their choice may be arbitrary).

By (**C**) and Lemma 1 (6), it follows that there are $x_1, ..., x_n (n > 0)$ in $h(G)$ such that $\wedge_{i=1}^{n} x_i \vdash x$. Let $a_1, ..., a_n$ be the bodies of the rules in question. Since $\text{Incon}\, S$, $S \vdash \wedge_{i=1}^{n} a_i$. By (**C**) and Lemma 1 (6), $\wedge_{i=1}^{m} s_i \vdash \wedge_{i=1}^{n} a_i$ for $s_1, ..., s_m \in S$. The following indicates how a derivation of (S, x) may be obtained from G, contradicting the result $x \notin deriv_2(G, S)$ previously established.

$$\text{(SI, AND)}\ \cfrac{\cfrac{\cfrac{(a_1, x_1) \ldots (a_n, x_n)}{(\wedge_{i=1}^{n} a_i, \wedge_{i=1}^{n} x_i)}}{(\wedge_{i=1}^{n} a_i, x)}\ \text{WO}}{(\wedge_{i=1}^{m} s_i, x)}\ \text{SI}$$

For (ii), let $S \vdash y$ and $y \notin S$. From the former, $\wedge_{i=1}^{n} s_i \vdash y$, for $s_1, ..., s_n \in S$ by (**C**) and Lemma 1(6). From the latter, $x \in deriv_2(G, S \cup \{y\})$, by maximality of S. This means that the pair $(\wedge_{i=1}^{m} a_i \wedge y, x)$ can be derived from G, with $a_1, ..., a_m \in S$. From this, the contradiction $x \in deriv_2(G, S)$ follows:

$$
\dfrac{\begin{array}{c}\vdots\\ (\wedge_{i=1}^{m} a_i \wedge y, x)\end{array}
\qquad
\dfrac{
\dfrac{\dfrac{\wedge_{i=1}^{n} s_i \vdash y}{\wedge_{i=1}^{n} s_i, \wedge_{i=1}^{m} a_i \vdash y}\,(\mathbf{Mon})}{\wedge_{i=1}^{n} s_i \wedge (\wedge_{i=1}^{m} a_i) \vdash y}\,\text{Lem } 1(6)
\qquad
\dfrac{\dfrac{\wedge_{i=1}^{m} a_i \vdash \wedge_{i=1}^{m} a_i}{\wedge_{i=1}^{n} s_i, \wedge_{i=1}^{m} a_i \vdash \wedge_{i=1}^{m} a_i}\,(\mathbf{Mon})}{\wedge_{i=1}^{n} s_i \wedge (\wedge_{i=1}^{m} a_i) \vdash \wedge_{i=1}^{m} a_i}\,\text{Lem } 1(6)
}{\wedge_{i=1}^{n} s_i \wedge (\wedge_{i=1}^{m} a_i) \vdash \wedge_{i=1}^{m} a_i \wedge y}\,(\wedge{:}\mathbf{I})
}{(\wedge_{i=1}^{n} s_i \wedge (\wedge_{i=1}^{m} a_i), x)}\,\text{SI}
$$

For (iii), let $a \vee b \in S$. Assume $a \notin S$ and $b \notin S$. Any such a and b are such that $a = x_n$, for some $n \geq 1$, and $b = x_m$, for some $m \geq 1$. By construction, $x \in deriv_2(G, S_n \cup \{a\})$ and $x \in deriv_2(G, S_m \cup \{b\})$. By Claim 4 either $S_n \subseteq S_m$ or $S_m \subseteq S_n$. Suppose the first applies (the argument for the other case is similar). By monotony in the right argument for $deriv_2$, $x \in deriv_2(G, S_m \cup \{a\})$. This means that the pair $(\wedge_{i=1}^{l} s_i \wedge a, x)$ can be derived from G, with $s_1, \ldots, s_l \in S_m$, and that the pair $(\wedge_{i=1}^{p} a_i \wedge b, x)$ can be derived from G, with $a_1, \ldots, a_p \in S_m$. Note that, by Lemma 1 (10), we have

$$
\wedge_{i=1}^{l} s_i \wedge (\wedge_{i=1}^{p} a_i) \wedge (a \vee b) \vdash (\wedge_{i=1}^{l} s_i \wedge a) \vee (\wedge_{i=1}^{p} a_i \wedge b)
$$

Thus,

$$
\dfrac{\dfrac{\begin{array}{c}\vdots\\ (\wedge_{i=1}^{l} s_i \wedge a, x)\end{array} \qquad \begin{array}{c}\vdots\\ (\wedge_{i=1}^{p} a_i \wedge b, x)\end{array}}{((\wedge_{i=1}^{l} s_i \wedge a) \vee (\wedge_{i=1}^{p} a_i \wedge b), x)}\,\text{OR}}{(\wedge_{i=1}^{l} s_i \wedge (\wedge_{i=1}^{p} a_i) \wedge (a \vee b), x)}\,\text{SI}
$$

So, $x \in deriv_2(G, S_m \cup \{a \vee b\})$. Since $a \vee b \in S$ and $S_m \subseteq S$, $S_m \cup \{a \vee b\} \subseteq S$. So by monotony $x \in deriv_2(G, S)$, a contradiction.

This concludes the main step of the proof. The final step is as in the original proof. We have $out_1(G, S) = deriv_1(G, S) \subseteq deriv_2(G, S)$. From $x \notin deriv_2(G, S)$, it then follows that $x \notin out_1(G, S) = Cn(G(Cn(S)))$. Since S is saturated, $S = Cn(S)$ by Theorem 9 (1). So $x \notin Cn(G(S))$. Furthermore, A is consistent, and $S \supseteq A$. By Definition 3, $x \notin out_2(G, A)$, and the proof may be considered complete. □

It is noteworthy that the argument for saturatedness uses Lemma 1 (10), which in turn appeals to the deduction theorem.

It is also interesting to see what happens if, in Definition 3, maximal consistency is used in place of saturatedness. We would need to establish that S is maximal consistent. However, the latter fact does not follow from the stated hypotheses, and what has already been established. Therefore, the proof does not go through. If maximal consistency* is used instead, then the proof does not go through either unless the base logic has the law of excluded middle. The latter law is needed to establish that S is $\neg$-complete. To see why, suppose there is some a such that $a \notin S$ and $\neg a \notin S$. A similar argument as in the proof for saturatedness yields $x \in deriv_2(G, S_m \cup \{a\})$ and $x \in deriv_2(G, S_m \cup \{\neg a\})$, from which $x \in deriv_2(G, S_m \cup \{a \vee \neg a\})$ follows.

However, in the absence of excluded middle, there is no guarantee that $a \vee \neg a \in S$, and so the required contradiction $x \in deriv_2(G, S)$ does not follow any more.

We now turn to the reusable operation out_3. In Makinson and van der Torre (2000), the soundness and completeness results are established for a singleton input set A. We extend the argument to an input set A of arbitrary cardinality.

Theorem 16 (Soundness and completeness, simple-minded reusable output). $out_3(G, A) = deriv_3(G, A)$.

Proof. For the soundness part, we only verify that the new rule (CT) is still valid. Let $x \in out_3(G, a)$ and $y \notin out_3(G, a)$. To show: $y \notin out_3(G, a \wedge x)$. From the second hypothesis, there is some $B = Cn(B) \supseteq G(B)$ with $a \in B$ and $y \notin Cn(G(B))$. From the first hypothesis, $x \in Cn(G(B))$. By (**Mon**), $Cn(G(B)) \subseteq Cn(B)$. So $x \in Cn(B)$. By (**Ref**), $a \in Cn(B)$. By ($\wedge$:**I**), $a \wedge x \in Cn(B)$. So, $a \wedge x \in B$. This means that $y \notin out_3(G, a \wedge x)$ as required.

For the completeness part, we argue contrapositively. Let $x \notin deriv_3(G, A)$. To show: $x \notin out_3(G, A)$.

Let $B = Cn(A \cup deriv_3(G, A))$. By (**Ref**), $A \subseteq B$. By Lemma 1 (11), $Cn(B) = B$. To show: (i) $G(B) \subseteq B$; (ii) $x \notin Cn(G(B))$.

For (i). Let $y \in G(B)$. So $(b, y) \in G$ for some $b \in B$. Hence $A \cup deriv_3(G, A) \vdash b$. By (**C**), $a_1, \ldots, a_m, x_1, \ldots, x_n \vdash b$ for $a_1, \ldots, a_m \in A$ and $x_1, \ldots, x_n \in deriv_3(G, A)$. For all $i \leq n$, $x_i \in deriv_3(G, b_i)$, where b_i is the conjunction of elements in A. By Lemma 1 (6), $\wedge_{i=1}^{m} a_i, \wedge_{i=1}^{n} x_i \vdash b$. By ($\rightarrow$:**I**), $\wedge_{i=1}^{n} x_i \vdash \wedge_{i=1}^{m} a_i \rightarrow b$. Based on this, one might argue that $y \in deriv_3(G, A)$ as follows. By Lemmas 1 (2) and (9), $\wedge_{i=1}^{n} b_i \wedge (\wedge_{i=1}^{m} a_i) \wedge (\wedge_{i=1}^{m} a_i \rightarrow b) \vdash b$ and $\wedge_{i=1}^{n} b_i \wedge (\wedge_{i=1}^{m} a_i) \vdash \wedge_{i=1}^{n} b_i$. We, thus, have:

$$\dfrac{\dfrac{\dfrac{\begin{array}{ccc}\vdots & & \vdots \\ (b_1, x_1) & \cdots\ \cdots & (b_n, x_n)\end{array}}{(\wedge_{i=1}^{n} b_i, \wedge_{i=1}^{n} x_i)}\,(\text{SI,AND})}{(\wedge_{i=1}^{n} b_i \wedge (\wedge_{i=1}^{m} a_i), \wedge_{i=1}^{m} a_i \rightarrow b)}\,(\text{SI,WO}) \qquad \dfrac{(b, y)}{(\wedge_{i=1}^{n} b_i \wedge (\wedge_{i=1}^{m} a_i) \wedge (\wedge_{i=1}^{m} a_i \rightarrow b), y)}\,\text{SI}}{(\wedge_{i=1}^{n} b_i \wedge (\wedge_{i=1}^{m} a_i), y)}\,\text{CT}$$

Once this established, the conclusion $y \in B$ immediately follows from (**Ref**).

The argument for (ii) uses a *reductio ad absurdum*, and invokes the completeness result for out_1. Assume $x \in Cn(G(B))$. This amounts to assuming that $x \in out_1(G, A \cup deriv_3(G, A))$. By the above completeness result, $x \in deriv_1(G, A \cup deriv_3(G, A))$. By definition of $deriv_1$, $x \in deriv_1(G, \wedge_{i=1}^{n} a_i \wedge (\wedge_{i=1}^{m} x_i))$, where $a_1, \ldots, a_n \in A$ and $x_1, \ldots, x_m \in deriv_3(G, A)$. So, *a fortiori*, $x \in deriv_3(G, \wedge_{i=1}^{n} a_i \wedge (\wedge_{i=1}^{m} x_i))$. For all $i \leq m$, $x_i \in deriv_3(G, b_i)$, where b_i is the conjunction of elements in A. The following indicates how a derivation of (A, x) may be obtained, contradicting the opening assumption.

$$\dfrac{\dfrac{\dfrac{\vdots}{(b_1, x_1)} \quad \cdots \cdots \quad \dfrac{\vdots}{(b_m, x_m)}}{(\wedge_{i=1}^m b_i \wedge (\wedge_{i=1}^n a_i), \wedge_{i=1}^m x_i)}\text{(SI, AND)} \quad \dfrac{\dfrac{\vdots}{(\wedge_{i=1}^n a_i \wedge (\wedge_{i=1}^m x_i), x)}}{(\wedge_{i=1}^m b_i \wedge (\wedge_{i=1}^n a_i) \wedge (\wedge_{i=1}^m x_i), x)}\text{SI}}{(\wedge_{i=1}^m b_i \wedge (\wedge_{i=1}^n a_i), x)}\text{CT}$$

This shows that $x \notin out_3(G, A)$. □

It is noteworthy that the argument for (i) uses both (→:**I**) and (→:**E**).

We end with a few remarks on out_4. The reader may easily verify that the four rules of $deriv_4$ are still validated. However, the original completeness proof breaks down when going intuitionistic. This is because what is described as Lemma 11 a in the original argument is no longer provable. This is the lemma that states that $deriv_4(G, a \wedge (b \to x)) \subseteq deriv_4(G, a)$ whenever $(b, x) \in G$. Its proof uses the following law, which is no longer available:

$$(\dagger) \qquad a \vdash (a \wedge b) \vee (a \wedge (b \to x))$$

This brings to the fore what may be considered the bottom-line between the two approaches. In Makinson and van der Torre (2000) there is a brief mention of the rule of (as they call it) "ghost contraposition". This is the rule

$$\text{(GC)}\ \dfrac{(a, b) \qquad (\neg a, c)}{(\neg c, b)}$$

Intuitively: although we cannot contrapose the rightmost premisse $(\neg a, c)$, we can still use its contrapositive $(\neg c, a)$ for an application of the rule of transitivity. If $(\neg a, c)$ is rewritten as $(\neg c, a)$, then the conclusion $(\neg c, b)$ follows by plain transitivity.

It is not difficult to see that although none of the output operations validate plain contraposition (GC) holds for the classically-based reusable basic output out_4. In this respect, contraposition still plays a 'ghostly' role for out_4. The following shows how to derive (GC).

$$\dfrac{\dfrac{\dfrac{\dfrac{(\neg a, c)}{(\neg c \wedge \neg a, c)}\text{SI} \quad \dfrac{(a, b)}{(\neg c \wedge \neg a \wedge c, b)}\text{SI}}{(\neg c \wedge \neg a, b)}\text{CT} \quad \dfrac{(a, b)}{(\neg c \wedge a, b)}\text{SI}}{((\neg c \wedge \neg a) \vee (\neg c \wedge a), b)}\text{OR}}{(\neg c, b)}\text{SI}$$

In an intuitionistic setting the last move is blocked, because (‡) does not hold:

$$(\ddagger) \qquad \neg c \vdash (\neg c \wedge \neg a) \vee (\neg c \wedge a)$$

If (‡) was valid, then we would get the law of excluded middle. This can be seen as follows. Substituting, in (‡), $x \wedge \neg x$ for c, we get by Theorem 2:

Fig. 1 A counter-model to (‡)

$$\neg(x \wedge \neg x) \vdash (\neg(x \wedge \neg x) \wedge \neg a) \vee (\neg(x \wedge \neg x) \wedge a)$$

The reader may easily verify that $\vdash \neg(x \wedge \neg x)$. So,

$$\vdash (\neg(x \wedge \neg x) \wedge \neg a) \vee (\neg(x \wedge \neg x) \wedge a)$$

This in turn can be simplified into[4]:

$$\vdash a \vee \neg a$$

Example 2 provides a counter-model to (‡), which is *mutatis mutandis* a counter-model to the law of excluded middle.

Example 2 Consider a model $M = (W, \geq, V)$ with $W = \{s, t\}$, $s \geq s$, $t \geq t$, $t \geq s$, $V(a) = \{t\}$, and $V(c) = \emptyset$. This can be depicted as in Fig. 1. Here the general convention is that $v \geq u$ iff either $u = v$ or v is above u. And each world is labelled with the atoms it makes true. Thus, a missing atom indicates falsehood. s forces $\neg c$, because neither s nor t forces c. s does not force $\neg c \wedge \neg a$, because it does not force $\neg a$ (witness: t). And neither does s force $\neg c \wedge a$. Therefore, $(\neg c \wedge \neg a) \vee (\neg c \wedge a)$ is not a semantic consequence of $\neg c$, and thus (by soundness) the former is not derivable from the latter.

As the reader may see, the fact that $\neg$ occurs in the premiss in (‡) plays no role. This is also a counter-model to the law of excluded middle. Indeed, $a \vee \neg a$ is not forced at s.

Example 3 shows that (GC) fails in an intuitionistic setting.

Example 3 (Ghost contraposition) Put $G = \{(a, b), (\neg a, c)\}$, where a, b and c are atomic formulae. Suppose out_4 is based on intuitionistic logic.

- Assume the input is $A = \{a\}$. Since a is consistent, the top clause in Definition 4 applies. Consider any saturated set S meeting the requirements mentioned in this clause. Any such S contains a. And, for any such S, $b \in G(S) \subseteq S$, so that $G(S) \vdash b$, by (**Ref**). This implies that $b \in out_4(G, a)$.
- Assume the input is $A = \{\neg a\}$. For a similar reason, $c \in out_4(G, \neg a)$.
- Assume the input is $A = \{\neg c\}$. Since $\neg c$ is consistent, the top clause in Definition 4 applies again. Consider $Cn(\neg c)$. By Theorem 7, $Cn(\neg c)$ is saturated. By (**Ref**),

[4] Using Theorem 3, $\top \dashv\vdash \neg(x \wedge \neg x)$, Lemma 1(7) and $a \vee b \dashv\vdash b \vee a$.

$\neg c \in Cn(\neg c)$. And $Cn(\neg c) \supseteq G(Cn(\neg c)) = \emptyset$. Furthermore, $b \notin Cn(\emptyset)$. So there is a saturated set S such that $A \subseteq S \supseteq G(S)$ but $b \notin Cn(G(S))$, which suffices for $b \notin out_4(G, \neg c)$.

The counter-example is blocked, if out_4 is defined in terms of maximal consistent* rather than saturated sets. Given input $\neg c$, there is only one maximal consistent* set S meeting the requirement $A \subseteq S \supseteq G(S)$. It is $Cn(a, \neg c, b)$. So $b \in out_4(G, \neg c)$.

Of course, this leaves open the question of whether a representation result may, or may not, be obtained for intuitionistic out_4.

At first sight, it would seem that out_4 is more suitable to model normative reasoning, because of the presence in its proof-theory of rules that are quite attractive. Indeed the operation supports both reasoning by cases (OR) and chaining (CT). But one would like to have the core rules of $deriv_4$ without getting (GC) by the same way. Indeed, the following natural language example, due to Hansen (2008), suggests that failure of ghost contraposition is a desirable feature. Let a, b, and c stand for the propositions that it is raining, that I wear my rain coat, and that I wear my best suit, respectively. It makes sense for my mum to order me to wear my rain coat if it rains, and my best suit if it does not. It does not follow that I am obliged to wear my rain coat given that I cannot wear my best suit (e.g. it is in the laundry).

4 Conclusion and Future Research

This chapter has demonstrated that, for three unconstrained output operations, the use of intuitionistic logic as base logic does not affect the axiomatic characterization of the resulting framework. This shows that it would be a mistake to think that I/O logic, in one way or another, 'presupposes' classical logic.

The proofs given in the chapter help to appreciate what elementary rules are required besides the Tarski conditions. It is natural to ask if the full power of intuitionistic logic is needed for the completeness results to hold.

We said that intuitionistic logic is the smallest logic that allows for both the deduction theorem and the *ex falso* principle. The completeness proofs for out_2 and out_3 both invoke the former. But (as far as we can see) they do not appeal to the latter, and neither do the arguments for soundness. Therefore, it looks as if our results carry over to the minimal logic of Johansson (1936), where the *ex falso* principle goes away. It is striking that the latter principle is used to show the hard half of the compactness theorem in terms of consistency. This is the right-to-left direction of the biconditional appearing in the statement of Theorem 5(3). It states that a (possibly infinite) set of wffs is consistent if every finite subset of it is consistent. However, this property does not seem to play any role in the arguments.

For out_1, neither the deduction theorem nor the *ex falso* principle are used. Instead the argument makes essential use of $(\wedge{:}\mathbf{I})$, $(\wedge{:}\mathbf{E})$, and $(\top{:}\mathbf{I})$. These rules can be viewed as the cornerstone of a natural deduction system, and they appear in logics weaker than intuitionistic logic, like the aforementioned minimal logic. What is clear is that some substructural logic like relevance logic would not do the job. ($\top$:**I**) goes away along with (**Mon**).

Several directions for future research can be taken. First, the question of whether a representation result may, or may not, be obtained for out_4 remains an open problem. Second, there are known embeddings of the classically-based out_2 and out_4 into modal logic. It would also be interesting to know if their intuitionistic analogs can be similarly mapped onto intuitionistic modal logic. Third, the question naturally arises as to whether one can move to the algebraic level by using lattices. Both classical and intuitionistic logics can be given an algebraic treatment. The interesting thing about lattices is that they give a geometrical flavor to I/O logic. A lattice is mainly about moving from points to points along a relation $\geq$. The travelling always goes in the upwards direction. We can think of the set G of generators used in I/O logic as "jump" points or bridges. A pair $(a, x) \in G$ is an instruction to deviate from a path that would be normally taken. Once we have reached node a, instead of continuing up we jump to an unrelated node x and continue your journey upwards.

Acknowledgments We would like to thank two anonymous referees and David Makinson for valuable comments.

References

Dummett, M. (1973). The philosophical basis of intuitionistic logic. *Truth and other enigmas* (pp. 215–247). Cambridge: Harvard University Press.

Gabbay, D. (1981). *Semantical investigations in Heyting's intuitionistic logic*. Dordrecht: Reidel.

Gabbay, D. (2007). *Logic for artificial intelligence and information technology*, volume 9 of texts in computer science. London: College Publications.

Hansen, J. (2008). Prioritized conditional imperatives: problems and a new proposal. *Autonomous Agents and Multi-Agent Systems, 17*(1), 11–35.

Harrop, R. (1956). On disjunctions and existential statements in intuitionistic systems of logic. *Mathematische Annalen, 132*(4), 347–361.

Johansson, I. (1936). Der Minimalkalkul, ein reduzierter intuitionistischer Formalismus. *Compositio Mathematica, 4*, 119–136.

Kripke, S. (1965). Semantical analysis of intuitionistic logic I. In M. Dummett & J. N. Crossley (Eds.), *Formal systems and recursive functions* (pp. 92–130). Amsterdam: North-Holland Publishing Co.

Makinson, D. (2012). *Sets, logic and maths for computing* (2nd ed.). London: Springer.

Makinson, D., & van der Torre, L. (2000). Input/output logics. *Journal of Philosophical Logic, 29*(4), 383–408.

Makinson, D., & van der Torre, L. (2001). Constraints for input/output logics. *Journal of Philosophical Logic, 30*(2), 155–185.

Makinson, D., & van der Torre, L. (2003). Permission from an input/output perspective. *Journal of Philosophical Logic, 32*(3), 391–416.

Nonnengart, A., & Weidenbach, C. (2001). Computing small clause normal forms. In *Handbook of automated reasoning* (Vol. I, pp. 335–367). Cambridge: Elsevier Science and MIT Press.

Restall, G. (2000). *An introduction to substructural logics*. New York: Routledge.

Thomason, R. H. (1968). On the strong semantical completeness of the intuitionistic predicate calculus. *Journal of Symbolic Logic, 33*(1), 1–7.

Troelstra, A. S., & van Dalen, D. (1988). *Constructivism in mathematics* (Vol. 1). Amsterdam: Elsevier Science & Technology.

van Dalen, D. (1994). *Logic and structure* (3rd ed.). Berlin: Springer.

van Dalen, D. (2002). Intuitionistic logic. In D. Gabbay & F. Guenthner (Eds.), *Handbook of philosophical logic* (Vol. 5, pp. 1–114). (2nd ed.). Dordrecht: Kluwer Academic Publishers.

Reasoning About Permission and Obligation

Jörg Hansen

Abstract How is deontic logic possible on a positivistic philosophy of norms? D. Makinson (1999) considered this question the 'fundamental problem of deontic logic', and called to reconstruct deontic logic as a logic of reasoning about norms. A solution is to use a semantics which defines the truth of monadic and dyadic deontic sentences with respect to an explicitly modelled set of norms. Here, I explore how such a norm-based semantics can be adapted to include not just mandatory but also permissive norms that possibly conflict with the first, and describe how this may affect the validity of well-known theorems. All studied proposals are based on a definition of consistency from the later theory of G. H. von Wright. The approach may shed new light on the problem of 'free choice permission,' while D. Lewis's 'problem of permission' persists. Finally, I question a persistent belief about permissions: that unlike obligations they must be considered one by one, and not collectively.

Keywords Deontic logic · Logic of norms · Free choice permission

1 Introduction

In 1986, E. Bulygin (1986) stated: "It is a well known fact that permissive norms are not very popular among legal philosophers." D. Makinson and L. van der Torre echoed this remark 16 years later when they wrote:

> In formal deontic logic, permission is studied less frequently than obligation. For a long time it was naively assumed that it can simply be taken as a dual of obligation, just as possibility is the dual of necessity in modal logic. As time passed, more and more researchers realized how subtle and multi-faceted the concept is. Nevertheless, they continued focussing on obligation because there it is easier to make progress within existing paradigms. Consequently the understanding of permission is still in a less satisfactory state. (Makinson and van der Torre 2003, p. 391)

J. Hansen (✉)
Beethovenstraße 4, 99817 Eisenach, Germany
e-mail: jhansen@uni-leipzig.de

S. O. Hansson (ed.), *David Makinson on Classical Methods for Non-Classical Problems*, Outstanding Contributions to Logic 3, DOI: 10.1007/978-94-007-7759-0_14,

One of the difficult features of permissions that does not seem to agree with "existing paradigms" is that explicitly permitting an action makes little sense when this action has not been generally prohibited before and will never be so in the foreseeable future. But when the action is already regulated, a permission may limit an obligation or make an exception to an existing prohibition, and it may even block or limit the effect of prohibiting the action in the future. So instead of merely mirroring the absence of a prohibition, it belongs to the nature of a permissive norm to be in disagreement with the mandatory norms, or to come to be so during the time of its existence. This may be called the 'dynamic view' of permissions, according to which their particular role is "that of cancelling or derogating prohibitions which could not be possibly performed by mandatory norms" (Bulygin 1986, p. 213).

Deontic logic, by contrast, defines its operator of permission by the absence of a prohibition, which again is defined as an obligation to the contrary: $PA =_{def} \neg O \neg A$. This treatment is supported by the use of deontic modalities in ordinary language: it is the same whether I am informed by some fellow member that 'it is not permitted to enter the club in street clothing' or whether I am told that such an action is prohibited, and if I am told that it is not prohibited I can conclude that I may enter as I am. This does not mean ignoring the dynamic aspect of permitting an action. Rather, by stating truly or falsely that something is permitted or obligatory according to some regulations, the description is of the picture that obtains *after* all norms, obligating and permissive, have been taken into account. If a permission has limited the range of a general obligation, then only what the limited obligation (still) makes obligatory is (truly) so described. If it has provided an exception to a prohibition, then the exception can no longer (truly) be described as prohibited. Then to state that an action is permitted does indeed mean the same as that it is not prohibited. The resulting 'static' picture of the relationship between obligation and permission resembles what J. Bentham in 1782 called the "universal law of liberty":

> Every efficient [mandatory] law whatever may be considered as a limitation or exception, grafted on a pre-established universal law of liberty [...]: a boundless expanse in which the several efficient laws appear as so many spots; like islands and continents projecting out of the ocean: or like material bodies scattered over the immensity of space. (Bentham 1782, pp. 119–120)

In what follows the truth and falsity of sentences that use deontic modal operators are modelled by a 'norm-based' logical semantics. Such a semantics mimics a legal code or any other set of regulations, with respect to which something may be described as obligatory or forbidden, by a formal representation of the norms. Here, only simple obligating and permissive norms are considered. It will obviously make a difference if just obligating norms are present, or if there are additional explicit permissions with which they might conflict, and thus the dynamic view of permission is captured. This will not, however, force us to work outside of the 'existing paradigms' of deontic logic. Even though there is no duality among norms—the existence of a prohibition does not rule out the existence of an explicit permission for the same action—, we can still describe in the usual way what these norms make obligatory or permitted in the end, by using deontic O- and P-operators that behave as duals, and thus the static view of permission is also accommodated.

Section 2 describes and motivates the formal framework. Here a definition is borrowed from the later theory of G. H. von Wright of what it means for two sets of obligating and permissive norms to be 'consistent', and first steps are made towards a deontic logic that describes what is obligatory or permitted according to such sets. Here we will also reencounter the 'static' and 'dynamic' definitions of strong permission that were proposed by Makinson and van der Torre (2003) for input/output logic.[1] Section 3 contains the bulk of this work. I will compare two proposals of how dyadic deontic operators can be defined with respect to sets of possibly conflicting obligating and permissive norms, one that rests again on von Wright's notion of consistency and was used for a similar purpose by G. Boella and L. van der Torre (2003), and a variant that rests on a proposal by C. E. Alchourrón and Bulygin (1981) and was recently supported by A. Stolpe (2010). Additionally, a unifying framework is presented where priority relations may hold between the norms and of which the previous proposals turn out to be special cases. Section 4 addresses some problems: free choice permission—it turns out that explicit disjunctive permissions already come with a strong free choice flavor—, D. Lewis's (1979) notorious 'problem of permission', and J. F. Horty's (1994) 'asparagus puzzle' for which I present a fresh solution. In the concluding Sect. 5 I return to von Wright's definition of consistency and call into question the belief that unlike mandatory norms permissions must be considered only one by one, and not collectively. All proofs are delegated to an appendix.

2 Basic Concepts

2.1 Norm-Based Semantics and its Motivation

For reasoning about what is obligatory, prohibited or forbidden according to some normative system, or what 'deontic propositions' it makes true,[2] here a 'norm-based semantics' is employed that uses explicitly represented norms. The idea to explain the truths of deontic logic not by some set of possible worlds among which some are ideal or at least better than others, but with reference to an explicit set of given norms or existing moral standards, dates back to the very beginning of deontic logic as a purely axiomatic theory.[3] Indeed, it seems very natural to say that A is obligatory if A is necessary to satisfy all of some set of commands, or some or one of them, and permitted if it does not violate any command, and such definitions will then

---

[1] Note, however, that the terms used for their definitions have little to do with the 'dynamic' and 'static' views of permission as described above.

[2] Cf. von Wright (1951): "The system of Deontic Logic, which we are outlining in this paper, studies propositions (and truth-functions of propositions), about the obligatory, permitted, forbidden [...]. We call the propositions which are the object of study deontic propositions."

[3] Cf. von Wright (1951): "Instead of dealing with propositions of the type 'A is permitted', we might consider propositions of the types 'A is permitted according to the moral code C'." For authors of an imperatival tradition of deontic logic that models deontic truth by means of a given set of imperatives or commands cf. Hansen (2008a) Chap. 3.

inevitably link the truth of a deontic statement to a consequence relation between the 'contents' of the norms (what has been commanded) and the 'A' in a deontic statement of the form $O\,A$. The resulting validities may then be compared to theorems of existing axiomatic systems of deontic logic. Thus, a branch of modal logic may be linked to operations of classical consequence, and deontic logic may be semantically 'reconstructed' with respect to given sets of norms.[4]

My exploration of such reconstructions, in a project begun in Hansen (2001) and continued in Hansen (2004, 2005, 2006, 2008b), was triggered by Makinson's presentation at the ΔEON'98 workshop in Bologna, where he criticized that work in deontic logic goes on "as if a distinction between norms and normative propositions has never been heard of," and called it a "fundamental problem" of deontic logic to justify and explain the view that deontic logic is indeed a logic of reasoning about norms, positive and thus arbitrary linguistic entities that do not bear truth values:

> It is thus a central problem—we would say, a fundamental problem—of deontic logic to reconstruct it in accord with the philosophical position that norms are devoid of truth values. In other words: to explain how deontic logic is possible on a positivistic philosophy of norms. (Makinson 1999, p. 30)

The norm-based semantics used here is closely related to Makinson's own response to the "fundamental problem" by co-developing input/output logic in Makinson and van der Torre (2000, 2001). Input/output logic does not commit itself to an interpretation of what it means to obtain an output from some input set G and some fact expressed by a propositional formula C, so—skipping formalities—we are free to define

$$G \models O(A/C) \textit{ iff } A \in out(G, C)$$

and thereby obtain, from any output operation *out*, a dyadic deontic operator that claims that according to a code G, A is obligatory in the context of C. If I do not use the input/output framework here, but prefer to state the truth conditions for the traditional (monadic or dyadic) deontic operators more directly, this is also owed to the fact that I treat only unconditional norms in what follows,[5] and so the whole power of input/output logic is not required.

Norm-based semantics, like the one employed here or a similarly interpreted input/output logic, should be considered an alternative tool to explain deontic logic. An analysis of how explicit permissions modify what is obligatory or permitted according to a code seems more natural or at least more perspicuous with a semantics that models explicit obligations and permissions directly, even though a similar analysis could probably also be carried out using possible worlds semantics.[6]

---

[4] As dyadic deontic logic is a nonmonotonic logic *avant la lettre*, this provides another bridge between classical and nonmonotonic logic in the sense of Makinson (2005).

[5] In the framework studied below, the only way to express a conditionality of a norm is to use a material conditional '$\rightarrow$' for its content. For an extension to conditional imperatives cf. Hansen (2008b).

[6] A sketch must suffice for the claim: Consider Goble's (2003, 2004) multiplex preference semantics, which characterizes a deontic logic that allows for conflicts by a multitude of Lewis-type 'systems

2.2 The Basic Framework

Let I be a set of objects, meant to be simple imperatives used for commanding some addressee and which the addressee must obey or at least believes must be obeyed. A function f associates with each imperative $i \in I$ a sentence $f(i)$ in the language of some basic logic, meant to be a formalization of the sentence that is true iff the imperative is satisfied (traditionally this sentence is called the 'content' of the imperative). I use '$!A$' as a name for an imperative $i \in I$ with $f(i) = A$. As we want to model the dynamic view of permission, according to which explicit permissions are normative acts distinct from obligating acts, we also have a set L of objects, meant to represent simple explicit permissions which the same addressee has been granted, or believes to have been granted. I shall call these *licenses*, as a reminder of the *licentium* of the older theories before the advent of the deontic square.[7] Let now $f : I \cup L \rightarrow \mathscr{L}_{BL}$ associate a sentence in the language of a basic logic with each of the imperatives and with each of the licenses. Obviously, an addressee must be able to understand 'what' has been allowed, and so there must be such an associated descriptive sentence also for each license.[8] This sentence will be called the content of the license, and I use $l(A)$ as a name for a license $l \in L$ with $f(l) = A$.

I, L, and f will then be used to define the truth of a deontic sentence.

For all purposes here, the basic logic will be propositional logic *PL*. Moreover, to avoid formal clutter, for most sections we can simply represent the imperatives and licenses by their contents, let $I \cup L \subseteq \mathscr{L}_{PL}$, and drop the reference to f. I might switch between the notations for convenience. Things become more strict when we consider a priority relation between the norms, since some may have the same

(Footnote 6 continued)
of spheres', each built from its own connected preference relation over the set of possible worlds. Associate with each such relation subsets of possible worlds (representing what is explicitly permitted) and require that any of the 'spheres' produced by the relation must have non-empty intersections with any of the subsets, except if a subset is empty or lies completely within an inner sphere. Deontic operators might then behave similarly to the ones defined by semantics three below.

[7] The linguistic distinction between what has been explicitly permitted and what the law merely "lets go through" seems to have almost vanished since Leibniz sought to "transfer all the complications, transpositions and oppositions of the modalities from Aristotle's *De Interpretatione* to the *iuris modalia*" (Leibniz (1930) pp. 480–481, also cf. Hruschka (1986)). An exception is Bentham (1782) who employs the deontic square as we know it (i.e. the analogue of the Aristotelian square of the alethic modalities), but distinguishes a 'non-command' from an 'original permission'.

[8] Cf. Hansen (2001, 2008a) for 'Weinberger's principle' that allows us to define the content of an explicit obligating norm via the conditions for its satisfaction or violation. For a permissive norm Weinberger (1996) p. 174 proposes to define its content as the content of a corresponding prohibition that would result if the words of the permissive sentence were changed, e.g. from "I permit you to ..." to "I forbid you to ...". The treatment of licenses by the semantics discussed below suggests that, informally, we may think of the content of a license as a disjunction of all the states it permits, just as we may think of the content of an imperative as a conjunction of all the states it commands. However, there are no formal restrictions on contents, and so L or $f(L)$ are the same as the set of 'P-norms' used in von Wright's later theory and by Alchourrón and Bulygin in Alchourrón and Bulygin (1981), and the set Z employed by Makinson and van der Torre (2003) in input/output logic.

content but different ranks. Yet no matter how convenient and formally legitimate this representation is, fundamentally, norms are different objects than propositions, and licenses are different from imperatives (though their contents may overlap), so I will drop this clutch whenever a more accurate analysis is called for.

2.3 Separability and Inseparability

There are no further formal requirements for I and L: the sets may be finite or infinite or empty, and contain tautologies or contradictions. There are, however, two informal assumptions for the norms in the sets: (i) that they are *not separable*, and (ii) that they are *independent*. I will motivate these by examples, but for further analysis the reader should refer to Hamblin (1987). The assumptions have the effect that we should not assume that I and L are closed under some consequence relation.

Sometimes doing only a part of what has been requested or commanded is seen as something good and (partly) following the order, while at other times failing a part means that satisfying what remains no longer makes sense.

Example 1 (The shopping list).
Suppose I am told: "Buy apples and walnuts!" If the walnuts are needed for biscuits and the apples for dessert, then it makes sense to get the apples even if walnuts are out. If, on the other hand, both are required for a Waldorf salad, then it may be unwanted and a waste of money to buy the apples regardless.

In the first case the imperative is called separable, and in the second case inseparable. The same reasoning applies if not one but several sentences were used for commanding, like when I am told first to buy the apples and some time later to get walnuts too. Then the sentences are called independent (when the addressee still has to satisfy some commands even if all cannot be satisfied), or they combine inseparably (when the commands were only meant to be satisfied collectively). Shopping requirements are good examples that there is no logical method to determine whether an imperative is separable or inseparable, or whether imperatives are independent or combine inseparably. Sometimes they are separable or independent (when the intention was to stock up supplies), and sometimes they are inseparable or combine inseparably (when all items were needed for a common purpose).[9] We cannot even hope that by formalizing some of the context the difference can be made explicit as the speaker is not forced to explain herself. She may simply add: "and you must do it all" (also cf. Goble 1990).

The assumption that the imperatives in I are inseparable and independent becomes important when conflicts and dilemmas are considered, that is, where two or more imperatives cannot be satisfied collectively, or in contrary-to-duty circumstances, where some imperative has been invariably violated:

[9] Medical treatments provide more morbid examples: sometimes you must take some pills even if others ran out, and sometimes you don't (I owe this example to M. Sergot, private correspondence).

Example 2 (The shopping list, continued I).
I have spent most of the money on ice cream, so I am not able to buy *both* the apples and the walnuts. I might even have spent so much that I cannot buy the walnuts with the change, though I still have enough to purchase the apples. I might then be obliged to buy the apples or the walnuts (in the first case), or just the apples (in the second), or I may be under no obligation to buy anything any more (if I was helping prepare a Waldorf salad).

Similarly, when more important duties override less important ones, the reasoning will depend on whether an imperative is separable or not:

Example 3 (The shopping list, continued II).
In the shop I remember that earlier I was told by the same speaker never to buy walnuts because some family member is allergic to them. I reason—safety first—that I mustn't buy walnuts. If both were for a Waldorf salad, it makes no sense to still buy the apples, and so I don't have to.

Questions of separability and inseparability have, in the form of 'paradoxes,' plagued proposals for a 'logic of imperatives' and later deontic logic.[10] By representing imperatives semantically, and requiring that they have been properly formalized (combined if they are not independent, unseparated if they are not separable, and separate otherwise), the problems seem adequately addressed.

Turning to licenses, they too may be separable or inseparable:

Example 4 (Working at home).
There is a general requirement to work at the office. However, today my boss told me: "Take the company laptop and work at home tomorrow, if you like!" On my way home I drop the laptop. It is now broken and cannot be used for work. Am I still allowed to stay at home tomorrow (I could do some work-related phone calls)? The situations permitted by my boss were those where I stay at home *and* work *and* use the company laptop, for whatever reason. If I want to work at home regardless, I must ask again.

Example 5 (The hotel voucher).
A hotel gives vouchers to important people like you. The voucher includes a free night in a suite and free use of the swimming pool which normally incurs extra charges. You can't swim and so won't use the pool. Do you have to pay for the suite? I think not. (Things might be different if the hotel was using the vouchers to promote their new pool.)

A special case of separable norms are *general norms*, like common or standing obligations that require a similar behavior by several subjects or by the same subject in similar situations ("all members of the staff come to tomorrow's meeting," "play the trumpet every day at 10 a.m. and 5 p.m."), or like open invitations or, in the

---

[10] Examples have included 'write a letter and post it' (Ross 1941), 'take the parachute and jump' (Hare 1967), 'pay the bill and file it' (Weinberger 1958), 'close the window and play the piano' (Weinberger 1991), and 'fill up the boiler with water and heat it' (Weinberger 1999).

law, temporally limited licenses or general freedoms.[11] If we want to consider such general norms they must be split into all their instances. (To make things more natural, we can alternatively assume that I and L are generated by a set of general norms in some appropriate manner.)

2.4 Formal Languages and Logic

The analysis in the following sections will be based on the following formal languages (unlike in Hansen (2004) they do not include nested deontic operators or mixed sentences that combine a propositional sub-formula with a deontic one):

- The language of propositional logic $\mathscr{L}_{PL}$ is based on an alphabet that has a set $Prop = \{p_1, p_2, ...\}$ of proposition letters, Boolean operators '$\neg$', '$\wedge$', '$\vee$', '$\rightarrow$', and brackets '(', ')'. If $p \in Prop$ then $p \in \mathscr{L}_{PL}$, and if $A, B \in \mathscr{L}$ then all of $\ulcorner\neg A\urcorner$, $\ulcorner(A \wedge B)\urcorner$, $\ulcorner(A \vee B)\urcorner$ and $\ulcorner(A \rightarrow B)\urcorner$ are in $\mathscr{L}_{PL}$. Nothing else is in $\mathscr{L}_{PL}$.
- The language of (monadic) deontic logic $\mathscr{L}_{DL}$ is based on an alphabet that is like that of $\mathscr{L}_{PL}$ but additionally includes an operator 'O'. If $A \in \mathscr{L}_{PL}$ then $\ulcorner OA\urcorner \in \mathscr{L}_{DL}$, and if $A, B \in \mathscr{L}_{DL}$ then so are $\ulcorner\neg A\urcorner$, $\ulcorner(A \wedge B)\urcorner$, $\ulcorner(A \vee B)\urcorner$, $\ulcorner(A \rightarrow B)\urcorner$. Nothing else is in $\mathscr{L}_{DL}$. For any $A \in \mathscr{L}_{PL}$, $\ulcorner PA\urcorner$ means $\ulcorner\neg O\neg A\urcorner$.
- The language of dyadic deontic logic $\mathscr{L}_{DDL}$ is based on an alphabet that is like that of $\mathscr{L}_{DL}$ but additionally contains a stroke '/'. If $A, C \in \mathscr{L}_{PL}$ then $\ulcorner O(A/C)\urcorner \in \mathscr{L}_{DDL}$, and if $A, B \in \mathscr{L}_{DDL}$ so are $\ulcorner\neg A\urcorner$, $\ulcorner(A \wedge B)\urcorner$, $\ulcorner(A \vee B)\urcorner$, $\ulcorner(A \rightarrow B)\urcorner$. Nothing else is in $\mathscr{L}_{DDL}$. For any $A, C \in \mathscr{L}_{PL}$, $\ulcorner P(A/C)\urcorner$ means $\ulcorner\neg O(\neg A/C)\urcorner$, and $\ulcorner OA\urcorner$ means $\ulcorner O(A/\top)\urcorner$ in $\mathscr{L}_{DDL}$.

The usual conveniences for brackets apply. The metalanguage uses $A, B, ...$ for sentences, $p, q, ...$ for proposition letters, and $\Gamma, \Delta, ...$ for sets of sentences. Corner quotes around mixed expressions are mostly omitted. Variants of the languages will be the obvious ones: when it is stated that a new operator P^+ for $\mathscr{L}_{DL}$ behaves grammatically like O, then $\mathscr{L}_{DL+P^+}$ is like $\mathscr{L}_{DL}$ except that the alphabet also has '+' and 'P', if $A \in \mathscr{L}_{PL}$ then also $P^+A \in \mathscr{L}_{DL+P^+}$, and the language is again closed.

A theory of classical propositional logic PL is assumed that defines a classical consequence relation $\vdash_{PL}$ between a set of sentences $\Gamma \subseteq \mathscr{L}_{PL}$ and a sentence $A \in \mathscr{L}_{PL}$. When the sentences on both sides of the turnstile are obviously propositional, or meant to be so, the index may be dropped as in '$\Gamma \vdash A$'. As usual, '$\top$' denotes an arbitrary tautology, and '$\bot$' an arbitrary contradiction.

When some axiomatic theory DL is said to be defined by some axiom schemes and rules, then DL means the set of all sentences from the referenced language

[11] In this respect my use of the term 'license' might be misleading. In legal terminology a license usually refers to standing exceptions from general prohibitions for a number of cases and a longer duration of time, like a license to drive, to publish, to carry out a certain trade, etc.

$\mathscr{L}_{DL}$ that are $\mathscr{L}_{DL}$-instances of tautologies, or $\mathscr{L}_{PL}$-instances of the given axiom schemes, and closed under all $\mathscr{L}_{PL}$-instances of the given rules and *modus ponens*. '$\vdash_{DL} A$' means that $A \in DL$. A set $\Gamma \subseteq \mathscr{L}_{DL}$ is called DL-inconsistent if there are $A_1, ..., A_n \in \Gamma$ such that $\vdash_{DL} \neg(A_1 \wedge ... \wedge A_n)$, and DL-consistent otherwise. $A \in \mathscr{L}_{DL}$ is called DL-derivable from Γ iff $\Gamma \cup \{\neg A\}$ is DL-inconsistent (we write '$\Gamma \vdash_{DL} A$').

Semantically, it suffices to provide a truth definition for any deontic operator (Boolean operators are defined as usual). Such definitions may appeal to some or all of I, L, and f to interpret some sentence of deontic logic B. If B is defined true with respect to some I, L, and f, I write '$\langle I, L, f\rangle \models B$', and '$\models B$' if B is true for all such tuples (and call B valid). An axiomatic theory that only contains valid sentences is called sound, and (weakly) complete if it contains all valid sentences. It is strongly complete if for any consistent set Γ there is a tuple that makes all of Γ true.

2.5 First Steps and Von-Wright-Consistency

For first steps in reasoning about imperatives and licenses, let the setting be monadic and assume for simplicity $I, L \subseteq \mathscr{L}_{PL}$ (so we identify the norms with their contents). Let us experiment with truth definitions for when a statement that something is permitted is true for an ordered pair of sets of imperatives and licenses $\langle I, L\rangle$. A truth definition for 'explicit permission' $P^x A$ (that behaves grammatically like O) may simply state that there is a license that permits A,

(td-1) $\langle I, L\rangle \models P^x A$ iff $\exists l \in L: \ \vdash l \leftrightarrow A$,

which produces a classic modal operator P^x that just validates the rule of extensionality: if $\vdash A \leftrightarrow B$ then $\models P^x A \leftrightarrow P^x B$. Similarly, a monotonic operator P^m, that additionally validates the axiom of monotonicity: $P^m(A \wedge B) \rightarrow (P^m A \wedge P^m B)$, can be defined as follows:

(td-2) $\langle I, L\rangle \models P^m A$ iff $\exists l \in L: \ \vdash l \rightarrow A$.

More interesting are truth definitions for what is 'strongly' permitted that make use of not only the set L, but also of the set I. For this purpose, a proposal from G. H. von Wright's later theory[12] may be adapted as to when a set of O-norms and a set of P-norms 'entails' another norm.[13] The proposal rests on a concept of 'consistency' (cited from von Wright 1999b, pp. 5–6, my numbering):

[12] Von Wright's later theory has its roots in von Wright (1963) Chap. 8, Sect. 7–10, emerged in vonWright (1980, 1983) and was stated and restated many times later (e.g. in von Wright 1991, 1993, 1999a,b).

[13] It is an adaptation since here the question is how to define descriptive deontic operators, not 'entailment' relations between norms. However, it should be noted that von Wright's later theory is one of 'normological scepticism.' Entailment relations are meant as advice to the norm giver who contemplates the addition of a norm to a system: entailed norms need not be added, and norms that make the system inconsistent should not be added.

1. A set of O-norms is consistent if, and only if, the conjunction of the contents of all members of the set is doable. (One could also say: if all the member-norms are jointly satisfiable.)

2. The notion of consistency does not apply to sets of P-norms.

3. A joint set of O- and P-norms is consistent if, and only if, the conjunction of all the contents of all O-norms with the content of any one of the P-norms individually is doable.

4. By the negation-norm of a O-norm I understand a P-norm, the content of which is the negation of the content of the O-norm. Similarly, the negation norm of a P-norm is a O-norm, the content of which is the negation of the content of the P-norm.

5. A consistent set of norms will be said to entail a given norm if, and only if, adding the negation of the given norm to the set makes the set inconsistent.

Let $\langle I, L\rangle$ be likened to a 'joint set' of O-norms and P-norms. Let 'doability' of contents be understood as PL-consistency. Then von Wright's definition of consistency translates as follows[14]:

Definition 1 (Genuine von-Wright-consistency).
$\langle I, L\rangle$ is (genuinely) vW-consistent iff $I \nvdash \bot$ and $\forall l \in L : I \cup \{l\} \nvdash \bot$.

We can then easily translate von Wright's idea of an 'entailment-relation' between norms into truth definitions for a deontic O-operator and an additional operator of 'positive' or 'strong' permission P^+ as follows:

(td-3) $\langle I, L\rangle \models OA$ iff $I \cup \{\neg A\} \vdash \bot$.

(td-4) $\langle I, L\rangle \models P^+A$ iff either $I \cup \{\neg A\} \vdash \bot$ *or* $\exists l \in L : I \cup \{\neg A\} \cup \{l\} \vdash \bot$.

Note that the second clause in (td-4) for P^+A is similar to the definition of 'dynamic permission' of Makinson and van der Torre (2003), also called 'antithetic permission' (cf. Stolpe 2010): P^+A is true because the law giver cannot prohibit A in a way consistent with existing obligating and permissive norms. Obviously, both definitions are equivalent to the following:

(td-5) $\langle I, L\rangle \models OA$ iff $I \vdash A$.

(td-6) $\langle I, L\rangle \models P^+A$ iff either $I \vdash A$ *or* $\exists l \in L : I \cup \{l\} \vdash A$.

The second clause in (td-6) for P^+A is now similar to what Makinson and van der Torre call 'static permisson': in an unconditional setting, 'static' and 'dynamic' permissions coincide (cf. Makinson and van der Torre 2003, p. 397).

Von Wright's later theory deliberately lacks a concept of negative permission since with such a concept "one could derive arbitrary permissions from a set of norms, which have nothing to do with the contents of the norms in the set."[15] Here, an operator of negative permission PA is defined as $\neg O \neg A$ as usual.

---

[14] Note that the first clause by von Wright is used to cover the case where $L = \varnothing$.

[15] Private correspondence, letter dated 4 July 1999.

Let standard deontic logic *SDL* be an axiomatic theory in the language of $\mathscr{L}_{DL}$ that is based on the following axiom schemes and rules:

$$\begin{array}{rl} \text{(Ext)} & \text{If } \vdash A \leftrightarrow B \text{ then } \vdash_{SDL} OA \leftrightarrow OB \\ \text{(M)} & O(A \wedge B) \rightarrow (OA \wedge OB) \\ \text{(C)} & (OA \wedge OB) \rightarrow O(A \wedge B) \\ \text{(N)} & O\top \\ \text{(P)} & P\top \end{array}$$

(C) and (P) derive the typical 'deontic' theorem (D):

$$\text{(D)} \quad OA \rightarrow PA$$

SDL is then sound and (strongly) complete for a norm-based semantics that defines the truth of the O-operator by (td−5) and where the additional restriction holds that I is consistent: $I \nvdash \bot$ (cf. Theorem 4.1.5 in Hansen (2004)).

Consider now the language $\mathscr{L}_{DL+P^+}$ that has an additional operator of strong permission P^+ which behaves grammatically like O. Let $SDL+P^+$ be an axiomatic theory in this language that is based on the axiom schemes and rules of *SDL* and the following additional schemes and rules:

$$\begin{array}{rl} \text{(Ext}^+\text{)} & \text{If } \vdash A \leftrightarrow B \text{ then } \vdash_{SDL+P^+} P^+A \leftrightarrow P^+B \\ \text{(M}^+\text{)} & P^+(A \wedge B) \rightarrow (P^+A \wedge P^+B) \\ \text{(C}^+\text{)} & (OA \wedge P^+B) \rightarrow P^+(A \wedge B) \\ \text{(N}^+\text{)} & \neg P^+\bot \\ \text{(P}^+\text{)} & P^+\top \end{array}$$

Theorem 1 (Soundness and completeness of $SDL + P^+$).
The system $SDL+P^+$ is sound and (strongly) complete for a norm-based semantics that defines the truth of its deontic operators by (td-5) and (td-6) and where the additional restriction holds that $\langle I, L\rangle$ is (genuinely) vW-consistent.

Note that (C$^+$) corresponds closely to the second clause in the truth definition of P^+. (C$^+$) with (P$^+$) or (N$^+$) respectively derive the following two theorems:

$$\begin{array}{rl} \text{(D}^+\text{)} & OA \rightarrow P^+A \\ \text{(P}^+\text{P)} & P^+A \rightarrow PA \end{array}$$

(D^+) expresses the first half of (td-6), and (P^+P) expresses that if A is strongly permitted, then A is not forbidden. One might use these operators to mark what can consistently be prohibited and define $\mathit{Prohibitable}(A) =_{def} (P\neg A \wedge \neg P^+A)$.

3 Dyadic Deontic Reasoning About Norms

To any subscriber of the dynamic view of permission, the just discussed definitions of P-operators—whether they be called positive, strong, dynamic or antithetic—must seem anemic. If it is the purpose of an explicit permission to make a change in what would be obligatory and prohibited without it, then the above picture is insufficient. Namely, if a pair $\langle I, L\rangle$ must be consistent in von Wright's sense, then a set of conflicting imperatives and licenses, for example one that contains $\{!\neg k, l(k)\}$—a prohibition to kill and a license to kill—is not even admitted for consideration, and so we do not get to find out how a resolution mechanism might work. So such a restriction should not be employed.

To describe what possibly conflicting imperatives and licenses make, in the end, obligatory and permitted, a dyadic deontic setting seems natural.[16] All norms may be consistent at first, but then some situation emerges in which an explicit permission conflicts with an explicit prohibition, and we want to know what is then obligatory or permitted for an agent. The setting should be one that also allows for conflicts between the imperatives alone. Otherwise, the set of imperatives would have to be unduly restricted: unless the contents of all imperatives are logically chained and for any two imperatives $!A, !B \in I$, either $\vdash A \rightarrow B$ or $\vdash B \rightarrow A$, the situation $\neg A \vee \neg B$ will create a dilemma. A conflict-tolerant semantics for dyadic deontic logic was investigated in Hansen (2005), it is explained first before licenses are added.

3.1 Semantics 1: Conflicting Imperatives

Let I be a set of imperatives. To avoid clutter, I is again identified with the set of contents of the imperatives and so $I \subseteq \mathscr{L}_{PL}$. We are back to where there are no restrictions on the contents, so there might be conflicts[17] or the agent might find that in the given circumstances, described by $C \in \mathscr{L}_{PL}$, not all of the imperatives are still satisfiable. Then, the best the agent can do is to satisfy a maximal set of the commands. One might say with Goble (2000) that each such set determines a consistent 'normative standard' which an agent may adopt when faced with conflicting norms. No one in the same situation can do more but the agent will still be culpable for doing less. So what the agent must do is determined by the remainder set $I \perp \neg C$ of I by $\neg C$, where as usual (cf. Alchourrón and Makinson 1981, Hansson 1999)

$$I \perp \neg C$$

[16] The dyadic setting does not mean we are suddenly concerned with conditional norms. $O(A/C)$ expresses not a conditional norm, but what must obtain in the context of C to satisfy the norms, which here are unconditional. Conditional imperatives, that cannot be satisfied or violated unless their conditions are true, or permissives that require a similar 'triggering', play no role here at all.

[17] We do not have to assume an unreasonable normgiver. The agent might just find herself the addressee of imperatives from different sources.

is the set of all subsets $I' \subseteq I$ such that

(i) $I' \nvdash \neg C$, and
(ii) there is no $I'' \subseteq I$ that meets (i) and $I' \subset I''$.

As there may be several remainders, we can define two kinds of dyadic deontic O-operators that claim A is obligatory in the context of C: an 'existential' O^e-operator pronouncing as obligatory what is necessary, given the context, to satisfy all imperatives in one of the remainders, and a 'universal' O^a-operator pronouncing as obligatory what for any remainder is necessary, given the context, to satisfy its imperatives.[18] So instead of an unmarked operator O I use two grammatically similar operators O^e and O^a in the language $\mathscr{L}_{DDL^{ea}}$, and define their truth as follows:

(td-7) $I \models O^e(A/C)$ iff $\exists I' \in I\bot\neg C : I' \cup \{C\} \vdash A$

(td-8) $I \models O^a(A/C)$ iff $\forall I' \in I\bot\neg C : I' \cup \{C\} \vdash A$

For semantics with these two truth definitions there is a neat axiomatic theory DDL^{ea} that is sound and (weakly) complete and related to well-known systems of dyadic deontic logic and nonmonotonic reasoning in general (cf. Hansen 2005). Its a-fragment (the sentences in which 'e' does not occur) coincides with the KLM (Kraus et al. 1990) system P with the additional deontic axiom (DD) $O^a(A/C) \rightarrow P^a(A/C)$ for consistent C. The e-fragment of DDL^{ea} corresponds to Bochman's (1999) credulous nonmonotonic inference to which the axiom (DP) $P(\top/C)$ is added. The complete system is similar to Goble's dyadic deontic logic for multipreference semantics in Goble (2004).[19]

3.2 Semantics 2: Competing Licenses

In semantics 1, the obligations that remain in a conflict are defined with respect to subsets of I that are maximally consistent with the situation. One deontic operator tells what the agent must do to satisfy the requirements of some such set, and the other tells the agent what is required to be done according to all such sets (and so the agent will, by necessity, satisfy the requirements of at least one, if one exists). To apply a similar reasoning to pairs of imperatives I and permissive norms L, we need some sort of analogue—a definition of when such a pair is consistent.

Now we already encountered such a definition in von Wright's later theory. According to it, $\langle I, L\rangle$ is (genuinely) vW-consistent iff $I \nvdash \bot$ and $\forall l \in L : I \cup \{l\} \nvdash \bot$ (Definition 1 above). For the present purposes, I propose a minor change regarding

[18] The terminology varies—similar 'existential' operators are also called 'credulous' or 'full join,' and 'universal' operators called 'skeptical' or 'meet'. Note that the dual P^e will be universal, and P^a existential, so the indexes are somewhat misleading.

[19] Goble's $SDDL_a DP_e$ only differs on the question which of $O(\top/\bot)$ or $P(\top/\bot)$ should hold, of which Lewis (1974) p. 5 has said: "The mind boggles."

licenses for the impossible, like ones with contradictions as their contents.[20] Due to the definition's second clause, $\langle I, L\rangle$ is pronounced inconsistent if L includes such a license. But consider $\langle \varnothing, \{\bot\}\rangle$: to call it inconsistent runs counter to von Wright's intuition that the notion of consistency does not apply to sets of P-norms (point 2 in Sect. 2.5). In von Wright's own theory this does not pose a problem, as he only considers 'genuine' norms and not 'spurious' ones whose contents are 'not doable' (cf. von Wright 1999b, p. 5).[21] Here, the sets I and L are not restricted in this way. In accord with von Wright's intuition, I suggest to treat such licenses as inoperative, and to test the set of imperatives only against licenses with consistent contents. Von Wright's (modified) concept of normative consistency is then easily generalized to a 'consistency with the circumstances' as follows[22]:

Definition 2 (Von-Wright-consistency in a situation).
$\langle I, L\rangle$ is vW-consistent with $C \in \mathscr{L}_{PL}$ iff $I \nvdash \neg C$ and $\forall l \in L \overset{n}{-} \neg C : I \cup \{l\} \nvdash \neg C$.

Here, C represents the circumstances, and '$\overset{n}{-}$' is a naïve subtraction that eliminates single elements which derive the subtract: $\Gamma \overset{n}{-} A = \{B \in \Gamma \mid B \nvdash A\}$. If $\langle I, L\rangle$ is vW-consistent with $\top$, I henceforth call $\langle I, L\rangle$ 'arbitrarily vW-consistent,' or vW-consistent simpliciter. Applying this definition of consistency, let

$$\langle I, L\rangle \bot^{vW} \neg C$$

be the set of all pairs $\langle I', L'\rangle$, $I' \subseteq I$, $L' \subseteq L$ such that

(i) $\langle I', L'\rangle$ is vW-consistent with C, and
(ii) there is no such pair $\langle I'', L''\rangle \neq \langle I', L'\rangle$ that meets (i) and $I' \subseteq I''$, $L' \subseteq L''$.

Existential and universal dyadic deontic O-operators are then defined as follows:

(td-9) $\langle I, L\rangle \models O^e(A/C)$ iff $\exists \langle I', L'\rangle \in \langle I, L\rangle \bot^{vW} \neg C : I' \cup \{C\} \vdash A$

(td-10) $\langle I, L\rangle \models O^a(A/C)$ iff $\forall \langle I', L'\rangle \in \langle I, L\rangle \bot^{vW} \neg C : I' \cup \{C\} \vdash A$

The construction has the effect that imperatives and licenses which conflict may be seen to compete for a place in a maximal pair of a set of obligating and a set of permissive norms that is vW-consistent. If there is such a pair and we want to include an additional license, then at least one of the imperatives has to go, and vice versa. Suppose that $I = \{\neg k\}$ and $L = \{k\}$: there is a prohibition to kill and a license to kill. In the default circumstances $\top$, $\langle I, L\rangle \bot^{vW} \neg\top = \{\langle\{\neg k\}, \varnothing\rangle, \langle\varnothing, \{k\}\rangle\}$: So while $O^e(\neg k/\top)$ is true, because according to some maximal and vW-consistent

[20] None of the theorems and validities presented later depend on this modification, but it will make it much easier to compare the semantics proposed in this paper.

[21] This was partly reflected in the semantic constraint for system $SDL + P^+$ in Sect. 2.5 above. There seems to be no need to take similar special care of norms with necessary (e.g. tautologous) contents, which von Wright also considers spurious.

[22] A similar definition (without the modification) was employed by Boella and van der Torre (2003) to limit the maximal subsets of given sets of obligating and permitting norms in an input/output setting of hierarchically ordered norms.

normative standard it is still forbidden to kill, $O^a(\neg k/\top)$ is false, for killing is no longer prohibited overall, and $P^a(k/\top)$ is true: at least some other normative standard permits killing.

The following observation links the two semantics provided so far:

Observation 1 (Relation of semantics 1 to semantics 2)
If $I' \in I \bot \neg C$ then $\langle I', \{l \in L \mid \{l\} \vdash \neg C \text{ or } I' \cup \{l\} \nvdash \neg C\}\rangle$ is in $\langle I, L\rangle \bot^{vW} \neg C$.

So far, we do not have information about whether some license ranks higher than some imperative, and so its inclusion in a normative standard may be preferred over the inclusion of the imperative. The above observation may be interpreted as stating that then a standard that requires doing what a maximally consistent set of the imperatives alone demands, without regard to the licenses, must also be considered. In this sense, semantics 1 is a special case of semantics 2: by disregarding licenses, it 'solves' conflicts between permissive and mandatory norms always in favor of the latter. I now turn to an alternative to semantics 2 that does just the opposite.

3.3 Semantics 3: Overriding Licenses

At the same time that G. H. von Wright's later theory of norms emerged, Alchourrón and Bulygin (1981) proposed an 'expressive conception of norms' that defines the truth of normative propositions relative to a given normative system. Obligating norms play a very similar role as they do here. Permission, however, is treated as derogation, the act of rejecting a prohibiting norm. The authors leave it open whether explicit permissions should be treated as (meta-) operations of derogation (rejection) on the set of 'O-norms' **A**, or need to be represented as separate entities and therefore by a set of permissive norms **B**. For the second case, they remark:

> But then we must unify somehow both sets, if we want a non-ambivalent system. [...] Therefore the operation of unification requires subtracting from the commanded set the negations of the propositions that are members of the permitted set. So if p is permitted, $\neg p$ must be subtracted (eliminated from **A**) and vice versa. Thus the permission of p gives rise to the same operation as the rejection of $\neg p$. (Alchourrón and Bulygin 1981, p. 118)

Alchourrón and Bulygin's proposal was echoed by Stolpe (2010) in a recent critique of Makinson's and van der Torre's (2003) treatment of explicit permissions in an input/output setting. According to Stolpe, positive (explicit) permission must be treated as "a particular instance of the more general concept of derogation" and must "denote the elimination of a norm from a normative system". Stolpe therefore uses the contents of explicit permissions to contract the set of contents of obligating norms to capture what he considers an essential feature of permissions:

> What a permissive provision does is to render a prohibition null and void for a particular case. [...] Permissive provisions override mandatory norms, and are therefore to a certain extent protected—in cases of conflict the permission prevails. (Stolpe 2010, p. 103)

The present framework makes it quite straightforward to model the intuition behind Stolpe's and Alchourrón's and Bulygin's proposals: simply 'maximize' the set of licenses first. Since $\langle \emptyset, L \rangle$ is always vW-consistent with any C, this means keeping L fixed, and maximizing just the set of imperatives until adding a further imperative makes the pair vW-inconsistent with C. Formally, let

$$\langle I, L \rangle \bot^{AB} \neg C$$

be the set of all pairs $\langle I', L \rangle$, $I' \subseteq I$, such that

(i) $\langle I', L \rangle$ is vW-consistent with C, and
(ii) there is no $I'' \subseteq I$ such that $\langle I'', L \rangle$ meets (i) and $I' \subset I''$.

So for each $\langle I', L' \rangle \in \langle I, L \rangle \bot^{AB} \neg C$, $L' = L$, and I' is a maximal subset of the imperatives that is consistent with the situation, and with the situation and any of the licenses in $L^{\underline{n}} \neg C$ (all but those licenses that contradict the situation and are therefore treated as inoperative by Definition 2).

Existential and universal dyadic O-operators are then defined as before:

(td-11) $\langle I, L \rangle \models O^e(A/C)$ iff $\exists \langle I', L \rangle \in \langle I, L \rangle \bot^{AB} \neg C : I' \cup \{C\} \vdash A$

(td-12) $\langle I, L \rangle \models O^a(A/C)$ iff $\forall \langle I', L \rangle \in \langle I, L \rangle \bot^{AB} \neg C : I' \cup \{C\} \vdash A$

The construction makes sure that in cases of conflict between imperatives and an explicit permission, the permission prevails. Suppose that $I = \{\neg k\}$ and $L = \{k\}$, so there is a prohibition to kill and a license to kill. In the default circumstances $\top$ we have $\langle I, L \rangle \bot^{AB} \neg \top = \{\langle \emptyset, \{k\} \rangle\}$: So $P^e(k/\top)$ and $P^a(k/\top)$ are both true—the inclusion of explicit permissions in a normative standard is always more important than the inclusion of imperatives, and then no admissible standard prohibits killing. Explicit permissions need not compete with imperatives—they override them.

It is easy to see that any pair of imperatives and licenses that is admitted by semantics 3 is also admitted by the wider semantics 2:

Observation 2 (Relation between semantics 3 and 2)
If $\langle I', L \rangle \in \langle I, L \rangle \bot^{AB} \neg C$ then $\langle I', L \rangle \in \langle I, L \rangle \bot^{vW} \neg C$.

3.4 Logical Comparison

Semantics 2 and 3 have no special deontic permission operators, like explicit or strong permission P^x or P^+ from Sect. 2.5. Instead there are two types of negative permission: $P^a A$ means that some admissible standard does not commit the agent to doing $\neg A$, and $P^e A$ means that no such standard does (A is overall permitted). Other P-operators are easily defined. But perhaps Bentham (1782) p. 98 is right who, when considering whether to distinguish an 'original permission' from a 'non-command', remarked that "where there can be no difference in the conduct of the subject, it is to no purpose to mark out any difference in the mind of the legislator."

As observations 1 and 2 shows, semantics 1 and 3 may be considered special cases of semantics 2. Semantics 1 pays no regard to licenses and just maximizes the set of imperatives; from any such set $I' \in I \bot \neg C$ a maximal pair $\langle I', L' \rangle$ in the sense of semantics 2 can then be obtained by filling L' maximally with $l \in L$ such that vW-consistency with C is maintained. Semantics 3 'maximizes' the set of licenses (includes them all), and only then a maximal subset of imperatives is added such that the resulting pair is vW-consistent. Semantics 2, by contrast, considers all ways to obtain maximal vW-consistent pairs from the given sets I and L equally. The duality of semantics 1 and 3 suggests looking for definitions in between, which may be achieved by adding priority orderings as will be discussed later.

For the logical properties of the deontic operators defined by the three semantics, consider the following table:

		Semantics 1–3		
CExt	If $\vdash C \to (A \leftrightarrow B)$ then $\models O(A/C) \leftrightarrow O(B/C)$	e, a		
ExtC	If $\vdash C \leftrightarrow D$ then $\models O(A/C) \leftrightarrow O(A/D)$	e, a		
DM	$O(A \wedge B/C) \to (O(A/C) \wedge O(B/C))$	e, a		
DC	$(O(A/C) \wedge O(B/C)) \to O(A \wedge B/C)$	a, aee		
DN	$O(\top/C)$	a, if $\nvdash \neg C$ then e		
DP	$P(\top/C)$	e, if $\nvdash \neg C$ then a		
CCMon	$O(A \wedge D/C) \to O(A/C \wedge D)$	e, a		
Cut	$O(D/C) \to (O(A/C \wedge D) \to O(A/C))$	a, aee, eae		
Cond	$O(A/C \wedge D) \to O(D \to A/C)$	e, a	e, –	–, –
RMon	$P(D/C) \to (O(A/C) \to O(A/C \wedge D))$	e, eaa, aae	e, –, aae	–, –, –
Or	$(O(A/C) \wedge O(A/D)) \to O(A/C \vee D)$	a, aee, eae	–, aee, eae	–, –, –
DR	$(O(A/C \vee D)) \to (O(A/C) \vee O(A/D))$	e, aea, aae	e, aea, aae	–, –, –
Loop	$(X(A_2/A_1) \wedge \ldots \wedge X(A_n/A_{n-1}) \wedge X(A_1/A_n))$ $\to X(A_n/A_1)$	P^e, O^a	P^e, O^a	–, O^a
		Sem.1	*Sem.2*	*Sem.3*

Theorem 2 (Validities for the three semantics).
All of the formulas in the table are valid in the semantics as indicated in the columns to the right, where a single 'a' or 'e' means the deontic operators are uniformly indexed by 'a' or 'e', and a string containing 'a' and 'e' means the operators are indexed in the order of the string. If the entry is '–', then the formula indexed as in a previous column is not valid. (Loop) means the formula where X is uniformly replaced by one of the indicated operators.

All formulas are well known from dyadic deontic logic and nonmonotonic logic. (CExt) and (ExtC) establish (contextual) extensionality for antecedents and consequents, and (DM)–(DP) correspond to the usual monadic theorems. The other names are from the study of nonmonotonic logic, namely *conjunctive cautious monotony*, *cumulative transitivity*, *conditionalization*, and *disjunctive reasoning*. Given (CExt), (DN) and (DP) are equivalent to (similarly constrained) O- and P-*reflexivity* $O(A/A)$, $P(A/A)$. (Loop) was crucial in Spohn's (1975) completeness proof for Hansson's (1969) system of dyadic standard deontic logic *DSDL*3, and was later rediscovered by Kraus et al. (1990).

For the monadic fragments, the following theorem is easily established:

Theorem 3 (Soundness and completeness of $SDL^a P^e$).
For all the semantics 1–3, the monadic fragment—the valid formulas in which each occurrence of a stroke '/' is followed by '⊤)'— equals the axiomatic theory $SDL^a P^e$ defined by the axiom schemes and rules above the middle line of the table that are indexed as in Theorem 2 and where all occurrences of C and D are replaced by ⊤ (and $\models$ is replaced by $\vdash_{SDL^a P^e}$).

In Hansen (2005), I proved that for semantics 1, the axiomatic theory based on the given validities is weakly complete (in fact, a subset suffices). If this also holds for semantics 2 and 3 remains open.[23] However, it is clear that if $A \in \mathscr{L}_{DDL^{ea}}$ is invalid in semantics 1, and so there is a set I such that $I \models \neg A$, then $\langle I, \varnothing \rangle \models \neg A$ in semantics 2 and 3, so A is invalid in these as well. This establishes the following observation:

Observation 3 (DDL^{ea} as upper limit logic for semantics 2, 3)
Let DDL^{ea} be the axiomatic theory based on the given validities for semantics 1. Suppose $DDL2^{ea}$, $DDL3^{ea}$ are axiomatic theories sound and complete with respect to semantics 2 and 3. Then $DDL2^{ea} \subseteq DDL^{ea}$, $DDL3^{ea} \subseteq DDL^{ea}$.

3.5 Discussion

Semantics 2 and 3 may equally claim to explain how a resolution of conflicts between explicit permissions and obligations works. But which semantics captures our intuitions about such a resolution better?

In both semantics, the addition of a license $l(A)$ will remove obligations to the contrary: it makes $O^a(\neg A/C)$ generally false and so A will not be prohibited overall, except if $\neg A$ is a fact (then by reflexivity $O^a(\neg A/\neg A)$). However, the import of this exception is not negligible: it means that the effect of a conflicting obligating norm is never eliminated completely. Let $I = \{p \wedge q\}$ and $L = \{\neg p\}$. Then both semantics make true $O^a(q/p)$: in the situation p realizing q remains overall obligatory. We might interpret this as meaning that if it has become a fact that a license is *not* used then the obligations remain the same. There are two competing intuitions: One wants to make an agent (legally, morally) free to do a certain act, and for this purpose opposing duties must be suspended. If the freedom is not used, then the suspended duties come into play again. The other only observes the clash of authority that is embodied in the existence of two conflicting norms, and wants to totally remove ("void") one of them for reasons of principle or tacitly assumed rank. Then this norm should not be considered in other, more special situations either. Only the first intuition is modelled here, not the second.

---

[23] The *a*-fragment of semantics 2 and 3 is similar to the KLM-system ***CL*** to which (DP) is added for consistent antecedents. The *e*-fragment of semantics 2 seems the same as that of semantics 1.

To motivate why permissions should act as rejections and so prevail over prohibitions, Alchourrón and Bulgyin employ a legal example:

> This is what happens with constitutional rights and guarantees: the constitution rejects in advance certain norm-contents (that would affect the basic rights), preventing the legislature from promulgating those norm-contents, for if the legislature promulgates such a norm-content, it can be declared unconstitutional by the courts and will not be added to the system. (Alchourrón and Bulygin 1981, p. 108)

Stolpe (2010) also motivates his proposal by a legal example: that of Article 19 of the Universal Declaration of Human Rights guaranteeing the freedom to publish cartoons depicting the prophet Mohammad even when this might be forbidden by local laws. Seemingly, the idea that permissions always prevail is most convincing when considering fundamental rights and freedoms that rank high in some normative hierarchy. Then a lower-ranking, conflicting prohibition is 'materially derogated' (cf. Weinberger 1996, p. 175). However, such a ranking may not always exist:

Example 6 (Partying with friends).
A couple you know is having a party. One of them leaves a message: "I am sorry, you can't come, it's close friends only!" The other one also leaves a message: "You may come if you bring a bottle of champagne." You may consistently reason that you are prohibited to come as one of the hosts told you not to, and also that you are allowed to come bearing a bottle of champagne, since one of the hosts permitted it.

Even if such a ranking exists, it may not be known:

Example 7 (Partying with friends, continued).
The same person left the messages. *Case 1:* Both came by email, first the apology, then the conditional invite. *Case 2:* The apology arrived later. *Case 3:* One arrived by email, the other was left on your answering machine which has no time stamp. Using *lex posterior* you may reason that, after all, you are allowed to come (case 1) or prohibited to come (case 2). In case 3, both seem reasonable. If the host asks why you weren't there you may reply: "I wasn't allowed to come, you disinvited me!" If you go and are looked at strangely you may say: "You allowed me to come bringing a bottle of champagne, here it is!"

So sometimes there are rules of thumb that help to resolve conflicts, like *lex superior*, *lex specialis* and *lex posterior*. But even these may lead to contradictory results, like when a more specific command or permission comes from a lower authority. If there are no such guidelines we are at a loss, and there is no reason to prefer permissions over obligations.

Some have understood the deontic D-axiom as stating that obligating norms always contain a permission by the authority to carry out what they make obligatory (cf. Weinberger 1996, p. 173). Reasonable or not, we can easily incorporate this assumption in our framework by letting $I \subseteq L$. Consider then the following Buridan's ass type example provided by Barcan Marcus (1980):

Example 8 (Saving a twin).
Identical twins are in danger of being crushed by a rock. Only one can be saved. You are liable for the lives of both. What are your obligations? Let A mean you save the first twin and B you save the second. $I = \{A, B\}$, and so by the above assumption $L = \{A, B\}$. The situation C equals $\neg(A \wedge B)$. Semantics 1, developed in van Fraassen (1973), Horty (1997) to solve such puzzles, makes true $O^e(A/C)$, $O^e(B/C)$, and $O^a(A \vee B/C)$. Semantics 2 makes true $O^e(A/C)$ and $O^e(B/C)$, but not $O^a(A \vee B/C)$ as you may use both licenses. In semantics 3 the 'implied' licenses cancel all imperatives, making any action consistent with C by default permitted. So you may just walk away.

I will not pursue this assumption.[24] But quite similarly, in semantics 3 additional licenses may make *more* things overall obligatory as they can clear conflicts out of the way. Suppose $I = \{p, \neg p\}$ and let $L = \{p\}$. Then the only pair in $\langle I, L\rangle \bot^{AB} \neg\top$ is $\langle\{p\}, \{p\}\rangle$, so $O^a p$ is true and doing p overall obligatory. But if a command to realize p did not prevail against a command to the contrary, why should it do so with an additional permission? Note that a command may override an explicit permission as easily as vice versa (e.g., if uttered later).

So I think the proposal by Alchourrón and Bulgyin, endorsed by Stolpe, rests on an error. It mistakes the situation in which the explicit permissions all have higher ranks than the obligating norms, and so may be addressed in this fashion, for the general case. But obligations and permissions may also conflict on the same level or when they are without discernable ranking, and then the treatment of permissions as always overriding seems unconvincing. However, the idea that licenses *can* override imperatives, and that these are then no longer considered in the situation, is embodied by the priority semantics discussed next.

3.6 Semantics 4: Adding Priorities

Let I, L and f be defined as before. As we consider prioritized norms, and two norms may have the same contents but different priority, the convenience to identify them with their contents is no longer available. Also, I will assume that I and L are disjoint (but not, necessarily, the sets $f(I)$ and $f(L)$ of contents). In addition to I, L and f, there is a priority relation $<$ on $I \cup L$, where '$n < m$' is read as 'norm n takes priority over norm m'. Let $<$ be a strict partial order (irreflexive and transitive), and additionally be well-founded (infinite descending subchains are excluded). Note that $<$ may be empty (e.g., if there is no priority relation between the norms).

For conflict resolution I make use of a proposal developed for the purpose of theory revision in Brewka (1989, 1991).[25] Brewka uses 'full prioritizations' $\prec$ obtainable from $<$, which are all strict well-orders on the given set that preserve $<$: for all

[24] As the example demonstrates, there is a difficulty of ensuring that the 'implied' license is only used in connection with actually satisfying the imperative that 'implied' it.

[25] The following expands—with regard to licenses—the framework used in Hansen (2006).

$i, j \in I \cup L$, if $i < j$ then $i \prec j$. Intuitively, from each prioritization one obtains a maximally vW-consistent pair $\langle I', L'\rangle$ as follows: Work 'from top to bottom'. Include the next[26] imperative or license to the respective set of the so far formed pair if it preserves vW-consistency. Otherwise, disregard it as overridden. The formal definition is a bit more complex since infinite ascending subchains are not excluded.

I use '+' for the addition of an element from some set of a pair of disjoint sets $\langle I, L\rangle$ to a pair of subsets: Let $\Gamma \subseteq I, \Delta \subseteq L$, and $n \in I \cup L$. Then $\langle \Gamma, \Delta\rangle + n$ is $\langle \Gamma \cup \{n\}, \Delta\rangle$ if $n \in I$ and $\langle \Gamma, \Delta \cup \{n\}\rangle$ otherwise. By a union of ordered pairs I mean the pair that results from uniting the respective sets: $\langle \Gamma, \Delta\rangle \cup \langle \Gamma', \Delta'\rangle = \langle \Gamma \cup \Gamma', \Delta \cup \Delta'\rangle$. $f[\langle \Gamma, \Delta\rangle]$ means $\langle f(\Gamma), f(\Delta)\rangle$. Finally, let $I, L, f, <$, and C be as described, and

$$\langle I, L\rangle \Downarrow^{vW} \neg C$$

be the set of all pairs $\langle \Gamma, \Delta\rangle$ such that $\nvdash \neg C$ and $\langle \Gamma, \Delta\rangle$ is obtained from some full prioritization $\prec$ of $<$ by defining

$$\langle \Gamma, \Delta\rangle_n = \begin{cases} \bigcup_{m \prec n} \langle \Gamma, \Delta\rangle_m + n & \text{if } f\left[\langle \Gamma, \Delta\rangle_n\right] \text{is then } vW\text{-consistent with } C, \text{ and} \\ \bigcup_{m \prec n} \langle \Gamma, \Delta\rangle_m & \text{otherwise,} \end{cases}$$

for any $n \in I \cup L$, and letting $\langle \Gamma, \Delta\rangle = \bigcup_{n \in I \cup L} \langle \Gamma, \Delta\rangle_n$.

Neglecting again the difference between norms and their contents, if $<= \emptyset$, then $\langle I, L\rangle \Downarrow^{vW} \neg C = \langle I, L\rangle \bot^{vW} \neg C$, and so semantics 4 coincides with semantics 2. If the imperatives take priority and $<= \{i < l \mid i \in I, l \in L\}$, we obtain semantics 1: $\langle I', L'\rangle \in \langle I, L\rangle \Downarrow^{vW} \neg C$ for some L' iff $I' \in I \bot \neg C$. If the licenses take priority and $<= \{l < i \mid i \in I, l \in L\}$, we obtain semantics 3: $\langle I, L\rangle \Downarrow^{vW} \neg C = \langle I, L\rangle \bot^{AB} \neg C$. In this sense, semantics 4 contains all previous semantics.

Truth definitions for dyadic deontic operators are then as now usual:

(td-13) $\langle I, L, f, <\rangle \models O^e(A/C)$ iff $\exists \langle I', L'\rangle \in \langle I, L\rangle \Downarrow^{vW} \neg C : f(I') \cup \{C\} \vdash A$

(td-14) $\langle I, L, f, <\rangle \models O^a(A/C)$ iff $\forall \langle I', L'\rangle \in \langle I, L\rangle \Downarrow^{vW} \neg C : f(I') \cup \{C\} \vdash A$

It is immediate that all theorems above the line in the earlier table are still valid, and that the monadic fragment also equals $SDL^a P^e$. Also, all versions of (CCMon) and (Cut) indicated there are still valid.[27] But it is equally clear that all sentences that are invalid with respect to semantics 3 are still so, and hence no more validities are obtainable: if A is invalid in semantics 3, then there is a pair $\langle I, L\rangle$ such that $\langle I, L\rangle \models \neg A$, but then $\langle I, L, f, <\rangle \models \neg A$ in semantics 4, where f maps the respective contents to some disjoint sets and $<$ just ranks all licenses highest.

Regarding the earlier discussion, semantics 4 seems to capture the best of all worlds. If there is a conflict between a license and a prohibition, and neither takes priority, then both compete for a place in a vW-consistent pair as in semantics 2. If the license ranks higher, then the prohibition is overridden as in semantics 3. And if the license ranks lower then the prohibition prevails as in semantics 1. The semantics

---

[26] Well-foundedness of $<$ guarantees the existence of a 'next' ($\prec$-minimal) norm in the chain $\prec$.

[27] For a sketch of the inductive proof cf. that of Theorem 5 in Hansen (2006), p. 25.

preserves the 'paradigms' of deontic logic by describing the normative situation in a language where P is defined as the dual of O. But *what* is described as obligatory, permitted or prohibited depends not only on the presence of explicit obligations but also that of explicit permissions and the priority ordering. It puts permissive norms in a dynamic context, as pictured by E. Bulygin:

> Permissive norms are in an important sense system dependent: it is only in a dynamic perspective of a hierarchically structured normative system [...] where the concept of permissive norm becomes really fruitful. (Bulygin 1986, p. 216)

4 Problems of Permission and Obligation

After these presentations of logical semantics for reasoning about explicit obligations and permissions, I now turn to some problems: we look at 'free choice permission', consider D. Lewis's famous 'problem of permission' (Lewis 1979), and revisit J. F. Horty's (1994) 'asparagus puzzle' for a new solution by use of explicit permissions.

4.1 Free Choice Permission

Free choice permission is probably the most discussed problem in the literature on permissions in deontic logic.[28] Here I will examine it in the semantic framework discussed above. For the problem, consider the following example:

Example 9 (Tea or coffee?).
You visit a colleague. Her gracious secretary points at a coffee table with some silverware and says: "Have some tea or coffee, if you like!" Being a deontic logician you wonder what exactly has been permitted. Let p mean you have tea, and q that you have coffee. The natural sentence suggests $P(p \vee q)$—surely the secretary did not imagine you might want to have both? But just as if she had permitted both, it seems now true that you are permitted to have tea and permitted to have coffee, so her permission makes true Pp and Pq. But $P(p \vee q)$ does not derive these in standard deontic logic.[29] So it seems as if either a new deontic logic is needed (e.g. one that adds a new permission operator), or the one natural sentence must be split into two permissive sentences. (Finally, you settle for tea.)

To model the example, let a license set L contain a single disjunctive license $l(p \vee q)$. Of course, if nothing is prohibited then any consistent course of action is permitted. So let us assume that there are some background prohibitions to the effect that having

[28] For more details on the problem and proposed solutions cf. Hansson (2013).

[29] Otherwise, with the usual axiom of monotonicity, a statement that posting a certain letter is permitted would derive the statement that burning the letter is permitted.

tea or having coffee is normally somehow forbidden.[30] It seems natural to assume that it lies within the secretary's power to override these background prohibitions, and so we may use semantics 3 to describe the situation. The next table compares the results for different backgrounds:

I	L	True	False
$!\neg(p \wedge q)$	$l(p \vee q)$	$P^e p,\ P^e q$	$P^a(p \wedge q)$
$!\neg p$	– " –	$P^e q,\ P^e(p/\neg q)$	$P^e p,\ P^a p$
$!\neg(p \vee q)$	– " –	$P^e p,\ P^e q,\ P^e(p \wedge q)$	
$!\neg p, !\neg q$	– " –	$P^a p,\ P^a q,\ P^e(p \vee q),\ P^e(p/\neg q),\ P^e(q/\neg p)$	$P^e p,\ P^e q,\ P^a(p \wedge q)$

As can be seen, adding the license $l(p \vee q)$ to *any* of the backgrounds results in *both*, p and q, becoming permitted, in some way or other. Cases: If just having both tea and coffee is forbidden in the background, then there is no conflict and having both remains forbidden while having either remains overall permitted.[31] If the background is not symmetric and there is just a prohibition against having tea, then only having coffee is permitted by default. But if you make up your mind and won't have the coffee (it is not forbidden) and so $\neg q$ is a fact, then you are also overall permitted to have tea.[32] If the background prohibition is the opposite of the license, then it is overridden and any consistent action again permitted. If there are two prohibitions, against having tea and against having coffee, then there are two normative standards: one where you may have tea and one where you may have coffee. Without further information, it is up to you which standard you decide to follow, so from an addressee's perspective, again either is permitted.[33]

This result may be compared to that of the proposal to translate a free choice permission into two separate permissions for each disjunct (cf. Hansson 2013): As

I	L	True	False
$!\neg(p \wedge q)$	$l(p), l(q)$	$P^e p,\ P^e q,\ P^e(p/q),\ P^e(q/p)$	$P^a(p \wedge q)$
$!\neg p$	– " –	$P^e p,\ P^e q,\ P^e(p \wedge q)$	
$!\neg(p \vee q)$	– " –	– " –	
$!\neg p, !\neg q$	– " –	– " –	

30 The example suggests that such prior prohibitions are, in fact, in place: people don't just help themselves to tea or coffee when visiting, if they do they risk being called rude, and then most likely will try to excuse themselves ("I thought it was free for all"). This is all what is required for the existence of a norm (cf. Hart 1961, pp. 54–56).

31 Such a background prohibition seems to be the main culprit in the case of free choice permission. It may express a social norm against appearing greedy or wasteful and thus come with overriding strength. It explains why one may hesitate to have tea *and* coffee (or whatever the options are) even when expressly permitted to have both.

32 It is perhaps not very intuitive to assume an asymmetric background, since there is no indication for it in the example.

33 Moreover, as listed in the table, not only does the license make having tea or coffee overall permitted, but also having one once you have declined the other (again, this is not forbidden).

the tables demonstrate, having two distinct licenses, one for having coffee and one for having tea, makes a difference. The two licenses override almost all of the background prohibitions and then any consistent course of action is permitted, including your having both tea and coffee. The only case where you cannot immediately[34] have both is the first, since each license is individually consistent with the background prohibition, and so the norms are vW-consistent.

So if there are very strong disjunctive permissions, ones that aim at creating overall permissions for each disjunct for all accommodating circumstances and against all normative backgrounds, then they are obviously better modelled using distinct licenses for each disjunct. However, the above analysis shows that even if they are modelled using ordinary disjunction,[35] they, and in fact all overriding licenses,[36] still behave very much like 'free choice permissions': they make each of their disjuncts somehow permitted in a variety of normative settings. This seems to greatly diminish the problem. It suggests that mostly,[37] disjunctive licenses are indeed correctly modelled using ordinary disjunctions, as seems natural.

But what about a supposed need to have a new 'free choice' permission operator in the deontic object language? As the table demonstrates, to sufficiently describe what becomes permitted when a disjunctive license $l(A \vee B)$ is added to a set of norms, often deontic statements of all the forms PA, PB and $P(A \vee B)$ will be required.[38] But this is an effect of the normative dynamics, and not a deficiency of the logical

[34] As the truth of $P^e(p/q)$ and $P^e(q/p)$ demonstrates, once you are having one (it is not forbidden) then you may also have the other. Also note that if the concept of vW-consistency is changed as will be discussed in the final section, this case becomes the same as the others in this table.

[35] Of course, syntactical form does not matter in the semantics, so it is the same if the license is $l(p \vee q)$, $l(\neg p \rightarrow q)$, or $l(\neg(\neg p \wedge \neg q))$.

[36] As noted above (fn. 8), the content of a license may be considered a disjunction of the states it permits. In this sense, *any* license is disjunctive, and 'free choice' when it is overriding. Consider $\langle\{!q\}, \{l(p)\}\rangle$. Syntactical form does not matter, so $l(p)$ behaves like $l((p \wedge q) \vee (p \wedge \neg q))$. Semantics 3 makes true $P^e(p \wedge q)$, and also $P^e(p \wedge \neg q/C)$ is true for some C, namely for $C = \neg(p \wedge q)$. Note that C is not forbidden: both $O^a(p \wedge q)$ and $O^e(p \wedge q)$ are false. So again both disjuncts are somehow overall permitted. Whenever licenses are treated as absolute, some permitted circumstances may pit a license and an obligation against each other in a way that forces a choice. Then, being overriding, the license wins. (This may show that, purely by definition, absolute rights create loopholes for unwanted behavior. I owe this point to M. Beirlaen, private correspondence.)

[37] Much effort has been spent on trying to prove that there are acts of using disjunctions for permitting that are *not* free choice. There are two types of examples: One adds an explicit permission to a background with a higher-ranking obligating norm (cf. Kamp 1973, p. 67, Hilpinen 1982, p. 189). E.g. there is a permission to buy apples or pears and a (stronger) background obligation to always buy the cheaper product when there is a choice. It seems clear from my description how this case may be accommodated. The other considers natural sentences of the form "You may take an apple or a pear, but I don't know which (yet)" (cf. Kamp 1978, p. 271). I think that at best such a sentence expresses not a permissive but a deontic statement, uttered e.g. when the speaker only knows that there is or will be a prohibition for one of the actions but not both (and then it is true).

[38] This provides an argument that—contrary to what Lewis (1979) and, following him, Kamp (1978) and Schwager (2006) have suggested—the performative of making something permitted cannot be identified with one basic deontic statement that describes what has thus been made permitted.

language. It just shows how complex the transition from an act of permitting into a description of what it makes permitted may be.

4.2 *Lewis's Problem of Permission*

Above it was explained that natural language imperatives might be separable or inseparable, and if they are inseparable, satisfying only a part makes little sense. Shopping lists are good examples since they may be interpreted either way. For the semantics it was assumed that the problem has been taken care of by the formalization of natural language, and by not using logical closure. In the case of conflicts between imperatives this means that if an imperative is overridden, by default nothing it (alone) pronounced obligatory remains so. If there are two imperatives $!(p \wedge q)$ and $!\neg p$, and the second takes priority, then Oq is false.

A conflict between an imperative and a license was treated just in the same way: if there is an imperative $!(p \wedge q)$ and a license $l(\neg p)$ and the second takes priority over the first, then again Oq is false. But often, in natural language, an explicit permission is used to do exactly that, to modify a prior inseparable command to make only a part of it still obligatory. Consider the following example:

Example 10 (Waldorf salad).
I have been told: "Buy celery, apples and walnuts, I want to make a Waldorf salad!" A little later: "You don't have to buy the walnuts, I found some."

In the example, the original imperative is inseparable: if, in the supermarket, I find out that apples are out, I best call home and ask if there is some different dish I might help prepare. Let p mean I buy the celery, q mean I buy the apples, and r mean I buy the walnuts, so the set of imperatives should contain $!(p \wedge q \wedge r)$. The permissive sentence frees me from having to buy the walnuts, so the licenses should include $l(\neg r)$, and this license should take priority. Our semantics then makes Op and Oq false: I am freed from my shopping chores, contrary to what the speaker intended.

So one might want to modify the semantics and let a license eliminate some conjuncts of an imperative, but the other conjuncts should remain untouched: I must still buy the celery *and* the apples, and so an inseparable imperative $!(p \wedge q)$ should remain in the revised set. But how are the 'remaining conjuncts' to be determined? An equivalent description of the original imperative is $!(((p \wedge q) \leftrightarrow r) \wedge r)$, but removing the conjunct to the right leaves $!((p \wedge q) \leftrightarrow r)$—I may not buy the walnuts, but then I mustn't buy one of the other items either, which again is clearly not what the speaker intended. This is Lewis's (1979) 'problem of permission': There are right ways and wrong ways to describe what has become permissible after something has been permitted, and there is no obvious logical method that helps choose which ones are right, though in ordinary language "somehow that is understood."

To solve the problem, some theory contraction operator seems required—we want to break up individual formulas (imperatives) in a way that makes them consistent with another formula (the license). Full meet theory contraction might be considered (cf. Stolpe 2010). But when do we use it? We must also take care of conflicts and dilemmas created by the situation. If we first construct the remainder sets of imperatives that are consistent with the situation, some imperative might not get added to a remainder even though, after having been modified by a license, it could have been. If we apply the contraction first, then the resulting set of imperatives will not just contain, as desired, an inseparable imperative $!(p \wedge q)$, but also $!q$ or $!(p \leftrightarrow q)$, and then Oq or $O(p \leftrightarrow q)$ might be falsely described as true even when, in the second step, a conflicting imperative $!\neg p$ overrides the first. So we need to somehow pick the now inseparable rests of the originally inseparable formulas from the set obtained by the contraction, and are back to Lewis's problem full circle.

There is an obvious problem to Lewis's presentation of the 'problem of permission' as such. Consider the following variant of the example:

Example 11 (Waldorf salad, continued).
I was told: "Buy celery, apples and walnuts, I want to make a Waldorf salad!" A little later: "Don't buy the walnuts! I found some, so don't waste the money!"

Again I am no longer required to buy the walnuts, and an inseparable part of the imperative remains (obliging me to buy celery and apples). But now the original imperative is modified not by a permission, but by a second, later imperative. So not only permissions but also imperatives can do this. But then the same problem arises: Lewis's problem, it turns out, is not just one of permission, but also of commanding.

Curiously, Lewis seems to have recognized this problem. In his setting a 'Master' continually restricts the freedom of a 'Slave' by issuing commands, but sometimes the Master also issues permissions and gives the Slave some of his freedom back. Lewis considers conflicting orders, when "having commanded at dawn that the Slave devote all his energies all day to carrying rocks, the Master may decide at noon that it would be better to have the Slave spend the afternoon on some lighter or more urgent task." His solution is to shift the burden back to the act of permitting:

> What he [the Master] should have done was first to permit and then to command that p. He should say to the Slave, in quick succession, first $l(p)$ and then $!p$; that way, he would be commanding not the impermissible but the newly permissible. We could indeed have equipped the language game with a labor-saving device: whenever Pp is false, a command that p is deemed to be preceded by a tacit permission that p. (Lewis 1979, p. 168)[39]

So Lewis's solution to our case is that in order to forbid me to buy the walnuts when I was originally obliged to buy them, I should first be permitted and then commanded not to buy them. Lewis assumes that the act of permission separates the previous obligations to the contrary in the 'right' way, and then the act of commanding leads to no oddities. But similarly, in order to permit me not to buy the walnuts when

[39] I replaced Lewis' symbolism with mine. Lewis employs '$\varnothing$' as propositional variable, and—being notoriously sceptic of the distinction between performatives and constatives—'¡p' both for the description that p is permitted (I write 'Pp') and the act that permits it (I use '$l(p)$').

I was obliged to buy them, I could first be forbidden and then permitted to buy them. We assume that the prohibition separates the previous obligations in the 'right' way, and since the original situation implied my having to buy the walnuts, adding a permission for it leads to no further disagreement, except that it eliminates the new prohibition in the simple, overriding manner defined before. Both solutions do not 'solve' anything, of course: one places the burden to find the 'right' remaining obligations on the concept of permission, and the other places the same burden on the concept of prohibition (commanding).

It seems hard to formally distinguish cases in which subsequent permissions (or commands) are meant to modify a previous inseparable command, from cases where they are meant to eliminate or suspend it completely. Consider the following variant:

Example 12 (Winter shopping).
It is winter and very cold. I have been told: "Put on your warmest coat, walk to the shop and buy some walnuts!" A little later: "You don't have to buy the walnuts, I found some." You don't have to wear your coat now.

The original command seems inseparable: the speaker would not have the agent buy the walnuts in a different way. But walking to the shop or wearing a coat without buying the walnuts is pointless. Therefore, the later permission not only frees from having to buy the walnuts but also from all prescribed accessories to this action.[40] Pending better ideas, I suggest that if a speaker wants to use a license or a command for modifying a previous inseparable command, this should be treated as an act of replacing the command by a new one, and not as separate norm-giving.

4.3 Eating Asparagus With Your Fingers

Being able to use not just mandatory norms but also explicit permissions to model normative reasoning is a game changer. It provides a fresh look at seemingly unrelated problems, which were believed solvable only by means of specially formalized conditional norms. Such a problem is the 'asparagus puzzle' (Horty 1993, 1994):

Example 13 (Asparagus puzzle).
An agent decides that his behavior should be governed by the following three oughts: You ought not to eat with your fingers; you ought to put your napkin on your lap; if you are served asparagus, you ought to eat it with your fingers.

Let A mean you are eating asparagus, N that you put a napkin on your lap, and F that you are eating with your fingers. All of $O\neg F$, ON and $O(F/A)$ should be true. This is a puzzle because at the time there was no account of dyadic deontic logic which allowed to model these sentences without undesired effects. Either $O(\neg F/A)$

---

[40] Such accessorial actions occur in examples like that of Weinberger's piano player (cf. fn. 10). It is tempting to just split the commands: 'play the piano,' 'if you play the piano, close the window.' But then playing the piano with the window open would satisfy a norm, something the speaker would most likely object to. So like the single-purpose shopping lists, they are cases of inseparability.

is true, so eating asparagus with your fingers is also forbidden, but it should not be. Or $O\neg A$ is true, so you must not eat asparagus in the first place, which the example seems not to imply. Or $O(N/A)$ is false, so when served asparagus you don't have to use the napkin, but the example did not suggest it can be discarded with the cutlery.

Horty's solution was to model the underlying norms by default conditionals, where an imperative not to eat with your fingers is overridden by a conditional one to eat with your fingers when eating asparagus, once in fact you do. But there is also a solution in the present framework, without recourse to such conditionals.

First we translate the three obligations from the example into imperatives, so from $O\neg F$, ON and $O(F/A)$ we obtain the set $I = \{!\neg F, !N, !(A \rightarrow F)\}$. From the above discussion we pick up that eating asparagus with your fingers should not be forbidden, and also that eating asparagus should not be forbidden. So we include these desiderata in the license set $L = \{l(A \rightarrow F), l(A)\}$. Finally, just like Horty, we use a priority ordering of the norms which may be as follows:

$$l(A \rightarrow F) < l(A) < !\neg F < !N < !(A \rightarrow F)$$

It is easily checked that by use of semantics 4 we obtain

$$\langle I, L\rangle \Downarrow^{vW} \neg\top = \{\langle\{!\neg F, !N\}, L\rangle\},$$
$$\langle I, L\rangle \Downarrow^{vW} \neg A = \{\langle\{!N, !(A \rightarrow F)\}, L\rangle\},$$

which makes all forms of $O^*\neg F$, O^*N, $O^*(F/A)$ and $O^*(N/A)$ true, and all forms of $O^*\neg A$ and $O^*(\neg F/A)$ false. Thus the puzzle is solved.

A related puzzle by Prakken and Sergot (1997) requires to model the following sentences: You must not have a fence (F), a white fence (W) if you do, and may have a fence if you live near the sea (S). It has a similar solution in semantics 4: if the norms are $l(S \rightarrow F) < !\neg F < !(F \rightarrow W)$, then all of $O^*\neg F$, $O^*(W/F)$, $P^*(F/S)$ and $O^*(W/S \wedge F)$ are true, as desired by the authors.

All this demonstrates that instead of reasoning with default conditionals to limit a mandatory norm by ones that apply in more special circumstances, often simple licenses can be used to provide the same exceptions.

5 Beyond Von-Wright-Consistency

In the spirit of D. Makinson's call to reconstruct deontic logic as a logic of reasoning about norms, proposals were examined for what can be described as obligatory or permitted if a considered set of norms contains not only mandatory norms ('imperatives') but also explicit permissions ('licenses'). Some typical theorems of dyadic deontic logic were examined for the proposals. It remains an open question what axiomatic theories are complete with respect to the proposed semantic definitions. Such a relation was proved only for fragments of the language.

All proposals rest on a definition of normative consistency taken from G. H. von Wright's later theory of norms. Here, a combined set of 'O-norms' and 'P-norms' was defined as consistent iff the contents of all O-norms are jointly satisfiable by themselves and together with the content of any single P-norm. This definition was used to define monadic or dyadic deontic operators with respect to maximal subsets of the norms that are consistent (either simpliciter or with a given context) in von Wright's sense. The proposals differ only on the strategy to obtain such subsets: semantics 1 maximizes the set of O-norms, semantics 3 prefers the P-norms, while semantics 2 considers all norms equally. Semantics 4 allows to specify which norms should be included rather than some others by means of a priority relation.

As far as I can see, von Wright's definition was never questioned. It has its roots in his *Norm and Action* of 1963.[41] In Alchourrón and Bulygin's logic of normative systems *LNS*, which has an operator of strong permission P^+, it appears in the form that a derivation of the formula $\text{IN}(A) =_{def} (P^+A \wedge O\neg A)$, which serves as a "definition of inconsistent regulation," never involves more than one P^+-premiss since P^+ satisfies just monotonicity.[42] More recently, von Wright's definition was accepted by Makinson and van der Torre (2003),[43] by Stolpe (2010),[44] and Boella and Torre (2003).[45] But is it reasonable? Why should obligations be taken into account collectively, but not permissions? Why should a set of norms $\{!\neg(p \wedge q), l(p), l(q)\}$, consisting of a prohibition against doing both p and q and two explicit permissions for doing p and for doing q, be considered consistent when one can only fulfill one's duties by not making use of at least one of the two permissions that have been granted? Are permissions lesser norms? There seem to be two main objections against taking permissions into account collectively:

1. There may be two explicit permissions with opposite contents: "You may stay at home tonight. Or join us at our practice, if you like." Likewise, $\{!p, l(q), l(\neg q)\}$ should be consistent, even though the two licenses cannot be used collectively.
2. It is common that permissions cannot be used collectively. Consider the set $\{!\neg(p \wedge q), l(p), l(q)\}$, and let p mean the agent drinks tonight, and q mean

---

[41] Cf. von Wright (1963) p. 152: "The inconsistency of a set of commands and permissions means this: [...] *Something* which the authority lets the subject(s) do or forbear is, however, at least in some circumstances, logically impossible to do or forbear together with *everything* which he wants them to do or forbear."(My emphasis.)

[42] Cf. Alchourrón and Bulygin (1993) p. 291, also Alchourrón (1969, 1972), Alchourrón and Bulygin (1984). *LNS* is like $SDL + P^+$ from Sect. 2.5, except that it lacks (N), (P), (N^+) and (P^+), and (D^+) replaces (C^+). Note that IN(A) is the negation of (P^+P).

[43] Cf. Makinson and van der Torre (2003) p. 396 on their sets of obligations A and permissions Z: "The elements of A may be used jointly, while the elements of Z can only be used one by one," and continuing in a footnote: "We do not consider here the contractions or revisions that one might wish to make to the code when A is inconsistent with some $z \in Z$."

[44] Cf. Stolpe (2010) p. 104: "While obligations proper may be used jointly, permissions may only be applied one by one. This restriction is intended to capture the fact that two actions may be permitted under a common condition without being jointly so."

[45] Cf. Boella and van der Torre (2003) Sect. 4, where consistency of a set of obligations G and a set of permissions P with an antecedent formula a is defined to mean that $out(G \cup Q, a) \cup \{a\}$ "is consistent for every singleton or empty $Q \subseteq P$."

the agent drives tonight. The agent's mother has permitted her to take the car to the party, and her father has permitted her to drink at the party. But there is a background obligation not to drink and drive, and then having to choose between drinking and driving does not seem out of the ordinary.

Consider, however, tightening the requirements of vW-consistency as follows[46]:

Definition 3 (Strong von-Wright-consistency in a situation).
$\langle I, L\rangle$ is vW^+-consistent with $C \in \mathcal{L}_{PL}$ iff $I \nvdash \neg C$ and $\forall L' \in L \bot \neg C : I \cup L' \nvdash \neg C$.

So 'strong von-Wright-consistency' tests the imperatives not against each license individually, as before, but against sets of licenses: against any maximal subset of licenses whose contents are jointly consistent with the situation C. Clearly, if $\langle I, L\rangle$ is vW^+-consistent then it is also vW-consistent: if the collective test is passed, so is the individual test. Objection (1) is accommodated by only testing against sets of licenses whose contents are jointly consistent, so there may be licenses with opposite contents like $l(q)$ and $l(\neg q)$ among a vW^+-consistent set of norms, just as before.

Semantics 2, 3, and 4 may then be strengthened to semantics 2^+, 3^+, and 4^+: in the definitions of $\langle I, L\rangle \bot^{vW} \neg C$, $\langle I, L\rangle \bot^{AB} \neg C$, and $\langle I, L\rangle \Downarrow^{vW} \neg C$, replace the expression 'vW-consistent with C' with 'vW^+-consistent with C'.[47] We obtain:

Theorem 4 (Relation between semantics 2 [3] and 2^+ [3^+]).
1. If $A \in \mathcal{L}_{DDL^{ea}}$ is valid in semantics $2^+[3^+]$ then A is valid in semantics 2 [3].
2. If $A \in \mathcal{L}_{DDL^{ea}}$ is valid in semantics 3 then A is valid in semantics 3^+.
3. If $A \in \mathcal{L}_{DDL^{e}}$ is valid in semantics 2 then A is valid in semantics 2^+.
4. All validities listed in Sect. 3.4 for semantics 2 remain valid in semantics 2^+.

But what about objection (2)? Let $I = \{!\neg(p \wedge q)\}$ and $L = \{l(p), l(q)\}$. The contents of the licenses are jointly consistent, but not with that of the imperative. So the norms in the story about drinking and driving are not vW^+-consistent. First look at semantics 2^+: we have $\langle I, L\rangle \bot^{vW^+} \neg\top = \{\langle I, \{l(p)\}\rangle, \langle I, \{l(q)\}\rangle, \langle \varnothing, L\rangle\}$. As before, semantics 2^+ considers all norms equally, which means a standard $\langle \varnothing, L\rangle$, which maximizes the set of licenses first, is also accepted. So $P^a(p \wedge q)$ is true: by some standard, the agent may drink and drive. Semantics 3^+ again always maximizes the set of licenses (accepts them all), so for it the just considered standard is the only standard and $\langle I, L\rangle \bot^{AB^+} \neg\top = \{\langle \varnothing, L\rangle\}$. So $P^e(p \wedge q)$ is true: the agent is overall permitted to drink and drive. But I believe there is a tacit understanding that the prohibition against drinking and driving ranks higher than the two permissions: DUI is against the law, it endangers innocent third parties, and the life and health of the agent herself. Assuming $!\neg(p \wedge q) < l(p)$ and $!\neg(p \wedge q) < l(q)$ we obtain, in semantics 4^+, $\langle I, L\rangle \Downarrow^{vW^+} \neg\top = \{\langle I, \{l(p)\}\rangle, \langle I, \{l(q)\}\rangle\}$. So $O^a\neg(p \wedge q)$ is true: the agent remains overall forbidden to drink and drive.[48]

[46] I owe this proposal to L. Goble (private discussion on a draft of this contribution).

[47] For clarity, I add a '+' to the names, and write $\langle I, L\rangle \bot^{vW^+} \neg C$, $\langle I, L\rangle \bot^{AB^+} \neg C$, $\langle I, L\rangle \Downarrow^{vW^+} \neg C$.

[48] Stolpe (2010) uses almost the same objection: "For example, whereas it is usually the case that drinking is permitted and that driving is permitted according to the same legal system, it is usually

Perhaps such tacit assumptions of priority always play a part in objections of this type, and then von Wright's definition should be set right,[49] as described.[50]

Acknowledgments A draft of this contribution was presented at a workshop held in Ghent, Belgium, on 25 May 2012 in honor of the Ph.D. obtained by F. van de Putte, and I thank D. Makinson, D. Batens and M. Beirlaen for critical remarks at this occasion and later on. I thank A. Stolpe and M. Sergot for their criticism and help in overcoming countless mistakes. I am indebted to L. Goble, whose comments and suggestions improved anything of value in this contribution considerably.

Appendix: Proofs

Theorem 1 (Soundness and completeness of $SDL + P^+$).
The system $SDL + P^+$ is sound and (strongly) complete for a norm-based semantics that defines the truth of its deontic operators by (td-5) and (td-6) and where the additional restriction holds that $\langle I, L\rangle$ is (genuinely) vW-consistent.

Proof. *Soundness* is trivial with respect to the truth definitions employed. For *completeness*, we prove equivalently that if Γ is $SDL + P^+$-consistent then there is a (genuinely) vW-consistent pair $\langle I, L\rangle$ such that all of Γ are true. Let $A_1, A_2, \ldots$ be a fixed enumeration of $\mathscr{L}_{DL+P^+}$. Let $\Delta = \bigcup_n \Delta_n$, where $\Delta_0 = \Gamma$, and

$$\Delta_{n+1} = \begin{cases} \Delta_n \cup \{A_{n+1}\} \text{ if this is } SDL + P^+ - \text{consistent, and} \\ \Delta_n \cup \{\neg A_{n+1}\} \text{ otherwise} \end{cases}$$

Clearly each Δ_n must be $SDL+P^+$-consistent with either A_{n+1} or $\neg A_{n+1}$, hence (i) Δ is $SDL + P^+$-consistent, and (ii) for all $A \in \mathscr{L}_{DL+P^+}$ either $A \in \Delta$ or $\neg A \in \Delta$. Now we define:

$$I = \{A \mid OA \in \Delta\} \qquad L = \left\{A \mid P^+A \in \Delta\right\}$$

Coincidence: $\langle I, L\rangle \models A$ iff $A \in \Delta$ is proved by induction on A. I only give the case for $A = OB$ and $A = P^+B$; the others are trivial:

not the case that drinking *and* driving is permitted according to the same legal system" (p. 104). But the law rarely permits drinking explicitly. It may just not be forbidden. Or it is implied by some general freedom that can be overridden by more specific regulation; then the solution is similar.

[49] An operator of explicit permission P^x may be added to Alchourrón and Bulygin's *LNS* (cf. fn. 42), with (Ext^x), (C^{x+}): if $\nvdash \neg(A \wedge B)$ then $(P^x A \wedge P^x B) \rightarrow P^+(A \wedge B)$, and—since *LNS* admits that a contradiction may be strongly permitted, i.e. $P^+\bot$ is not inconsistent—(P^{x}P$^+$): $P^x A \rightarrow P^+A$. Then IN(A) may derive from more than one P^x-premiss, as in $(O\neg(p \wedge q) \wedge P^x p \wedge P^x q) \rightarrow$ IN($\neg(p \wedge q)$).

[50] That if $\{l(p), l(q)\} \subseteq L$ then $P^a(p \wedge q)$ and $P^e(p \wedge q)$ are true in semantics 2^+ and 3^+, respectively, appears like a late vindication of Weinberger's intuition that $P(p \wedge q)$ should be derivable from Pp and Pq in a logic of norms (Weinberger 1989, p. 249).

Suppose $OB \in \Delta$. Then $B \in I$ and $\langle I, L\rangle \models OB$. Suppose $OB \notin \Delta$, so $\neg OB \in \Delta$. Assume $\langle I, L\rangle \models OB$, so $I \vdash_{PL} B$. If $I = \varnothing$, $\vdash_{PL} B$ and so $\vdash_{SDL+P^+} OB$ by (Ext) and (N), so Δ is $SDL+P^+$-inconsistent. If $I \neq \varnothing$, then by compactness of *PL* there is a non-empty finite set $\{C_1, ..., C_n\} \subseteq I$ such that $\vdash_{PL} (C_1 \wedge ... \wedge C_n) \rightarrow B$. So $\{OC_1, ..., OC_n\} \subset \Delta$. But then we obtain $\vdash_{SDL+P^+} (OC_1 \wedge ... \wedge OC_n) \rightarrow OB$ from (Ext), (C), and (M), so Δ is SDL+P^+-inconsistent.

Suppose $P^+B \in \Delta$. Then $B \in L$ and $\langle I, L\rangle \models P^+B$. Suppose $P^+B \notin \Delta$, so $\neg P^+B \in \Delta$. Assume $\langle I, L\rangle \models P^+B$, so either $I \vdash_{PL} B$ or $\exists l \in L$: $I \cup \{l\} \vdash_{PL} B$. Suppose $I \vdash_{PL} B$. If $I = \varnothing$ then $\vdash_{PL} B$ and so $\vdash_{SDL+P^+} P^+B$ by (Ext) and (P^+), so Δ is SDL+P^+-inconsistent. If $I \neq \varnothing$ then by compactness of *PL* there is a non-empty finite set $\{C_1, ..., C_n\} \subseteq I$ such that $\vdash_{PL} (C_1 \wedge ... \wedge C_n) \rightarrow B$. So $\{OC_1, ..., OC_n\} \subset \Delta$. But from (Ext), (C), and (M) we obtain $\vdash_{SDL+P^+} (OC_1 \wedge ... \wedge OC_n) \rightarrow OB$, and from ($C^+$) and ($P^+$) also $\vdash_{SDL+P^+} (OC_1 \wedge ... \wedge OC_n) \rightarrow P^+B$. So Δ is SDL+P^+-inconsistent. Suppose $I \cup \{D\} \vdash_{PL} B$ for some $l \in L$ with $l = D$. Then, by compactness of *PL* there is a non-empty finite set $\{C_1, ..., C_n\} \subseteq I$ with $\vdash_{PL} (C_1 \wedge ... \wedge C_n \wedge D) \rightarrow B$. So $\{OC_1, ..., OC_n\} \subset \Delta$ and $P^+D \in \Delta$. Then $\vdash_{SDL+P^+} \left(OC_1 \wedge ... \wedge OC_n \wedge P^+D\right) \rightarrow P^+B$ is obtained with (C), (C^+), (Ext) and (M^+), and so Δ is SDL+P^+-inconsistent.

Verification: We must prove that $\langle I, L\rangle$ is (genuinely) vW-consistent. Suppose $I \vdash \bot$. Then by strong completeness of *PL* there is a non-empty finite set $\{C_1, ..., C_n\} \subseteq I$ such that $\vdash_{PL} (C_1 \wedge ... \wedge C_n) \rightarrow \bot$. So $\{OC_1, ..., OC_n\} \subset \Delta$. From (Ext), (C), and (M), we obtain $\vdash_{SDL+P^+} (OC_1 \wedge ... \wedge OC_n) \rightarrow O\bot$. But then Δ is inconsistent since due to (Ext) and (P) $\vdash_{SDL+P^+} \neg O\bot$. Suppose $I \cup \{D\} \vdash_{PL} \bot$ for some $l \in L$ with $l = D$. Then by strong completeness of *PL* there is a non-empty finite set $\{C_1, ..., C_n\} \subseteq I$ such that $\vdash_{PL} (C_1 \wedge ... \wedge C_n \wedge D) \rightarrow \bot$. So $\{OC_1, ..., OC_n\} \subset \Delta$ and $P^+D \in \Delta$. Then with (C), (C^+), (Ext) and (M^+), $\vdash_{SDL+P^+} \left(OC_1 \wedge ... \wedge OC_n \wedge P^+D\right) \rightarrow P^+\bot$, and so Δ is SDL+P^+-inconsistent since due to (N^+), $\vdash_{SDL+P^+} \neg P^+\bot$. □

Theorem 2 (Validities for the three semantics).
All of the formulas in the table are valid in the semantics as indicated in the columns to the right, where a single 'a' or 'e' means the deontic operators are uniformly indexed by 'a' or 'e', and a string containing 'a' and 'e' means the operators are indexed in the order of the string. If the entry is '–', then the formula indexed as in a previous column is not valid. (Loop) means the formula where X is uniformly replaced by one of the indicated operators.

Proof. Validity of all rules and 'monadic' axiom schemes above the line in the table is immediate (cf. Hansen (2005) Theorem 2). Note that due to (CExt) and (DM), in all semantics the following rule of monotonicity for consequents, sometimes called *right weakening*, holds for all dyadic operators X:

$$\text{(RW)} \quad \text{If} \vdash A \rightarrow B \text{ then} \models X(A/C) \rightarrow X(B/C).$$

Semantics 1: Validity of all versions of (Cond), (CCMon) and (RMon) was proved in Hansen (2005) p. 495–496. All other formulas are easily derived in the system DDL^{ea} characterized by the already given valid formulas (Theorem 1 in 2005).
Semantics 2: I first state a lemma, then validities, and finally give counterexamples.

Lemma 1 (Determination of licenses in vW-consistent pairs).
If $\langle \Gamma, \Delta \rangle \in \langle I, L \rangle \perp^{vW} \neg C$, then $\Delta = \Delta^{\underline{n}} \neg C \cup \{l \in L \mid \{l\} \vdash \neg C\}$.

Proof. Immediate from Definition 2 of vW-consistency with C. □

(CCMon^e) Assume $O^e(A \wedge D/C)$, so there is a $\langle \Gamma, \Delta \rangle \in \langle I, L \rangle \perp^{vW} \neg C$ such that $\Gamma \cup \{C\} \vdash A \wedge D$. Then $\Gamma \nvdash \neg(C \wedge D)$ for otherwise $\Gamma \vdash \neg C$, which contradicts vW-consistency of $\langle \Gamma, \Delta \rangle$ with C. So by construction there is a $\langle \Gamma', \Theta \rangle \in \langle I, L \rangle \perp^{vW} \neg(C \wedge D)$ such that $\Gamma \subseteq \Gamma'$, so $\Gamma' \cup \{C\} \vdash A \wedge D$ and $\Gamma' \cup \{C \wedge D\} \vdash A$, hence $O^e(A/C \wedge D)$ is true.

(CCMon^a) Assume $O^a(A \wedge D/C)$, so for all $\langle \Gamma, \Delta \rangle \in \langle I, L \rangle \perp^{vW} \neg C$, $\Gamma \cup \{C\} \vdash A \wedge D$. Suppose $\langle \Gamma', \Delta' \rangle \in \langle I, L \rangle \perp^{vW} \neg(C \wedge D)$. Then $\Gamma' \nvdash \neg C$, for otherwise $\Gamma' \vdash \neg(C \wedge D)$, which contradicts vW-consistency of $\langle \Gamma', \Delta' \rangle$ with $C \wedge D$. Similarly, for all $l \in \Delta'^{\underline{n}} \neg(C \wedge D)$, $\Gamma' \cup \{l\} \nvdash \neg C$, for otherwise $\Gamma' \cup \{l\} \vdash \neg(C \wedge D)$, which is likewise excluded. So by construction there is a $\langle \Gamma, \Delta \rangle \in \langle I, L \rangle \perp^{vW} \neg C$ such that $\Gamma' \subseteq \Gamma$ and $\Delta'^{\underline{n}} \neg(C \wedge D) \subseteq \Delta$. $\Gamma \cup \{C\} \vdash A \wedge D$, so $\Gamma \nvdash \neg(C \wedge D)$ for otherwise $\Gamma \vdash \neg C$, which is excluded by vW-consistency of $\langle \Gamma, \Delta \rangle$ with C. For all $l \in \Delta'^{\underline{n}} \neg(C \wedge D) \subseteq \Delta$, $l \in \Delta^{\underline{n}} \neg C$, for otherwise $\{l\} \vdash \neg(C \wedge D)$, and so similarly $\Gamma \cup \{l\} \nvdash \neg(C \wedge D)$, for otherwise $\Gamma \cup \{l\} \vdash \neg C$, which is excluded by vW-consistency of $\langle \Gamma, \Delta \rangle$ with C. So by construction there is a $\langle \Gamma'', \Delta'' \rangle \in \langle I, L \rangle \perp^{vW} \neg(C \wedge D)$ such that $\Gamma \subseteq \Gamma''$ and $\Delta'^{\underline{n}} \neg(C \wedge D) \subseteq \Delta''$. But then by maximality and Lemma 1, $\Gamma' = \Gamma = \Gamma''$ and $\Delta' = \Delta''$. So $\Gamma' \cup \{C\} \vdash A \wedge D$ and also $\Gamma' \cup \{C \wedge D\} \vdash A$, and so $O^a(A/C \wedge D)$ is true since $\langle \Gamma', \Delta' \rangle$ was arbitrary.

(Cut^a) Assume $O^a(D/C)$ so for all $\langle \Gamma, \Delta \rangle \in \langle I, L \rangle \perp^{vW} \neg C$, $\Gamma \cup \{C\} \vdash D$. Assume $O^a(A/C \wedge D)$ so for all $\langle \Gamma', \Delta' \rangle \in \langle I, L \rangle \perp^{vW} \neg(C \wedge D)$, $\Gamma' \cup \{C \wedge D\} \vdash A$. Suppose $\langle \Gamma, \Delta \rangle \in \langle I, L \rangle \perp^{vW} \neg C$. Since $\Gamma \cup \{C\} \vdash D$ we have $\Gamma \nvdash \neg(C \wedge D)$, for otherwise $\Gamma \vdash \neg C$ which is excluded by vW-consistency of $\langle \Gamma, \Delta \rangle$ with C, and similarly for all $l \in \Delta^{\underline{n}} \neg C$, $\Gamma \cup \{l\} \nvdash \neg(C \wedge D)$, for otherwise $\Gamma \cup \{l\} \vdash \neg C$ which is likewise excluded. So by construction there is a $\langle \Gamma', \Delta' \rangle \in \langle I, L \rangle \perp^{vW} \neg(C \wedge D)$ such that $\Gamma \subseteq \Gamma'$ and $\Delta^{\underline{n}} \neg C \subseteq \Delta'$. Then $\Gamma' \nvdash \neg C$, for otherwise $\Gamma' \vdash \neg(C \wedge D)$, which contradicts vW-consistency of $\langle \Gamma', \Delta' \rangle$ with $C \wedge D$. For all $l \in \Delta^{\underline{n}} \neg C \subseteq \Delta'$, $\{l\} \nvdash \neg(C \wedge D)$, for otherwise $\Gamma \cup \{l\} \vdash \neg C$, which is excluded by vW-consistency of $\langle \Gamma, \Delta \rangle$ with C, and so similarly it must be that $\Gamma' \cup \{l\} \nvdash \neg C$, for otherwise $\Gamma' \cup \{l\} \vdash \neg(C \wedge D)$, which is excluded by vW-consistency of $\langle \Gamma', \Delta' \rangle$ with $C \wedge D$. So by construction there is a $\langle \Gamma'', \Delta'' \rangle \in \langle I, L \rangle \perp^{vW} \neg C$ such that $\Gamma' \subseteq \Gamma''$ and $\Delta^{\underline{n}} \neg C \subseteq \Delta''$. But then by maximality and Lemma 1, $\Gamma = \Gamma' = \Gamma''$ and $\Delta = \Delta''$. So $\Gamma \cup \{C \wedge D\} \vdash A$, and since also $\Gamma \cup \{C\} \vdash D$, we obtain $\Gamma \cup \{C\} \vdash A$, and so $O^a(A/C)$ is true since $\langle \Gamma, \Delta \rangle$ was arbitrary.

(Cut^{aee}) Immediate from (Cond^e) below, (DC^{aee}), (DM), and (ExtC):

$$\dfrac{\dfrac{O^a(D/C) \quad \dfrac{O^e(A/C \wedge D)}{O^e(D \to A/C)}\,\text{Cond}^e}{O^e(D \wedge (D \to A)/C)}\,\text{DC}^{aee}}{\dfrac{O^e(A \wedge D/C)}{O^e(A/C)}\,\text{DM}}\,\text{ExtC}$$

(Cut^{eae}) Assume $O^e(D/C)$, so there is a $\langle \Gamma, \Delta \rangle \in \langle I, L \rangle \bot^{vW} \neg C$ such that $\Gamma \cup \{C\} \vdash D$. Assume $O^a(A/C \wedge D)$, so for all $\langle \Gamma', \Theta \rangle \in \langle I, L \rangle \bot^{vW} \neg(C \wedge D)$, $\Gamma' \cup \{C \wedge D\} \vdash A$. Since $\Gamma \cup \{C\} \vdash D$ we have $\Gamma \nvdash \neg(C \wedge D)$, for otherwise $\Gamma \vdash \neg C$, which is excluded by vW-consistency of $\langle \Gamma, \Delta \rangle$ with C. So by construction there is a $\langle \Gamma', \Theta \rangle \in \langle I, L \rangle \bot^{vW} \neg(C \wedge D)$ such that $\Gamma \subseteq \Gamma'$. Then $\Gamma' \nvdash \neg C$, for otherwise $\Gamma' \vdash \neg(C \wedge D)$, which contradicts vW-consistency of $\langle \Gamma', \Theta \rangle$ with $C \wedge D$. So by construction there is a $\langle \Gamma'', \Lambda \rangle \in \langle I, L \rangle \bot^{vW} \neg C$ such that $\Gamma' \subseteq \Gamma''$. Then $\Gamma'' \cup \{C \wedge D\} \vdash A$, and, since also $\Gamma'' \cup \{C\} \vdash D$, we obtain $\Gamma'' \cup \{C\} \vdash A$, and so $O^e(A/C)$ is true.

(Cond^e) Assume $O^e(A/C \wedge D)$, so there is a $\langle \Gamma, \Delta \rangle \in \langle I, L \rangle \bot^{vW} \neg(C \wedge D)$ s.t. $\Gamma \cup \{C \wedge D\} \vdash A$. So $\Gamma \cup \{C\} \vdash D \to A$. Since $\Gamma \nvdash \neg(C \wedge D)$ by vW-consistency of $\langle \Gamma, \Delta \rangle$, also $\Gamma \nvdash \neg C$. By maximality there is a $\langle \Gamma', \Theta \rangle \in \langle I, L \rangle \bot^{vW} \neg C$ such that $\Gamma \subseteq \Gamma'$, so $\Gamma' \cup \{C\} \vdash D \to A$ and $O^e(D \to A/C)$ is true.

(RMon^e) Assume $O^e(A/C)$, so there is a $\langle \Gamma, \Delta \rangle \in \langle I, L \rangle \bot^{vW} \neg C$ such that $\Gamma \cup \{C\} \vdash A$. Assume $P^e(D/C)$, so for all $\langle \Gamma, \Delta \rangle \in \langle I, L \rangle \bot^{vW} \neg C$, $\Gamma \cup \{C\} \nvdash \neg D$. So also $\Gamma \nvdash \neg(C \wedge D)$. So by construction there is a $\langle \Gamma', \Theta \rangle \in \langle I, L \rangle \bot^{vW} \neg(C \wedge D)$ with $\Gamma \subseteq \Gamma'$. Then $\Gamma' \cup \{C\} \vdash A$, so $\Gamma' \cup \{C \wedge D\} \vdash A$ and $O^e(A/C \wedge D)$ is true.

(RMon^{aae}) Assume $O^a(A/C)$, so for all $\langle \Gamma, \Delta \rangle \in \langle I, L \rangle \bot^{vW} \neg C$, $\Gamma \cup \{C\} \vdash A$. Assume $P^a(D/C)$, so there is a $\langle \Gamma, \Delta \rangle \in \langle I, L \rangle \bot^{vW} \neg C$ such that $\Gamma \cup \{C\} \nvdash \neg D$. So also $\Gamma \nvdash \neg(C \wedge D)$. So by construction there is a $\langle \Gamma', \Theta \rangle \in \langle I, L \rangle \bot^{vW} \neg(C \wedge D)$ such that $\Gamma \subseteq \Gamma'$. Then $\Gamma' \cup \{C\} \vdash A$, so $\Gamma' \cup \{C \wedge D\} \vdash A$ and $O^e(A/C \wedge D)$ is true.

(Or^{aee}) Assume $O^a(A/C)$, so for all $\langle \Gamma, \Delta \rangle \in \langle I, L \rangle \bot^{vW} \neg C$, $\Gamma \cup \{C\} \vdash A$. Assume $O^e(A/D)$ is true, so with (ExtC) also $O^e(A/(C \vee D) \wedge D)$, and with (Cond^e) also $O^e(D \to A/C \vee D)$ is true. So there is a $\langle \Gamma', \Theta \rangle \in \langle I, L \rangle \bot^{vW} \neg(C \vee D)$ with $\Gamma' \cup \{C \vee D\} \vdash D \to A$. Suppose $\Gamma' \vdash \neg C$. Then $\Gamma' \cup \{C \vee D\} \vdash D$ and $\Gamma' \cup \{C \vee D\} \vdash A$. Otherwise $\Gamma' \nvdash \neg C$. Then by construction there is a $\langle \Gamma, \Delta \rangle \in \langle I, L \rangle \bot^{vW} \neg C$ such that $\Gamma' \subseteq \Gamma$. By the first assumption, $\Gamma \cup \{C\} \vdash A$. $\Gamma \nvdash \neg(C \vee D)$, for otherwise $\Gamma \vdash \neg C$, which contradicts vW-consistency of $\langle \Gamma, \Delta \rangle$ with C. So by construction there is a $\langle \Gamma'', \Lambda \rangle \in \langle I, L \rangle \bot^{vW} \neg(C \vee D)$ such that $\Gamma \subseteq \Gamma''$. Since $\Gamma' \subseteq \Gamma \subseteq \Gamma''$, $\Gamma'' \cup \{C \vee D\} \vdash D \to A$ and $\Gamma'' \cup \{C\} \vdash A$, so $\Gamma'' \vdash (C \to A) \wedge (D \to A)$ and $\Gamma'' \cup \{C \vee D\} \vdash A$. So $O^e(A/C \vee D)$ is true.

(DR^e) The proof is immediate from (CCMon^e) and (RMon^e):

$$\dfrac{\dfrac{\dfrac{O^e(A/C \vee D) \qquad \dfrac{\dfrac{\dfrac{P^e(\neg A/C)}{P^e(C \to \neg A/C \vee D)}\ \text{ExtC, CCMon}^e}{P^e((C \to \neg A) \wedge ((C \wedge \neg A) \vee D)/C \vee D)}\ \text{CExt}}{P^e((C \wedge \neg A) \vee D/C \vee D)}\ \text{RW}}{O^e(A/(C \wedge \neg A) \vee D)}\ \text{RMon}^e\text{, ExtC}}{O^e(A \wedge D/(C \wedge \neg A) \vee D)}\ \text{CExt}}{O^e(A/D)}\ \text{CCMon}^e\text{, ExtC}$$

(DR^{aea}) In the sketched proof for (DR^e) replace the top formulas by their a-versions, replace the upper application of (CCMon^e) with (CCMon^a), and replace the application of (RMon^e) with (RMon^{aae}).

(P^e-Loop) (P^e-Loop) can be derived with (RMon^e), (CCMon^e), and (Cond^e), whose validity was already demonstrated. I leave the proof for the reader (cf. Hansen (2005), Spohn (1975)).

(O^a-Loop) I will show that (O^a-Loop) is valid by first proving that a stronger version of (Cut), named (Cut$^+$), is valid for the operator O^a in semantics 2, which provides a limited version of (Cond):

$$(\text{Cut}^+)\ \mathrm{O^a(C/D) \to (O^a(A/C) \to O^a(C \to A/C \vee D))}$$

Proof. Assume $O^a(C/D)$, so for all $\langle \Gamma, \Delta \rangle \in \langle I, L \rangle \bot^{vW} \neg D$, $\Gamma \cup \{D\} \vdash C$. Assume $O^a(A/C)$, so for all $\langle \Gamma', \Delta' \rangle \in \langle I, L \rangle \bot^{vW} \neg C$, $\Gamma' \cup \{C\} \vdash A$. Suppose $\langle \Theta, \Lambda \rangle \in \langle I, L \rangle \bot^{vW} \neg(C \vee D)$. If $\Theta \vdash \neg C$, then $\Theta \cup \{C \vee D\} \vdash C \to A$ holds trivially. So suppose $\Theta \nvdash \neg C$. If there are $l \in \Lambda^{\underline{n}} \neg(C \vee D)$ such that $\Theta \cup \{l\} \vdash \neg C$, then $\Theta \cup \{l\} \nvdash \neg D$ due to vW-consistency of $\langle \Theta, \Lambda \rangle$ with $C \vee D$. So then there is by construction a $\langle \Gamma, \Delta \rangle \in \langle I, L \rangle \bot^{vW} \neg D$ such that $\Theta \subseteq \Gamma$ and $l \in \Delta$. But by the first assumption $\Gamma \cup \{D\} \vdash C$, so then there would be a $l \in \Delta^{\underline{n}} \neg D$ such that $\Gamma \cup \{l\} \vdash \neg D$, contrary to vW-consistency of $\langle \Gamma, \Delta \rangle$ with D. So for all $l \in \Lambda^{\underline{n}} \neg(C \vee D)$, $\Theta \cup \{l\} \nvdash \neg C$. So by construction there must be a $\langle \Gamma', \Delta' \rangle \in \langle I, L \rangle \bot^{vW} \neg C$ such that $\Theta \subseteq \Gamma'$, $\Lambda^{\underline{n}} \neg(C \vee D) \subseteq \Delta'$. Then $\Gamma' \nvdash \neg(C \vee D)$, for otherwise $\Gamma' \vdash \neg C$, contrary to vW-consistency of $\langle \Gamma', \Delta' \rangle$ with C. For all $l \in \Lambda^{\underline{n}} \neg(C \vee D) \subseteq \Delta'$, $\{l\} \nvdash \neg C$ (as just proved), so similarly $\Gamma \cup \{l\} \nvdash \neg(C \vee D)$ due to vW-consistency of $\langle \Gamma', \Delta' \rangle$ with C, and so by construction there is a $\langle \Theta', \Lambda' \rangle \in \langle I, L \rangle \bot^{vW} \neg(C \vee D)$ such that $\Gamma' \subseteq \Theta'$ and $\Lambda^{\underline{n}} \neg(C \vee D) \subseteq \Lambda'$. Then by maximality and Lemma 1, $\Theta = \Gamma' = \Theta'$ and $\Lambda = \Lambda'$, and so $\Theta \cup \{C\} \vdash A$ and $\Theta \cup \{C \vee D\} \vdash C \to A$. So $O^a(C \to A/C \vee D)$ is true since $\langle \Theta, \Lambda \rangle$ was arbitrary. □

Returning to (O^a-Loop), first we prove that $O^a(A_i \to A_j/A_1 \vee \ldots \vee A_n)$ for any i, j such that $O^a(A_j/A_i)$ is in the loop, as demonstrated below for $O^a(A_1/A_n)$.

Lines that apply (Cut$^+$) are marked '+'.

$$\text{RW}\ \dfrac{\dfrac{\dfrac{O^a(A_{n-1}/A_{n-2})}{O^a(A_{n-1} \vee A_n/A_{n-2})}\ \text{RW} \qquad \dfrac{O^a(A_n/A_{n-1}) \qquad O^a(A_1/A_n)}{O^a(A_n \rightarrow A_1/A_{n-1} \vee A_n)}\ +}{O^a((A_{n-1} \vee A_n) \rightarrow (A_n \rightarrow A_1)/A_{n-2} \vee A_{n-1} \vee A_n)}\ +}{O^a(A_n \rightarrow A_1/A_{n-2} \vee A_{n-1} \vee A_n)}$$

$$\vdots$$

$$\vdots$$

$$\dfrac{\dfrac{\dfrac{O^a(A_2/A_1)}{O^a(A_2 \vee A_3 \vee \ldots \vee A_n/A_1)}\ \text{RW} \qquad \vdots \atop O^a(A_n \rightarrow A_1/A_2 \vee A_3 \vee \ldots \vee A_n)}{O^a((A_2 \vee A_3 \vee \ldots \vee A_n) \rightarrow (A_n \rightarrow A_1)/A_1 \vee \ldots \vee A_n)}\ +}{O^a(A_n \rightarrow A_1/A_1 \vee \ldots \vee A_n)}\ \text{RW}$$

Then use (DC) to obtain $O^a(A_1 \wedge \ldots \wedge A_n/A_1 \vee \ldots \vee A_n)$:

$$\dfrac{\dfrac{O^a(A_1 \rightarrow A_2/A_1 \vee \ldots \vee A_n) \qquad \ldots \qquad O^a(A_n \rightarrow A_1/A_1 \vee \ldots \vee A_n)}{O^a((A_1 \rightarrow A_2) \wedge \ldots \wedge (A_{n-1} \rightarrow A_n) \wedge (A_n \rightarrow A_1)/A_1 \vee \ldots \vee A_n)}\ \text{DC}}{O^a(A_1 \wedge \ldots \wedge A_n/A_1 \vee \ldots \vee A_n)}\ \text{CExt}$$

From this we then get $O^a(A_j/A_i)$ for any i, j such that $1 \leq i \leq n$, $1 \leq j \leq n$ as desired:

$$\dfrac{\dfrac{O^a(A_1 \wedge \ldots \wedge A_n/A_1 \vee \ldots \vee A_n)}{O^a(A_1 \wedge \ldots \wedge A_n/A_i)}\ \text{CCMon, ExtC}}{O^a(A_j/A_i)}\ \text{DM}$$

Counterexamples that refute the remaining formulas are listed below:

Scheme	Refuted instance for semantics 2	I	L
(Cond)	$O^a(p/q) \rightarrow O^a(q \rightarrow p/\top)$	$\{p \wedge q\}$	$\{\neg q\}$
(RMon)	$P^e(p/\top) \rightarrow (O^a(q/\top) \rightarrow O^a(q/p))$	$\{q\}$	$\{q \rightarrow \neg p\}$
(Or)	$(O^a(r/p) \wedge O^a(r/q)) \rightarrow O^a(r/p \vee q)$	$\{r \wedge p, r \wedge q\}$	$\{\neg p, \neg q\}$

Semantics 3: I first state two lemmas, then validities, and finally counterexamples.

Lemma 2 (Properties of naïve contraction).
(1) $\Gamma \underline{n} \neg(A \vee B) = \Gamma \underline{n} \neg A \cup \Gamma \underline{n} \neg B$
(2) $\Gamma \underline{n} \neg(A \wedge B) \subseteq \Gamma \underline{n} \neg A \cap \Gamma \underline{n} \neg B$

Proof. (1) Left to right: If $l \notin \Gamma \underline{n} \neg A \cup \Gamma \underline{n} \neg B$, then $l \notin \Gamma \underline{n} \neg A$ and $l \notin \Gamma \underline{n} \neg B$. So $\{l\} \vdash \neg A$ and $\{l\} \vdash \neg B$, and so $\{l\} \vdash \neg(A \vee B)$. So $l \notin \Gamma \underline{n} \neg(A \vee B)$. Right to left: If $l \notin \Gamma \underline{n} \neg(A \vee B)$, then $\{l\} \vdash \neg A$ and $\{l\} \vdash \neg B$. Hence $l \notin \Gamma \underline{n} \neg A$ and $l \notin \Gamma \underline{n} \neg B$ and so $l \notin \Gamma \underline{n} \neg A \cup \Gamma \underline{n} \neg B$. (2) Assume $l \notin \Gamma \underline{n} \neg A \cap \Gamma \underline{n} \neg B$. So either

$l \notin \Gamma \underline{n} \neg A$ and $\{l\} \vdash \neg A$, or $l \notin \Gamma \underline{n} \neg B$ and $\{l\} \vdash \neg B$. In either case $\{l\} \vdash \neg(A \wedge B)$ and so $l \notin \Gamma \underline{n} \neg(A \wedge B)$. □

Lemma 3 (Property of obligation in semantics 3).
If $\langle I, L\rangle \models O^a(D/C)$ or $\langle I, L\rangle \models O^e(D/C)$, then $L \underline{n} \neg C \subseteq L \underline{n} \neg D$.

Proof. If $\vdash \neg C$, then $L \underline{n} \neg C = \varnothing$ and the inclusion is trivial. Otherwise there exists some $\langle \Gamma, L \underline{n} \neg C\rangle \in \langle I, L\rangle \perp^{AB} \neg C$ such that $\Gamma \cup \{C\} \vdash D$. If there exists a $l \in L \underline{n} \neg C$ such that $\{l\} \vdash \neg D$, then $\Gamma \cup \{l\} \vdash \neg C$. But this violates the second half of clause (i) in the definition of $\langle I, L\rangle \perp^{AB} \neg C$. □

(CCMon^e) Assume $O^e(A \wedge D/C)$, so there is a $\langle \Gamma, L\rangle \in \langle I, L\rangle \perp^{AB} \neg C$ such that $\Gamma \cup \{C\} \vdash A \wedge D$. Due to (M) $O^e(D/C)$ is true, so $L \underline{n} \neg C \subseteq L \underline{n} \neg(C \wedge D)$ by Lemma 3, and hence by Lemma 2 $L \underline{n} \neg C = L \underline{n} \neg(C \wedge D)$. Furthermore, $\Gamma \nvdash \neg(C \wedge D)$. for otherwise $\Gamma \vdash \neg C$ with the assumption and contrary to νW-consistency. For the same reason, for all $l \in L \underline{n} \neg C = L \underline{n} \neg(C \wedge D)$, $\Gamma \cup \{l\} \nvdash \neg(C \wedge D)$. So by construction there is a $\langle \Delta, L\rangle \in \langle I, L\rangle \perp^{AB} \neg(C \wedge D)$ such that $\Gamma \subseteq \Delta$, and then $\Delta \cup \{C\} \vdash A \wedge D$ and also $\Delta \cup \{C \wedge D\} \vdash A$ and hence $O^e(A/C \wedge D)$ is true.

(CCMon^a) Assume $O^a(A \wedge D/C)$, so for all $\langle \Gamma, L\rangle \in \langle I, L\rangle \perp^{AB} \neg C$, $\Gamma \cup \{C\} \vdash A \wedge D$. Due to (M) $O^a(D/C)$ is true, so by Lemma 3 $L \underline{n} \neg C \subseteq L \underline{n} \neg(C \wedge D)$ and hence by Lemma 2 $L \underline{n} \neg C = L \underline{n} \neg(C \wedge D)$. Suppose $\langle \Delta, L\rangle \in \langle I, L\rangle \perp^{AB} \neg(C \wedge D)$. $\Delta \nvdash \neg C$, for otherwise $\Delta \vdash \neg(C \wedge D)$. Similarly, for all $l \in L \underline{n} \neg(C \wedge D) = L \underline{n} \neg C$, $\Delta \cup \{l\} \nvdash \neg C$, for otherwise $\Delta \cup \{l\} \vdash \neg(C \wedge D)$. So by construction there is a $\langle \Gamma, L\rangle \in \langle I, L\rangle \perp^{AB} \neg C$ such that $\Delta \subseteq \Gamma$. Then by the same reasoning as in the proof of (CCMon^e) there must be a $\langle \Delta', L\rangle \in \langle I, L\rangle \perp^{AB} \neg(C \wedge D)$ such that $\Gamma \subseteq \Delta'$. By maximality $\Delta = \Gamma = \Delta'$. So $\Delta \cup \{C\} \vdash A \wedge D$ and $\Delta \cup \{C \wedge D\} \vdash A$. Since Δ was arbitrary, $O^a(A/C \wedge D)$ is true.

(Cut^a) Assume $O^a(D/C)$, so for all $\langle \Gamma, L\rangle \in \langle I, L\rangle \perp^{AB} \neg C$, $\Gamma \cup \{C\} \vdash D$. By Lemma 3 $L \underline{n} \neg C \subseteq L \underline{n} \neg(C \wedge D)$ and hence $L \underline{n} \neg C = L \underline{n} \neg(C \wedge D)$ by Lemma 2. Assume $O^a(A/C \wedge D)$, so for all $\langle \Delta, L\rangle$ in $\langle I, L\rangle \perp^{AB} \neg(C \wedge D)$, $\Delta \cup \{C \wedge D\} \vdash A$. Suppose $\langle \Gamma, L\rangle \in \langle I, L\rangle \perp^{AB} \neg C$. The first assumption ensures $\Gamma \cup \{C\} \vdash D$ and so $\Gamma \nvdash \neg(C \wedge D)$, for otherwise $\Gamma \vdash \neg C$, contrary to νW-consistency. Similarly, for all $l \in L \underline{n} \neg C = L \underline{n} \neg(C \wedge D)$, $\Gamma \cup \{l\} \nvdash \neg(C \wedge D)$, for otherwise $\Gamma \cup \{l\} \nvdash \neg C$. So by construction there is a $\langle \Delta, L\rangle \in \langle I, L\rangle \perp^{AB} \neg(C \wedge D)$ such that $\Gamma \subseteq \Delta$. $\Delta \nvdash \neg C$ for otherwise $\Delta \vdash \neg(C \wedge D)$, and similarly for all $l \in L \underline{n} \neg(C \wedge D) = L \underline{n} \neg C$, $\Delta \cup \{l\} \nvdash \neg C$, for otherwise $\Delta \cup \{l\} \vdash \neg(C \wedge D)$. So by construction there is a $\langle \Gamma', L\rangle \in \langle I, L\rangle \perp^{AB} \neg C$ such that $\Delta \subseteq \Gamma'$. But then by maximality $\Gamma = \Delta = \Gamma'$ and so $\Gamma \cup \{C \wedge D\} \vdash A$ and $\Gamma \cup \{C\} \vdash D$, and so $\Gamma \cup \{C\} \vdash A$. Since Γ was arbitrary, $O^a(A/C)$ is true.

(Cut^{aee}) Assume $O^a(D/C)$, so for all $\langle \Gamma, L\rangle \in \langle I, L\rangle \perp^{AB} \neg C$, $\Gamma \cup \{C\} \vdash D$. By Lemma 3 $L \underline{n} \neg C \subseteq L \underline{n} \neg(C \wedge D)$ and hence by Lemma 2 $L \underline{n} \neg C = L \underline{n} \neg(C \wedge D)$. Assume $O^e(A/C \wedge D)$, so there is a $\langle \Delta, L\rangle \in \langle I, L\rangle \perp^{AB} \neg(C \wedge D)$ such that $\Delta \cup \{C \wedge D\} \vdash A$. By νW-consistency, $\Delta \nvdash \neg(C \wedge D)$ and so also $\Delta \nvdash \neg C$. Likewise $\Delta \cup \{l\} \nvdash \neg(C \wedge D)$ for all $l \in L \underline{n} \neg(C \wedge D) = L \underline{n} \neg C$ and so also

$\Delta \cup \{l\} \nvdash \neg C$ for any such l. So by construction there is a $\langle \Gamma, L \rangle \in \langle I, L \rangle \bot^{AB} \neg C$ such that $\Delta \subseteq \Gamma$, so $\Gamma \cup \{C \wedge D\} \vdash A$ and $\Gamma \cup \{C\} \vdash D$. So $\Gamma \cup \{C\} \vdash A$ and $O^e(A/C)$ is true.

(Cut^{eae}) Assume $O^a(A/C \wedge D)$, so for all $\langle \Delta, L \rangle \in \langle I, L \rangle \bot^{AB} \neg(C \wedge D)$, $\Delta \cup \{C \wedge D\} \vdash A$. Assume $O^e(D/C)$, so for some $\langle \Gamma, L \rangle \in \langle I, L \rangle \bot^{AB} \neg C$, $\Gamma \cup \{C\} \vdash D$. By Lemma 3 $L^{\underline{n}} \neg C \subseteq L^{\underline{n}} \neg(C \wedge D)$, and hence by Lemma 2 $L^{\underline{n}} \neg C = L^{\underline{n}} \neg(C \wedge D)$. $\Gamma \nvdash \neg(C \wedge D)$, for otherwise $\Gamma \vdash \neg C$, contrary to vW-consistency, and similarly for all $l \in L^{\underline{n}} \neg C = L^{\underline{n}} \neg(C \wedge D)$, $\Gamma \cup \{l\} \nvdash \neg(C \wedge D)$, for otherwise $\Gamma \cup \{l\} \vdash \neg C$. So by construction, for some $\langle \Delta, L \rangle \in \langle I, L \rangle \bot^{AB} \neg(C \wedge D)$, $\Gamma \subseteq \Delta$. By definition $\Delta \nvdash \neg(C \wedge D)$ and so also $\Delta \nvdash \neg C$, and likewise $\Delta \cup \{l\} \nvdash \neg(C \wedge D)$ for all $l \in L^{\underline{n}} \neg(C \wedge D) = L^{\underline{n}} \neg C$ and therefore also $\Delta \cup \{l\} \nvdash \neg C$ for these l. So again by the construction, for some $\langle \Gamma', L \rangle \in \langle I, L \rangle \bot^{AB} \neg C$, $\Delta \subseteq \Gamma'$. Then by maximality $\Gamma = \Delta = \Gamma'$, so $\Gamma \cup \{C \wedge D\} \vdash A$ and since also $\Gamma \cup \{C\} \vdash D$ we obtain $\Gamma \cup \{C\} \vdash A$ and so $O^e(A/C)$ is true.

(O^a-Loop) I only consider $n = 3$, expansions to higher n are immediate[51]: Assume (a) $O^a(A_2/A_1)$, so for all $\langle \Gamma, L \rangle \in \langle I, L \rangle \bot^{AB} \neg A_1$, we have $\Gamma \cup \{A_1\} \vdash A_2$. Assume (b) $O^a(A_3/A_2)$, so for all $\langle \Delta, L \rangle \in \langle I, L \rangle \bot^{AB} \neg A_2$, we have $\Delta \cup \{A_2\} \vdash A_3$. Assume (c) $O^a(A_1/A_3)$, so for all $\langle \Theta, L \rangle \in \langle I, L \rangle \bot^{AB} \neg A_3$, we have $\Theta \cup \{A_3\} \vdash A_1$. By (a), (b), (c) and Lemma 3, $L^{\underline{n}} \neg A_1 \subseteq L^{\underline{n}} \neg A_2$, $L^{\underline{n}} \neg A_2 \subseteq L^{\underline{n}} \neg A_3$, and $L^{\underline{n}} \neg A_3 \subseteq L^{\underline{n}} \neg A_1$. Hence $L^{\underline{n}} \neg A_1 = L^{\underline{n}} \neg A_2 = L^{\underline{n}} \neg A_3$.
Suppose $\langle \Gamma, L \rangle \in \langle I, L \rangle \bot^{AB} \neg A_1$. Then $\Gamma \nvdash \neg A_2$, since otherwise $\Gamma \vdash \neg A_1$ due to (a), which is excluded by vW-consistency. Likewise, for all $l \in L^{\underline{n}} \neg A_2$, $\Gamma \cup \{l\} \nvdash \neg A_2$, for otherwise $\Gamma \cup \{l\} \vdash \neg A_1$ for some $l \in L^{\underline{n}} \neg A_1$, which is excluded by vW-consistency. By construction, $\Gamma \subseteq \Delta$ for some $\langle \Delta, L \rangle \in \langle I, L \rangle \bot^{AB} \neg A_2$.
Consider this $\langle \Delta, L \rangle \in \langle I, L \rangle \bot^{AB} \neg A_2$. Then $\Delta \nvdash \neg A_3$, since otherwise $\Delta \vdash \neg A_2$ due to (b), which is excluded by vW-consistency. Again for all $l \in L^{\underline{n}} \neg A_3$, $\Delta \cup \{l\} \nvdash \neg A_3$, for otherwise $\Delta \cup \{l\} \vdash \neg A_2$ for some $l \in L^{\underline{n}} \neg A_2$, which is excluded by vW-consistency. By construction, $\Delta \subseteq \Theta$ for some $\langle \Theta, L \rangle \in \langle I, L \rangle \bot^{AB} \neg A_3$.
Consider this $\langle \Theta, L \rangle \in \langle I, L \rangle \bot^{AB} \neg A_3$. Then $\Theta \nvdash \neg A_1$, since otherwise $\Theta \vdash \neg A_3$ due to (c), which is excluded by vW-consistency. Again for all $l \in L^{\underline{n}} \neg A_1$, $\Theta \cup \{l\} \nvdash \neg A_1$, for otherwise $\Theta \cup \{l\} \vdash \neg A_3$ for some $l \in L^{\underline{n}} \neg A_3$, which is excluded by vW-consistency. By construction, $\Theta \subseteq \Gamma'$ for some $\langle \Gamma', L \rangle \in \langle I, L \rangle \bot^{AB} \neg A_1$.
But then $\Gamma = \Delta = \Theta = \Gamma'$ by maximality. So we have $\Gamma \cup \{A_1\} \vdash A_2$ and $\Gamma \cup \{A_2\} \vdash A_3$ and therefore $\Gamma \cup \{A_1\} \vdash A_3$ by transitivity of classical consequence. Since Γ was arbitrary, $O^a(A_3/A_1)$ is true.

[51] (Cut$^+$), used above to prove validity of (O^a-Loop), is not valid here. Its instance $O^a(p/q) \rightarrow (O^a(r/p) \rightarrow O^a(p \rightarrow r/p \vee q))$ is refuted by $I = \{p \wedge \neg s, r \wedge \neg s, s\}$ and $L = \{s \rightarrow \neg q, s \rightarrow \neg p\}$.

Counterexamples that refute the remaining formulas are listed below:

Scheme	Refuted instance for semantics 3	I	L
(Cond)	$O^e(p/q) \rightarrow O^e(q \rightarrow p/\top)$	$\{p \wedge q\}$	$\{\neg q\}$
	$O^a(p/q) \rightarrow O^a(q \rightarrow p/\top)$	–″–	–″–
(RMon)	$P^e(p/\top) \rightarrow (O^e(q/\top) \rightarrow O^e(q/p))$	$\{q\}$	$\{q \rightarrow \neg p\}$
	$P^e(p/\top) \rightarrow (O^a(q/\top) \rightarrow O^a(q/p))$	–″–	–″–
	$P^a(p/\top) \rightarrow (O^a(q/\top) \rightarrow O^e(q/p))$	–″–	–″–
(Or)	$(O^a(r/p) \wedge O^a(r/q)) \rightarrow O^a(r/p \vee q)$	$\{r \wedge p, r \wedge q\}$	$\{\neg p, \neg q\}$
	$(O^a(r/p) \wedge O^e(r/q)) \rightarrow O^e(r/p \vee q)$	–″–	–″–
(DR)	$O^e(r/p \vee q) \rightarrow (O^e(r/p) \vee O^e(r/q))$	$\{r\}$	$\{r \rightarrow \neg p, r \rightarrow \neg q\}$
	$O^a(r/p \vee q) \rightarrow (O^e(r/p) \vee O^a(r/q))$	–″–	–″–
(Loop)	$P^e(q/p) \wedge P^e(r/q) \wedge P^e(p/r) \rightarrow P^e(r/p)$	$\{\neg r\}$	$\{q \rightarrow r\}$

□

Theorem 3 (Soundness and Completeness of SDL^aP^e).
For all the semantics 1–3, the monadic fragment—the valid formulas in which each occurrence of a stroke '/' is followed by '$\top$)'—equals the axiomatic theory SDL^aP^e defined by the axiom schemes and rules above the middle line of the table that are indexed as in Theorem 2 and where all occurrences of C and D are replaced by $\top$ (and $\models$ is replaced by $\vdash_{SDL^aP^e}$).

Proof. Soundness of all axiom schemes and rules of SDL^aP^e for semantics 1–3 is immediate. The completeness proof is a straightforward adaptation of the similar proof for DDL^{ea} in Hansen (2005). Sketch: Let δ be the SDL^aP^e-consistent disjunct in a disjunctive normal form of the sentence A we need to model. Let the fragment $\mathcal{L}^{\delta}_{PL}$ of $\mathcal{L}_{PL}$ contain only proposition letters occurring in δ. Let $r(\mathcal{L}^{\delta}_{PL})$ be a (finite) set of mutually non-equivalent representatives in $\mathcal{L}^{\delta}_{PL}$. By a propositional formula we now mean its equivalent representative. Construct a SDL^aP^e-consistent set Δ that includes all conjuncts of δ and additionally either $P^e\neg B$ or $O^e\neg B$, and either $P^a\neg B$ or $O^a\neg B$, for any $B \in r(\mathcal{L}^{\delta}_{PL})$. The 'deontic bases' in Δ are then identified:

- $\mathcal{O}^a = \bigwedge\{A \in r(\mathcal{L}^{\delta}_{PL}) \mid O^a A \in \Delta\}$,
- $\mathbb{O}^e = min\,\{A \in r(\mathcal{L}^{\delta}_{PL}) \mid O^e A \in \Delta\}$.

where $min\,\Gamma = \{A \in \Gamma \mid \forall B \in \Gamma, \text{if} \vdash_{PL} B \rightarrow A, \text{then} \vdash_{PL} B \leftrightarrow A\}, \Gamma \subseteq \mathcal{L}_{PL}$.

The last step is to make all deontic bases inconsistent with each other. Let $\phi : \{\mathcal{O}^a\} \cup \mathbb{O}^e \rightarrow \mathcal{L}_{PL}$ be a function that associates with each of the bases a sentence built only from proposition letters not occurring in δ such that for each $B \neq C$, $\{\phi(B)\} \nvdash_{PL} \bot$ and $\{\phi(B), \phi(C)\} \vdash_{PL} \bot$. We then define

$$I = \{\ulcorner \mathcal{O}^a \wedge \phi(\mathcal{O}^a)\urcorner\} \cup \{\ulcorner \mathcal{O}^e \wedge \phi(\mathcal{O}^e)\urcorner \mid \mathcal{O}^e \in \mathbb{O}^e\}.$$

For *coincidence*, prove first that $I\bot\neg\top = \{\{\mathcal{O} \wedge \phi(\mathcal{O})\} \mid \mathcal{O} \in \{\mathcal{O}^a\} \cup \mathbb{O}^e\}$. Non-vacuity of I follows easily from (DN), singularity from construction of $\phi(\mathcal{O})$, and consistency of $\mathcal{O} \wedge \phi(\mathcal{O})$ from (CExt), (DM), and (DP). Show for semantics 1 that for all $B \in r(\mathcal{L}^{\delta}_{PL})$, $I \models O^e B$ iff $O^e B \in \Delta$, and $I \models O^a B$ iff $O^a B \in \Delta$. The left-to-right version for O^e is immediate from the construction, and for O^a goes through with (DC^{aee}). For the right to left versions assume $O^* B \notin \Delta$ and suppose for *r.a.a.* that $I \models O^* B$. Inconsistency of Δ follows then from (CExt) and (DM) (for $*$=e), and from (CExt), (DM), and (DC) (for $*$=a). For semantics 2 and 3 (or 4) use $\langle I, \varnothing \rangle$ (or $\langle I, \varnothing, f, < \rangle$, with $f(i) = i$ and $<= \varnothing$). □

Theorem 4 (Relation between semantics 2 [3] and 2$^+$ [3$^+$]).
1. If $A \in \mathcal{L}_{DDL^{ea}}$ is valid in semantics 2$^+$ [3$^+$] then A is valid in semantics 2 [3].
2. If $A \in \mathcal{L}_{DDL^{ea}}$ is valid in semantics 3 then A is valid in semantics 3$^+$.
3. If $A \in \mathcal{L}_{DDL^{e}}$ is valid in semantics 2 then A is valid in semantics 2$^+$.
4. All validities listed in Sect. 3.4 for semantics 2 remain valid in semantics 2$^+$.

Proof. Ad (1): We prove equivalently that if there is a $\langle I, L \rangle$ such that $\langle I, L \rangle \models A$ for some $A \in \mathcal{L}_{DDL^{ea}}$ in semantics 2 [3], then we can construct a $\langle I, L^* \rangle$ such that $\langle I, L^* \rangle \models A$ in semantics 2$^+$[3$^+$]. The basic idea is simple: we can force semantics 2$^+$[3$^+$] to consider licenses in L individually making them inconsistent with each other. The trick is to do so without changing the evaluation of A.

To make licenses inconsistent with each other, let $\mathcal{L}^{A}_{PL}$ be the fragment of the language of propositional logic that contains only sentences composed of proposition letters occurring in A, and let $\mathcal{L}^{I,L,A}_{PL}$ be the similar fragment for proposition letters occurring in $I \cup L \cup \{A\}$. Let a function

$$\phi : \mathcal{L}^{I,L,A}_{PL} \rightarrow \mathcal{L}_{PL}$$

associate with each sentence $B \in \mathcal{L}^{I,L,A}_{PL}$ a sentence $\phi(B)$ composed only of proposition letters *not* occurring in $I \cup L \cup \{A\}$, such that $\{\phi(B)\} \nvdash \bot$ and for $B \neq C \in \mathcal{L}^{I,L,A}_{PL}$, $\{\phi(B), \phi(C)\} \vdash \bot$.

The construction presupposes that there is a (possibly infinite) supply of proposition letters available for the range of ϕ that are not occurring in A or any sentence in I or L. Of course A has finite length due to the definition of $\mathcal{L}_{PL}$. But may not $I \cup L$, together with A, 'use up' all proposition letters in *Prop*? Based on compactness of *PL* one might conjecture that since A is finite only a finite number of sentences in I, L are required by the truth definitions to make A true. But presently we can always uniformly replace in all sentences of I and L all proposition letters p_n not occurring in A by, say, p_{n*2}. Let the thus modified sets be the new I, L. Obviously still $\langle I, L \rangle \models A$. This leaves all odd-numbered proposition letters of *Prop* not occurring in A at our disposal. Let $q_1, q_2, q_3 ...$ be an enumeration of these. Then, for example, the sequence $q_1, \ulcorner \neg q_1 \wedge q_2 \urcorner, \ulcorner \neg q_1 \wedge \neg q_2 \wedge q_3 \urcorner, ...$ suffices for the range of ϕ.

For any $\Delta \subseteq L$, let

$$\Delta^* = \{\ulcorner l \wedge \phi(l) \urcorner \mid l \in \Delta\}.$$

We have to prove that $\langle I, L^* \rangle \models A$ in semantics 2^+ [3^+], which is done by induction on the construction of A. The only interesting cases are the deontic operators (B, C are in $\mathscr{L}^A_{PL}$, and '$\models$' is indexed by the respective semantics for clarity):

Semantics $2/2^+$:

$O^e(B/C)$ Assume $\langle I, L \rangle \models_2 O^e(B/C)$, so there is a $\langle \Gamma, \Delta \rangle \in \langle I, L \rangle \bot^{vW} \neg C$ such that $\Gamma \cup \{C\} \vdash B$. $\Gamma \nvdash \neg C$, so by construction there is a $\langle \Gamma', \Theta \rangle \in \langle I, L^* \rangle \bot^{vW^+} \neg C$ with $\Gamma \subseteq \Gamma'$, so $\Gamma' \cup \{C\} \vdash B$ and so $\langle I, L \rangle \models_{2^+} O^e(B/C)$. The opposite direction that if $\langle I, L \rangle \models_{2^+} O^e(B/C)$, then $\langle I, L \rangle \models_2 O^e(B/C)$, is proved in exactly the same way.

$O^a(B/C)$ Assume $\langle I, L \rangle \models_2 O^a(B/C)$, so for all $\langle \Gamma, \Delta \rangle \in \langle I, L \rangle \bot^{vW} \neg C$, $\Gamma \cup \{C\} \vdash B$. Suppose $\langle \Theta, \Lambda^* \rangle \in \langle I, L^* \rangle \bot^{vW^+} \neg C$. $\Theta \nvdash \neg C$, and for all $\{l \wedge \phi(l)\} \in \Lambda^* \bot \neg C$, $\Theta \cup \{l \wedge \phi(l)\} \nvdash \neg C$, so also $\Theta \cup \{l\} \nvdash \neg C$ for all $l \in \Lambda$ unless $\vdash l \to \neg C$, so there is a $\langle \Gamma, \Delta \rangle \in \langle I, L \rangle \bot^{vW} \neg C$ such that $\Theta \subseteq \Gamma$ and $\Lambda \subseteq \Delta$. Suppose there is an $i \in \Gamma$ such that $i \notin \Theta$. Since $\Theta \cup \{i\} \subseteq \Gamma \nvdash \neg C$ there must be a $\{l \wedge \phi(l)\} \in \Lambda^* \bot \neg C$ such that $\Theta \cup \{i\} \cup \{l \wedge \phi(l)\} \vdash \neg C$. Since $\{l \wedge \phi(l)\} \in \Lambda^* \bot \neg C$, also $\{l\} \nvdash \neg C$, so there is a $l \in \Lambda \underline{n} \neg C$ such that $\Gamma \cup \{l\} \vdash \neg C$, which contradicts vW-consistency of $\langle \Gamma, \Delta \rangle$. So $\Gamma \subseteq \Theta$, hence $\Theta \cup \{C\} \vdash B$ by the assumption, and $\langle I, L^* \rangle \models_{2^+} O^a(B/C)$ since $\langle \Theta, \Lambda^* \rangle$ was arbitrary.

Assume $\langle I, L^* \rangle \models_{2^+} O^a(B/C)$, so for all $\langle \Gamma, \Delta^* \rangle \in \langle I, L^* \rangle \bot^{vW^+} \neg C$, $\Gamma \cup \{C\} \vdash B$. Suppose $\langle \Theta, \Lambda \rangle \in \langle I, L \rangle \bot^{vW} \neg C$. $\Theta \nvdash \neg C$, and for all $l \in \Lambda \underline{n} \neg C$, $\Theta \cup \{l\} \nvdash \neg C$, so also $\Theta \cup \{l \wedge \phi(l)\} \nvdash \neg C$, so there is a $\langle \Gamma, \Delta^* \rangle \in \langle I, L^* \rangle \bot^{vW^+} \neg C$ such that $\Theta \subseteq \Gamma$ and $\Lambda^* \subseteq \Delta^*$. Suppose there is an $i \in \Gamma$ such that $i \notin \Theta$. Since $\Theta \cup \{i\} \subseteq \Gamma \nvdash \neg C$ there must be a $l \in \Lambda \underline{n} \neg C$ such that $\Theta \cup \{i\} \cup \{l\} \vdash \neg C$. Since $\langle \Theta, \Lambda \rangle$ is vW-consistent we have $\{l \wedge \phi(l)\} \nvdash \neg C$ and so $\{l \wedge \phi(l)\} \in \Delta^* \bot \neg C$. But then $\Gamma \cup \{l \wedge \phi(l)\} \vdash \neg C$ contradicts vW^+-consistency of $\langle \Gamma, \Delta^* \rangle$ with C. So $\Gamma \subseteq \Theta$, hence $\Theta \cup \{C\} \vdash B$ by the assumption, and $\langle I, L \rangle \models_2 O^a(B/C)$ since $\langle \Theta, \Lambda \rangle$ was arbitrary.

Semantics $3/3^+$:

$O^e(B/C)$ Assume $\langle I, L \rangle \models_3 O^e(B/C)$, so there is a $\langle \Gamma, L \rangle \in \langle I, L \rangle \bot^{AB} \neg C$ such that $\Gamma \cup \{C\} \vdash B$. $\Gamma \nvdash \neg C$. Suppose there is a $\{l \wedge \phi(l)\} \in L^* \bot \neg C$ such that $\Gamma \cup \{l \wedge \phi(l)\} \vdash \neg C$. Then $\{l\} \nvdash \neg C$, so $l \in L \underline{n} \neg C$, and also $\Gamma \cup \{l\} \vdash \neg C$. But then $\langle \Gamma, L \rangle$ is vW-inconsistent. So for all $\{l \wedge \phi(l)\} \in L^* \bot \neg C$, $\Gamma \cup \{l \wedge \phi(l)\} \nvdash \neg C$, so $\langle \Gamma, L^* \rangle \in \langle I, L^* \rangle \bot^{AB^+} \neg C$, hence $\langle I, L^* \rangle \models_{3^+} O^e(B/C)$.

Assume $\langle I, L^* \rangle \models_{3^+} O^e(B/C)$, so there is a $\langle \Gamma, L^* \rangle \in \langle I, L^* \rangle \bot^{AB^+} \neg C$ such that $\Gamma \cup \{C\} \vdash B$. $\Gamma \nvdash \neg C$. Suppose there is a $l \in L \underline{n} \neg C$ with $\Gamma \cup \{l\} \vdash \neg C$. Then $\{l \wedge \phi(l)\} \in L^* \bot \neg C$ since $\{l\} \nvdash \neg C$. But then also $\Gamma \cup \{l \wedge \phi(l)\} \vdash \neg C$ and so $\langle \Gamma, L^* \rangle$ is vW^+-inconsistent. So for all $l \in L \underline{n} \neg C$, $\Gamma \cup \{l\} \nvdash \neg C$, so $\langle \Gamma, L \rangle \in \rangle \bot^{AB} \neg C$, hence $\langle I, L \rangle \models_3 O^e(B/C)$.

$O^a(B/C)$ Assume $\langle I, L \rangle \models_3 O^a(B/C)$, so for all $\langle \Gamma, L \rangle \in \langle I, L \rangle \bot^{AB} \neg C$, $\Gamma \cup \{C\} \vdash B$. Suppose $\langle \Delta, L^* \rangle \in \langle I, L^* \rangle \bot^{AB^+} \neg C$. So $\Delta \nvdash \neg C$ and there is no $\{l \wedge \phi(l)\} \in L^* \bot \neg C$ such that $\Delta \cup \{l \wedge \phi(l)\} \vdash \neg C$. So also for no $l \in L \underline{n} \neg C$, $\Gamma \cup \{l\} \vdash \neg C$. So there is a $\langle \Gamma, L \rangle \in \langle I, L \rangle \bot^{AB} \neg C$ such that $\Delta \subseteq \Gamma$. Suppose there is a $i \in \Gamma$ such that $i \notin \Delta$. Since $\Gamma \nvdash \neg C$, there must be a $\{l \wedge \phi(l)\} \in L^{*}\bot\neg C$ such that $\Delta \cup \{i\} \cup \{l \wedge \phi(l)\} \vdash \neg C$. Then $\{l\} \nvdash \neg C$, so $l \in L^{\underline{n}}\neg C$, and $\Gamma \cup \{l\} \vdash \neg C$. But then $\langle \Gamma, L\rangle$ is vW-inconsistent with C. So $\Gamma \subseteq \Delta$, hence $\Delta \cup \{C\} \vdash B$ by the assumption, and so $\langle I, L^{*}\rangle \models_{3^{+}} O^{a}(B/C)$ since $\langle \Delta, L^{*}\rangle$ was arbitrary.

Assume $\langle I, L^{*}\rangle \models_{3^{+}} O^{a}(B/C)$, so for all $\langle \Gamma, L^{*}\rangle \in \langle I, L^{*}\rangle\bot^{AB^{+}}\neg C$, $\Gamma \cup \{C\} \vdash B$. Suppose $\langle \Delta, L\rangle \in \langle I, L\rangle\bot^{AB}\neg C$. So $\Delta \nvdash \neg C$ and there is no $l \in L^{\underline{n}}\neg C$ such that $\Delta \cup \{l\} \vdash \neg C$. So also for no $\{l \wedge \phi(l)\} \in L^{*}\bot\neg C$, $\Gamma \cup \{l \wedge \phi(l)\} \vdash \neg C$. So there is a $\langle \Gamma, L^{*}\rangle \in \langle I, L^{*}\rangle\bot^{AB^{+}}\neg C$ such that $\Delta \subseteq \Gamma$. Suppose there is a $i \in \Gamma$ such that $i \notin \Delta$. Since $\Gamma \nvdash \neg C$, there must be a $l \in L^{\underline{n}}\neg C$ such that $\Delta \cup \{i\} \cup \{l\} \vdash \neg C$. Then since $\{l\} \nvdash \neg C$, $\{l \wedge \phi(l)\} \in L^{*}\bot\neg C$, and $\Gamma \cup \{l \wedge \phi(l)\} \vdash \neg C$. But then $\langle \Gamma, L^{*}\rangle$ is vW^{+}-inconsistent. So $\Gamma \subseteq \Delta$, hence $\Delta \cup \{C\} \vdash B$ by the assumption, and so $\langle I, L\rangle \models_{3} O^{a}(B/C)$ since $\langle \Delta, L\rangle$ was arbitrary.

Ad (2/3): I prove equivalently that if there is a $\langle I, L\rangle$ such that $\langle I, L\rangle \models A$ for some $A \in \mathscr{L}_{DDL^{e}}$ [$A \in \mathscr{L}_{DDL^{ea}}$] in semantics 2^{+} [3^{+}], then there is a $\langle I, L^{*}\rangle$ such that $\langle I, L^{*}\rangle \models A$ in semantics 2 [3]. For any $\Delta \subseteq L$, let Δ^{*} be the smallest set such that

(i) if $l \in \Delta$ then $l \in \Delta^{*}$, and
(ii) if $\{l_1, \dots, l_n\} \subseteq \Delta$ then $(l_1 \wedge \dots \wedge l_n) \in \Delta^{*}$.

$\langle I, L^{*}\rangle \models A$ in semantics 2 [3] is then proved by induction on the construction of A. Again, I only give the deontic cases:

Semantics $2/2^{+}$:

$O^{e}(B/C)$ Assume $\langle I, L\rangle \models_{2^{+}} O^{e}(B/C)$, so there is a $\langle \Gamma, \Delta\rangle \in \langle I, L\rangle\bot^{vW^{+}}\neg C$ such that $\Gamma \cup \{C\} \vdash B$. So $\Gamma \nvdash \neg C$, so by the construction there is a $\langle \Theta, \Lambda\rangle \in \langle I, L^{*}\rangle\bot^{vW}\neg C$ such that $\Gamma \subseteq \Theta$ and so $\Theta \cup \{C\} \vdash B$, hence $\langle I, L^{*}\rangle \models_{2} O^{e}(B/C)$. The opposite direction is proved similarly.

$O^{a}(B/C)$ I note for completeness that the left-to-right direction does *not* work: sometimes $\langle I, L\rangle \models_{2^{+}} O^{a}(B/C)$ but not $\langle I, L^{*}\rangle \models_{2} O^{a}(B/C)$: Let $I = \{(p_1 \vee p_2) \wedge \neg p_3, p_3 \wedge p_6, p_4 \wedge p_6, p_5 \wedge p_6\}$, and $L = \{\neg p_1 \wedge \neg p_4, \neg p_2 \wedge \neg p_5\}$. It is easily checked that $\langle I, L\rangle \models_{2^{+}} O^{a}(p_6/\top)$, but not $\langle I, L^{*}\rangle \models_{2} O^{a}(p_6/\top)$ since $\langle\{(p_1 \vee p_2) \wedge \neg p_3\}, L\rangle \in \langle I, L^{*}\rangle\bot^{vW}\neg\top$ and $\{(p_1 \vee p_2) \wedge \neg p_3\} \nvdash p_6$. So the converse of claim 1 of the theorem (which would imply claims 2–4) cannot be proved with the present construction. If other ones may succeed remains an open question.

Semantics $3/3^{+}$:

$O^{e}(B/C)$ Assume $\langle I, L\rangle \models_{3^{+}} O^{e}(B/C)$, so there is a $\langle \Gamma, L\rangle \in \langle I, L\rangle\bot^{AB^{+}}\neg C$ such that $\Gamma \cup \{C\} \vdash B$. $\Gamma \nvdash \neg C$. Suppose there is a $l \in L^{*\underline{n}}\neg C$ such that $\Gamma \cup \{l\} \vdash \neg C$. Then by the construction either $l \in L$ or $l = l_1 \wedge \dots \wedge l_n$ for some $\{l_1, \dots, l_n\} \subseteq L$. Since $\{l\} \nvdash \neg C$ there is then a $\Delta \in L\bot\neg C$ such that either $l \in \Delta$ or $\{l_1, \dots, l_n\} \subseteq \Delta$, but then there is a $\Delta \in L\bot\neg C$ such that $\Gamma \cup \Delta \vdash \neg C$ and so $\langle \Gamma, L\rangle$ is vW^{+}-inconsistent. So for all $l \in L^{*\underline{n}}\neg C$, $\Gamma \cup \{l\} \nvdash \neg C$, so there is a $\langle \Theta, L^{*}\rangle \in \langle I, L^{*}\rangle\bot^{AB}\neg C$ such that $\Gamma \subseteq \Theta$, by the assumption $\Theta \cup \{C\} \vdash B$

and hence $\langle I, L^* \rangle \models_3 O^e(B/C)$.

Assume $\langle I, L^* \rangle \models_3 O^e(B/C)$, so there is a $\langle \Gamma, L^* \rangle \in \langle I, L^* \rangle \bot^{AB} \neg C$ such that $\Gamma \cup \{C\} \vdash B$. $\Gamma \nvdash \neg C$. Suppose there is a $\Delta \in L \bot \neg C$ such that $\Gamma \cup \Delta \vdash \neg C$. Then by compactness of PL there are $l_1, ..., l_n \in \Delta$ such that $\Gamma \cup \{l_1, ..., l_n\} \vdash \neg C$. But then $l_1 \wedge ... \wedge l_n \in L^{*\underline{n}} \neg C$, and so $\langle \Gamma, L^* \rangle$ is vW-inconsistent. So for all $\Delta \in L \bot \neg C$, $\Gamma \cup \Delta \nvdash \neg C$, so there is a $\langle \Theta, L \rangle \in \langle I, L \rangle \bot^{AB^+} \neg C$ such that $\Gamma \subseteq \Theta$, by the assumption $\Theta \cup \{C\} \vdash B$ and hence $\langle I, L \rangle \models_{3^+} O^e(B/C)$.

$O^a(B/C)$ Assume $\langle I, L \rangle \models_{3^+} O^a(B/C)$, so for all $\langle \Gamma, L \rangle \in \langle I, L \rangle \bot^{AB^+} \neg C$, $\Gamma \cup \{C\} \vdash B$. Suppose $\langle \Theta, L^* \rangle \in \langle I, L^* \rangle \bot^{AB} \neg C$. $\Theta \nvdash \neg C$. Suppose there is a $\Delta \in L \bot \neg C$ such that $\Theta \cup \Delta \vdash \neg C$. Then by compactness of PL there are $l_1, ..., l_n \in \Delta$ such that $\Theta \cup \{l_1, ..., l_n\} \vdash \neg C$. But then $l_1 \wedge ... \wedge l_n \in L^{*\underline{n}} \neg C$, which contradicts vW-consistency of $\langle \Theta, L^* \rangle$. So for all $\Delta \in L \bot \neg C$, $\Theta \cup \Delta \nvdash \neg C$, so there is a $\langle \Gamma, L \rangle \in \langle I, L \rangle \bot^{AB^+} \neg C$, such that $\Theta \subseteq \Gamma$. Suppose there are $i \in \Gamma$: $i \notin \Theta$. Then by maximality there is a $l \in L^{*\underline{n}} \neg C$ such that $\Theta \cup \{i\} \cup \{l\} \vdash \neg C$. Then either $l \in L$ or there are $l_1, ..., l_n \in L$ such that $l = l_1 \wedge ... \wedge l_n$, and there is a $\Delta \in L \bot \neg C$ such that either $l \in \Delta$ or $\{l_1, ..., l_n\} \subseteq \Delta$. But then $\Gamma \cup \Delta \vdash \neg C$ and so $\langle \Gamma, L \rangle$ is vW^+-inconsistent. So $\Gamma \subseteq \Theta$, by the assumption $\Theta \cup \{C\} \vdash B$, hence $\langle I, L^* \rangle \models_3 O^a(B/C)$ since $\langle \Theta, L^* \rangle$ was arbitrary.

Assume $\langle I, L^* \rangle \models_3 O^a(B/C)$, so for all $\langle \Gamma, L^* \rangle \in \langle I, L^* \rangle \bot^{AB} \neg C$, $\Gamma \cup \{C\} \vdash B$. Suppose $\langle \Theta, L \rangle \in \langle I, L \rangle \bot^{AB^+} \neg C$. $\Theta \nvdash \neg C$. Suppose there is a $l \in L^{*\underline{n}} \neg C$ such that $\Theta \cup \{l\} \vdash \neg C$. Then either $l \in L$ or there are $l_1, ..., l_n \in L$ such that $l = l_1 \wedge ... \wedge l_n$, so there is a $\Delta \in L \bot \neg C$ with either $l \in \Delta$ or $\{l_1, ..., l_n\} \subseteq \Delta$. But then $\langle \Theta, L \rangle$ is vW^+-inconsistent. So for all $l \in L^{*\underline{n}} \neg C$, $\Theta \cup \{l\} \nvdash \neg C$, so there is a $\langle \Gamma, L^* \rangle \in \langle I, L^* \rangle \bot^{AB} \neg C$ such that $\Theta \subseteq \Gamma$. Suppose there are $i \in \Gamma$: $i \notin \Theta$. Then by maximality there is a $\Delta \in L \bot \neg C$ such that $\Theta \cup \{i\} \cup \Delta \vdash \neg C$. By compactness of PL there are $l_1, ..., l_n \in \Delta$ such that $\Theta \cup \{i, l_1, ..., l_n\} \vdash \neg C$. But $l_1 \wedge ... \wedge l_n \in L^{*\underline{n}} \neg C$, and so $\langle \Gamma, L^* \rangle$ is vW-inconsistent with C. So $\Gamma \subseteq \Theta$, by the assumption $\Theta \cup \{C\} \vdash B$, hence $\langle I, L \rangle \models_{3^+} O^a(B/C)$ since $\langle \Theta, L \rangle$ was arbitrary.

Ad (4): 2^+-validity of all rules and 'monadic' axiom schemes is immediate. For all versions of (Cond), (RMon), (Or), and (DR) valid in semantics 2, as for the *aee* and *eae*-versions of (Cut), the proofs for semantics 2^+ are straightforward adaptations of those provided in the proof of Theorem 2. 2^+-validity of all *e*-versions follows from claim 3 just proved. For the remaining proofs a lemma is helpful:

Lemma 4 (Property of universal obligation in semantics 2^+).
If $\langle I, L \rangle \models O^a(D/C)$, then $\forall \langle \Gamma, \Delta \rangle \in \langle I, L \rangle \bot^{vW^+} \neg C$: $\Delta \bot \neg C \subseteq \Delta \bot \neg (C \wedge D)$.

Proof. Suppose $\langle \Gamma, \Delta \rangle \in \langle I, L \rangle \bot^{vW^+} \neg C$. $\langle I, L \rangle \models O^a(D/C)$, so $\Gamma \cup \{C\} \vdash D$. If there is a $X \in \Delta \bot \neg C$ with $X \vdash \neg D$, then $\Gamma \cup X \vdash \neg C$, contrary to vW^+-consistency of $\langle \Gamma, \Delta \rangle$. So $X \nvdash \neg D$. If $X \vdash \neg C \vee \neg D$, then likewise $\Gamma \cup X \vdash \neg C$. So $X \nvdash \neg C \vee \neg D$ and $X \in \Delta \bot \{\neg C, \neg D, \neg C \vee \neg D\} = \Delta \bot \neg (C \wedge D)$ (Hansson (1999) p. 42). □

Corollary 1. If $\langle I, L\rangle \models O^a(D/C)$, then $\forall \Delta \subseteq L$: $\Delta\bot\neg C \subseteq \Delta\bot\neg(C \wedge D)$.

Proof. If $\vdash \neg C$, then $\Delta\bot\neg C = \varnothing$ and the inclusion is trivial. Otherwise by construction $\langle \Gamma, L\rangle \in \langle I, L\rangle\bot^{\nu W^+}\neg C$ for some $\Gamma \subseteq I$. Let $X \in \Delta\bot\neg C$. Since $X \nvdash \neg C$, there is a $Y \in L\bot\neg C$ such that $X \subseteq Y$. By lemma 4, $Y \in L\bot\neg(C \wedge D)$, so $X \nvdash \neg D$ and $X \nvdash \neg(C \wedge D)$, so $X \in \Delta\bot\{\neg C, \neg D, \neg C \vee \neg D\} = \Delta\bot\neg(C \wedge D)$. □

(CCMon^a) Assume $\langle I, L\rangle \models O^a(A \wedge D/C)$, so also $\langle I, L\rangle \models O^a(D/C)$ by validity of (DM^a). Suppose $\langle \Theta, \Lambda\rangle \in \langle I, L\rangle\bot^{\nu W^+}\neg(C \wedge D)$. $\Theta \nvdash \neg C$, since otherwise $\Theta \vdash \neg(C \wedge D)$, which is excluded by νW^+-consistency with $C \wedge D$. Suppose there is a $X \in \Lambda\bot\neg C$ such that $\Theta \cup X \vdash \neg C$. But $X \in \Lambda\bot\neg(C \wedge D)$ by Corollary 1, so then $\langle \Theta, \Lambda\rangle$ is νW^+-inconsistent with $C \wedge D$. So for all $X \in \Lambda\bot\neg C$, $\Theta \cup X \nvdash \neg C$, so there is a $\langle \Gamma, \Delta\rangle \in \langle I, L\rangle\bot^{\nu W^+}\neg C$ such that $\Theta \subseteq \Gamma$ and $\Lambda \subseteq \Delta$. $\Gamma \nvdash \neg(C \wedge D)$ for by the assumption $\Gamma \cup \{C\} \vdash D$, so otherwise $\Gamma \vdash \neg C$, contrary to νW^+-consistency of $\langle \Gamma, \Delta\rangle$ with C. Suppose there is a $X \in \Delta\bot\neg(C \wedge D)$ such that $\Gamma \cup X \vdash \neg(C \wedge D)$. Since $X \nvdash \neg(C \wedge D)$, also $X \nvdash \neg C$, but then $X \subseteq Y \in \Delta\bot\neg C$ and $\Gamma \cup Y \vdash \neg C$ again contrary to νW^+-consistency of $\langle \Gamma, \Delta\rangle$ with C. So for all $X \in \Delta\bot\neg(C \wedge D)$, $\Gamma \cup X \nvdash \neg(C \wedge D)$, so there is a $\langle \Theta', \Lambda'\rangle \in \langle I, L\rangle\bot^{\nu W^+}\neg(C \wedge D)$ such that $\Gamma \subseteq \Theta'$ and $\Delta \subseteq \Lambda'$. Then by maximality $\Theta = \Gamma = \Theta'$ and $\Lambda = \Delta = \Lambda'$, so by the assumption $\Theta \cup \{C\} \vdash A$ and so also $\Theta \cup \{C \wedge D\} \vdash A$, and $\langle I, L\rangle \models O^a(A/C \wedge D)$ since $\langle \Theta, \Lambda\rangle$ was arbitrary.

(Cut^a) Assume $\langle I, L\rangle \models O^a(D/C)$ and $\langle I, L\rangle \models O^a(A/C \wedge D)$. Suppose $\langle \Theta, \Lambda\rangle \in \langle I, L\rangle\bot^{\nu W^+}\neg C$. $\Theta \nvdash \neg(C \wedge D)$, for by the assumption $\Theta \cup \{C\} \vdash D$, so otherwise $\Theta \vdash \neg C$, contrary to νW^+-consistency of $\langle \Theta, \Lambda\rangle$ with C. Suppose there is a $X \in \Lambda\bot\neg(C \wedge D)$ such that $\Theta \cup X \vdash \neg(C \wedge D)$. Since $X \nvdash \neg(C \wedge D)$, also $X \nvdash \neg C$, but then there is a $Y \in \Theta\bot\neg C$ such that $X \subseteq Y$, but then $\Theta \cup Y \vdash \neg C$, again contrary to νW^+-consistency of $\langle \Theta, \Lambda\rangle$ with C. So for all $X \in \Lambda\bot\neg(C \wedge D)$, $\Theta \cup X \nvdash \neg(C \wedge D)$, so there is a $\langle \Gamma, \Delta\rangle \in \langle I, L\rangle\bot^{\nu W^+}\neg(C \wedge D)$ such that $\Theta \subseteq \Gamma$ and $\Lambda \subseteq \Delta$. $\Gamma \nvdash \neg C$, for otherwise $\Gamma \vdash \neg(C \wedge D)$, contrary to νW^+-consistency of $\langle \Gamma, \Delta\rangle$ with C. Suppose there is a $X \in \Delta\bot\neg C$ such that $\Gamma \cup X \vdash \neg C$. So also $\Gamma \cup X \vdash \neg(C \wedge D)$. But $X \in \Delta\bot\neg(C \wedge D)$ due to Corollary 1, so then $\langle \Gamma, \Delta\rangle$ is νW^+-inconsistent with $C \wedge D$. So for all $X \in \Delta\bot\neg C$, $\Gamma \cup X \nvdash \neg C$, so there is a $\langle \Theta', \Lambda'\rangle \in \langle I, L\rangle\bot^{\nu W^+}\neg C$ such that $\Gamma \subseteq \Theta'$ and $\Delta \subseteq \Lambda'$. Then by maximality $\Theta = \Lambda = \Theta'$ and $\Lambda = \Delta = \Lambda'$. By the second assumption, $\Theta \cup \{C \wedge D\} \vdash A$ and so $\Theta \cup \{C\} \vdash D \rightarrow A$, so $\langle I, L\rangle \models O^a(D \rightarrow A/C)$ since $\langle \Theta, \Lambda\rangle$ was arbitrary.

(O^a-Loop) I prove that (Cut$^+$) $O^a(C/D) \rightarrow (O^a(A/C) \rightarrow O^a(C \rightarrow A/C \vee D))$ is 2^+-valid, which implies 2^+-validity of (O^a-Loop) (proof of Theorem 2): Assume $\langle I, L\rangle \models O^a(C/D)$ and $\langle I, L\rangle \models O^a(A/C)$. Suppose $\langle \Gamma, \Delta\rangle \in \langle I, L\rangle\bot^{\nu W^+}\neg(C \vee D)$. If $\Gamma \vdash \neg C$ then $\Gamma \cup \{C \vee D\} \vdash C \rightarrow A$ holds trivially. So suppose $\Gamma \nvdash \neg C$. If there is a $X \in \Delta\bot\neg C$ such that $\Gamma \cup X \vdash \neg C$, then $X \subseteq Y \in \Delta\bot\neg(C \vee D)$. Then $\Gamma \cup Y \nvdash \neg D$, for otherwise $\Gamma \cup Y \vdash \neg(C \vee D)$ contrary to νW^+-consistency of $\langle \Gamma, \Delta \rangle$ with $C \vee D$. So then there is by construction a $\langle \Theta, \Lambda \rangle \in \langle I, L \rangle \bot^{\nu W^+} \neg D$ such that $\Gamma \subseteq \Theta$ and $Y \subseteq \Lambda$. Since $Y \nvdash \neg D$, $Y \subseteq Z \in \Lambda \bot \neg D$, so $\Theta \cup Z \vdash \neg C$. But also $\Theta \cup \{D\} \vdash C$ by the first assumption, so $\Theta \cup Z \vdash \neg D$, which contradicts νW^+-consistency of $\langle \Theta, \Lambda \rangle$ with D. So for all $X \in \Delta \bot \neg C$, $\Gamma \cup X \nvdash \neg C$, so there is a $\langle \Theta, \Lambda \rangle \in \langle I, L \rangle \bot^{\nu W^+} \neg C$ such that $\Gamma \subseteq \Theta$ and $\Delta \subseteq \Lambda$. Since $\Theta \nvdash \neg C$, also $\Theta \nvdash \neg(C \vee D)$. Suppose there is a $X \in \Lambda \bot \neg(C \vee D)$ such that $\Theta \cup X \vdash \neg(C \vee D)$. Either $X \nvdash \neg C$ or $X \nvdash \neg D$. If $X \nvdash \neg C$ then $X \subseteq Y \in \Lambda \bot \neg C$, but then $\Theta \cup Y \vdash \neg C$, contrary to νW^+-consistency of $\langle \Theta, \Lambda \rangle$ with $\neg C$. So $X \vdash \neg C$ and $X \nvdash \neg D$. But then $X \subseteq Y \in \Lambda \bot \neg D$, and by Corollary 1 $Y \in \Lambda \bot \neg(C \wedge D)$, which contradicts $X \vdash \neg C$. So for all $X \in \Lambda \bot \neg(C \vee D)$, $\Theta \cup X \nvdash \neg(C \vee D$, so there is a $\langle \Gamma', \Delta' \rangle$ such that $\Theta \subseteq \Gamma'$ and $\Lambda \subseteq \Delta'$. Then by maximality $\Gamma = \Theta = \Gamma'$ and $\Delta = \Lambda = \Delta'$, so by the second assumption $\Gamma \cup \{C\} \vdash A$, so $\Gamma \cup \{C \vee D\} \vdash C \rightarrow A$ and $\langle I, L \rangle \models O^a(C \rightarrow A / C \vee D)$ since $\langle \Gamma, \Delta \rangle$ was arbitrary. □

References

Alchourrón, C. E. (1969). Logic of norms and logic of normative propositions. *Logique and Analyse, 12*, 242–268.

Alchourrón, C. E. (1972). The intuitive background of normative legal discourse and its formalization. *Journal of Philosophical Logic, 1*, 447–463.

Alchourrón, C. E., Bulygin, E. (1981). The expressive conception of norms. In R. Hilpinen (Ed.), *New studies in deontic logic* (pp. 95–124). Dordrecht: Reidel.

Alchourrón, C. E., & Bulygin, E. (1984). Permission and permissive norms. In W. Krawietz, H. Schelsky, O. Weinberger, & G. Winkler (Eds.), *Theorie der Normen* (pp. 349–371). Berlin: Duncker & Humblot.

Alchourrón, C. E., & Bulygin, E. (1993). On the logic of normative systems. In H. Stachowiak (Ed.), *Handbuch pragmatischen Denkens* (pp. 273–293). Hamburg: Meiner.

Alchourrón, C. E., & Makinson, D. (1981). Hierarchies of regulations and their logic. In R. Hilpinen (Ed.), *New studies in deontic logic* (pp. 125–148). Dordrecht: Reidel.

Barcan Marcus, R. (1980). Moral dilemmas and consistency. *Journal of Philosophy, 77*, 121–136.

Bentham, J. (1970). Of laws in general. In H. L. A. Hart (Ed.). London: Athlone Press (first appeared 1782).

Bochman, A. (1999). Credulous nonmonotonic inference. In T. Dean (Ed.), *Proceedings of the 16th International Joint Conference on Artificial Intelligence, IJCAI '99, Stockholm, August 1999* (pp. 30–35). San Francisco: Morgan Kaufmann.

Boella, G., van der Torre, L. (2003). Permissions and obligations in hierarchical normative systems. In *Proceedings of the 9th International Conference on Artificial Intelligence and Law, ICAIL '03* (pp. 109–118) (republ. w. minor changes as Boella (2008)).

Boella, G., van der Torre, L. (2008). Institutions with a hierarchy of authorities in distributed dynamic environments. *Artificial Intelligence and Law, 16*, 53–71.

Brewka, G. (1989). Preferred subtheories: An extended logical framework for default reasoning. In N. S. Sridharan (Ed.), *Proceedings of the Eleventh International Joint Conference on Artificial Intelligence IJCAI-89, Detroit, Michigan, USA, August 20–25, 1989* (pp. 1043–1048). San Mateo: Kaufmann.

Brewka, G. (1991). Belief revision in a framework for default reasoning. In A. Fuhrmann & M. Morreau (Eds.), *The Logic of Theory Change, Workshop, Konstanz, Germany, October 13–15, 1989* (pp. 206–222). Berlin: Springer.

Bulygin, E. (1986). Permissive norms and normative systems. In A. A. Martino & F. Socci Natali (Eds.), *Automated Analysis of Legal Texts* (pp. 211–218). North-Holland: Elsevier.

van Fraassen, B. (1973). Values and the heart's command. *Journal of Philosophy, 70*, 5–19.

Goble, L. (1990). A logic of good, should, and would. *Journal of Philosophical Logic, 19*, 169–199.

Goble, L. (2000). Multiplex semantics for deontic logics. *Nordic Journal of Philosophical Logic, 5*, 113–134.

Goble, L. (2003). Preference semantics for deontic logics: Part I—simple models. *Logique et Analyse, 46*, 383–418.

Goble, L. (2004). Preference semantics for deontic logics: Part II—multiplex models. *Logique et Analyse, 47*, 335–363.

Hamblin, C. L. (1987). *Imperatives*. Oxford: Blackwell.

Hansen, J. (2001). Sets, sentences, and some logics about imperatives. *Fundamenta Informaticae, 48*, 205–226.

Hansen, J. (2004). Problems and results for logics about imperatives. *Journal of Applied Logic, 2*, 39–61.

Hansen, J. (2005). Conflicting imperatives and dyadic deontic logic. *Journal of Applied Logic, 3*, 484–511.

Hansen, J. (2006). Deontic logics for prioritized imperatives. *Artificial Intelligence and Law, 14*, 1–34.

Hansen, J. (2008a). Imperatives and deontic logic: On the semantic foundations of deontic logic. Ph.D. thesis, Universität Leipzig, Inst. f. Philosophie, Leipzig.

Hansen, J. (2008b). Prioritized conditional imperatives: Problems and a new proposal. *Autonomous Agents and Multi-Agent Systems, 17*, 11–35.

Hansson, B. (1969). An analysis of some deontic logics. *Noûs, 3*, 373–398 (reprinted in Hilpinen (1971)).

Hansson, S. O. (1999). *A textbook of belief dynamics: Theory change and dabase updating*. Dordrecht: Kluwer.

Hansson, S. O. (2013). The varieties of permission. In D. Gabbay, J. F. Horty, R. van der Meyden, & L. van der Torre (Eds.), *Handbook of deontic logic and normative systems* (Vol. 1). London: College Publications.

Hare, R. M. (1967). Some alleged differences between imperatives and indicatives. *Mind, 76*, 309–326.

Hart, H. L. A. (1961). *The concept of law*. Oxford: Clarendon Press.

Hilpinen, R. (1971). *Deontic logic: Introductory and systematic readings*. Dordrecht: Reidel.

Hilpinen, R. (1982). Disjunctive permissions and conditionals with disjunctive antecedents. *Acta Philosophica Fennica, 35*, 175–194.

Horty, J. F. (1993). Deontic logic as founded on nonmonotonic logic. *Annals of Mathematics and Artificial Intelligence, 9*, 69–91.

Horty, J. F. (1994). Moral dilemmas and nonmonotonic logic. *Journal of Philosophical Logic, 23*, 35–65.

Horty, J. F. (1997). Nonmonotonic foundations for deontic logic. In D. Nute (Ed.), *Defeasible deontic logic* (pp. 17–44). Dordrecht: Kluwer.

Hruschka, J. (1986). *Das deontologische Sechseck bei Gottfried Achenwall im Jahre 1767*. Göttingen: Vandenhoek & Ruprecht

Kamp, H. (1973). Free choice permission. *Proceedings of the Aristotelian Society, 74*, 57–74.

Kamp, H. (1978). Semantics versus pragmatics. In F. Guenthner & S. J. Schmidt (Eds.), *Formal semantics and pragmatics for natural languages* (pp. 255–287). Dordrecht: Reidel.

Kraus, S., Lehmann, D., & Magidor, M. (1990). Nonmonotonic reasoning, preferential models and cumulative logics. *Artificial Intelligence, 44*, 167–207.

Leibniz, G. W. (1930). Elementa juris naturalis [2nd half 1671, holographic concept B (Hannover)]. In Preussische Akademie der Wissenschaften (Ed.), *Gottfried Wilhelm Leibniz: Sämtliche Schriften und Briefe, 6. Reihe: Philosophische Schriften* (Vol. 1, pp. 465–485). Darmstadt: Otto Reichl.
Lewis, D. (1974). Semantic analyses for dyadic deontic logic. In S. Stenlund (Ed.), *Logical theory and semantic analysis* (pp. 1–14). Dordrecht: Reidel.
Lewis, D. (1979). A problem about permission. In E. Saarinen, R. Hilpinen, I. Niiniluoto, & M. Provence Hintikka (Eds.), *Essays in honour of Jaakko Hintikka* (pp. 163–175). Dordrecht: D. Reidel.
Makinson, D. (1999). On a fundamental problem of deontic logic. In P. McNamara & H. Prakken (Eds.), *Norms, logics and information systems* (pp. 29–53). Amsterdam: IOS.
Makinson, D. (2005). *Bridges from classical to nonmonotonic logic*. London: King's College.
Makinson, D., & van der Torre, L. (2000). Input/output logics. *Journal of Philosophical Logic, 29*, 383–408.
Makinson, D., & van der Torre, L. (2001). Constraints for input/output logics. *Journal of Philosophical Logic, 30*, 155–185.
Makinson, D., & van der Torre, L. (2003). Permissions from an input/output perspective. *Journal of Philosophical Logic, 32*, 391–416.
Prakken, H., Sergot, M. (1997). Dyadic deontic logic and contrary-to-duty obligations. In D. Nute (Ed.), *Defeasible deontic logic* (pp. 223–262). Dordrecht: Kluwer.
Ross, A. (1941). Imperatives and logic. *Theoria, 7*, 53–71.
Schwager, M. (2006). Interpreting imperatives. Ph.D. thesis, Johann-Wolfgang-von-Goethe Universität, Institut für Kognitive Linguistik, Frankfurt am Main.
Spohn, W. (1975). An analysis of Hansson's dyadic deontic logic. *Journal of Philosophical Logic, 4*, 237–252.
Stolpe, A. (2010). A theory of permission based on the notion of derogation. *Journal of Applied Logic, 8*, 97–113.
Weinberger, O. (1958). Die Sollsatzproblematik in der modernen Logik: Können Sollsätze (Imperative) als wahr bezeichnet werden? Prague: Nakladatelství Ceskoslovenské Akademie Ved. Page numbers refer to the reprint in Weinberger (1974) pp. 59–186.
Weinberger, O. (1974). *Studien zur Normlogik und Rechtsinformatik*. Berlin: Schweitzer.
Weinberger, O. (1989). *Rechtslogik* (2nd ed.). Berlin: Duncker & Humblot.
Weinberger, O. (1991). The logic of norms founded on descriptive language. *Ratio Juris, 4*, 284–307.
Weinberger, O. (1996). *Alternative Handlungstheorie*. Wien: Böhlau.
Weinberger, O. (1999). Logical analysis in the realm of law . In G. Meggle (Ed.), *Actions, norms, values: Discussions with Georg Henrik von Wright* (pp. 291–304). Berlin: De Gruyter.
von Wright, G. H. (1951). Deontic logic. *Mind, 60*, 1–15.
von Wright, G. H. (1963). *Norm and action*. London: Routledge & Kegan Paul.
von Wright, G. H. (1980). Problems and prospects of deontic logic. In E. Agazzi (Ed.), *Modern logic: A survey* (pp. 399–423). Dordrecht: D. Reidel.
von Wright, G.H. (1983). Norms, truth and logic. In G. H. von Wright (Ed.), *Practical reason: Philosophical papers* (Vol. 1, pp. 130–209). Oxford: Blackwell.
von Wright, G. H. (1991). Is there a logic of norms? *Ratio Juris, 4*, 265–283.
von Wright, G. H. (1993). A pilgrim's progress. In G. H. von Wright (Ed.), *The tree of knowledge and other essays* (pp. 103–113). Leiden: Brill.
von Wright, G. H. (1999a). Deontic logic—as I see it. In P. McNamara & H. Prakken (Eds.), *Norms, logics and information systems* (pp. 15–25). Amsterdam: IOS.
von Wright, G. H. (1999b). Ought to be—ought to do. In G. Meggle (Ed.), *Actions, norms, values: Discussions with Georg Henrik von Wright* (pp. 3–9). Berlin: De Gruyter.

Norm Change in the Common Law

John Horty

Abstract An account of legal change in a common law system is developed. Legal change takes place incrementally through court decisions that are constrained by previous decisions in other courts. The assumption is that a court's decision has to be consistent with the rules set out in earlier court decisions. However, the court is allowed to make add new distinctions and therefore make a different decision based on factors not present in the previous decision. Two formal models of this process are presented. The first model is based on refinement of (the set of factors taken into account in) the set of previous cases on which a decision is based. In the second model the focus is on a preference ordering on reasons. The court is allowed to supplement, but not to revise the preference ordering on reasons that can be inferred from previous cases. The two accounts turn out to be equivalent. A court can make a consistent decision even if the case base is not consistent; the important requirement is that no new inconsistencies should be added to the case base.

Keywords Norm change · Common law · Legal code · Derogation · Legal reasoning · Legal factor · Precedent case · Rule · Refinement

1 Introduction

Among David Makinson's many achievements in logic, none is more important than his development, along with Carlos Alchourrón and Peter Gärderfors, of the AGM theory of belief change.

The origin of that work has now been documented—in David's obituary of Alchourrón, in Gärdenfors's brief history, and in David's own reflections—and it

J. Horty (✉)
Philosophy Department, Institute for Advanced Computer Studies,
University of Maryland, College Park, MD, USA
e-mail: horty@umiacs.umd.edu
http://www.umiacs.umd.edu/users/horty

S. O. Hansson (ed.), *David Makinson on Classical Methods for Non-Classical Problems*, Outstanding Contributions to Logic 3, DOI: 10.1007/978-94-007-7759-0_15,

is, in many ways, a dramatic saga.[1] From David's perspective, it began with the problem of norm change in the law, or more specifically, with Alchourrón's interest, together with that of his colleague Eugenio Bulygin, in the concept of derogation: the removal of a norm from a system of norms, such as a legal code.[2] The difficulty is that the individual norm to be derogated might not simply be listed in the legal code, but instead, or in addition, implied by other individual norms from the code, or by sets of other norms taken together. In the latter case, it will be possible for the derogation of a particular norm to be achieved in a number of ways, depending on which adjustments are made to the set of norms supporting it; the result is, therefore, indeterminate.

David reports that he did not, at first, see much of interest in the concept of derogation for exactly this reason, the indeterminacy of its result, which he viewed as "just an unfortunate fact of life ... about which formal logic could say little or nothing". By the end of the 1970s, however, he and Alchourrón had managed to frame the issue in a way that was amenable to formal analysis, and published the outcome in the second of Risto Hilpinen's two influential collections on deontic logic.[3] Just as they were completing this paper, they realized that both the issues under consideration and their logical analysis could be seen in a more general light—as a matter of belief revision in general, not just norm revision. This perspective was adopted in a second paper, submitted to *Theoria*.[4]

As it happens, the editor of that journal was then Peter Gärdenfors, who was working on formally similar problems, though with a distinct philosophical motivation—Gärdenfors had been exploring the semantics of conditionals, not norm change—and a collaboration was joined. Of course, there would have been differences: Gärdenfors's approach had been largely postulational, while the approach of Alchourrón and Makinson was definitional; Alchourrón and Makinson had focused on derogation, now called contraction, as the fundamental operation, with revision defined through the Levi identity, while Gärdenfors took the opposite route, treating revision as fundamental with contraction defined through the Harper identity. Nevertheless, David describes the collaboration as "a dream", with differences resolved and further progress achieved, in those days before email, through a series of longhand letters "circulating incessantly between Buenos Aires, Lund, Beirut, and Paris". The result was the initial AGM paper, which, taken together with subsequent work on the topic by the original authors and many others—in fields including philosophy, computer science, economics, and psychology—stands as one of the great success stories from the past 25 years of philosophical logic.[5]

---

[1] See Makinson (1996, 2003) and Gärdenfors (2011).

[2] The term "derogation" is often used to refer only to limitation of a norm, while its full removal is described as an "abrogation". My terminology here follows that of Alchourrón and Makinson (1981).

[3] Alchourrón and Makinson (1981).

[4] Alchourrón and Makinson (1982).

[5] Alchourrón, Gärdenfors, and Makinson (1985).

I will not try to advance this story here. Instead, I want to return to its roots: the problem of norm change in the law. It is natural that Alchourrón, as an Argentinian, from a civil law country, would explore this problem in the context of a changing legal canon, an evolving body of rules. But any jurist working in the United Kingdom, America, Canada, Australia, New Zealand, or any of the other common law countries, if asked about norm change in the law, would think first, not about explicit modifications to a legal code, but about the common law itself. And here, the process of norm change is typically more gradual, incremental, and mediated by the common law doctrine of precedent, according to which the decisions of earlier courts generalize to constrain the decisions of later courts, while still allowing these later courts a degree freedom in responding to fresh circumstances.

On what is, perhaps, the standard view, the constraints of precedent are themselves carried through rules: a court facing a particular problem situation either invokes a previous common law rule or articulates a new one to justify its decision in that case, and this rule is then generally thought to determine the decisions that might be reached in any future case to which it applies. There are, however, two qualifications. Some courts, depending on their place in the judicial hierarchy, have the power to *overrule* the decisions of earlier courts. The effect of overruling is much like that of derogation: the normative force of a case that has been overruled is removed entirely. Overruling is, therefore, radical, but it is also rare, and not a form of norm change that I will discuss here.

Although only certain courts have the power to overrule earlier decisions, all courts are thought to have the power of *distinguishing* later cases—the power, that is, to point out important differences between the facts present in some later case and those of earlier cases, and so modifying the rules set out in those earlier cases to avoid what they feel would be an inappropriate application to the later case. Of course, later courts cannot modify the rules set out by earlier courts entirely at will, in any way whatsoever. There must be some restrictions on this power, and the most widely accepted restrictions are those first set out explicitly by Joseph Raz, although, as Raz acknowledges, the account owes much previous work of A. W. B. Simpson.[6] According to this account, any later modification of an earlier rule must satisfy two conditions: first, the modification can consist only in the addition of further qualifications, which will thus narrow the original rule; and second, the modified rule must continue to yield the original outcome in the case in which it was introduced, as well as in any further cases in which this rule was applied.

In recent work, motivated in part by research from the field of Artificial Intelligence and Law, as well as by a previous proposal due to Grant Lamond, I developed an account of precedent in the common law according to which constraint is not a matter of rules at all, but of reasons.[7] More exactly, I suggested that what is important about a precedent case is the previous court's assessment of the balance of reasons

---

[6] See Raz (1979, pp. 180–209) and Simpson (1961).

[7] See Horty (2011), and then Horty and Bench-Capon (2012) for a development of this account within the context of related research from Artificial Intelligence and Law; see Lamond (2005) for his earlier proposal.

presented by that case; later courts are then constrained, not to follow some rule set out by the earlier court, but to reach a decision that is consistent with the earlier court's assessment of the balance of reasons.

The account I propose is precise and allows, I believe, for a good balance between the constraints imposed by previous decisions and the freedoms granted to later courts for developing the law. But my account is also unusual, especially in abstaining from any appeal to rules in its treatment of the common law, and the question immediately arises: what is the relation between this account of precedential constraint, developed in terms of reasons, and the standard account, relying on rules?

The goal of the current paper is to answer this question. More precisely, what I show is that, even though the account of precedential constraint developed in terms of reasons was introduced as an alternative to the standard account, in terms of rules, it turns out that these two accounts are, in an important sense, equivalent. Establishing this result requires a precise statement of the notion of constraint at work in the standard account, which is offered in Sect. 3 of this paper, after basic concepts are introduced in Sect. 2. The account of precedential constraint in terms of reasons is reviewed in Sects. 4, and 5 establishes its equivalence with the standard account. Section 6 mentions some of the formal issues raised by this work; a discussion of the philosophical motivation is reserved for a companion paper.

2 Factors, Rules, and Cases

I follow the work of Edwina Rissland, Kevin Ashley, and their colleagues in supposing that the situation presented to the court in a legal case can usefully by represented as a set of *factors*, where a factor stands for a legally significant fact or pattern of facts.[8] Cases in different areas of the law will be characterized by different sets of factors, of course. In the domain of trade secrets law, for example, where the factor-based analysis has been developed most extensively, a case will typically concern the issue of whether the defendant has gained an unfair competitive advantage over the plaintiff through the misappropriation of a trade secret; and here the factors involved might turn on, say, questions concerning whether the plaintiff took measures to protect the trade secret, whether a confidential relationship existed between the plaintiff and the defendant, whether the information acquired was reverse-engineerable or in some other way publicly available, and the extent to which this information did, in fact, lead to a real competitive advantage for the defendant.[9]

[8] See Rissland and Ashley (1987) and then Ashley (1989, 1990) for an introduction to the model; see also, Rissland (1990) for an overview of research in Artificial Intelligence and Law that places this work in a broader context.

[9] Aleven (1997) has analyzed 147 cases from trade secrets law in terms of a factor hierarchy that includes 5 high-level issues, 11 intermediate-level concerns, and 26 base-level factors. The resulting knowledge base is used in an intelligent tutoring system for teaching elementary skills in legal argumentation, which has achieved results comparable to traditional methods of instruction in controlled studies; see Aleven and Ashley (1997).

We will assume, as usual, that factors have polarities, always favoring one side or another. In the domain of trade secrets law, once again, the presence of security measures favors the plaintiff, since it strengthens the claim that the information secured was a valuable trade secret; reverse-engineerability favors the defendant, since it suggests that the product information might have been acquired through proper means. The paper is based, furthermore, on the simplifying assumption that the reasoning under consideration involves only a single step, proceeding from the factors present in a case immediately to a decision—in favor of the plaintiff or the defendant—rather than moving through a series of intermediate legal concepts.[10]

Formally, then, we will let $F^{\pi} = \{f_1^{\pi}, \ldots, f_n^{\pi}\}$ represent the set of factors favoring the plaintiff and $F^{\delta} = \{f_1^{\delta}, \ldots, f_m^{\delta}\}$ the set of factors favoring the defendant. Since each factor favors one side of the other, we can suppose that the entire set F of legal factors is exhausted by these two sets: $F = F^{\pi} \cup F^{\delta}$. A *fact situation* X, of the sort presented in a legal case, can then be defined as some particular subset of the overall set of factors: $X \subseteq F$.

A *precedent case* will be represented as a fact situation together with an outcome as well as a rule through which that outcome is reached. Such a case can be defined as a triple of the form $c = \langle X, r, s \rangle$, where X is a fact situation containing the legal factors present in the case, r is the rule of the case, and s is its outcome.[11] We define three functions—$Factors$, *Rule*, and *Outcome*—to map cases into their component parts, so that, in the case c above, for example, we would have $Factors(c) = X$, $Rule(c) = r$, and $Outcome(c) = s$.

Given our assumption that reasoning proceeds in a single step, we can suppose that the *outcome* s of a case is always either a decision in favor of the plaintiff or a decision in favor of the defendant, with these two outcomes represented as π or δ respectively; and where s is a particular outcome, a decision for some side, we suppose that $\overline{s}$ represents a decision for the opposite side, so that $\overline{\pi} = \delta$ and $\overline{\delta} = \pi$. Where X is a fact situation, we let X^s represent the factors from X that support the side s; that is, $X^{\pi} = X \cap F^{\pi}$ and $X^{\delta} = X \cap F^{\delta}$.

Rules are to be defined in terms of reasons, where a *reason for a side* is a set of factors favoring that side. A *reason* can then be defined as a set of factors favoring one side or another. To illustrate: $\{f_1^{\pi}, f_2^{\pi}\}$ is a reason favoring the side π, and so a reason, while $\{f_1^{\delta}\}$ is a reason favoring δ, and likewise a reason; but the set $\{f_1^{\pi}, f_1^{\delta}\}$ is not a reason, since the factors it contains do not uniformly favor one side or another.

A statement of the form $X \models R$ indicates that the fact situation X *satisfies* the reason R, or that the reason *holds* in that situation; this idea can be defined by stipulating that

$$X \models R \text{ just in case } R \subseteq X,$$

[10] Both of the assumptions mentioned in this paragraph are discussed in Horty (2011).

[11] For the purpose of this paper, I simplify by assuming that the rule underlying a court's decision is plain, ignoring the extensive literature on methods for determining the rule, or *ratio decidendi*, of a case. I will also assume that a case always contains a single rule, ignoring situations in which a judge might offer several rules for a decision, or in which a court reaches a decision by majority, with different judges offering different rules, or in which a judge might simply render a decision in a case without setting out any general rule at all.

and then extended in the usual way to statements ϕ and ψ formed by closing the reasons under conjunction and negation:

$$X \models \neg\phi \text{ if and only if it fails that } X \models \phi,$$
$$X \models \phi \wedge \psi \text{ if and only if } X \models \phi \text{ and } X \models \psi.$$

We stipulate, in the usual way, that ϕ implies a statement $\phi \vdash \psi$—that is, ϕ implies ψ—just in case $X \models \psi$ whenever $X \models \phi$.

Given this notion of a reason, a rule can now be defined as a pair whose premise is a certain kind of conjunction of reasons and their negations and whose conclusion is an outcome, a decision favoring one side or the other. More specifically, where R^s is a single reason for the side s and $R^{\overline{s}}_1, \ldots, R^{\overline{s}}_i$ are zero or more reasons for the opposite side, then a *rule for the side s* has the form

$$R^s \wedge \neg R^{\overline{s}}_1 \wedge \ldots \wedge \neg R^{\overline{s}}_i \rightarrow s$$

and a *rule* is simply a rule for one side or the other; the idea, of course is that, when the reason R^s favoring s holds in some situation, and none of the reasons $R^{\overline{s}}_1, \ldots, R^{\overline{s}}_i$ favoring the opposite side hold, then r requires a decision for the side s. Given a rule r of this form, we define functions *Premise*, *Premise*^{s}, *Premise*$^{\overline{s}}$, and *Conclusion* picking out its premise, the positive part of its premise, the negative part, and its conclusion, all as follows:

$$Premise(r) = R^s \wedge \neg R^{\overline{s}}_1 \wedge \ldots \wedge \neg R^{\overline{s}}_i,$$
$$Premise^s(r) = R^s,$$
$$Premise^{\overline{s}}(r) = \neg R^{\overline{s}}_1 \wedge \ldots \wedge \neg R^{\overline{s}}_i,$$
$$Conclusion(r) = s.$$

We can then say that r *applies* in a fact situation X just in case $X \models Premise(r)$.

Let us return, now, to the concept of a precedent case $c = \langle X, r, s\rangle$, containing a fact situation X along with a rule r leading to the outcome s. In order for this concept to make sense, we impose two coherence constraints. The first is that the rule contained in the case must actually apply to the facts of the case, or that $X \models Premise(r)$. The second is that the conclusion of the precedent rule must match the outcome of the case itself, or that $Conclusion(r) = Outcome(c)$.

These various concepts and constraints can be illustrated through the concrete case $c_1 = \langle X_1, r_1, s_1\rangle$, containing the fact situation $X_1 = \{f^\pi_1, f^\pi_2, f^\pi_3, f^\delta_1, f^\delta_2, f^\delta_3, f^\delta_4\}$, with three factors favoring the plaintiff and four favoring the defendant, where r_1 is the rule $\{f^\pi_1, f^\pi_2\} \wedge \neg\{f^\delta_5\} \wedge \neg\{f^\delta_4, f^\delta_6\} \rightarrow \pi$, and where the outcome s_1 is π, a decision for the plaintiff. Since we have both $X_1 \models Premise(r_1)$ and $Conclusion(r_1) = Outcome(c_1)$, it is clear that the case satisfies our two coherence constraints: the precedent rule is applicable to the fact situation: and the conclusion of the precedent rule matches the outcome of the case. This particular precedent, then, represents a case in which the court decided for the plaintiff by applying or introducing a rule according to which the presence of the factors f^π_1 and f^π_2, together with the absence of the factor f_5^δ, as well as the absence of the pair of factors f_4^δ and f_6^δ, leads to decision for the plaintiff.

With this notion of a precedent case in hand, we can now define a *case base* as a set Γ of precedent cases. It is a case base of this sort that will be taken to represent the common law in some area, and to constrain the decisions of future courts.

3 Constraint by Rules

We now turn to the standard account of precedential constraint, in terms of rules that can be modified. I motivate this account by tracing a three simple examples of legal development according to the standard view, generalizing from these examples, and then characterizing what I take to be the standard notion of precedential constraint in terms of this generalization.

To begin with, then, suppose that the background case base is $\Gamma_1 = \{c_2\}$, containing only the single precedent case $c_2 = \langle X_2, r_2, s_2 \rangle$, with $X_2 = \{f_1^\pi, f_1^\delta\}$, where $r_2 = \{f_1^\pi\} \rightarrow \pi$, and where $s_2 = \pi$; this precedent represents a situation in which a prior court, confronted with the conflicting factors f_1^π and f_1^δ, decided for π on the basis of f_1^π. Now imagine that, against the background of this case base, a later court is confronted with the new fact situation $X_3 = \{f_1^\pi, f_2^\delta\}$, and takes the presence of the new factor f_2^δ as sufficient to justify a decision for δ. Of course, the previous rule r_2 applies to the new fact situation, apparently requiring a decision for π. But according to the standard account, the court can decide for δ all the same by distinguish the new fact situation from that of the case in which r_2 was introduced—pointing out that the new situation, unlike that of the earlier case, contains the factor f_2^δ, and so declining to apply the earlier rule on that basis.

The result of this decision, then, is that the original case base is changed in two ways. First, by deciding the new situation for δ on the basis of f_2^δ, the court supplements this case base with the new case $c_3 = \langle X_3, r_3, s_3 \rangle$, where X_3 is as above, where $r_3 = \{f_2^\delta\} \rightarrow \delta$, and where $s_3 = \delta$. And second, by declining to apply the earlier r_2 to the new situation due to the presence of f_2^δ, the court, in effect, modifies this earlier rule so that it now carries the force of $r_2' = \{f_1^\pi\} \wedge \neg\{f_2^\delta\} \rightarrow \pi$. The new case base is thus $\Gamma_1' = \{c_2', c_3\}$, with $c_2' = \langle X_2', r_2', s_2' \rangle$ where $X_2' = X_2$, where r_2' is as above, and where $s_2' = s_2$, and with c_3 as above.

The process could continue, of course. Suppose now that, against the background of the modified case base $\Gamma_1' = \{c_2', c_3\}$, another court is confronted with the further fact situation $X_4 = \{f_1^\pi, f_3^\delta\}$, and again takes the new factor f_3^δ as sufficient to justify a decision for δ, in spite of the fact that even the modified rule r_2' requires a decision for π. Once again, this decision changes the current case base in two ways: first, supplementing this case base with a new case representing the current decision, and second, further modifying the previous rule to avoid a conflicting result in the current case. The resulting case base is therefore $\Gamma_1'' = \{c_2'', c_3, c_4\}$, with $c_2'' = \langle X_2'', r_2'', s_2'' \rangle$ as a modification of the previous c_2', where $X_2'' = X_2'$, where $s_2'' = s_2'$, and now where $r_2'' = \{f_1^\pi\} \wedge \neg\{f_2^\delta\} \wedge \neg\{f_3^\delta\} \rightarrow \pi$, with c_3 is as above, and with $c_4 = \langle X_4, r_4, s_4\rangle$ representing the current decision, where X_4 is as above, where $r_4 = \{f_3^\delta\} \rightarrow \delta$, and where $s_4 = \delta$.

As our second example, suppose that the background case base is $\Gamma_2 = \{c_2, c_5\}$, with c_2 as above, and with $c_5 = \langle X_5, r_5, s_5\rangle$, where $X_5 = \{f_1^\pi, f_2^\delta\}$, where $r_5 = \{f_1^\pi\} \rightarrow \pi$, and where $s_5 = \pi$. This case base represents a pair of prior decisions for π on the basis of f_1^π, in spite of the conflicting factors f_1^δ, in one case, and f_2^δ, in the other. Now suppose that, against this background, a later court confronts the new situation $X_6 = \{f_1^\pi, f_1^\delta, f_2^\delta\}$, and decides that, although earlier cases favored f_1^π over the conflicting f_1^δ and f_2^δ presented separately, the combination of f_1^δ and f_2^δ together justifies a decision for δ. Again, this decision supplements the existing case base with the new case $c_6 = \langle X_6, r_6, s_6\rangle$, where X_6 as above, where $r_6 = \{f_1^\delta, f_2^\delta\} \rightarrow \delta$, and where $s_6 = \delta$. But here, the rules from both of the existing cases, c_2 and c_5, must be modified to block application to situations in which f_1^δ and f_2^δ appear together, and so now carry the force of $r_2' = r_5' = \{f_1^\pi\} \wedge \neg\{f_1^\delta, f_2^\delta\} \rightarrow \pi$. The case base resulting from this decision is thus $\Gamma_2' = \{c_2', c_5', c_6, \}$, with $c_2' = \langle X_2', r_2', s_2'\rangle$ where $X_2' = X_2$, where r_2' as above, and where $s_2' = s_2$, with $c_5' = \langle X_5', X_5', s_5'\rangle$ where $X_5' = X_5$, where X_5' is as above, and where $s_5' = s_5$, and with c_6 as above.

Finally, suppose the background case base is $\Gamma_3 = \{c_2, c_7\}$ again with c_2 as above, but with $c_7 = \langle X_7, r_7, s_7\rangle$, where $X_7 = \{f_2^\pi, f_2^\delta\}$, where $r_7 = \{f_2^\pi\} \rightarrow \pi$, and where $s_7 = \pi$. This case base represents a pair of previous decisions for π, one on the basis of f_1^π in spite of the conflicting f_1^δ, and one on the basis of f_2^π in spite of the conflicting f_2^δ. Now imagine that a later court confronts the new situation $X_8 = \{f_1^\pi, f_2^\delta\}$, containing two factors that have not yet been compared, and concludes that f_2^δ is sufficient to justify a decision for δ, in spite of the conflicting f_1^π. Once again, the earlier rule r_2 must be taken to have the force of $r_2' = \{f_1^\pi\} \wedge \neg\{f_2^\delta\} \rightarrow \pi$, in order not to conflict with the current decision. In this case, however, the new rule cannot be formulated simply as $\{f_2^\delta\} \rightarrow \delta$, but must now take the form of $r_8 = \{f_2^\delta\} \wedge \neg\{f_2^\pi\} \rightarrow \delta$, in order not to conflict with the decision for π previously reached in c_7. This scenario, then, is one in which modifications are forced in both directions: a previous rule must be modified to avoid conflict with the current decision, while at the same time, the rule of the current case must be hedged to avoid conflict with a previous decision. The resulting case base is $\Gamma_3' = \{c_2', c_7, c_8, \}$, with $c_2' = \langle X_2', r_2', s_2'\rangle$, where $X_2' = X_2$, where r_2' is as above, and where $s_2' = s_2$, with c_7 as above, and with $c_8 = \langle X_8, r_8, s_8\rangle$, where X_8 as above, where r_8 as above, and where $s_8 = \delta$.

Each of these examples describes a scenario in which a sequence of fact situations are confronted, decisions are reached, rules are formulated to justify the decisions, and rules are modified to accommodate later, or earlier, decisions. It is interesting, and somewhat surprising, to note that, as long as all decisions can be accommodated, with rules properly modified to avoid conflicts, then the order in which cases are confronted is irrelevant. To put this point precisely, let us stipulate that, where $c = \langle X, r, s\rangle$ is a precedent case decided for the side s, the *reason* for this decision is $Premise^s(r)$, the positive part of the premise of the case rule; and suppose that a case base has been constructed through the process of considering fact situations in some particular sequence, in each case rendering a decision for some particular reason and modifying other rules accordingly. It then turns out that, as long as the same decisions are rendered for the same reasons, the same case base will be constructed, with all rules modified in the same way, regardless of the sequence in which the fact situations are considered. Indeed, the fact situations need not be considered in any sequence at all: as long as the set of decisions in these situations is capable of being accommodated through appropriate rule modifications, then all the rules can be modified at once, through a process of case base *refinement*.

This process of transforming a case base Γ into its refinement can be described informally as follows: first, for each case c belonging to Γ, decided for some side and for some particular reason, collect together into Γ_c all of the cases in which that reason hold, but which were decided for the other side; next, for each such case c' from Γ_c, take the negation of the reason for which that case was decided, and then conjoin all of these negated reasons together; finally, replace the rule from the original case c with the new rule that results when this complex conjunction is itself conjoined with the reason for the original decision. And this informal description can be transformed at once into a formal definition.

Definition 1 (Refinement of a case base) Where Γ is a case base, its refinement—written, Γ^+—is the set that results from carrying out the following procedure. For each case $c = \langle X, r, s\rangle$ belonging to Γ:

1. Let
$$\Gamma_c = \{c' \in \Gamma : c' = \langle Y, r', \overline{s}\rangle \;\&\; Y \models \mathit{Premise}^s(r)\}$$

2. For each case $c' = \langle Y, r', \overline{s}\rangle$ from Γ_c, let
$$d_{c,c'} = \neg \mathit{Premise}^{\overline{s}}(r')$$

3. Define
$$D_c = \bigwedge_{c' \in \Gamma_c} d_{c,c'}$$

4. Replace the case $c = \langle X, r, s\rangle$ from Γ with $c'' = \langle X, r'', s\rangle$, where r'' is the new rule
$$\mathit{Premise}^s(r) \wedge D_c \rightarrow s$$

It is easy to verify that, in each of our three examples, the case base resulting from our sequential rule modification is identical with the case base that would have resulted simply from deciding the same fact situations for the same reasons, and then modifying all rules at once, through refinement. Focusing only on the first of our examples, we can see that $\Gamma_1' = (\Gamma_1 \cup \{c_3\})^+$, and then that $\Gamma_1'' = (\Gamma_1' \cup \{c_4\})^+$—or, considering the two later decisions together, that $\Gamma_1'' = (\Gamma_1 \cup \{c_3, c_4\})^+$.

In these situations, then, where a decision can be accommodated against the background of a case base through an appropriate modification of rules, the same outcome can be achieved, the rules modified in the same way, simply by supplementing the background case base with that decision and then refining the result. But of course, there are some decisions that cannot be accommodated against the background of certain case bases—the rules simply cannot be modified appropriately. What does refinement lead to in a situation like this? As it turns out, the result of refining the case base supplemented with the new decision will not then be a case base at all, since refinement will produce rules that fail to apply to their corresponding fact situations. And this linkage between accommodation and refinement, I suggest, works in both directions, and can be taken as a formal explication for the concept of accommodation: a decision can be accommodated against the background of a case base just in case the result of supplementing that case base with the decision can itself be refined into a case base.

We can now turn to the notion of precedential constraint itself according to the standard model, in terms of rules that can be modified. The initial idea is that a court is constrained to reach a decision that can be accommodated within the context of a background case base through an appropriate modification of rules—or, given our formal explication of this concept, a decision that can be combined with the background case base to yield a result whose refinement is itself a case base.

Definition 2 (Rule constraint) Let Γ be a case base and X a new fact situation confronting the court. Then the rule constraint requires the court to base its decision on some rule r leading to an outcome s such that $(\Gamma \cup \{\langle X, r, s\rangle\})^+$ is a case base.

This definition can be illustrated by taking as background the case base $\Gamma_4 = \{c_9\}$, containing the single case $c_9 = \langle X_9, r_9, s_9\rangle$, where $X_9 = \{f_1^\pi, f_2^\pi, f_1^\delta, f_2^\delta\}$, where $r_9 = \{f_1^\pi\} \rightarrow \pi$, and where $s_9 = \pi$. Now suppose the court confronts the new situation $X_{10} = \{f_1^\pi, f_1^\delta, f_2^\delta, f_3^\delta\}$, and considers finding for δ on the basis of f_1^δ and f_2^δ, leading to the decision $c_{10} = \langle X_{10}, r_{10}, s_{10}\rangle$, where X_{10} is as above, where $r_{10} = \{f_1^\delta, f_2^\delta\} \rightarrow \delta$, and where $s_{10} = \delta$. According to current view, this decision is ruled out by precedent, since the result of supplementing the background case base Γ_4 with c_{10} cannot itself be refined into a case base. Indeed, we have $(\Gamma_4 \cup \{c_{10}\})^+ = \{c_9', c_{10}'\}$ with $c_9' = \langle X_9', r_9', s_9'\rangle$, where $X_9' = X_9$, where $r_9' = \{f_1^\pi\} \wedge \neg\{f_1^\delta, f_2^\delta\} \rightarrow \pi$, and where $s_9 = \pi$, and with $c_{10}' = \langle X_{10}', r_{10}', s_{10}'\rangle$, where $X_{10}' = X_{10}$, where $r_{10}' = \{f_1^\delta, f_2^\delta\} \wedge \neg\{f_1^\pi\} \rightarrow \delta$, and where $s_{10} = \delta$. But it is easy to see that neither c_9' nor c_{10}' is a case, in our technical sense, since the rule r_9' fails to apply to X_9', and the rule r_{10}' fails to apply to X_{10}'.

4 Constraint by Reasons

Having provided a formal reconstruction of what I take to be the standard account of precedential constraint, in terms of rules that can be modified, I now want to review my own account, developed in terms of an ordering relation on reasons.

In order to motivate this concept, let us return to the case $c_9 = \langle X_9, r_9, s_9\rangle$—where again $X_9 = \{f_1^\pi, f_2^\pi, f_1^\delta, f_2^\delta\}$, where $r_9 = \{f_1^\pi\} \rightarrow \pi$, and where $s_9 = \pi$—and ask what information is actually carried by this case; what is the court telling us with its decision? Well, two things, at least. First of all, by appealing to the rule r_9 as justification, the court is telling us that the reason for the decision—that is, $Premise^{\beta}(r_9)$, or $\{f_1^{\pi}\}$—is actually sufficient to justify a decision in favor of π. But second, with its decision for π, the court is also telling us that this reason is preferable to whatever other reasons the case might present that favor the δ.

To put this precisely, let is first stipulate that, if X and Y are reasons favoring the same side, then Y is *at least as strong* a reason as X for that side whenever $X \subseteq Y$. Returning to our example, then, where $X_9 = \{f_1^{\pi}, f_2^{\pi}, f_1^{\delta}, f_2^{\delta}\}$, it is clear that the strongest reason present for δ is $X_9^{\delta} = \{f_1^{\delta}, f_2^{\delta}\}$, containing all those factors from the original fact situation that favor δ. Since the c_9 court has decided for π on the grounds of the reason $Premise^{\beta}(r_9)$, even in the face of the reason X_9^{δ}, it seems to follow as a consequence of the court's decision that the reason $Premise^{\beta}(r_9)$ for π is preferred to the reason X_9^{δ} for the δ—that is, that $\{f_1^{\pi}\}$ is preferred to the reason $\{f_1^{\delta}, f_2^{\delta}\}$. If we introduce the symbol $<_{c_9}$ to represent the preference relation on reasons that is derived from the particular case c_9, then this consequence of the court's decision can be put more formally as the claim that $\{f_1^{\delta}, f_2^{\delta}\} <_{c_9} \{f_1^{\pi}\}$, or equivalently, that $X_9^{\delta} <_{c_9} Premise^{\beta}(r_9)$.

As far as the preference ordering goes, then, the earlier court is telling us at least that $X_9^{\delta} <_{c_9} Premise^{\beta}(r_9)$, but is it telling us anything else? Perhaps not explicitly, but implicitly, yes. For if the reason $Premise^{\beta}(r_9)$ for π is preferred to the reason X_9^{δ} for δ, then surely any reason for π that is at least as strong as $Premise^{\beta}(r_9)$ must likewise be preferred to X_9^{δ}, and just as surely, $Premise^{\beta}(r_9)$ must be preferred to any reason for δ that is at least as weak as X_9^{δ}. As we have seen, a reason Z for π is at least as strong as $Premise^{\beta}(r_9)$ if it contains all the factors contained by $Premise^{\beta}(r_9)$—that is, if $Premise^{\beta}(r_9) \subseteq Z$. And we can conclude, likewise, that a reason W for δ is at least as weak as X_9^{δ} if it contains no more factors than X_9^{δ} itself—that is, if $W \subseteq X_9^{\delta}$. It therefore follows from the earlier court's decision in c_9, not only that $X_9^{\delta} <_{c_9} Premise^{\beta}(r_9)$, but that $W <_{c_9} Z$ whenever W is at least as weak a reason for δ as X_9^{δ} and Z is at least as strong a reason for π as $Premise^{\beta}(r_9)$—whenever, that is, $W \subseteq X_9^{\delta}$ and $Premise^{\beta}(r_9) \subseteq Z$. To illustrate: from the court's explicit decision that $\{f_1^{\delta}, f_2^{\delta}\} <_{c_9} \{f_1^{\pi}\}$, we can conclude also that $\{f_1^{\delta}\} <_{c_9} \{f_1^{\pi}, f_3^{\pi}\}$, for example.

This line of argument leads to the following definition of the preference relation among reasons that can be derived from a single case.

Definition 3 (Preference relation derived from a case) Let $c = \langle X, r, s \rangle$ be a case, and suppose W and Z are reasons. Then the relation $<_c$ representing the preferences on reasons derived from the case c is defined by stipulating that $W <_c Z$ if and only if $W \subseteq X^{\overline{s}}$ and $Premise^{s}(r) \subseteq Z$.

Once we have defined the preference relation derived from a single case, we can introduce a preference relation $<_{\Gamma}$ derived from an entire case base Γ in the natural way, by stipulating that one reason is stronger than another according to the entire case base if that strength relation is supported by some particular case in the case base.

Definition 4 (Preference relation derived from a case base) Let Γ be a case base, and suppose W and Z are reasons. Then the relation $<_\Gamma$ representing the preferences on reasons derived from the case base Γ is defined by stipulating that $W <_\Gamma Z$ if and only if $W <_c Z$ for some case c from Γ.

And we can then define a case base as inconsistent if it provides conflicting information about the preference relation among reasons—telling us, for any two reasons, the each is preferred to the other—and consistent otherwise.

Definition 5 (Reason consistent case bases) Let Γ be a case base with $<_\Gamma$ the derived preference relation. Then Γ is reason inconsistent if and only if there are reasons X and Y such that $X <_\Gamma Y$ and $Y <_\Gamma X$. Γ is reason consistent if and only if it is not reason inconsistent.

Given this notion of consistency, we can now turn to the concept of precedential constraint itself, according to he reason account. The intuition could not be simpler: in deciding a case, a constrained court is required to preserve the consistency of the background case base. Suppose, more exactly, that a court constrained by a background case base Γ is confronted with a new fact situation X. Then the court is required to reach a decision on X that is itself consistent with Γ—that is, a decision that does not result in an inconsistent case base.

Definition 6 (Reason constraint) Let Γ be a case base and X a new fact situation confronting the court. Then reason constraint requires the court to base its decision on some rule r leading to an outcome s such that the new case base $\Gamma \cup \{\langle X, r, s\rangle\}$ is reason consistent.

This idea can be illustrated by assuming as background the previous case base $\Gamma_4 = \{c_9\}$, containing only the previous case c_9, supposing once again that, against this background, the court confronts the fresh situation $X_{10} = \{f_1^\pi, f_1^\delta, f_2^\delta, f_3^\delta\}$ and considers finding for δ on the basis of f_1^δ and f_2^δ, leading to the decision $c_{10} = \langle X_{10}, r_{10}, s_{10}\rangle$, where X_{10} is as above, where $r_{10} = \{f_1^\delta, f_2^\delta\} \rightarrow \delta$, and where $s_{10} = \delta$. We saw in the previous section that such a decision would fail to satisfy the rule constraint, and we can see now that it fails to satisfy the reason constraint as well. Why? Because the new case c_{10} would support the preference relation $\{f_1^\pi\} <_{c_{10}} \{f_1^\delta, f_2^\delta\}$, telling us that the reason $\{f_1^\delta, f_2^\delta\}$ for δ outweighs the reason $\{f_1^\pi\}$ for π. But Γ_4 already contains the case c_9, from which we can derive the preference relation $\{f_1^\delta, f_2^\delta\} <_{c_9} \{f_1^\pi\}$, telling us exactly the opposite. As a result, the augmented case base $\Gamma_4 \cup \{c_{10}\}$ would be reason inconsistent.

5 An Equivalence

The two accounts presented in Sects. 3 and 4 of this paper offer strikingly different pictures of precedential constraint, and of legal development and norm change. According to the standard account from Sect. 3, what is important about a background case base is the set of rules it contains, together with the facts of the cases in which they were formulated. In reaching a decision concerning a new fact situation, the court is obliged by modify the existing set of rules appropriately, to accommodate this decision. Precedential constraint derives from the fact that such accommodation is not always possible; legal development is due to the modification of existing rules, together with the addition to the case base of the new rule from the new decision. According to the reason account from Sect. 4, what is important about a background case base is, not the set of rules it contains, but a derived preference ordering on reasons. In confronting a new fact situation, then, a court is obliged only to reach a decision that is consistent with the existing preference ordering on reasons. Constraint derives from the fact that not all such decisions are consistent; legal development is due to the supplementation of the existing preference ordering on reasons with the new preferences derived from the new decision.

Given the very different pictures presented by these two accounts of precedential constraint, it is interesting to note that the accounts are in fact equivalent, in the sense that, given a background case base Γ and a new fact situation X, a decision on the basis of a rule r is permitted by the rule constraint just in case it is permitted by the reason constraint. This observation—the chief result of the paper—can be established very simply, after showing, first, that any reason consistent case base has a case base as its refinement, and second, that any case base with a case base as its refinement must be reason consistent.

Observation 1 If Γ is a reason consistent case base, then its refinement Γ^+ is a case base.

Proof Suppose Γ is a reason consistent case base. Γ^+ is constructed from Γ by replacing each case $c = \langle X, r, s\rangle$ from Γ with the new $c'' = \langle X, r'', s\rangle$, where the new rule r'' has the form $\mathit{Premise}^s(r) \wedge D_c \rightarrow s$, as specified as in Definition 1. Since all of the new rules involved in moving from Γ to Γ^+ support the same outcomes as the original, we can verify that Γ^+ is a case base as well simply by establishing that, for each $c'' = \langle X, r'', s\rangle$ from Γ^+, the new rule r'' continues to be applicable to the fact situation X—that is, that $X \models \mathit{Premise}(r'')$, or that $X \models \mathit{Premise}^s(r) \wedge D_c$. We know, of course, that $X \models \mathit{Premise}^s(r)$, since Γ is a case base, and so need only show that $X \models D_c$.

It follows from Steps 2 and 3 of the construction that establishing that $X \models D_c$ amounts to showing, for each $c' = \langle Y, r', \bar{s}\rangle$ from Γ_c, where $c = \langle X, r, s\rangle$, that $X \models \neg\mathit{Premise}^{\bar{s}}(r')$. So suppose the contrary—that $X \not\models \neg\mathit{Premise}^{\bar{s}}(r')$, or $X \models \mathit{Premise}^{\bar{s}}(r')$, from which we can conclude that (1) $\mathit{Premise}^{\bar{s}}(r') \subseteq X^{\bar{s}}$. Since $c' = \langle Y, r', \bar{s}\rangle$ belongs to Γ_c, we know from Step 1 of the construction that $Y \models \mathit{Premise}^s(r)$, from which we can conclude that (2) $\mathit{Premise}^s(r) \subseteq Y^s$. From (1), we can then conclude by Definition 3 that (3) $\mathit{Premise}^{\bar{s}}(r') <_c \mathit{Premise}^s(r)$, and from (2), that (4) $\mathit{Premise}^s(r) <_{c'} \mathit{Premise}^{\bar{s}}(r')$. But since both c and c' belong to Γ, the combination of (3) and (4) contradicts the stipulation that Γ is reason consistent. Hence, our assumption fails, from which we can conclude that $X \models D_c$. ■

Observation 2 If Γ is a case base whose refinement Γ^+ is also a case base, then Γ is reason consistent.

Proof Suppose Γ is a case base whose refinement Γ^+ is a case base, but that Γ itself is not reason consistent. Since Γ is not reason consistent, there are reasons A and B such that (1) $A <_c B$ and (2) $B <_{c'} A$ for cases $c = \langle X, r, s\rangle$ and $c' = \langle Y, r', \overline{s}\rangle$ from Γ. From (1) we have (3) $A \subseteq X^{\overline{s}}$ and (4) $Premise^s(r) \subseteq B$, and from (2) we have (5) $B \subseteq Y^s$ and (6) $Premise^{\overline{s}}(r') \subseteq A$. Together, (4) and (5), along with the fact that $Y^s \subseteq Y$, yield $Premise^s(r) \subseteq Y$, or (7) $Y \models Premise^s(r)$. In the same way, (3) and (6), together with the fact that $X^{\overline{s}} \subseteq X$, yield $Premise^{\overline{s}}(r') \subseteq X$, or (8) $X \models Premise^{\overline{s}}(r')$.

Γ^+ is constructed from the case base Γ by replacing each case $c = \langle X, r, s\rangle$ with the new $c'' = \langle X, r'', s\rangle$, where the new rule r'' has the form $Premise^s(r) \wedge D_c \rightarrow s$, as specified in Definition 1. Step 1 of this construction, together with (7), tells us that c' belongs to Γ_c, and then Steps 2, 3, and 4 allow us to conclude that $\neg Premise^{\overline{s}}(r')$ is one of the conjuncts of D_c, and so of the new rule r''. From (8), however, we know that $X \models Premise^{\overline{s}}(r')$, from which it follows that $X \not\models \neg Premise(r'')$. As a result, the rule of c'' does not apply to its facts, from which it follows that c'' is not a case, and so Γ^+ not a case base, contrary to our assumption. ■

Observation 3 Let Γ be a case base, and let X be a new fact situation. Then a decision on the basis of a rule r leading to an outcome s is permitted by the reason constraint just in case that decision is permitted by the rule constraint.

Proof Suppose a decision on the basis of r and leading to the outcome s is permitted by the reason constraint, so that $\Gamma \cup \{\langle X, r, s\rangle\}$ is reason consistent. Then $(\Gamma \cup \{\langle X, r, s\rangle\})^+$ is a rule coherent case base, by Observation 1, so that the same decision is permitted by the rule constraint. Or suppose a decision on the basis of r and leading to the outcome s is permitted by the rule constraint, so that $(\Gamma \cup \{\langle X, r, s\rangle\})^+$ is a case base. Then $\Gamma \cup \{\langle X, r, s\rangle\}$ is reason consistent by Observation 2, so that the same decision is permitted by the reason constraint. ■

6 Discussion

The goal of this paper has been to establish the equivalence between two accounts of precedential constraint, the standard account from Sect. 3 and the reason account from Sect. 4. I discuss what I take to be the philosophical significance of this equivalence elsewhere.[12] Here I simply want to close with two technical remarks, one concerning the standard account and one concerning the reason account.

Beginning with the standard account, developed in terms of rules that can be modified, it is natural to ask *why* these rules should be modified—what property of the overall case base can we suppose courts are trying to establish, or guarantee, through the modification of rules? That natural answer to this natural question is that, by modifying rules, courts are trying to guarantee a kind of consistency. More exactly, suppose we take

[12] See Horty (2013).

$$Rule(\Gamma) = \{Rule(c) : c \in \Gamma\}$$

as the set of rules derived from a case base Γ. We can then define a case base as rule consistent just in case, whenever a rule derived from that case base applies to the facts of some case from the case base, the rule yields an outcome identical to the outcome that was actually reached in that case.

Definition 7 (Rule consistent case base) A case base Γ is rule consistent if and only if, for each r in $Rule(\Gamma)$ and for each c in Γ such that $Factors(c) \models Premise(r)$, we have $Outcome(c) = Conclusion(r)$.

And it is natural to conjecture that, by modifying rules to accommodate later decisions, the property that courts are trying to preserve is the property of rule consistency.

This conjecture is supported by reflection on the examples we used to motivate the standard account. Recall our initial example. Here, the background case base was $\Gamma_1 = \{c_2\}$, with $c_2 = \langle X_2, r_2, s_2 \rangle$, where $X_2 = \{f_1^\pi, f_1^\delta\}$, where $r_2 = \{f_1^\pi\} \rightarrow \pi$, and where $s_2 = \pi$; and we imagined that the court, confronting the new fact situation $X_3 = \{f_1^\pi, f_2^\delta\}$, wishes to decide for δ on the basis of f_2^δ, leading to the decision $c_3 = \langle X_3, r_3, s_3 \rangle$, where X_3 is as above, where $r_3 = \{f_2^\delta\} \rightarrow \delta$, and where $s_3 = \delta$. Now suppose the rule from the original case had not been modified, so that the result of this decision was that the original case base was simply supplemented with the new decision, leading to the revised case base $\Gamma_1 \cup \{c_3\}$. It is easy to see that the revised case base would not be rule consistent, since r_2 belongs to $Rule(\Gamma_1 \cup \{c_3\})$ and $Factors(c_3) \models Premise(r_2)$, yet $Outcome(c_3) \neq Conclusion(r_2)$—the rule from the original case applies to the facts of the new case, but supports a different result from that actually reached in the new case. By modifying the original rule r_2 to have the force of $r_2{}' = \{f_1^\pi\} \wedge \neg\{f_2^\delta\} \rightarrow \pi$, the court can thus be seen as guaranteeing rule consistency, blocking application of the rule to a case with a conflicting outcome.

Our other motivating examples have the same form: the later decisions would introduce rule inconsistency on their own, but modification of the earlier rules restores consistency. Is it, then, rule consistency that we should see a court as attempting to guarantee by modifying rules? Surprisingly, perhaps, I would say No—for some case bases are peculiar even though they are rule consistent. Consider, for example, the case base $\Gamma_5 = \{c_{11}\}$ with $c_{11} = \langle X_{11}, r_{11}, s_{11} \rangle$, where $X_{11} = \{f_1^\pi, f_2^\pi, f_1^\delta\}$, where $r_{11} = \{f_1^\pi\} \rightarrow \pi$, and where $s_{11} = \pi$. And suppose that, against the background of this case base, the court confronts the new fact situation $X_{12} = \{f_1^\pi, f_1^\delta, f_2^\delta\}$ and wishes to decide for δ on the basis of f_1^δ. There is, then, the risk of rule inconsistency, since the previous rule r_{11} applies to the fact situation X_{12}, but leads to π as an outcome, rather than δ. But now, imagine that the court distinguishes in the following way: first, by modifying the previous r_{11} to have the force of $r_{11}{}' = \{f_1^\pi\} \wedge \neg\{f_2^\delta\} \rightarrow \pi$, and second, by hedging its rule for the new case to read $r_{12} = \{f_1^\delta\} \wedge \neg\{f_2^\pi\} \rightarrow \delta$. The resulting case base would then be $\Gamma_5{}' = \{c_{11}{}', c_{12}\}$, with $c_{11}{}' = \langle X_{11}{}', r_{11}{}', s_{11}{}' \rangle$, where $X_{11}{}' = X_{11}$, where $r_{11}{}'$ is as above, and where $s_{11}{}' = s_{11}$, and with

$c_{12} = \langle X_{12}, r_{12}, s_{12}\rangle$, where $X_{12} = \{f_1^\pi, f_1^\delta, f_2^\delta\}$, where $r_{12} = \{f_1^\delta\} \wedge \neg\{f_2^\pi\} \rightarrow \delta$, and where $s_{12} = \delta$.

It is easy to see that the new case base $\Gamma_5{}'$ is rule consistent, since neither of the rules involved applies to the other case. It is, nevertheless, a peculiar case base. One way of seeing this is by noting that, although rule consistent, the case base is not reason consistent: we have both $\{f_1^\delta\} <_{c_{11}{}'} \{f_1^\pi\}$ and $\{f_1^\pi\} <_{c_{12}} \{f_1^\delta\}$. Another way—which gets to the root of the problem—is by noting that, in each of the two case rules, the exception clause, which blocks applicability to the other case, has nothing to do with the reason for which that other case was decided.

Because a case base can be peculiar even if it is rule consistent, I do not think that mere rule consistency is the property that courts are concerned to guarantee, as they modify rules. Instead, I believe, courts must be seen as trying to avoid, not just rule inconsistency, but also peculiarity in the sense illustrated above, by guaranteeing the property of rule coherence.

Definition 8 (Rule coherent case base) Let Γ be a case base. Then Γ is rule coherent just in case, for each $c = \langle X, r, s\rangle$ and $c' = \langle Y, r', \overline{s}\rangle$ in Γ, if $Y \models \mathit{Premise}^s(r)$, then $\mathit{Premise}(r) \vdash \neg \mathit{Premise}^{\overline{s}}(r')$.

What this property requires is that, whenever the reason for a decision in some particular case holds in another case where the opposite outcome was reached, then the negation of the reason for the latter decision must be entailed by the premise of the rule supporting the original.[13] The property of rule coherence is thus supposed to be explanatory in a way that mere rule consistency is not: when the original reason holds in a latter case but fails to yield the appropriate outcome, the rule putting forth the original reason must help us understand why, by containing the information that it does not apply when the reason from the latter case is present.

We can verify that rule coherence is a stronger property than mere rule consistency, in the sense that a rule coherent case must be rule consistent.

Observation 4 Any rule coherent case base is rule consistent.

Proof Suppose a case base Γ is rule coherent but not rule consistent. Since Γ is not rule consistent, $\mathit{Rule}(\Gamma)$ contains some rule r, derived from some case $c = \langle X, r, s\rangle$ belonging to Γ, for which there is another case $c' = \langle Y, r', \overline{s}\rangle$ from Γ such that $Y \models \mathit{Premise}(r)$ and $\mathit{Conclusion}(r) \neq \mathit{Outcome}(c)$. Since $Y \models \mathit{Premise}(r)$, we know that $Y \models \mathit{Premise}^s(r)$, of course. By rule coherence, we then have $\mathit{Premise}(r) \vdash \neg\mathit{Premise}^{\overline{s}}(r')$, from which it follows that $Y \not\models \mathit{Premise}^{\overline{s}}(r')$, so that $Y \not\models \mathit{Premise}(r')$, which contradicts the requirement that the rule of $c' = \langle Y, r', \overline{s}\rangle$ must be applicable to the facts of the case. ■

And we can also see that, in contrast with case bases that are merely rule consistent, a case base that is rule coherent must be reason consistent as well.

---

[13] The most straightforward way in which the negation of the reason for the opposite decision would be entailed by the rule supporting the original, of course, is by being contained explicitly among the exceptions to that rule; but speaking more generally of entailment allows for other encodings as well.

Observation 5 Any rule coherent case base is reason consistent.

Proof Suppose a case base Γ is rule coherent but not reason consistent. Since Γ is not reason consistent, there are reasons A and B such that (1) $A <_c B$ and (2) $B <_{c'} A$ for cases $c = \langle X, r, s\rangle$ and $c' = \langle Y, r', \overline{s}\rangle$ from Γ. From (1) we have (3) $A \subseteq X^{\overline{s}}$ and (4) $Premise^s(r) \subseteq B$, and from (2) we have (5) $B \subseteq Y^s$ and (6) $Premise^{\overline{s}}(r') \subseteq A$. Together, (4) and (5), along with the fact that $Y^s \subseteq Y$, yield $Premise^s(r) \subseteq Y$, or (7) $Y \models Premise^s(r)$. In the same way, (3) and (6), together with the fact that $X^{\overline{s}} \subseteq X$, yield $Premise^{\overline{s}}(r') \subseteq X$, or (8) $X \models Premise^{\overline{s}}(r')$. From (7), rule coherence tells us that $Premise(r) \vdash \neg Premise^{\overline{s}}(r')$, or that (9) $Premise^{\overline{s}}(r') \vdash \neg Premise(r)$. But then (8) and (9) tell us that $X \models \neg Premise(r)$, or that (10) $X \not\models Premise(r)$, which contradicts the requirement that the rule of $c = \langle X, r, s\rangle$ must be applicable to the facts of that case. ■

It follows from this last observation that, unlike case bases in general, any case base that is rule coherent must have a case base as its refinement. Why is this? Because the observation tells us that any rule coherent case base is reason consistent, and we know from Observation 1 that the refinement of a reason consistent case base is a case base. Of course, a case base might well be reason consistent without being rule coherent—though the case base is reason consistent, its rules may simply not have been modified properly. But it is easy to see that, once the rules of a reason consistent case base have been modified through refinement, the result will be a rule coherent case base.

Observation 6 The refinement of a reason consistent case base is a rule coherent case base.

Proof The proof of Observation 1 shows that the refinement Γ^+ of a reason consistent case base Γ is a case base. To see that Γ^+ is also rule coherent, we need only continue that proof by noting that, where $c'' = \langle X, r'', s\rangle$ is the new case replacing the original case $c = \langle X, r, s\rangle$ from Γ, it follows from the construction of Γ^+ that $Premise^s(r'')$ is identical with $Premise^s(r)$, so that, for any case $c' = \langle Y, r', \overline{s}\rangle$ from Γ^+, whenever $Y \models Premise^s(r'')$, the formula $\neg Premise^{\overline{s}}(r')$ is a conjunct of $Premise(r'')$. From this we can conclude that $Premise(r'') \vdash \neg Premise^{\overline{s}}(r')$ at once. ■

Turning now from the standard account to the reason account, it is worth noting that our definition of reason constraint makes sense only on the assumption that the background case base is itself consistent to begin with. This is, of course, an unrealistic assumption. Given the vagaries of judicial decision, with a body of case law developed by a number of different courts, at different places and different times, it would be surprising if any nontrivial case base were actually consistent. But in fact, this assumption is not essential. The notion of reason inconsistency at work here is not like logical inconsistency—it is local, not pervasive. A case base might be reason inconsistent in certain areas, providing conflicting information about the relative priority of particular reasons, while remaining consistent elsewhere. It is therefore possible to extend our account of reason constraint to apply also to inconsistent case bases, by requiring of a court, not necessarily that its decision should not yield an inconsistent case base, but only that its decision should not introduce any new inconsistencies, which were not present before, into a case base that may already be inconsistent.

To state this precisely, we can define an *inconsistency* in a case base Γ as a pair or reasons, Y and Z, such that $Y <_\Gamma Z$ and $Z <_\Gamma Y$. The idea that a court should introduce no new inconsistencies into a case base can then be captured through the requirement that every inconsistency present after the court's decision must already have been present prior to the decision, leading to the following definition.

Definition 9 (Reason constraint: general version) Let Γ be a case base and X a new fact situation confronting the court. Then reason constraint requires the court to base its decision on some rule r leading to an outcome s such that: whenever $Y <_{\Gamma\cup\{\langle X,r,s\rangle\}} Z$ and $Z <_{\Gamma\cup\{\langle X,r,s\rangle\}} Y$, we also have $Y <_\Gamma Z$ and $Z <_\Gamma Y$.

This more general definition of reason constraint can be illustrated by considering the case base $\Gamma_6 = \{c_{13}, c_{14}\}$, with $c_{13} = \langle X_{13}, r_{13}, s_{13}\rangle$, where $X_{13} = \{f_1^\pi, f_1^\delta, f_2^\delta\}$, where $r_{13} = \{f_1^\pi\} \rightarrow \pi$, and where $s_{13} = \pi$, and with $c_{14} = \langle X_{14}, r_{14}, s_{14}\rangle$, where $X_{14} = \{f_1^\pi, f_1^\delta, \}$, where $r_{14} = \{f_1^\delta\} \rightarrow \delta$, and where $s_{14} = \delta$. This case base is inconsistent, of course, since it tells us both that $\{f_1^\delta\} <_{\Gamma_6} \{f_1^\pi\}$ and that $\{f_1^\pi\} <_{\Gamma_6} \{f_1^\delta\}$. But now, suppose that, against the background of this case base, the court confronts the new fact situation $X_{15} = \{f_1^\pi, f_2^\delta\}$. According to our original Definition 6, nothing the court can do is right, since the case base is already inconsistent. According to our more general definition, however, there is nevertheless a right decision for the court to make, even though the background case base is inconsistent, and a wrong decision. The right decision in this new situation would be to find for π on the basis of f_1^π, since this introduces no new inconsistencies. The wrong decision would be to find for δ on the basis of f_2^δ, leading to the new case base $\Gamma_6 \cup \{c_{15}\}$, with $c_{15} = \langle X_{15}, r_{15}, s_{15}\rangle$, where $X_{15} = \{f_1^\pi, f_2^\delta\}$, where $r_{15} = \{f_2^\delta\} \rightarrow \delta$, and where $s_{15} = \delta$. This decision would introduce a new inconsistency, since we would then have both $\{f_2^\delta\} <_{\Gamma_6\cup\{c_{15}\}} \{f_1^\pi\}$ and $\{f_1^\pi\} <_{\Gamma_6\cup\{c_{15}\}} \{f_2^\delta\}$, even though we did not previously have both $\{f_2^\delta\} <_{\Gamma_6} \{f_1^\pi\}$ and $\{f_1^\pi\} <_{\Gamma_6} \{f_2^\delta\}$.

7 Conclusion

There are many other issues to explore, both technical and philosophical. The case-based priority ordering on reasons is not transitive. Of course, we could simply impose transitivity, by reasoning with the transitive closure of the basic relation; but the question of whether we should leads to a thicket of interesting problems concerning belief combination. In addition, the rules we work with are very simple in form—basically, nothing but a reason supporting a conclusion, and a list of contrary reasons that are required not to hold, if that conclusion is to be reached. Should we allow more complex rules, and if so, how complex? This question, likewise, has a technical side, concerning the ways in which more complex rules might be unwound into ordering relations on reasons, and a conceptual side, since in the common law, courts are not supposed to legislate, but simply to respond to the fact situations before them. How complex can the rules become before we are forced to say that courts are legislating, rather than ruling on cases?

These questions, and others, will have to wait for another occasion. The present paper has a more limited aim. The very influential work by David Makinson and his AGM collaborators began with reflections, by Alchourrón and Makinson, on norm change in the civil law—and my goal here has been simply to sketch one way in which a complimentary theory of norm change in the common law might be developed. In doing so, I find that the line of thought traced here conforms to several, though not all, of the maxims set out by David in a later article, in which he tries to explain what was different about the initial work in the AGM tradition.[14] The relevant maxims are:

Logic is not just about deduction
There is nothing wrong with classical logic
Don't internalize too quickly
Do some logic without logic

Concerning the first of these maxims, it was a notable feature of AGM that, while the overwhelming majority of contemporary philosophical logicians were exploring different consequence relations—extensions of or alternatives to classical logic—the authors of that work focused on the entirely separate topic of belief revision; the present paper, likewise, applies logical techniques to a topic other than the question of what follows from what. Concerning the second maxim, in moving into a new field, the AGM authors carried with them the familiar classical logic; and likewise the present paper. The point of the third maxim is that it is often useful to explore certain concepts in the metalanguage, before trying to represent these concepts through an explicit object language connective, not for Quinean or other philosophical reasons, but simply as a matter of research methodology—get the flat case right, before considering nesting or iteration. This maxim has particular relevance for the present work, since I had originally thought it best, in exploring precedential constraint, to concentrate on the obligations of later courts, with these obligations represented through an object-language deontic operator; it was only when that operator was removed that the shape of the current approach became clear. Finally, as to the fourth maxim, while there are some connectives in the present account—conjunction, negation—it should be clear that they are not contributing much: there is no nesting of formulas, for example. Just as with AGM, while there may be connectives present, the real interest lies elsewhere.

To these four maxims, I would add a fifth:

Sometimes, let the subject shape the logic

One frequently finds that a theorist brings an existing logic or set of logical techniques, particularly those already familiar to the theorist, to a new subject. There is nothing

[14] Makinson (2003).

wrong with that: it is important to develop existing theories, and to explore new applications. But sometimes, existing techniques do not fit the new subject, or do not fit it well. Then there is the opportunity to let the subject suggest new logical ideas, and it is important to be open to that opportunity. This is a path that David has taken more than once—not only in his AGM work, but also in other work that I am familiar with, such as his research on Hohfeld's rights relations, on the general theory of nonmonotonic consequence, on norms without truth values, and on input/output logics.[15] It is a path that has already led him to many vital contributions, and we can expect that it will lead to many more.

References

Alchourrón, C., Gärdenfors, P., & Makinson, D. (1985). On the logic of theory change: Partial meet contraction and revision functions. *Journal of Symbolic Logic*, *50*, 510–530.

Alchourrón, C., & Makinson, D. (1981). Hierarchies of regulations and their logic. In R. Hilpinen (Ed.), *New studies in deontic logic* (pp. 125–148). Dordrecht: D. Reidel Publishing Company.

Alchourrón, C., & Makinson, D. (1982). On the logic of theory change: Contraction functions and their associated revision functions. *Theoria*, *48*, 14–37.

Aleven, V. (1997). *Teaching case-based argumentation through a model and examples*. Unpublished PhD thesis, Intelligent Systems Program, University of Pittsburgh, Pittsburgh.

Aleven, V., & Ashley, K. (1997). Evaluating a learning environment for case-based argumentation skills. In *Proceedings of the 6th International Conference on Artificial Intelligence and Law (ICAIL-97)* (pp. 170–179). The Association for Computing Machinery Press.

Ashley, K. (1989). Toward a computational theory of arguing with precedents: accomodating multiple interpretations of cases. In *Proceedings of the 2nd International Conference on Artificial Intelligence and Law (ICAIL-89)* (pp. 93–110). The Association for Computing Machinery Press.

Ashley, K. (1990). *Modeling legal argument: Reasoning with cases and hypotheticals*. Cambridge: The MIT Press.

Gärdenfors, P. (2011). Notes on the history of ideas behind AGM. *Journal of Philosophical Logic*, *40*, 115–120.

Horty, J. (2011). Rules and reasons in the theory of precedent. *Legal Theory*, *17*, 1–33.

Horty, J. (2013). Common law reasoning. Manuscript.

Horty, J., & Bench-Capon, T. (2012). A factor based definition of precedential constraint. *Artificial Intelligence and Law*, *20*, 181–214.

Lamond, G. (2005). Do precedents create rules? *Legal Theory*, *11*, 1–26.

Makinson, D. (1986). On the formal representation of rights relations: Remarks on the work of stig kanger and lars lindahl. *The Journal of Philosophical Logic*, *15*, 403–425.

Makinson, D. (1989). General theory of cumulative inference. In M. Reinfrank, J. de Kleer, M. Ginsberg & E. Sandewall (Eds.), *Proceedings of the 2nd International Workshop on Nonmonotonic Reasoning, volume 346 of Lecture Notes in Artificial Intelligence* (pp. 1–18). New York: Springer.

Makinson, D. (1994). General patterns in nonmonotonic reasoning. In D. Gabbay, C. Hogger, & J. Robinson (Eds.), *Handbook of Logic in Artificial Intelligence and Logic Programming: Nonmonotonic Reasoning and Uncertain Reasoning* (Vol. 3, pp. 35–110). Oxford: Oxford University Press.

Makinson, D. (1996). In memoriam Carlos Eduardo Alchourrön. *Nordic Journal of Philosophical Logic*, *1*, 3–10.

[15] See Makinson (1986, 1989, 1994, 1998) and Makinson and van der Torre (2000).

Makinson, D. (1999). On a fundamental problem of deontic logic. In P. McNamara, H. Prakken. (Eds.), *Norms, Logics and Information Systems: New Studies in Deontic Logic and Computer Science, Series: Frontiers in Artificial Intelligence and Applications* (Vol. 49, pp. 29–53). Amsterdam: IOS Press.

Makinson, D. (2003). Ways of doing logic: What was different about AGM 1985? *Journal of Logic and Computation*, *13*, 5–15.

Makinson, D., & van der Torre, L. (2000). Input/output logics. *The Journal of Philosophical Logic*, *29*, 383–408.

Raz, J. (1979). *The Authority of Law*. Oxford: Oxford University Press.

Rissland, E. (1990). Artificial intelligence and law: Stepping stones to a model of legal reasoning. *Yale Law Journal*, *99*, 1957–1981.

Rissland, E., & Ashley, K. (1987). A case-based system for trade secrets law. In *Proceedings of the 1st International Conference on AI and Law (ICAIL-87)* (pp. 60–66). The Association for Computing Machinery Press.

Simpson, A. W. B. (1961). The ratio decidendi of a case and the doctrine of binding precedent. In A. G. Guest (Ed.), *Oxford essays in jurisprudence* (pp. 148–175). New York: Oxford University Press.

Part V
Classical Resources

Intelim Rules for Classical Connectives

David Makinson

Abstract We investigate introduction and elimination rules for truth-functional connectives, focusing on the general questions of the existence, for a given connective, of at least one such rule that it satisfies, and the uniqueness of a connective with respect to the set of all of them. The answers are straightforward in the context of rules using general set/set sequents of formulae, but rather complex and asymmetric in the restricted (but more often used) context of set/formula sequents, as also in the intermediate set/formula-or-empty context.

Keywords Intelim rules · Introduction rules · Elimination rules · Classical logic · Truth-functional connectives

1 Introduction

Introduction and elimination (briefly intelim) rules are, in a natural sense, the simplest of all Horn rules for propositional connectives. They have also played a prominent role in some philosophical discussions of the meaning of the connectives; for example, they are sometimes seen as giving grounds to fall back to a sub-classical logic. For this reason, their behaviour has more than a purely formal interest.

We consider two questions about intelim rules for truth-functional connectives: the *existence*, for a given connective, of at least one such rule that it satisfies, and the *uniqueness* of a connective with respect to the set of all of them. The answers are straightforward in the context of rules using *set/set* sequents of formulae, but rather complex and asymmetric for the restricted (but more often used) of *set/formula*

D. Makinson (✉)
Department of Philosophy, Logic and Scientific Method, London School of Economics, Houghton Street, London WC2A 2AE, UK
e-mail: david.makinson@gmail.com
URL: https://sites.google.com/site/davidcmakinson/

S. O. Hansson (ed.), *David Makinson on Classical Methods for Non-Classical Problems*, Outstanding Contributions to Logic 3, DOI: 10.1007/978-94-007-7759-0_16,

context—and differently so in that of set/formula-or-empty sequents, as will be illustrated in the final section.

We write classical consequence as $\vDash$, flanked with a set of formulae on the left and, on the right, either a set of formulae or an individual formula according to the context chosen. We also refer to arbitrary consequence relations, written $\vdash$. They are assumed to be closure relations (i.e. satisfying the Tarski conditions appropriate to the context, be it set/set or set/formula), and also to be closed under uniform substitution for sentence letters. Familiar definitions, e.g. of formula-schemes, sequents and rules understood as schemes, the categories of elementary, structural, introduction and elimination rules as well as the notion of satisfaction of a rule, are recalled with some comments in an appendix. For unimpeded reading, the reader is advised to plunge directly into the text, but to check definitions in the appendix at delicate moments of formulation and proof.

2 Questions of Existence

Under what conditions does a given truth-functional connective satisfy at least one intelim rule? First, we consider elimination rules, where the conditions for existence are the same in the set/formula and set/set contexts. Then we look at introduction rules, where the existence requirements turn out to be much tighter in the set/formula context than in the set/set one.

Throughout the section, $*$ will be a truth-functional connective in a classical language L, and satisfaction means satisfaction wrt $\vDash$. We omit explicit reference to L since, as explained in the appendix, its identity will not matter so long as it is classical and can express at least one tautology and at least one contradiction. In Sect. 3, when considering questions of uniqueness, we will need to compare $*$ with other connectives # that may not be truth-functional, and more care will be needed when specifying the language.

2.1 Existence of Elimination Rules Satisfied by a Given Connective

Trivially, any connective $*$ satisfies at least one set/formula elimination rule (which is thus also a set/set elimination rule) with respect to any consequence relation. We need only take the (zero-premise) elimination rule $p, \circ(p_1, \ldots, p_k) \Rightarrow p$. But of course this rule is degenerate, being an instance of the structural rule $p, q \Rightarrow p$. So the interesting question is: when does $*$ satisfy at least one *non-degenerate* elimination rule?

For truth-functional connectives, the answer is straightforward. Recall that a k-ary truth-functional connective $*$ is called a *verum* iff $v(*(p_1, \ldots, p_k)) = 1$ for every valuation v, and is called a *falsum* iff $v(*(p_1, \ldots, p_k)) = 0$ for every valuation v.

Observation 1. Let $*$ be any truth-functional connective. Then the following are equivalent in both the set/set and set/formula contexts: (a) $*$ satisfies at least one non-degenerate elimination rule, (b) $*$ is not a verum.

Notation. To reduce overload, in this and the following proofs, we will often write K for $\{p_1, \ldots, p_k\}$, T_v for $\{p \in K : v(p) = 1\}$, and F_v for $K \backslash T_v$. When no ambiguity can arise, we will sometimes write just $\circ$ in place of $\circ(p_1, \ldots, p_k)$ where $\circ$ is a k-ary dummy connective and $p_1, \ldots, p_k$ are fixed distinct sentence letters in a rule, and likewise write $*$ instead of $*(p_1, \ldots, p_k)$. Thus, for example, when $*$ is a falsum we say that $v(*) = 0$ for every valuation v.

Proof. For the implication from (a) to (b), suppose that $*$ is a k-ary verum connective satisfying the elimination rule $A_i \Rightarrow B_i\,(i{\leq}n)/C, \circ \Rightarrow D$; we need to show that the rule is degenerate. Clearly it is a substitution instance of the rule $A_i \Rightarrow B_i\,(i{\leq}n)/C, q \Rightarrow D$, where q is a fresh letter. This rule contains no connectives and so is structural. For any instantiation σ taking $\circ$ to $*$ we have $\sigma(\circ(p_1, \ldots, p_k)) = *(\sigma(p_1), \ldots, \sigma(p_k))$ and since $*$ is a verum this gives $\sigma(C)$, $*(\sigma(p_1), \ldots, \sigma(p_k)) \nvDash \sigma(D)$ if $\sigma(C), \sigma(q) \nvDash \sigma(D)$. Thus, since the first rule is satisfied by $*$, the second one is too. Hence the first rule is degenerate.

For the converse, suppose that $*$ is not a verum, so that $v(*) = 0$ for some valuation v. We construct a non-degenerate set/formula elimination rule satisfied by $*$. We break the argument into two cases.

Case 1. Suppose $T_v = K$. Then $*$ satisfies the non-degenerate (zero-premise) set/formula elimination rule $K, \circ \Rightarrow q$ where q is a fresh sentence letter, since for any instantiation σ into classical logic we have $\sigma(p_1), \ldots, \sigma(p_k)$, $*(\sigma(p_1), \ldots, \sigma(p_k)) \vDash \sigma(q)$.

Case 2. Suppose $T_v \neq K$. We claim that $*$ satisfies the non-degenerate (multi-premise) set/formula elimination rule

$$\{p_i \Rightarrow p_j : p_i, p_j \in F_v\}/T_v, \circ \Rightarrow q,$$

where q is chosen in the non-empty set F_v. In other words, we claim that for any instantiation σ into classical logic:

$$\text{If}\ \sigma(p_i) \vDash \sigma(p_j)\,(\text{all}\ p_i, p_j \in F_v)\ \text{then}\ \sigma(T_v) \vDash \sigma(q), *(\sigma(p_1), \ldots, \sigma(p_k)) \vDash \sigma(q).$$

For suppose the left side holds, but for some valuation w we have $w(\sigma(T_v)) = 1$ while $w(\sigma(q)) = 0$ with $q \in F_v$. To complete the proof, it suffices to show that $w(*(\sigma(p_1), \ldots, \sigma(p_k))) = 0$. By the two suppositions, $w(\sigma(p)) = 0$ for all $p \in F_v$, so $w(\sigma(p)) = v(p)$ for all $p \in T_v \cup F_v = K$, so by the truth-functionality of $*$ we have $w(*(\sigma(p_1), \ldots, \sigma(p_k))) = v(*(p_1, \ldots, p_k)) = v(*) = 0$ as desired. □

2.2 Existence of Introduction Rules Satisfied by a Given Connective

In the set/set context, any connective $*$ trivially satisfies at least one introduction rule wrt any consequence relation. We need only take the (zero-premise) degenerate introduction rule $p \Rightarrow p, \circ$. So the interesting question again is: when does $*$ satisfy at least one non-degenerate introduction rule?

The answer differs according to whether we are in the set/set or set/formula context. For the former, it is a natural dual of Observation 1. Its proof is also a mirror image; nevertheless we give it in detail so as to pinpoint where it breaks down for the set/formula context.

Observation 2. In the set/set context, for any truth-functional connective $*$ the following are equivalent: (a) $*$ satisfies at least one non-degenerate introduction rule, (b) $*$ is not a falsum.

Proof. For the implication from (a) to (b), suppose that $*$ is a k-ary falsum connective such that $*$ satisfies the introduction rule $A_i \Rightarrow B_i\ (i \leq n)/C \Rightarrow D, \circ$; we need to show that the rule is degenerate. Clearly it is a substitution instance of the rule $A_i \Rightarrow B_i\ (i \leq n)/C \Rightarrow D, q$ where q is a fresh letter. This rule contains no connectives and so is structural. Moreover, since $*$ is a falsum, for any instantiation σ we have $\sigma(C), D \nvDash \sigma(*)$ if $\sigma(C), D \nvDash \sigma(q)$. So, since the first introduction rule above is satisfied by $*$, the second one is too. Hence the first rule is degenerate.

For the converse, suppose that $*$ is not a falsum, so that $v(*) = 1$ for some valuation v. We need to find a non-degenerate set/set introduction rule satisfied by $*$. We break the argument into two cases.

Case 1. Suppose Suppose $T_v = K$. Then $*$ satisfies the non-degenerate (zero-premise) set/set introduction rule $K \Rightarrow \circ$ and we are done.

Case 2. Suppose $T_v \neq K$. We claim that $*$ satisfies the non-degenerate (multi-premise) set/formula introduction rule:

$$\{p_i \Rightarrow p_j : p_i, p_j \in F_v\}/T_v, \Rightarrow q, \circ$$

where q is chosen in the non-empty set F_v. In other words, we claim that for any instantiation σ into classical logic:

$$\text{If } \sigma(p_i) \vDash \sigma(p_j)\ (\text{all} p_i, p_j \in F_v) \text{ then } \sigma(T_v) \vDash \sigma(q), *(\sigma(p_1), \ldots, \sigma(p_k)).$$

For suppose the left side holds, but for some valuation w we have $w(\sigma(T_v)) = 1$ while $w(\sigma(q)) = 0$ with $q \in F_v$. To complete the proof, it suffices to show that $w(*(\sigma(p_1), \ldots, \sigma(p_k))) = 1$. By the two suppositions, $w(\sigma(p)) = 0$ for all $p \in F_v$, so $w(\sigma(p)) = v(p)$ for all $p \in T_v \cup F_v = K$, so by the truth-functionality of $*$, we have $w(*(\sigma(p_1), \ldots, \sigma(p_k))) = v(*(p_1, \ldots, p_k)) = v(*) = 1$. □

This proof breaks down in the set/formula context because the rule appealed to in Case 2 has two formulae on the right of its conclusion-sequent. We will however use a set/formula tweak of it in Observation 6.

As one would expect, in the set/formula context the existence conditions for introduction rules turn out to be much tighter. Call a k-ary truth-functional connective $*$ *contrarian* (Humberstone and Makinson 2011) iff $v(*) = 0$ when all $v(p) = 1$. Before using this concept in our next observation, we note some of its more obvious features.

All falsum connectives are contrarian, but not conversely. For each k, the notion splits the set of k-ary truth-functional connectives right down the middle. For example, of the 16 two-place connectives, there are 8 *non*-contrarian ones: $\top$, $\vee$, $\wedge$, *left*(p,q), *right*(p,q), $\rightarrow$, $\leftarrow$, $\leftrightarrow$; we will return to them in Corollary 9. The 8 contrarian ones are expressed by their negations, respectively: $\bot$, 'neither nor', nand ('not both'), $\neg$ *left*(p,q), $\neg$ *right*(p,q), $p\wedge\neg q$, $q\wedge\neg p$, and exclusive disjunction (which we write as $+$). Any functionally complete set of truth-functional connectives must contain at least one that is contrarian (otherwise the valuation that puts all sentence letters true will make all formulae true, so that e.g. negation becomes inexpressible); it must also contain a connective that is dual-contrarian, in the sense that it comes out true when all its letters are false (same reason, with 'false' in place of 'true'). Of course, some connectives, such as 'nand' and 'neither nor' are both contrarian and dual contrarian.

Observation 3. Suppose that our classical language L contains at least one tautology and at least one contradiction. Then, in the set/formula context, no truth-functional connective satisfies any degenerate introduction rule. Moreover, for any truth-functional connective $*$ the following are equivalent: (a) $*$ satisfies at least one introduction rule, (b) $*$ is not contrarian.

Proof. For the first part, suppose that the k-ary truth-functional connective $*$ satisfies an introduction rule $\{A_i \Rightarrow q_i : i \leq n\}/C \Rightarrow \mathrm{o}$; we need to show that this rule is not degenerate. Suppose that it is a substitution instance of a structural rule, which must have the form $\{A_i' \Rightarrow q_i' : i \leq n\}/C' \Rightarrow r$ where all mentioned sets contain only sentence letters. It suffices to show that this latter rule is not satisfied by $\models$. Since o is not in C or any $A_i \cup \{q_i\}$, it follows that r is not in C' or any $A_i' \cup \{q_i'\}$, so we may without conflict substitute a contradiction (call it $\bot$) for r and a tautology (call it $\top$) for all other letters in the structural rule. Then each instantiated premise becomes $\top \models \top$ or $\varnothing \models \top$ (depending on whether A_i was non-empty), while the conclusion is $\top \models \bot$ or $\varnothing \models \bot$, and thus the structural rule is not satisfied by $\models$.

For the second part of the Observation, right to left, suppose that the truth-functional connective $*$ is not contrarian, so that for every valuation v we have $v(*) = 1$ if $v(p) = 1$ for all $p \in K$. Then $*$ satisfies the (zero-premise) introduction rule $K \Rightarrow \mathrm{o}$. For the converse, we recall the following verification from Humberstone and Makinson (2011). Suppose that $*$ is contrarian. Consider any introduction rule $\{A_i \Rightarrow q_i : i \leq n\}/C \Rightarrow \mathrm{o}$, noting that all mentioned sets contain only sentence letters. Instantiate all sentence letters by $\top$. Then each $\sigma(A_i) \models \sigma(q_i)$; while $\sigma(C) = \{\top\}$ and $\sigma(*(p_1, \ldots, p_k)) = *(\sigma(p_1), \ldots, \sigma(p_k)) = *(\top, \ldots,\top)$ which is equivalent to $\bot$ since $*$ is contrarian, so $\sigma(C) \nvDash *(\sigma(p_1), \ldots, \sigma(p_k))$ and thus $*$ does not satisfy the rule. □

It has long been appreciated by logicians assembling natural deduction systems and sequent calculi for classical logic that it is difficult to find any introduction rule for classical negation in a set/formula context. Observation 3 implies that the task is in fact *impossible*, since negation is contrarian. But the gap is not filled by falling back to intuitionistic logic for, as also shown by Humberstone and Makinson (2011), there are no intuitionistically acceptable introduction rules for negation either.

3 Questions of Uniqueness

We now consider the extent to which a truth-functional connective is uniquely determined by the intelim rules that it satisfies. It turns out that there are several ways of understanding such unique determination, with two important dimensions of choice.

The first choice is between what we call *local* and *global* intelim-uniqueness. Under the local option one compares a given truth-functional connective $*$ with other *truth-functional* connectives. Under the global option one reaches further, comparing $*$ with *arbitrary* connectives; precise definitions are given below. Within each of these options, one can also choose between *weak* and *strong* versions of the notion, according to whether the definition considers what happens given the identity, or inclusion, of suitable sets of intelim rules.

In Sect. 3.1 we begin by articulating these notions of intelim-uniqueness; in Sect. 3.2 we study them in the set/set context, and in Sect. 3.3 we consider the more complex situation arising in the set/formula context.

3.1 Concepts of Intelim-Uniqueness

The concept of local interim-uniqueness is defined solely in terms of truth-functional connectives. For this reason we can continue to be relaxed about the exact specification of the purely classical language (continuing to assume, however, that it can express at least one tautology and at least one contradiction). When $*$ is a k-ary truth-functional connective we write $\mathcal{R}_*$ for the set of all the intelim rules satisfied by $*$. Context will make it clear whether we are considering set/set or set/formula rules.

Local intelim-uniqueness in the strong sense: This holds of a truth-functional connective $*$ iff no other truth-functional connective # of the same arity satisfies all the intelim rules that are satisfied by $*$. Briefly, iff: $\mathcal{R}_* \subseteq \mathcal{R}_\#$ implies $* \dashv\vDash \#$.

Local intelim-uniqueness in the weak sense: Same definition, replacing *all* by *exactly the same*, and in the brief form replacing $\subseteq$ by $=$.

Clearly, the strong sense implies the weak one and, for both senses, if the property holds in the set/formula context, then it holds in the set/set context.

In contrast, the definition of global intelim-uniqueness of a truth-functional connective $*$ considers arbitrary connectives, in arbitrary languages equipped with arbitrary consequence relations. In this context, we need to be much more attentive to the choice of those parameters. We continue to write $\mathcal{R}_*$ for the set of all the intelim rules satisfied by $*$ in a classical language with $\models$, but when L contains a possibly non-truth-functional connective # and is equipped with a possibly non-classical consequence relation $\vdash$, we write $\mathcal{R}_{L,\#,\vdash}$ for the set of all the intelim rules that # satisfies wrt $(L, \vdash)$.

Global intelim-uniqueness in the strong sense: This holds of a truth-functional connective $*$ iff for any two propositional connectives #, #′ (of the same arity as $*$) that occur together in a language L equipped with a consequence relation $\vdash$, if #, #′ both satisfy wrt $(L, \vdash)$ all the intelim rules that are satisfied by $*$, then they are equivalent under $\vdash$. Briefly, iff in any such language L, if $\mathcal{R}_{L,\#,\vdash} \supseteq \mathcal{R}_* \subseteq \mathcal{R}_{L,\#',\vdash}$ then $\# \dashv\vdash \#'$.

Global intelim-uniqueness in the weak sense: Same definition, replacing *all* by *exactly the same*, and in the brief form replacing the inclusions by equalities.

Note that in these two definitions, it is required that #, #′ are in a common language L under a single consequence relation $\vdash$. Again, the strong sense implies the weak one, and the set/formula version implies the set/set one. Also, the global notions imply the corresponding local ones.

The eight notions thus form a cube in a Hasse diagram of relative power, with the strong global set/formula version at the bottom (most powerful) and the weak local set/set one at the top (weakest).[1]

---

[1] The reader should be warned that in some proof-theoretic studies of intelim rules, uniqueness is defined in a subtly different way. For example, in order to prove analogues of the interpolation theorem and Beth's theorem on definability, Došen and Schroeder-Heister (1998) define the uniqueness of connectives as follows. Where # is a connective of a propositional language $L_\#$ with consequence relation $\vdash_\#$, we make copies $L_{\#'}$ and $\vdash_{\#'}$ of them , replacing # by a fresh #′, and then consider $(L, \vdash)$ where L is the least propositional language including $L_\# \cup L_{\#'}$ and $\vdash$ is the least relation over L including $\vdash_\# \cup \vdash_{\#'}$ that is closed under the Tarski closure conditions. If then we have $\#(p_1, \ldots, p_k) \vdash \#'(p_1, \ldots, p_k)$ and conversely, the connective # is said to be unique.

Note that in this definition $\vdash$ is *not* required to be closed under all substitutions into L, even when its components were closed under substitution into their respective languages. For this reason, it can happen that while an intelim rule is satisfied by # wrt $(L_\#, \vdash_\#)$, and thus also by #′ wrt $(L_{\#'}, \vdash_{\#'})$, it is not satisfied (in our sense of the term) by either of these connectives wrt $(L, \vdash)$, since $\vdash_\#$ may not be closed under instantiation of #′-formulae for the sentence letters in an intelim rule for #, and vice versa. This is expressed by saying that applications of the intelim rule in the composite language do not allow 'mixing' connectives.

By thus restricting the range of allowed instantiations of a rule one reduces its power in the composite language, so that the definition of uniqueness ends up narrower. For example, while under our definitions classical disjunction is globally intelim-unique in the strong sense in the set/formula context (see Corollary 9), under the definitions (Došen and Schroeder-Heister 1998) it is not (although conjunction is, under both definitions).

The above comparisons prompt the question of whether, if one takes the (Došen and Schroeder-Heister 1998) definition of uniqueness and modifies it by also closing their composite consequence

3.2 Intelim-Uniqueness in the Set/Set Context

In the set/set context, the situation is straightforward, with the following very general result.

Observation 4. In the set/set context, every truth-functional connective is globally (and thus also locally) intelim-unique in the strong (and thus also the weak) sense.[2]

Proof. Let $f : \{0, 1\}^k \to \{0, 1\}$ be any k-ary truth-function, expressed by a connective $*$ in a language L equipped with the classical consequence relation $\models$. Consider any language L containing two k-ary connectives $\#$, $\#'$, equipped with a (set/set) consequence relation $\vdash$ such that $\mathcal{R}_{L,\#,\vdash} \supseteq \mathcal{R}_* \subseteq \mathcal{R}_{L,\#',\vdash}$. We show $\# \vdash \#'$; the converse is similar.

Let v be any assignment of truth-values to letters in K. It determines a set/set intelim rule R_v defined as follows. If $v(*) = 1$ then R_v is the rule $T_v \Rightarrow \circ, F_v$; if on the other hand $v(*) = 0$ then R_v is the rule $T_v, \circ \Rightarrow F_v$. Clearly $R_v \in \mathcal{R}_*$ so by supposition $R_v \in \mathcal{R}_{L,\#,\vdash}$ and $R_v \in \mathcal{R}_{L,\#',\vdash}$.That is, in the first case we have both $T_v \vdash \#, F_v$ and its $\#'$-counterpart, and in the second case both $T_v, \# \vdash F_v$ and its $\#'$-counterpart. In the first case, we obtain $T_v, \# \vdash \#', F_v$ by weakening on the left in the $\#'$-rule, and in the second case we obtain the same by weakening on the right in the $\#$-rule.

This shows that for each assignment v of truth-values to letters in K we have $T_v, \# \vdash \#', F_v$. To these 2^k items (one for each valuation v) we may finally apply the closure conditions for set/set closure relations (with many applications of cumulative transitivity) to obtain $\# \vdash \#'$ as desired. □

Evidently, the rules employed in this proof may have any number (up to k) of formulae on the right, and so are not available in the set/formula context.

3.3 Intelim-Uniqueness in the Set/Formula Context

We begin on the local level, where we have a simple and satisfying pattern; and then pass to the global level, where the situation is more complex. As it can be difficult to keep track of the various results, we summarise them in a tree in Sect. 3.4.

3.3.1 On the Local Level

The following negative fact shows the failure of Observation 4 in the set/formula context, and sets a boundary for the positive results that follow.

relation $\vdash$ under substitution, one would get exactly global intelim-uniqueness in the strong sense as defined in this paper. We will not attempt to answer that here.

[2] Observation 4 has connections with some results in Sect. 3.1 of Humberstone (2011), particularly with Observation 3.11.4.

Lemma 5. In the set/formula context, no contrarian truth-functional connectives that is not a falsum is locally (or globally) intelim-unique in the strong sense.[3]

Proof. Let $*$ be a k-ary contrarian truth-functional connective that is not a falsum; we want to show that it is not locally (so also not globally) intelim-unique in the strong sense. It suffices to show that the k-place falsum $\bot$ satisfies all intelim rules that are satisfied by $*$, briefly that $\mathcal{R}_* \subseteq \mathcal{R}_\bot$. Since $*$ is contrarian, we know from Observation 3 that in the set/formula context, it does not satisfy any introduction rules at all, so the only intelim rules that it can satisfy are elimination rules $A_i \Rightarrow q_i (i \leq n)/C, \circ \Rightarrow r$ where the displayed letters may be distinct or identical and no connectives occur in C or the A_i. These rules hold trivially for the falsum connectives, since $\bot \vDash r$. □

On the positive side, we have full characterizations at the local level, for both the weak and strong senses.

Observation 6. In the set/formula context, (a) every truth-functional connective is locally intelim-unique in the weak sense. Moreover, (b) a truth-functional connective is locally intelim-unique in the strong sense iff it is either a falsum or non-contrarian.[4]

Proof. We already have the left-to-right half of (b) from Lemma 5. It is also immediate that every falsum is globally (and thus also locally) intelim-unique in the strong sense: we have $\bot(p_1, \ldots, p_k) \vDash q$ and so both $\#(p_1, \ldots, p_k) \vdash \#'(p_1, \ldots, p_k)$ and conversely for $\#, \#'$ as considered in the definition. To complete the proof of (a) and (b), it remains to show that every k-ary truth-functional connective $*$ is locally intelim-unique in the weak sense, and also in the strong sense if it is non-contrarian.

Let $\#$ be any k-ary truth-functional connective distinct from $*$, so that there is a valuation v with $v(*) \neq v(\#)$. It suffices to show (a) $\mathcal{R}_* \neq \mathcal{R}_\#$ (b) if $*$ is non-contrarian then $\mathcal{R}_* \nsubseteq \mathcal{R}_\#$. Since $v(*) \neq v(\#)$, either $v(\#) = 1$ while $v(*) = 0$ or conversely. We consider the two cases separately.

Case 1. Suppose $v(\#) = 1$ while $v(*) = 0$. Since $v(*) = 0$, the argument used in the proof of Observation 1 from (b) to (a) tells us that if $T_v = K$ then $*$ satisfies the (zero-premise) elimination rule:

$$K, \circ \Rightarrow q, \text{ where } q \text{ is a fresh letter,}$$

while if $T_v \neq K$ then $*$ satisfies the (multi-premise) elimination rule:

---

[3] In its application to negation, Lemma 5 might seem to jar with exercise 4.31.1 (iii) of Humberstone (2011), stating that in a language L containing two connectives $\sim, \sim'$ and equipped with a consequence relation $\vdash$, if both of these connectives satisfy wrt $(L, \vdash)$ the rules $p, \circ(p) \Rightarrow q$ (explosion) and $p, q \Rightarrow r; p, q \Rightarrow \circ(r) \,/\, p \Rightarrow \circ(q)$ (a form of *reductio ad absurdum*), then they are equivalent under $\vdash$. But there is no conflict, since the second of these rules is *not intelim* in the sense of this paper: it has two occurrences of negation, one of which is, moreover, in a premise-sequent.

[4] Thanks to Frederik van de Putte for convincing me that Observation 6(b) should be true, contrary to my initial conjecture. Observation 6(a) has connections with Theorem 4 of Rautenberg (1981) and Theorem 3.13.15 of Humberstone (2011).

$$\{p_i \Rightarrow p_j : p_i, p_j \in F_v\}/T_v, \circ \Rightarrow q,$$

where q is chosen in the non-empty set F_v. In contrast, since $v(\#) = 1$, # fails the two displayed rules (for a failing instance of the latter one, instantiate the letters in T_v to $\top$ and those in F_v to $\bot$). Thus $\mathcal{R}_* \nsubseteq \mathcal{R}_\#$ (irrespectively, in this case, of whether $*$ is contrarian).

Case 2. Suppose $v(\#) = 0$ while $v(*) = 1$. Applying the same argument as in Case 1 but with $*$, # reversed, we get $\mathcal{R}_\# \nsubseteq \mathcal{R}_*$ so $\mathcal{R}_\# \neq \mathcal{R}_*$, completing the proof of Part (a). Suppose now that $*$ is non-contrarian; we show that also $\mathcal{R}_* \nsubseteq \mathcal{R}_\#$. Since $v(*) = 1$, we have that if $T_v = K$ then $*$ satisfies the (zero-premise) introduction rule:

$$K \Rightarrow \circ$$

while if $T_v \neq K$ and $*$ is non-contrarian, we claim that $*$ satisfies the (multi-premise) introduction rule:

$$\{p_i \Rightarrow p_j : p_i, p_j \in F_v\}/T_v \Rightarrow \circ$$

For suppose the left side holds for a substitution σ, and for some valuation w we have $w(\sigma(T_v)) = 1$. Choose any q in the non-empty set F_v. If $w(\sigma(q)) = 1$ then using the equivalences we can say that since $*$ is non-contrarian we have $w(*(\sigma(p_1), \ldots, \sigma(p_k))) = 1$ as needed. If, on the other hand, $w(\sigma(q)) = 0$ then using the equivalences again we have $w(\sigma(p)) = v(p)$ for all $p \in T_v \cup F_v = K$ and so by the truth-functionality of $*$ we get $w(*(\sigma(p_1), \ldots, \sigma(p_k))) = v(*(p_1, \ldots, p_k)) = v(*) = 1$ again.

In contrast, since $v(\#) = 0$, by instantiating the letters in T_v to $\top$ and those in F_v to $\bot$ we see that # fails the two displayed rules. Thus $\mathcal{R}_* \nsubseteq \mathcal{R}_\#$ and the proof is complete. □

Remark. The multi-premise introduction rule displayed in Case 2 of the proof is *almost* the same as that used in Case 2 of the proof of Observation 2. The difference is that here the conclusion-sequent $T_v \Rightarrow \circ$ has a single formula on the right, as required in the set/formula context, while the conclusion-sequent $T_v, \Rightarrow q, \circ$ of the previous rule had two formulae on the right, as allowed in the set/set context. Inspection of the proof here reveals that it is the assumption that $*$ is non-contrarian that permits us to do without q on the right.

3.3.2 On the Global Level

In order to state our results on this level, we need the concept of the *partner* of a finite set of zero-premise (set/formula) introduction resp. elimination rules.[5]

[5] Presentations of natural deduction systems for classical and other logics routinely gravitate around zero-premise intelim rules, sometimes accompanied by partners. The *general* notion of the partner

- Consider any finite collection of zero-premise introduction rules $A_i \Rightarrow \circ\,(i{\leq}n)$, all with the same dummy connective $\circ$. Its *partner* is the n-premise elimination rule $A_i \Rightarrow s$ (all $i{\leq}n$) / $\circ \Rightarrow s$, where s is a fresh letter, the same in all the n premise-sequents and in the conclusion-sequent of the rule.
- Likewise, consider any finite collection of zero-premise elimination rules $A_i, \circ \Rightarrow r_i\,(i{\leq}n)$, all with the same dummy connective $\circ$. Its *partner* is the n-premise introduction rule $A_i, s \Rightarrow r_i$ (all $i{\leq}n$) / $s \Rightarrow \circ$, where s is likewise a single fresh letter.

Note that this relation holds between an appropriate *set* of rules and an individual rule, rather than between individual rules. Even when the set is a singleton, the relation is not symmetric.

Observation 7. Let $*$ be any k-ary truth-functional connective in a classical language. Then, for the set/formula context, each of the following two conditions is sufficient for $*$ to be globally intelim-unique in the strong sense:

(i) $*$ satisfies some finite set of zero-premise introduction rules and its partner elimination rule.
(ii) $*$ satisfies some finite set of zero-premise elimination rules and its partner introduction rule.

Before giving the proof, we make some explanatory remarks on the two conditions and the notion of partner that they employ.

- On a conceptual level, satisfaction of conditions (i) and (ii) may be seen as a formal rendering of a notion of 'harmony' between introduction and elimination rules. But unlike previous accounts of that elusive notion, which have sought to express it in proof-theoretic terms (for example, of the normalizability of derivations in an axiomatic system using the rules), these two conditions, one dual to the other, express properties that hold/fail of a given truth-functional connective *independently* of the specification of any axiom system and its derivations. They are defined in terms of the satisfaction of the rules as closure conditions on classical consequence, and are thus essentially *semantic* and 'static' in nature.
- Roughly speaking, condition (i) says that $*$ is the strongest of the connectives that satisfy all its zero-premise introduction rules, while (ii) says roughly that $*$ is the weakest of the connectives satisfying all its zero-premise elimination rules.
- It may be asked why the conditions restrict attention to zero-premise rules and their (multi-premise) partners. The answer is that the partner of a multi-premise rule would not itself be a rule in the usual sense of the term (see the appendix) but rather a higher-level rule in the sense of Schroeder-Heister (1984). Such rules are certainly of interest, and there is a sense in which any set/set rule in the usual sense may be re-expressed as a higher-level set/formula rule in Schroeder-Heister's sense; but they are beyond our focus of attention.

(Footnote 5 continued)
of a finite collection of such rules must surely have been articulated explicitly somewhere in the literature, but the author has not found it anywhere.

- On a technical level, recall that by Observation 3, the contrarian connectives satisfy no introduction rules in the set/formula context, so for them the set mentioned in (i) is empty; but the empty set of zero-premise introduction rules still has a partner rule, namely $\circ \Rightarrow s$, which is satisfied when $\circ$ is taken to be the falsum. In contrast, no contrarian connective satisfies condition (ii).
- In Sect. 3.3.3 we will formulate a more general, but more complex, pair of conditions sufficient for (global) intelim-uniqueness (in the strong sense), that include (i) and (ii) as limiting cases.

Proof of Observation 7. Suppose $\mathcal{R}_{L,\#,\vdash} \supseteq \mathcal{R}_* \subseteq \mathcal{R}_{L,\#',\vdash}$. By symmetry we need only show $\# \vdash \#'$ under each of the two conditions.

Suppose condition (i). Since $*$ satisfies the introduction rules $A_i \Rightarrow \circ$, by the second inclusion we have $A_i \vdash \#'$ for all $i \leq n$. Since $*$ also satisfies the partner rule $A_i \Rightarrow s\ (i{\leq}n)\ /\ \circ \Rightarrow s$, the first inclusion tells us that # satisfies it too. Substituting $\#'$ for s in the partner rule, we have that if $A_i \vdash \#'$ (all $i \leq n$) then $\# \vdash \#'$. Thus $\# \vdash \#'$ as desired.

Suppose condition (ii). Since $*$ satisfies the elimination rules $A_i, \circ \Rightarrow r_i$, by the first inclusion we have $A_i, \# \vdash r_i$ for all $i \leq n$. Since $*$ also satisfies the partner rule $A_i, s \Rightarrow r_i\ (i{\leq}n)\ /\ s \Rightarrow \circ$, the second inclusion tells us that $\#'$ satisfies it too. Substituting # for s in the partner rule, we have that if $A_i, \# \vdash r_i$ (all $i{\leq}n$) then $\# \vdash \#'$. Thus $\# \vdash \#'$ as required. □

In the restricted arena of *two-place* truth-functional connectives, we can use Observation 7 to wrap up very neatly the story for global interim uniqueness.

Corollary 8. For *two-place* truth-functional connectives $*$, the following three conditions are equivalent: (a) $*$ is globally intelim-unique in the strong sense, (b) $*$ is locally intelim-unique in the strong sense, (c) $*$ is either the falsum or non-contrarian.

Proof. Condition (a) immediately implies (b); we know from Observation 6(b) that conditions (b) and (c) are equivalent; and we have already noted that the falsum is globally intelim-unique in the strong sense. So we need only show that each of the eight non-contrarian two-place connectives is globally intelim-unique in the strong sense. By Observation 7 it suffices to check that each of them satisfies condition (i) or (ii). The connectives T, $\vee$, $\wedge$, *left*, *right* satisfy condition (i) with the following rules:

T: $\{\varnothing \Rightarrow \mathrm{T}\}$ and partner $\varnothing \Rightarrow s\ /\mathrm{T} \Rightarrow s$
$\vee$: $\{p \Rightarrow p{\vee}q;\ q \Rightarrow p{\vee}q\}$ and partner $p \Rightarrow s;\ q \Rightarrow s/p{\vee}q \Rightarrow s$
$\wedge$: $\{p, q \Rightarrow p{\wedge}q\}$ and partner $p, q \Rightarrow s/p{\wedge}q \Rightarrow s$
left: $\{p \Rightarrow \mathit{left}(p,q)\}$ and partner $p \Rightarrow s\ /\ \mathit{left}(p,q) \Rightarrow s$
right: $\{q \Rightarrow \mathit{right}(p,q)\}$ and partner $q \Rightarrow s\ /\ \mathit{right}(p,q) \Rightarrow s$.

For the connectives $\rightarrow$, $\leftarrow$, $\leftrightarrow$ we can use condition (ii)[6]:

---

[6] At first sight, the global intelim-uniqueness of classical implication might seem to clash with the fact that it satisfies exactly the same set/formula intelim rules as does intuitionistic implication (see Humberstone and Makinson (2011)). But there is no conflict. The latter fact concerns a

->: $\{p\text{-}{>}\, q, p \Rightarrow q\}$ and partner $s, p \Rightarrow q / s \Rightarrow p\text{-}{>}\, q$
<-: $\{p{\leftarrow}q, q \Rightarrow p\}$ and partner $s, q \Rightarrow p / s \Rightarrow p <\text{-}q$
<->: $\{p <\text{-}> q, p \Rightarrow q; p{\leftrightarrow}q, q \Rightarrow p\}$ and partner $s, p \Rightarrow q\,;\, s, q \Rightarrow p / s \Rightarrow p{\leftrightarrow}q$. □

Remark. Of the five connectives verified via condition (i), four may also be checked using (ii) as follows. There is thus a great deal of overlap in their ranges of applicability.

T: The empty set of elimination rules and partner $s \Rightarrow \mathsf{T}$.
∧ : $\{p\wedge q \Rightarrow p; p\wedge q \Rightarrow q\}$ and partner $s \Rightarrow p; s \Rightarrow q / s \Rightarrow p\wedge q$
left: $\{left(p,q) \Rightarrow p\}$ and partner $s \Rightarrow p / s \Rightarrow left(p,q)$
right: $\{right(p,q) \Rightarrow q\}$ and partner $s \Rightarrow q / s \Rightarrow right(p,q)$.[7]

Conditions (i) and (ii) of Observation 7 are syntactic in character. Are there equivalent semantic ones? We have found a simple semantic condition for (i) and a rather less simple one for (ii). Writing $v \leq w$ for $v(p) \leq w(p)$ for all $p \in K = \{p_1, \ldots, p_k\}$, they are:

(I) For all valuations v,w, if $v \leq w$ then $v(*) \leq w(*)$.
(II) For every valuation v, if $v(*) = 0$ then there is a co-atomic $v' \geq v$ such that $w(*) = 0$ for all w with $v \leq w \leq v'$.

In the second condition, a valuation v' is called co-atomic iff $v'(p) = 0$ for exactly one letter $p \in K$ (no restrictions for letters outside K). Note that condition (II) immediately implies that $*$ is not contrarian.

In order to prove the equivalence transparently, it is convenient to begin by massaging the two syntactic conditions. Consider any zero-premise introduction rule $A \Rightarrow \circ$ and any truth-functional connective $*(p_1, \ldots, p_k)$. As far as satisfaction of the rule by $*$ is concerned, we may assume without loss of generality that $A \subseteq \{p_1, \ldots, p_k\} = K$, since $A, q \vDash *(p_1, \ldots, p_k)$ where $q \notin K$ iff $A \vDash *(p_1, \ldots, p_k)$. Similarly, for satisfaction of an elimination rule $A, \circ \Rightarrow r$ by $*$, we may assume wlog that $A \subseteq K \cup \{r\}$, since $A, q, *(p_1, \ldots, p_k) \vDash r$ where $q \notin K \cup \{r\}$ iff $A, *(p_1, \ldots, p_k) \vDash r$. In what follows, we make these assumptions, which ensure that all sets of formulae that we

(Footnote 6 continued)
shared property of two different logics using different consequence relations ⊨ and ⊢, with no opportunity for substituting formulae of one language in formulae of the other. In contrast, global intelim-uniqueness is defined by considering a *single* language with two connectives, under a *single* consequence relation that is closed under substitution. For further discussion of the significance of this distinction, see (Humberstone 2011, Sect. 4.3).

[7] Most of the rules with partners mentioned in the proof of Corollary 8 and the remark following it are perfectly well known (and are collected in Humberstone (2011, Sect. 4.3). To take one example, it has long been known that in classical (and also intuitionistic) logic → satisfies the intelim rule and partner mentioned in the proof, and also that any two connectives satisfying those rules wrt a common consequence relation ⊢ in a common language are equivalent under ⊢. However, the examples of *left* and *right* do not appear to have attracted attention, presumably because of their triviality; and the interesting example of ↔ may perhaps have passed unnoticed.

consider are finite so that the operations performed on them (such as conjoining their elements) are well-defined.

Note incidentally that in an elimination rule $A, \circ \Rightarrow r$ we cannot assume without loss of generality that $r \in K$. For example, the rule $\circ(p_1, \ldots, p_k) \Rightarrow r$ holds only for the falsum when $r \notin K$ but can hold for other truth-functional connectives when $r \in K$. Nevertheless, when r is outside K we can take it to be a fixed letter (distinct from the letter s in the partner rule) thus remaining finite.

Lemma 9. Let $*$ be any k-ary truth-functional connective. Then in the set/formula context, conditions (i) and (ii) are respectively equivalent to the following:

(i) $*'$ satisfies the partner of the set of all the zero-premise introduction rules that $*$ satisfies,

(ii) $*'$ satisfies the partner of the set of all the zero-premise elimination rules that $*$ satisfies.

Proof. The implication from the primed to the original version is immediate. The converse follows from the definition of a partner: given a set of introduction (resp. elimination) rules all of which are satisfied by $*$, simply fill out that set with all the remaining introduction (resp. elimination) rules satisfied by $*$. They become additional premises of the partner rule and so do not disturb its satisfaction. □

Observation 10. In the set/formula context, semantic conditions (I), (II) are respectively equivalent to syntactic conditions (i), (ii).

Remark. The following proof is rather intricate, but quite 'inevitable' once one has specified conditions (I) and (II). To follow the thread, it is important not to conflate the subscript i indexing zero-premise rules, with the subscript j indexing the argument places of the connective $*$.

Proof. By Lemma 9, it suffices in each instance to show equivalence to the primed syntactic conditions.

Suppose first that $*$ satisfies condition (I); we want to show that it satisfies (i)$'$. Let $\{A_i \Rightarrow \circ\, (i \leq n)\}$ be the set of all zero-premise introduction rules that $*$ satisfies, with $\{A_i \Rightarrow s\, (i \leq n)\} / \circ \Rightarrow s$ its partner elimination rule. Take any instantiation σ of it into L. Let u be a valuation with $u(\sigma(*)) = 1$ and $u(\sigma(s)) = 0$. It suffices to show that for some $i \leq n$ we have $u(\sigma(A_i)) = 1$.

Define a valuation v by putting $v(p) = u(\sigma(p))$ for each $p \in K$, and put $A = T_v = \{p \in K : v(p) = 1\}$. Then $v(A) = v(T_v) = 1$, and by the definition of v, $v(A) = u(\sigma(A))$, so that $u(\sigma(A)) = 1$. It remains to check that $A = A_i$ for some $i \leq n$, that is, to show that $\tau(A) \vDash \tau(*)$ for every substitution τ, for which it suffices to show that $A \vDash *$. Let w be any valuation and suppose $w(A) = 1$; we need to show that $w(*) = 1$. Since $A = T_v$ and $1 = w(A)$ we have $v \leq w$ so by condition (I) we have $v(*) \leq w(*)$. But using the definition of v and the truth-functionality of $*$ we have $1 = u(\sigma(*)) = v(*) \leq w(*)$ and the verification that (I) implies (i) is complete.

For the converse, suppose that $*$ does not satisfy (I); we show that it fails (i)$'$. Again, let $\{A_i \Rightarrow \circ\, (i \leq n)\}$ be the set of all zero-premise introduction rules that $*$ satisfies.

We need to show that $*$ does not satisfy its partner rule $\{A_i \Rightarrow s\,(i{\le}n)\} / \circ \Rightarrow s$. That is, we need to find a substitution σ such that $\sigma(A_i) \vDash \sigma(s)$ for all $i \le n$ but $\sigma(^*) \nvDash \sigma(s)$. Take σ to be the identity substitution on K and put $\sigma(s) = \vee\{\alpha_i\colon i \le n\}$ where each α_i is the conjunction of the finitely many elements of A_i. Then for each $i \le n$, $\sigma(A_i) = A_i \vDash \vee\{\alpha_i : i \le n\} = \sigma(s)$. It remains to show that $\sigma(^*) \nvDash \sigma(s)$, that is, $* \nvDash \vee\{\alpha_i : i \le n\}$.

Since by supposition $*$ does not satisfy (I), there are valuations v, w such that $v \le w$ while $v(*) = 1$, $w(*) = 0$. Since $v(*) = 1$ it suffices to show that $v(\vee\{\alpha_i\colon i \le n\}) = 0$, i.e. that $v(\alpha_i) = v(\wedge A_i) = v(A_i) = 0$ for every $i \le n$. Take any $i \le n$ and suppose that $v(A_i) = 1$; we seek a contradiction. Let u be the valuation putting $u(p) = 1$ iff $p \in A_i$, for all $p \in K$. Since $v(A_i) = 1$ we have $u \le v$, which combines with $v \le w$ to give $u \le w$, so $w(A_i) = 1$ while we already have $w(*) = 0$. But $A_i \vDash *$ since $*$ satisfies the zero-premise rule $A_i \Rightarrow \circ$, giving a contradiction as desired.

Next we consider the conditions (II) and (ii)$'$. The spirit of the argument is like the one above, but the details are rather different and a little more intricate.

Suppose first that $*$ satisfies condition (II); we show that it satisfies condition (ii)$'$. Let $\{A_i, \circ \Rightarrow r_i\,(i{\le}n)\}$ be the set of all zero-premise elimination rules that $*$ satisfies, with $\{A_i, s \Rightarrow r_i\,(i{\le}n)\} / s \Rightarrow \circ$ its partner introduction rule. Now each $r_i \in K$ (otherwise $*$ would be contrarian contrary to our supposition) and we may assume without loss of generality that that $r_i \notin A_i$ (otherwise the rule would be degenerate and satisfied by all connectives). Take any substitution σ into L and suppose $\sigma(s) \nvDash \sigma(*)$ so there is a valuation u with $u(\sigma(s)) = 1$ and $u(\sigma(*)) = 0$. We show that for some $i \le n$, $u(\sigma(A_i)) = 1 = u(\sigma(s))$ while $u(\sigma(r_i)) = 0$, so that $\sigma(A_i), \sigma(s) \nvDash \sigma(r_i)$. We already have $u(\sigma(s)) = 1$ so we need only check that for some $i \le n$, $u(\sigma(A_i)) = 1$ and $u(\sigma(r_i)) = 0$.

Define a valuation v by putting $v(p) = u(\sigma(p))$ for each $p \in K$. Since $u(\sigma(*)) = 0$, we have $v(*) = 0$, so by condition (II) there is a co-atomic $v' \ge v$ such that $w(*) = 0$ for all w with $v \le w \le v'$. Let p_j be the unique letter in K such that $v'(p_j) = 0$. Then $T_v, \neg p_j \vDash \neg *$, so $T_v, * \vDash p_j$, so $*$ satisfies the zero-premise elimination rule $T_v, \circ \Rightarrow p_j$. Putting $A_i = T_v$ and $r_i = p_j$, it remains to show that $u(\sigma(T_v)) = 1$ and $u(\sigma(p_j)) = 0$. For the former, $u(\sigma(T_v\})) = v(T_v) = 1$, while for the latter, $u(\sigma(p_j)) = v(p_j) \le v'(p_j) = 0$.

For the converse, suppose that $*$ does not satisfy (II); we wish to show that it fails (ii)$'$. Again let $\{A_i, \circ \Rightarrow r_i\,(i{\le}n)\}$ be the set of all zero-premise elimination rules that $*$ satisfies. We need to check that $*$ does not satisfy its partner introduction rule $\{A_i, s \Rightarrow r_i\,(i{\le}n)\} / s \Rightarrow \circ$. Now if $*$ is contrarian then by Observation 3 it does not satisfy any introduction rules, so we may suppose wlog that $*$ is not contrarian. That implies, as already noticed, that each $r_i \in K$. We need to find a substitution σ such that $\sigma(A_i), \sigma(s) \vDash \sigma(r_i)$ for all $i \le n$ but $\sigma(s) \vDash \sigma(*)$. Take σ to be the identity substitution on K and put $\sigma(s) = \wedge\{\alpha_i{\rightarrow}r_i : i{\le}n\}$ where each α_i is the conjunction of the finitely many elements of A_i. Then $\sigma(A_i), \sigma(s) = A_i, \wedge\{\alpha_i{\rightarrow}r_i : i{\le}n\} \vDash r_i = \sigma(r_i)$. It remains to show that $\sigma(s) \vDash \sigma(*)$, that is, $\wedge\{\alpha_i{\rightarrow}r_i : i{\le}n\} \nvDash *$.

Since by supposition $*$ does not satisfy (II) there is a valuation v with $v(*) = 0$ such that there is no co-atomic $v' \ge v$ with $w(*) = 0$ for all w with $v \le w \le v'$. It will suffice to show that $v(\wedge\{\alpha_i{\rightarrow}r_i\colon i{\le}n\}) = 1$. Choose any $i \le n$, suppose that

$v(\alpha_i) = 1$ and $v(r_i) = 0$; we derive a contradiction. Let v' be the valuation that puts all letters in K true except for r_i. Then v' is co-atomic and $v \leq v'$. So there is a w with $v \leq w \leq v'$ and $w(*) = 1$. Since $v(A_i) = v(\alpha_i) = 1$ and $v \leq w$ we have $w(A_i) = 1$, so since $A_i, * \models r_i$ we have $w(r_i) = 1$, so since $w \leq v'$ we have $v'(r_i) = 1$ giving us the desired contradiction. □

Although syntactic conditions (i) and (ii), likewise the equivalent semantic (I) and (II), are each sufficient for global intelim uniqueness in the strong sense, they are not collectively necessary for it. Consider the connective $+(p, q, r)$ of three place exclusive disjunction, defined from the familiar two-place one by putting $+(p, q, r) = p+q+r$ (no need for parentheses, by associativity) which, we recall, comes out true under a valuation just if an odd number of its three arguments are true. Then we have:

Observation 11. Three-place exclusive disjunction is globally intelim-unique in the strong sense, but fails conditions (i), (ii).

Proof. To show that + fails (i), it suffices by Observation 10 to show that it fails (I). That is, we need to find valuations v, w such that $v \leq w$, $v(+) = 1$ but $w(+) = 0$. Put $v = v_p$ and $w = v_{p,q}$ where the subscripts indicate which of the three letters are true. Clearly this does the job.

To show that + fails (ii), it suffices by Observation 10 to show that it fails (II). That is, we need to find a valuation v with $v(+) = 0$, such that for every co-atomic $v' \geq v$ there is a w with $v \leq w \leq v'$ and $w(+) = 1$. Put v to be the valuation making all three letters false; clearly $v(+) = 0$. There are just three co-atomic $v' \geq v$ namely $v_{p,q}$, $v_{p,r}$, $v_{q,r}$ (same subscript convention). Put w to be respectively v_p, v_r, v_q. Clearly these do the job.

It remains to show that + is globally intelim-complete in the strong sense. We articulate a bundle of intelim rules that it satisfies, and show that any connective satisfying those rules is globally intelim-complete in the strong sense. First, consider the following three zero-premise elimination rules, which + clearly satisfies:

$$(1p)\quad \circ, q, r \Rightarrow p;\ (1q)\quad \circ, p, r \Rightarrow q;\ (1r)\quad \circ, p, q \Rightarrow r.$$

Next, note that + also satisfies the following three introduction rules resembling (but not quite the same as) the partner of the above trio, differing from the partner (and from each other) only in their conclusion-sequent:

$$(2p)\quad s, q, r \Rightarrow p;\ s, p, r \Rightarrow q;\ s, p, q \Rightarrow r / s, p \Rightarrow \circ$$
$$(2q)\quad s, q, r \Rightarrow p;\ s, p, r \Rightarrow q;\ s, p, q \Rightarrow r / s, q \Rightarrow \circ$$
$$(2r)\quad s, q, r \Rightarrow p;\ s, p, r \Rightarrow q;\ s, p, q \Rightarrow r / s, r \Rightarrow \circ.$$

Finally, we check that + also satisfies the following three-premise elimination rule:

$$(3)\quad s, p \Rightarrow t;\ s, q \Rightarrow t;\ s, r \Rightarrow t / s, \circ \Rightarrow t.$$

Now consider a language containing three-place connectives #, #′, each of which satisfies the above seven rules under a common consequence relation $\vdash$. We show that $\# \vdash \#'$. By (1) taking $\circ$ as #, we have each of:

$$\#, q, r \vdash p; \#, p, r \vdash q; \#, p, q \vdash r,$$

so applying (2) three times, each time taking $\circ$ as #′ and substituting $s := \#$, we get:

$$\#, p \vdash \#'; \#, q \vdash \#'; \#, r \vdash \#',$$

so by (3) taking $\circ$ as # again and substituting $s := \#, t := \#'$ we conclude:

$$p\#q, p\#q \vdash p\#'q,$$

in other words:

$$p\#q \vdash p\#'q. \qquad \square$$

Remark on the proof. Intuitively, the rules of group (1) tell us that $\circ$ excludes the case that exactly two of its arguments are true, (3) tells us that it also excludes the case that none of them are true, and (2) that it is as weak as any connective satisfying such rules.

Remark on the Observation. The contrast between Observation 11 and Corollary 8 is rather curious: three-place exclusive disjunction is (globally) intelim-unique (in the strong sense, in the set/formula context) while its simpler two-place counterpart isn't. How is that possible? The essential reason is combinatorial: an n-place exclusive disjunction comes out true just when an odd number of its arguments are true. Thus, if we make all arguments true, the two-place connective comes out false (so that it is contrarian) while the three-place one comes out true (so it is not contrarian).

3.3.3 Generalized Sufficient Conditions

Observation 11 prompts the question whether there are more general sufficient conditions for intelim-uniqueness, extending (i) and (ii) and covering the rules used in the proof that three-place exclusive disjunction is intelim-unique. In this section we provide such a pair, although we still don't know whether they are collectively necessary for the uniqueness property.

To formulate the conditions we need to generalize the notion of the partner of a set of zero-premise introduction (resp. elimination) rules, as follows:

- For any set S of zero-premise introduction rules $A_i \Rightarrow \circ\, (i \leq n)$, all with the same dummy connective $\circ$, and any $X \subseteq K = \{p_1, \ldots, p_k\}$: the *X-partner* of S is the n-premise elimination rule $A_i \Rightarrow s$ (all $i \leq n$)$/\circ, X \Rightarrow s$, where s is a fresh letter, the same in all the n premise-sequents and in the conclusion-sequent of the rule.

- For any set S of zero-premise elimination rules $A_i, \circ \Rightarrow r_i\,(i{\leq}n)$, all with the same dummy connective $\circ$, and any $X \subseteq K = \{p_1, \ldots, p_k\}$: the *X-partner* of S is the n-premise introduction rule $A_i, s \Rightarrow r_i$ (all $i{\leq}n)/s, X \Rightarrow \circ$, where likewise s is a fresh letter, the same in all the n premise-sequents and in the conclusion-sequent of the rule.

Clearly, in both cases, the partner of S as defined earlier is just its $\varnothing$-partner. Now, let $\mathfrak{X}$ be any non-empty collection of subsets of K. We define the *$\mathfrak{X}$-platform rule* for a connective $\circ$ to be the multi-premise elimination rule, where s, t are distinct letters not in K:

$$s, X \Rightarrow t(\text{all}X \in \mathfrak{X})/s, \circ \Rightarrow t.$$

This rule says roughly there is an $X \in \mathfrak{X}$ such that the rules $\circ \Rightarrow p$ hold for all letters $p \in X$, which in the set/set context could be expressed more succinctly as $\circ \Rightarrow \{\wedge X : X \in \mathfrak{X}\}$. In the limiting case that $\mathfrak{X} = \{\varnothing\}$, the rule reduces to $s, \varnothing \Rightarrow t/s, \circ \Rightarrow t$, that is $s \Rightarrow t/s, \circ \Rightarrow t$, which is an instance of the valid structural rule $s \Rightarrow t/s, u \Rightarrow t$. Our two syntactic conditions generalizing (i) and (ii) may now be formulated as follows:

(i)$^+$ $*$ satisfies: (a) some finite set S of zero-premise introduction rules and (b) for some non-empty $\mathfrak{X} \in 2^K$, the X-partner elimination rules of S for all $X \in \mathfrak{X}$, plus (c) the $\mathfrak{X}$-platform rule.

(ii)$^+$ $*$ satisfies: (a) some finite set S of zero-premise elimination rules and (b) for some non-empty $\mathfrak{X} \in 2^K$, the X-partner introduction rules of S for all $X \in \mathfrak{X}$, plus (c) the $\mathfrak{X}$-platform rule.

Clearly, when $\mathfrak{X} = \{\varnothing\}$ then these conditions become conditions (i), (ii). The argument showing that 3-place exclusive disjunction is intelim-unique (Observation 11) in effect applies condition (ii)$^+$ to the case where $K = \{p, q, r\}$, $\mathfrak{X} = \{\{p\}, \{q\}, \{r\}\}$, and S consists of the three elimination rules: $\circ, q, r \Rightarrow p$ and $\circ, p, r \Rightarrow q$ and $\circ, p, q \Rightarrow r$.

There is an interesting residual asymmetry between (i)$^+$ and (ii)$^+$. In (i)$^+$ we are looking at a set of introduction rules, a set of partner X-elimination rules, and the $\mathfrak{X}$-platform rule, while in (ii)$^+$ we are dealing with a set of elimination rules, a set of partner X- introduction rules, and the *same* $\mathfrak{X}$-platform rule, which is an elimination rule. The reason for this asymmetry will be clear from the proof of the following sufficiency result.

Observation 12. Each of the conditions (i)$^+$, (ii)$^+$ suffices to guarantee (global) intelim-uniqueness of a truth-functional connective (in the strong sense, in the set/formula context).

Proof. We can consider the two conditions in parallel. Suppose that condition (i)$^+$ resp. (ii)$^+$ holds of a k-place truth-functional connective $*$, with $S = \{A_i \Rightarrow \circ\,(i{\leq}n)\}$ resp. $S = \{A_i, \circ \Rightarrow r_i\,(i{\leq}n)\}$. Consider a language with consequence relation $\vdash$ containing k-place connectives (not necessarily truth-functional) #, #′. Suppose that each of #, #′ satisfies wrt $\vdash$ all the intelim rules that are satisfied by $*$ wrt classical consequence, thus satisfying in particular the rules in condition (i)$^+$ resp. (ii)$^+$. We show that $\# \vdash \#'$ (the converse is similar). By part (a) of (i)$^+$ resp. (ii)$^+$ taking $\circ$ as $\#$, we have:

$$A_i \vdash \# \text{ (all } i \leq n) \qquad \text{resp.} \qquad A_i, \# \Rightarrow r_i \text{ (all } i \leq n)$$

For the proof concerning (i)$^+$ we apply n times part (b) of (i)$^+$ to the rules $A_i \vdash \#$ taking $\circ$ to be $\#$ and substituting $s := \#'$; while in the proof concerning (ii)$^+$ we apply part (b) of (ii)$^+$ to the rules $A_i, \# \Rightarrow r_i$ taking $\circ$ to be $\#'$ and substituting $s := \#$. Both steps give the same consequence, namely:

$$\#, X \vdash \#'.$$

So, by part (c) common to the two conditions, taking $\circ$ as $\#$ and substituting $s := \#, t := \#'$ we conclude:

$$\#, \# \vdash \#',$$

in other words:

$$\# \vdash \#'. \qquad \square$$

With reference to the asymmetry mentioned above, we see from this proof that in the two cases we needed the (eliminatory) platform rule for the same job, namely to pass from the derived consequence $\#, X \vdash \#'$ to the consequence $\#, \# \vdash \#'$, thus getting rid of X in favour of $\#$.

These results leave us with several open questions:

- Can we formulate equivalent *semantic* counterparts of conditions (i)$^+$, (ii)$^+$ as was done for their less general versions (i), (ii)?
- Are conditions (i)$^+$, (ii)$^+$ also collectively *necessary* for (global) intelim-uniqueness of a truth-functional connective (in the strong sense, in the set/formula context)? If not, can we characterize the dividing line semantically or syntactically?
- Can we say anything interesting about *weak* global intelim-uniqueness?

3.4 Summary Table and Open Questions for the Set/Formula Context

Table 1, which may be read as a tree with the title as root, summarizes the general results on existence and uniqueness that we have established for the set/formula context (thus omitting Corollary 8, proven only for the case $k = 2$, and the negative result in Observation 11).

In the table, the parentheses around 'non-degenerate' on the left recall the fact, noted in Observation 3, that in the set/formula context no truth-functional connective

Table 1 General results on existence and uniqueness in the set/formula context

Existence		Uniqueness			
Non-degenerate elimination	(Non-degenerate) Introduction	Local		Global	
		Weak	Strong	Strong	Weak
iff not a verum	iff not contrarian	all	Iff non-contrarian or falsum	(necessary condition) sufficient conditions: (i)/(I), (ii)/(II), $(i)^+$, $(ii)^+$	(sufficient conditions)
Obs 1	Obs 3	Obs 6	Obs 6	Lemma 9, Obs 7,10,12	

satisfies any degenerate introduction rule, so that it makes no difference whether or not we require non-degeneracy there. On the right, the inter-column arrows recall that the characteristic condition for the local strong property is *ipso facto* a necessary condition for the global strong one, while the sufficient conditions for the global strong property are likewise sufficient conditions for the global weak case.

4 The Intermediate Set/Formula-or-Empty Context

One may also ask what happens in the intermediate *set/formula-or-empty* context. In this final section we briefly note some basic facts about it.

Existence of elimination rules satisfied by a connective. Observation 1, which holds in the other two contexts, also holds in the intermediate one: a truth-functional connective satisfies at least one non-degenerate elimination rule iff it is not a verum.

Existence of introduction rules satisfied by a connective. For this property, the intermediate context behaves very much like the set/set one. First, a trivial point: every truth-functional connective satisfies some *degenerate* introduction rule, namely $K \Rightarrow \varnothing / K \Rightarrow \circ$, where $K = \{p_1, \ldots, p_k\}$. This contrasts with absence of any such rules in the set/formula situation (cf. Observation 3).

More significantly every k-ary truth-functional connective that is not a falsum satisfies at least one *non-degenerate* introduction rule, namely the k-premise introduction rule $\{T_v, p \Rightarrow \varnothing : p \in F_v\}/T_v \Rightarrow \circ$ where v is an arbitrarily chosen valuation with $v(*) = 1$ (at least one such v exists, as the connective is not a falsum; when $F_v = \varnothing$

the rule is understood as the zero-remise one $T_v \Rightarrow \circ$, i.e. $K \Rightarrow \circ$). This contrasts with the set/formula situation (cf. Observation 3), and agrees with the set/set one (cf. Observation 2, which can thus be given an alternative proof using empty-conclusion sequents in place of multi-conclusion ones).

Uniqueness of connectives given the set of intelim rules that they satisfy. On the *local* level, every truth-functional connective $*$ (of any number k of arguments) is now intelim-unique in the strong sense. To see this, associate each valuation v: $K \rightarrow \{0, 1\}$ with an intelim rule as follows: when $v(*) = 1$ take the introduction rule $\{T_v, p \Rightarrow \varnothing : p \in F_v\}/T_v \Rightarrow \circ$ mentioned in the preceding paragraph, and when $v(*) = 0$ take the elimination rule $\{T_v, p \Rightarrow \varnothing : p \in F_v\}/T_v, \circ \Rightarrow \varnothing$. To illustrate, in the case of negation this produces two rules, namely $p \Rightarrow \varnothing/\varnothing \Rightarrow \neg p$ and p, $\neg p \Rightarrow \varnothing$. It is not difficult to show that $*$ is the unique truth-functional connective (though perhaps not the unique connective) satisfying these 2^k intelim rules, giving us local intelim-uniqueness of $*$ in the strong sense.

This result suggests that methodologically, the set/formula-or-empty context is perhaps the most convenient one for treating classical consequence syntactically. It has enough expressive power to characterize truth-functions by means of intelim rules, without the rather unintuitive (though mathematically elegant) baggage of full set/set sequents.

For *global* intelim-uniqueness in the strong sense, we note that the notion of the *partner* of a set of zero-premise elimination (resp. introduction) rules extends meaningfully and naturally from the set/formula context to the set/formula-or-empty one, and each of syntactic conditions (i), (ii) continues to imply global intelim-uniqueness in the strong sense. Condition (i) covers no new connectives in the enlarged context, because zero-premise introduction rules remain of the form $A \Rightarrow \circ$ and so always have at least one formula on their right. However, condition (ii) does cover new connectives. Indeed, for the special case of the two-place connectives, it turns out that the condition is fulfilled by seven out of the eight contrarian ones, the exception being exclusive disjunction. Table 2 exhibits, for each of the seven, a set of zero-premise elimination rules with partner that it satisfies.

While $+$ fails condition (ii) in the set/formula-or-empty context, is it globally intelim-unique in the strong sense. These contrasting facts may be proven by essentially the same arguments as were used for three-place exclusive disjunction in Observation 11, with some simplifications. For example, to establish the positive part, we can consider the following three (instead of seven) rules:

$$p \circ q, p, q \Rightarrow \varnothing \tag{1}$$

$$s, p, q \Rightarrow \varnothing / s, p \Rightarrow p \circ q \tag{2}$$

$$s, p \Rightarrow r; s, q \Rightarrow r / s, p \circ q \Rightarrow r. \tag{3}$$

Thus, we can say that for the special case of the 16 two-place truth-functional connectives, exactly 15 (namely all except $+$) fulfill at least one of conditions (i) and (ii); but all 16 are globally intelim-unique in the strong sense. This agrees with the

Table 2 Seven two-place contrarian connectives satisfying condition (ii) in the set/formula-or-empty context

Connective	Zero-premise elimination rules	Partner
⊥ (p,q)	$p \circ q \Rightarrow \varnothing$	$s \Rightarrow \varnothing / s \Rightarrow p \circ q$
Neither p nor q	$p, p \circ q \Rightarrow \varnothing; q, p \circ q \Rightarrow \varnothing$	$p, s \Rightarrow \varnothing; q, s \Rightarrow \varnothing / s \Rightarrow p \circ q$
Not both p and q	$p \circ q, p, q \Rightarrow \varnothing$	$s, p, q \Rightarrow \varnothing / s \Rightarrow p \circ q$
¬ *left*(p, q)	$p, p \circ q \Rightarrow \varnothing$	$p, s \Rightarrow \varnothing / s \Rightarrow p \circ q$
¬ *right*(p, q)	$q, p \circ q \Rightarrow \varnothing$	$q, s \Rightarrow \varnothing / s \Rightarrow p \circ q$
$p \wedge \neg q$	$p \circ q \Rightarrow p; p \circ q, q \Rightarrow \varnothing$	$s \Rightarrow p; s, q \Rightarrow \varnothing / s \Rightarrow p \circ q$
$q \wedge \neg p$	$p \circ q \Rightarrow q; p \circ q, p \Rightarrow \varnothing$	$s \Rightarrow q; s, p \Rightarrow \varnothing / s \Rightarrow p \circ q$

Table 3 General results on existence and uniqueness in the set/formula-or-empty context

Existence		Uniqueness				
non-degenerate elimination	(non-degenerate) introduction	local		global		
		weak	strong	strong		weak
iff not a verum	iff not a falsum	all	all	sufficient conditions: (i)/(I), (ii)/(II), (i)$^+$, (ii)$^+$	⇒	(sufficient conditions)

situation in the set/set context (cf. Observation 4) and contrasts with the set/formula picture (cf. Corollary 9).

It is an open question whether this positive result extends from $k = 2$ to all values of k, that is, whether all truth-functional connectives are globally intelim-unique in the strong sense in the set/formula-or-empty context. If it does extend, then the main existence and uniqueness properties for the set/formula-or-empty context coincide with those for the set/set context. If not, then we may ask where the dividing line lies.

Table 3 summarizes the general results of this section on existence and uniqueness in the set/formula-or-empty context, serving as a counterpart to Table 1 for the set/formula context.

Acknowledgments Thanks to Lloyd Humberstone and to Frederik van de Putte for incisive comments and vigorous criticism over several drafts. Peter Schroeder-Heister's counsel from a proof-theoretic perspective was of great assistance, and the searching remarks of two anonymous referees also helped improve the text. Finally, thanks to the Editor for his patience with repeated post-final refinements.

Appendix

We recall familiar definitions of formula-schemes, sequents and rules, as well as elementary, structural, introduction and elimination rules, plus the central notion of satisfaction of a rule, with some comments on delicate points.

A *formula-scheme* is like a formula of a propositional logic except that it has 'dummy connectives', that is, place-markers $\circ$ in place of any connectives occurring in the formula; we assume that we have a countable supply of such dummy connectives for use as needed. A *set/set sequent* is an expression $A \Rightarrow B$ where A and B are finite (possibly empty) sets of formula-schemes. For brevity, when $A = \{\alpha_1, \ldots, \alpha_n\}$ and $B = \{\beta_1, \ldots, \beta_m\}$ we omit parentheses and write simply $\alpha_1, \ldots, \alpha_n \Rightarrow \beta_1, \ldots, \beta_m$. Similarly, $A \cup \{\alpha\} \Rightarrow B$ is written $A, \alpha \Rightarrow B$. *Set/formula-or-empty* sequents are those in which B is either a singleton or empty; when B is a singleton $\{\beta\}$ we have a *set/formula sequent* and we write it as $A \Rightarrow \beta$.

Expressions $A_1 \Rightarrow B_1; \ldots; A_n \Rightarrow B_n / C \Rightarrow D$ are called n-premise (*set/set*) *rules*, where $n \geq 0$ and the $A_i \Rightarrow B_i$, $C \Rightarrow D$ are set/set sequents. If all of the sequents are set/formula ones, it is a *set/formula rule*; similarly, if they are all set/formula-or-empty sequents, we have a *set/formula-or-empty rule*.

We say that a rule $A_1 \Rightarrow B_1; \ldots; A_n \Rightarrow B_n / C \Rightarrow D$ is *elementary* iff the following four conditions hold: (1) all formula-schemes in all its n+1 sequents are sentence letters except for at most one, in which case (2) it is of the form $\circ(p_1, \ldots, p_k)$ for some k-place dummy connective $\circ$ and sentence letters $p_1, \ldots, p_k$, and (3) it occurs exactly once, this occurrence being (4) in the conclusion-sequent $C \Rightarrow D$. In condition (2) it is convenient to require $p_1, \ldots, p_k$ to be distinct, but it makes no difference to the results of this paper.

When the connective does not appear at all in the rule, which thus consists entirely of sentence letters, the rule is *structural*. When the connective appears just once then, if it is in C we have an *elimination* rule, while if it is in D we have an *introduction* rule. Elimination and introduction rules are called briefly *intelim* rules. Note that when there is a unique occurrence of the connective but it is in one of the premise sequents, we do *not* count the rule as intelim in this paper. Some texts use the notion in a broader (and at times rather loose) way.

A *model* for a rule is a pair $(L, \vdash)$ where L is a propositional language and $\vdash$ is a consequence relation over it. In this paper $\vdash$ is always assumed to be a closure relation (i.e. satisfying the Tarski conditions appropriate to the set/set or set/formula context), and also to be closed under uniform substitution for sentence letters.

Consider any n-premise rule $A_1 \Rightarrow B_1; \ldots; A_n \Rightarrow B_n / C \Rightarrow D$ and model $(L, \vdash)$. A specification of connectives from L as interpretations of the dummy connectives in the rule is said to *satisfy* the rule in the model iff, after replacing the dummy connectives as specified, there is no instantiation σ of sentence letters to formulae of L such that $\sigma(A_i) \vdash \sigma(B_i)$ for all $i \leq n$ while $\sigma(C) \nvdash \sigma(D)$. For intelim rules, where there is just one dummy connective, we evidently need to specify just one connective $*$, and so we will say that $*$ satisfies the rule in the model. Satisfaction of a *sequent* $C \Rightarrow D$ in a model is identified with satisfaction of the zero-premise rule with conclusion $C \Rightarrow D$.

This is quite a complex definition, and it is important not to neglect the universal quantification over instantiations σ. It is also important to keep track of the parameters $L, \vdash$ in the model, since variation of either can change satisfaction into its failure or conversely. Nevertheless, it is not difficult to show that if a *truth-functional* connective $*$ satisfies an intelim rule in a classical model $(L, \models)$ that can express at least one tautology and at least one contradiction, then it does so in any other such classical model. This is because valuations of instantiations into one of the two languages may be mimicked by suitable substitutions of the tautology and the contradiction in the other. For this reason, when we are considering only truth-functional connectives (as in Sect. 2) we can be quite relaxed about the exact choice of the classical language.

An intelim rule that is satisfied by a model can be *degenerate* in the sense that it is a substitution instance of a structural rule that is satisfied by all models. For example, the zero-premise elimination rule $p, p \circ q \Rightarrow p$, which is satisfied by the model $(L, \wedge, \models)$ where L is a classical language, is also an instance of the structural rule $p, q \Rightarrow p$ satisfied by all models.

When one is dealing with a single propositional language, then it is possible to talk of rules without using formula-schemes and dummy connectives: one can simply identify the rule with the set of all its instances in the fixed language, or even with a representative underlying instance where the letters of the scheme are sent injectively to letters of the language. But if one wishes to consider the applications of a single rule to *multiple* languages, as in this paper, then such a simplified framework becomes cumbersome to operate with clarity and rigour.

References

Došen, K., & Schroeder-Heister, P. (1988). Uniqueness, definability and interpolation. *The Journal of Symbolic Logic, 53*, 554–570.

Humberstone, L. (2011). *The connectives*. Cambridge: MIT Press.

Humberstone, L., & Makinson, D. (2011). Intuitionistic logic and elementary rules. *Mind, 120*, 1035–1051.

Rautenberg, W. (1981). 2-element matrices. *Studia Logica, 40*, 315–353.

Schroeder-Heister, P. (1984). A natural extension of natural deduction. *The Journal of Symbolic Logic, 49*, 1284–1300.

Relevance Logic as a Conservative Extension of Classical Logic

David Makinson

Abstract Relevance logic is ordinarily seen as a subsystem of classical logic under the translation that replaces arrows by horseshoes. If, however, we consider the arrow as an additional connective alongside the horseshoe, then another perspective emerges: the theses of relevance logic, specifically the system R, may also be seen as the output of a conservative extension of the relation of classical consequence. We describe two ways in which this may be done. One is by defining a suitable closure relation out of the set of theses of relevance logic; the other is by adding to the usual natural deduction system for it further rules with 'projective constraints', whose application restricts the subsequent application of other rules. The significance of the two constructions is also discussed.

Keywords Relevance logic · System R · Consequence relations · Closure relations · Natural deduction · Projective constraints

1 Background and Goal

In a natural sense propositional relevance logic, as expressed by axiom systems such as R or E of Anderson and Belnap, is included in classical logic. If we systematically read its arrows (for relevant conditionals) as horseshoes (for material, i.e. truth-functional, conditionals) then every thesis of the relevance logic is a classical tautology. The inclusion is proper since notoriously there are also formulae that are

D. Makinson (✉)
Department of Philosophy, Logic and Scientific Method London School of Economics, Houghton Street, London WC2A 2AE, UK
e-mail: david.makinson@gmail.com

S. O. Hansson (ed.), *David Makinson on Classical Methods for Non-Classical Problems*, Outstanding Contributions to Logic 3, DOI: 10.1007/978-94-007-7759-0_17,

not accepted in relevance logic although they are classical tautologies when their arrows are read in this way.[1]

If, however, we think of the arrow as a connective additional to the truth-functional ones, it turns out that there is a sense in which relevance logics are conservative extensions of classical logic. To fix notation, let us take relevance logic as formulated as usual, with the connectives $\wedge$, $\vee$, $\neg$ plus the arrow $\rightarrow$, and classical propositional logic as expressed using just the first three. Then, as already recorded in Anderson and Belnap (1975) Sect. 24.1, we have the following basic fact.

Fact 1. A formula in the language of classical propositional logic is a tautology iff it is a thesis of R (indeed, of any of a range of neighbouring systems such as E, T, and R-Mingle) over the wider language with arrow added as a connective.

Meyer (1974) has described another way of isolating the classical tautologies, within a system that conservatively extends R by further non-classical connectives.[2] However, both of these results concern the *sets of formulae* serving as theses of classical and of relevance logic. The accepted wisdom has been that these results *do not extend to the corresponding consequence relations*; in other words, that classical consequence is not a subrelation of consequence for relevance logic.[3]

But *what is* the consequence relation for relevance logic? Things are not cut and dried, even when we have fixed a particular relevance logic, such as the system R. Historically, in both the 1975 and 1992 volumes of their treatise, Anderson and Belnap focussed attention on the *set* of formulae regarded as acceptable, rather than on a consequence *relation* between formulae. While they considered, in some detail, notions of derivation-from-assumptions within given Hilbertian axiomatizations of the set of acceptable formulae, and also articulated systems of natural deduction to generate the same set, these were not developed into a concept of consequence relation, between a set of formulae on the left and an individual formula on the right considered in abstraction from any particular path leading from one to the other.

By and large, those working in the wake of Anderson and Belnap have followed suit. As Avron (1992) put it (pp. 244–245): 'The relevantists have followed the [older] classical tradition of focusing on the set of logically valid sentences rather than on what is the really important aspect of a logic: the consequence relation(s) associated

[1] For axiomatizations of R and neighbouring systems as sets of formulae, see Anderson and Belnap (1975) Sects. 21.1 and 27.1, summarized in Anderson et al. (1992) Sect. R2; also e.g. Dunn (1986) Sect. 1.3 repeated in Dunn and Restall (2002) Sect. 1.3; Mares (2004) appendix A and Mares (2012). Following a tradition going back to Prior (1955), we use the term *thesis* for what is also commonly called a 'theorem' of the logic, reserving the latter word for whatever one can prove about the logic (instead of the rather cumbersome 'meta-theorem').

[2] To be specific, Meyer (1974) showed that if we introduce as further primitives a non-classical two-place 'fusion' connective and propositional constants f, t, writing $\sim\alpha$ as an abbreviation for $\alpha \rightarrow f$, then we can give a straightforward axiomatization of a conservative extension of R that neatly contains a classical axiom set for $\wedge$, $\vee$, $\neg$ alone.

[3] See for example the remarks in Wolf (1978) pp. 329–330; Dunn (1986) pp. 124 and 148–149 repeated in Dunn and Restall (2002) pp. 6–7 and 29–30; Restall (2000) Sect. 16.3; Mares (2004) p. 81.

with it'. Because of this, as Humberstone (2011) remarked (p. 1097), the latter topic has a 'somewhat neglected' status.

One may therefore ask: Is there any way in which we may define a consequence relation on formulae of relevance logic that leads to R, say, and is at the same time a conservative extension of classical consequence? In this paper we describe two ways of doing so.

2 A Closure Relation that outputs R

We work with relations $\vdash$ between sets of formulae of relevance logic on the left and individual formulae on the right. We do not consider generalizations in which the right position may be occupied by a set of formulae, nor those in which left position is occupied by a multi-set, sequence, or other such structure. Greek α, β, ... are formulae, upper case A, B, ... are for sets of formulae, and p, q, ... for elementary letters (aka variables) in formulae.

Following familiar terminology, we call $\vdash$ a *closure relation* iff it satisfies the three Tarski conditions of reflexivity ($A \vdash \alpha$ whenever $\alpha \in A$), cumulative transitivity ($A \vdash \gamma$ whenever both $A \vdash \beta$ for all $\beta \in B$ and $A \cup B \vdash \gamma$) and monotony ($A \vdash \alpha$ whenever $B \vdash \alpha$ and $B \subseteq A$). Our question may now be put in a sharper form: is there a closure relation in the language of relevance logic that is a conservative extension of classical consequence and *outputs* R in the sense that $\beta \in$ R iff $\emptyset \vdash \beta$ for all formulae β?

To construct one, write $\alpha \supset \beta$ as an abbreviation for $\neg\alpha \vee \beta$ and define $A \vdash_0 \beta$ to hold iff for some finite subset $\alpha_1,\ldots,\alpha_n \in A$ the formula $(\alpha_1 \wedge \ldots \wedge \alpha_n) \supset \beta$ is a thesis of R (with this formula understood to be β in the limiting case that $n = 0$). Then we have the following.

Theorem 2. $\vdash_0$ is a closure relation, is compact and closed under substitution, conservatively extends classical consequence, and outputs R. Indeed, it is the least such relation.

Proof. We give the proof in full detail, as we will also be considering how far it carries over to other contexts. There are seven points to verify.

(1) Compactness is immediate from the definition.

(2) To check that the relation is closed under substitution, suppose $A \vdash_0 \beta$. Then by definition, for some finite subset $\{\alpha_1,\ldots,\alpha_n\} \subseteq A$ the formula $(\alpha_1 \wedge \ldots \wedge \alpha_n) \supset \beta$ is a thesis of R. But R is closed under substitution, so $\sigma((\alpha_1 \wedge \ldots \wedge \alpha_n) \supset \beta) = (\sigma(\alpha_1) \wedge \ldots \wedge \sigma(\alpha_n)) \supset \sigma(\beta) \in$ R for any substitution function σ, so $\sigma(A) \vdash_0 \sigma(\beta)$.

(3) To show that $\vdash_0$ extends classical consequence, suppose $\alpha_1,\ldots,\alpha_n$, β are formulae of classical logic and $\{\alpha_1,\ldots,\alpha_n\}$ classically implies β. Then $(\alpha_1 \wedge \ldots \wedge \alpha_n) \supset \beta$ is a tautology. By Fact 1 (left to right) it is in R and so by definition $\{\alpha_1,\ldots,\alpha_n\} \vdash_0 \beta$.

(4) To show that the extension is conservative, suppose again that $\alpha_1,\ldots,\alpha_n,\beta$ are formulae of classical logic and $\{\alpha_1,\ldots,\alpha_n\} \vdash_0 \beta$. Then by the definition of $\vdash_0$ we have $(\alpha_1 \wedge \ldots \wedge \alpha_n) \supset \beta$ in R. Fact 1 (right to left) tells us that $(\alpha_1 \wedge \ldots \wedge \alpha_n) \supset \beta$ is a tautology, so that $\{\alpha_1,\ldots,\alpha_n\}$ classically implies β.
(5) That $\vdash_0$ outputs R is immediate from the definition with $n = 0$.
(6) To show that $\vdash_0$ is a closure relation, note first that monotony is immediate from the definition, while reflexivity follows from the fact that $\alpha \supset \alpha$ is in R. To shorten the verification of cumulative transitivity we introduce some notation. For a finite set $X = \{\alpha_1,\ldots,\alpha_n\}$ of formulae we write $\wedge X \supset \beta$ as shorthand for $(\alpha_1 \wedge \ldots \wedge \alpha_n) \supset \beta$, again reading this as β when $n = 0$. Clearly, for both classical and relevance logic, the order of the conjuncts in $\wedge X$ does not matter. Now suppose both $A \vdash_0 \beta$ for all $\beta \in B$ and $A \cup B \vdash_0 \gamma$; we need to show that $A \vdash_0 \gamma$. From the second supposition (with compactness) there are finite subsets $A_0 \subseteq A$, $B_0 \subseteq B$ such that $\wedge(A_0 \cup B_0) \supset \gamma$ is in R; and from the first supposition (with compactness) there is a finite subset $A_1 \subseteq A$ such that $A_1 \vdash_0 \beta$ for all $\beta \in B_0$, so that $\wedge A_1 \supset \beta$ is in R for all $\beta \in B_0$. Now, for $\{\beta_1,\ldots,\beta_n\} = B_0$ the following formula is a substitution instance of a tautology:

$$[\wedge A_1 \supset \beta_1] \supset [\ldots[\{\wedge A_1 \supset \beta_n\} \supset \{(\wedge(A_0 \cup B_0) \supset \gamma) \supset (\wedge(A_0 \cup A_1) \supset \gamma)\}]\ldots].$$

Since R is closed under substitution, Fact 1 tells us that this formula is in R. But it is known that the set of theses of R is closed under detachment for $\supset$; that is, whenever α and $\alpha \supset \beta$ are both theorems of R, so is β.[4] By $n + 1$ applications of detachment, $\wedge(A_0 \cup A_1) \supset \gamma$ is in R, so $A \vdash_0 \gamma$ as desired.
(7) For leastness, suppose $A \vdash_0 \beta$ and let $\vdash$ be any closure relation closed under substitution that extends classical consequence and outputs R; we show that $A \vdash \beta$. By the definition of $\vdash_0$, the formula $(\alpha_1 \wedge \ldots \wedge \alpha_n) \supset \beta$ is a thesis of R for some $\alpha_1,\ldots,\alpha_n \in A$. Since $\vdash$ outputs R we thus have $\emptyset \vdash (\alpha_1 \wedge \ldots \wedge \alpha_n) \supset \beta$, so by the monotony of $\vdash$ also $A \vdash (\alpha_1 \wedge \ldots \wedge \alpha_n) \supset \beta$. But since $\vdash$ extends classical consequence and is closed under substitution, also $A \cup \{(\alpha_1 \wedge \ldots \wedge \alpha_n) \supset \beta\} \vdash \beta$. So by cumulative transitivity for $\vdash$ we have $A \vdash \beta$ as desired. □

How should we interpret this? On the one hand, it can be seen as providing a motto for the relevantist policeman: *No arrests for classical connections between consenting formulae; they are perfectly legal under a suitably defined consequence relation for relevance logics!* On the other hand, it could be misleading to call $\vdash_0$ a *relevant consequence relation.* A candidate minimal condition for such a name to be appropriate might be that whenever $\alpha \vdash \beta$ then the formula $\alpha \to \beta$ is in R, where $\to$ is the arrow connective of R. Our consequence relation clearly fails that condition. For instance, since it extends classical consequence it tells us that $p \wedge \neg p \vdash_0 q$ and $p \vdash_0 q \vee \neg q$ while notoriously neither of $(p \wedge \neg p) \to q$, $p \to (q \vee \neg q)$ is in R. Likewise, since $p \supset (q \to q)$ is shorthand for $\neg p \vee (q \to q)$, which is a thesis of R, we have $p \vdash_0 q \to q$, while $p \to (q \to q)$ is not in R. We can however describe

[4] This is not a trivial result. It was first established by Meyer and Dunn (1969) and can also be found in Anderson and Belnap (1975), Sect. 25.2.3. In the context of relevance logic the rule of detachment for $\supset$ is also known as γ.

$\vdash_0$ as a consequence relation *for* relevance logic, in the sense that it outputs exactly the theses of the relevance logic R. By Theorem 2, it is also a closure relation and conservatively extends classical consequence.

Relevantists may well respond that $\vdash_0$ is not the relation they are thinking of. So what specific relation could they have in mind? Unfortunately, that is not at all clear from the literature, where at least eight candidates appear to be latent, differing from each other in three dimensions: whether to combine multiple premises by conjunction or by embedded arrows, whether or not to allow superfluous premises, and whether or not to countenance the empty premise set.

- One definition would put $A \vdash \beta$ iff for some $\alpha_1,\ldots,\alpha_n \in A$ the formula $(\alpha_1 \wedge \ldots \wedge \alpha_n) \to \beta$ is in R (with the formula again read as β when $n = 0$). This is like the definition of $\vdash_0$ above, but with $\to$ in place of $\supset$.
- A second puts $A \vdash \beta$ iff for some $\alpha_1,\ldots,\alpha_n \in A$ the formula $\alpha_1 \to (\alpha_2 \to \ldots(\alpha_n \to \beta)\ldots)$ is in R. Here the formula embeds arrows.
- The third and fourth are more stringent versions of the above obtained by requiring that A is a finite set, all of whose elements appear as antecedents of the certifying formula $(\alpha_1 \wedge \ldots \wedge \alpha_n) \to \beta$ or $\alpha_1 \to (\alpha_2 \to \ldots(\alpha_n \to \beta)\ldots)$ respectively.
- Four more are like the above but also requiring that the premise set A is non-empty. For brevity we call these the *non-empty counterparts* of the above four.

Historically, remarks in Sects. 22.2.1 and 23.6 of Anderson and Belnap (1975) seem to suggest an inclination towards one of the first two relations; on the other hand, their emphasis on the importance of conditionalization with respect to $\to$ also suggests the fourth one. But since the text does not conceptualize the notion of a consequence relation, it is difficult to give a firm reading; moreover it is not clear whether the non-empty counterparts might be preferred. Comments of Meyer (1974) pp. 54–55 seem to point to the fourth (or eighth) consequence relation, although again the text is open to interpretation. Avron (1992) pp. 247–248 appears to see relevance logicians as committed to the third (or seventh) relation. Mares, in his contribution to the present volume, in effect uses the fifth (in a context allowing multiple formulae on the right, and with a weaker relevance logic in place of R) to define his 'second relation' $\Vdash$.

Unlike our $\vdash_0$, *none of these eight extends classical consequence*—witness the failure in all of them of $\{p \wedge \neg p\} \vdash q$. They also differ considerably from each other in their behaviour. We will not treat them in detail, but note parenthetically a few salient differences on important properties; verifications are immediate.

Dependence on the background system. The second and fourth definitions (and their non-empty counterparts) use the formula scheme $\alpha_1 \to (\alpha_2 \to \ldots(\alpha_n \to \beta)\ldots)$, and so do not yield well-defined relations for a background Hilbertian system such as E, weaker than R, that lacks the principle of permutation for embedded arrow antecedents. More subtle formulations of the definition are then needed, say with the premises conceived as forming a sequence rather than a set. But in the case of R, with which we are concerned, all eight are well-defined for finite premise sets A.

Infinite premise sets. When the premise set A is infinite, the third and fourth relations (and their non-empty counterparts) are undefined. There are various ways of handling the gap; one may deliberately leave the relation undefined in that case, or adopt one or another reduction to the finite case.

Output sets. None of the non-empty counterpart relations output R, since they all output the empty set. Indeed, for those relations, a non-trivial notion of output is hard to define.

Relevant consequence relations. Neither the first nor the second relation satisfies the candidate minimal condition, suggested above, for being a 'relevant consequence relation', namely that whenever $\alpha \vdash \beta$ then the formula $\alpha \to \beta$ is in R. Counterexample: $p \vdash q \vee \neg q$ since (taking $n = 0$) we have $q \vee \neg q \in$ R, even though $p \to (q \vee \neg q) \notin$ R. This failure is remedied by their non-empty counterparts.

Monotony. The first two relations clearly satisfy monotony, as do their non-empty counterparts. In contrast, the fourth relation fails it (as also does its non-empty counterpart), since we may have $\alpha_1 \to (\alpha_2 \to \ldots (\alpha_n \to \beta) \ldots) \in$ R but $(\alpha_1 \to (\alpha_2 \to \ldots (\alpha_{n+1} \to \beta) \ldots) \notin$ R. The third relation goes close to satisfying monotony, since when $n \geq 1$ and $(\alpha_1 \wedge \ldots \wedge \alpha_n) \to \beta \in$ R then $(\alpha_1 \wedge \ldots \wedge \alpha_{n+1}) \to \beta \in$ R, but fails in the case $n = 0$, since again $q \vee \neg q \in$ R while $p \to (q \vee \neg q) \notin$ R; again this is remedied by its non-empty counterpart.

Cumulative transitivity. The first relation does not quite satisfy cumulative transitivity. A counterexample was given by Avron (1992) p. 248. Each of $p \to p$ and $((p \to p) \wedge q) \to ((p \to p) \wedge q)$ is in R, so that by the definition of $\vdash$ we have $q \vdash p \to p$ (using the case $n = 0$) and also $\{p \to p, q\} \vdash (p \to p) \wedge q$; but not $q \vdash (p \to p) \wedge q$, since neither $q \to ((p \to p) \wedge q)$ nor $(p \to p) \wedge q$ is in R. An even simpler counterexample was communicated by Peter Verdée in personal correspondence with the author: we have $p \to q \vdash p \to p$ (using the case $n = 0$ in the definition) as well as $p \to q \vdash p \to q$; we also have $\{p \to q, p \to p\} \vdash p \to (p \wedge q)$, but not $p \to q \vdash p \to (p \wedge q)$. Again, this is remedied by the non-empty counterpart, without damage to reflexivity or monotony, so that the fifth relation is a closure relation (as are the second and its non-empty counterpart).

A general conflict. Quite generally, for any consequence relation over the language (not only the eight defined above) there is an evident conflict between the three potential requirements of outputting R, being monotonic, and satisfying the candidate condition for being a 'relevant consequence relation'. For if $\vdash$ outputs R then $\emptyset \vdash q \vee \neg q$ so monotony gives $p \vdash q \vee \neg q$ while $p \to (q \vee \neg q) \notin$ R, violating the candidate condition.

Thus each of the eight consequence relations has its own peculiarities, which may be regarded as faults or as features, according to the eye of the beholder. We do not attempt to adjudicate between the attractions of the different relations, nor between them and our $\vdash_0$. We merely emphasize Theorem 2 as an *existence result*, telling us that alongside the various consequence relations that have been considered in the literature there is another, namely $\vdash_0$, that *does* conservatively extends classical consequence while outputting R (as well as being a compact closure relation closed under substitution). While it does not satisfy the candidate condition for being called a 'relevant consequence relation', it is nevertheless a well-behaved consequence relation *for* relevance logic.

3 The Idea of a Projective Constraint

We now consider the question of consequence relations for relevance logic from another angle. As is well known, Anderson and Belnap showed how to generate their system R using a natural deduction system with labels. It is natural to ask whether we can generate a conservative extension of classical consequence that outputs R in the same kind of way. In other words, can one obtain, via a natural deduction system with labels, a relation that outputs R and is a conservative extension of classical consequence? We show that the answer is positive.

Recall that the labels of the Anderson/Belnap system of natural deduction are supposed to record, for each line in a derivation, the undischarged assumptions that were *actually used* to get that line, so that when we come to apply the rule of arrow introduction ($\rightarrow +$) we may check the constraint that the formula being discharged has indeed been used. But, as is notorious, if the rules for conjunction and disjunction are not restricted in some way, they allow 'artificial employment' of assumptions. Quite trivially, with the rules for conjunction, one can apply the $\wedge +$ rule just to get an arbitrary assumption β on the payroll and then immediately apply the $\wedge -$ rule to send it on paid holiday. The same may be done using the disjunction rules $\vee +$, $\vee -$ with help from $\rightarrow -$: given α we can first use $\vee +$ to infer $\alpha \vee (\beta \rightarrow \alpha)$ with β chosen arbitrarily, then carry out two sub-proofs, one obtaining α from the first disjunct simply by reiteration, the other deriving α from the second disjunct together with additional assumption β by $\rightarrow -$, and finally apply $\vee -$ to emerge with α derived from α, β with both assumptions actually used.

In order to prevent such 'funny business',[5] Anderson and Belnap *restrict* the rules for $\wedge +$ and $\vee -$ to the special case that the assumptions actually used are the same for the two premises of the rule and, to compensate for the resulting overkill, they add to the system a rule expressing distribution of $\wedge$ over $\vee$. At the same time, for reasons quite independent of the 'funny business' problem, the classical rules for negation are truncated, leaving only their de Morgan versions. Not a very elegant dance, as all admit, but it works, at least in the sense of permitting derivation of all and only the theses of the system R from the empty set of assumptions. In particular, the natural deduction system allows derivation of all classical tautologies in $\wedge, \vee, \neg$ while, for example, not allowing arbitrary formulae to be derivable from explicit contradictions.

The idea behind our natural deduction system is to take exactly the same language, rules, and labelling regime as for Anderson and Belnap and supplement them with rules sufficient for the whole collection to yield classical consequence. However, applications of the additional rules are flagged and act as constraints on subsequent applications of the rule $\rightarrow +$ of arrow introduction. The flags are inherited as one passes along a derivation, and arrow introduction is given a second constraint: it cannot be applied when the conclusion of the sub-proof leading up to it is flagged. In

[5] The description of such detours as 'funny business' was popularized by R.J. Fogelin in the years before publication of Anderson and Belnap (1975), whose section 8.21 welcomes the epithet.

this way, the classical strength of the totality of rules is prevented from 'corrupting' the logic of the arrow.

Thus the system monitors the application of all rules by means of both labels and flags, but the flags constrain only the application of $\to+$. We get, one might say, a *projective constraint* version of the Anderson/Belnap natural deduction system. It provides a second motto for the relevantist policeman: *Take down incriminating classical steps in evidence now, but don't cite them until you are facing* $\to+$ *in the court-room!*

It must be said immediately that the general idea of projective constraint rules is not new to this paper, although it does not appear to have been given a name. In classical logic such rules have for long been used in some natural deduction systems for first-order reasoning. Specifically, some textbook versions of the rule known as existential instantiation (EI) permit us, under certain conditions, to strip off an existential quantifier and instantiate its free variable, while flagging the step to block certain other rules from subsequently being applied to formulae that inherit the flag. For details, see e.g. the overviews of Pelletier and Hazen (2001) or Pelletier and Hazen (2012). Indeed, the Anderson–Belnap labels may already be regarded as 'soft' projective constraints, since they restrict (but without banning) subsequent applications of the rules $\to+$, $\wedge+$, $\vee+$.

4 Formal Definition of our Projective Constraint System

When presenting their system of natural deduction, Anderson and Belnap write their labels as subscripts to formulae. But as they also make clear, the same formula may occur at several lines of a derivation with a different status at each line, requiring different labels. Strictly, a derivation in their system should be taken as a finite sequence of line numbers $1,\ldots,n$, with both formulae and labels attached to the line numbers. In turn, labels should be seen as sets of line numbers rather than sets of formulae. Our flag, written as #, will likewise be placed as a superscript to formulae but understood as attached to line numbers.

We recall from Sect. 1 that formulae of the language of relevance logic are generated using the connectives $\wedge$, $\vee$, $\neg$, $\to$ (thus without further connectives such as 'fusion'). The expression $\alpha \supset \beta$ is an abbreviation of $\neg\alpha \vee \beta$. There are two groups of rules.

Rules of Group I

These are all the rules of Anderson and Belnap for the system R in the connectives $\wedge, \vee, \neg, \to$.[6] The sole difference lies in the formulation of the rule $\to+$, where we impose a new flagging constraint alongside the existing labelling constraint, as italicized in the following.

---

[6] For presentations of the Anderson/Belnap system of natural deduction for R (and neighbouring systems) see Anderson and Belnap (1975) Sect. 27.2, summarized in Anderson et al. (1992) Sect. R3, and Mares (2004) appendix A. Unfortunately, the presentations in Dunn (1986) Sect. 1.5 and Dunn and Restall (2002) Sect. 1.5 lack the rules for negation.

$\to+$. Having inferred β_X from assumptions A,α, we may infer $(\alpha \to \beta)_{X-\{k\}}$ from A, where k is the line number of the assumption α, provided $k \in X$ *and* β_X *is not flagged.*

Rules of Group II

Unrestricted $\wedge+$. From α_X, β_Y infer $(\alpha \wedge \beta)^{\#}_{X\cup Y}$.

Modus Ponens for $\supset$. From α_X, $(\alpha \supset \beta)_Y$ infer $\beta^{\#}_{X\cup Y}$.

In both rules the conclusion is flagged. In primitive notation, the second one is a form of disjunctive syllogism, permitting passage from α_X, $(\neg\alpha \vee \beta)_Y$ to $\beta^{\#}_{X\cup Y}$, which is not permitted at all in the Anderson–Belnap natural deduction system, while $\wedge+$ is accepted there in restricted form only: from α_X, β_X infer $(\alpha \wedge \beta)_X$.[7]

We also need a general inheritance regime for flags. We need only mention the rules of Group I as the conclusions of rules in Group II are automatically flagged.

Flagging Inheritance Regime. In the application of any rule from Group I, if any premise carries a flag then it is inherited by the conclusion.

Note that rules of Group II supplement those of Group I. Thus, it can happen that a formula is obtainable in two different ways: by a derivation including applications of unrestricted but flagged rules from Group II, or by a derivation applying only restricted but unflagged ones from Group I. A savvy derivation-builder will follow the latter route whenever possible, so as to avoid unnecessarily blocking later attempts to apply $\to+$. We will return to this phenomenon in Sect. 6.

5 Behaviour of the Projective Constraint System

In what follows, it is important to keep in mind the distinction between a *closure relation* and the more general notion of a *consequence relation.* While the former, as defined in Sect. 2, satisfies all three Tarski conditions, the latter is understood rather loosely as any relation intended to reflect some notion of inference even if it fails one or more of the Tarski conditions. Both classical consequence and the relation $\vdash_0$ of Theorem 2 are genuine closure relations, but the relation emerging from our natural deduction system will not quite make it into that category, and will merely be a consequence relation.

We say that β is a *consequence of A in our system,* and write $A \vdash_1 \beta$, iff there is a derivation in the system with conclusion β, all of whose undischarged assumptions are in A, irrespective of the label and/or flag that may be attached to β.

---

[7] It doesn't really matter what label is placed on the conclusion of an application of a rule from Group II. We have used the natural label $X\cup Y$, but we could equally well have chosen any other subset of the currently undischarged assumptions, even $\emptyset$. For labels are needed only to constrain subsequent applications of $\to+$, and once we flag a line, as we do in the conclusions of the rules of Group II, such applications are barred anyway by the new proviso on $\to+$ and the flagging inheritance regime. One could even drop the label on the conclusions of Group II rules, keeping just the flags; but that would have the inconvenience of requiring notational adjustments to the labelling subscripts of the Anderson–Belnap rules in Group I, which we prefer to keep intact, and would complicate the proof of Corollary 9.

The consequence relation $\vdash_1$ should not be confused with the derivations that generate it. The relation is simply a set of ordered pairs (A, β), where β is a formula and A is a set of formulae, with no labels, flags, sub-proofs, or intermediate steps, whereas the derivations that generate those pairs will in general contain all those devices.

Checking the Tarski conditions, reflexivity is immediate. For monotony, the addition of unused premises does not modify the Anderson–Belnap labels, nor add to the flags, so that all steps that were correct before the addition remain so after it. However, cumulative transitivity fails, as is illustrated by the following simple example.

On the one hand, we have $\{\alpha, \beta\} \vdash_1 \alpha \wedge \beta$ by the following *flagged* derivation, where we omit the brackets separating a formula from its flag and label, and write the label without braces around the numerals, as no ambiguity is possible.

1. α_1	assumption
2. β_2	assumption
3. $\alpha \wedge \beta^{\#}_{1,2}$	unrestricted $\wedge$+ from Group II

We also have $\{\alpha, \beta, \alpha \wedge \beta\} \vdash_1 ((\alpha \wedge \beta) \rightarrow \gamma) \rightarrow \gamma$ by the following unflagged derivation with two premises unused:

1. $\alpha \wedge \beta_1$	assumption
2. $(\alpha \wedge \beta) \rightarrow \gamma_2$	assumption
3. $\gamma_{1,2}$	$\rightarrow -$
4. $((\alpha \wedge \beta) \rightarrow \gamma) \rightarrow \gamma_1$	$\rightarrow +$

But in the projective constraint system there is no way of constructing a derivation of $\{\alpha, \beta\} \vdash_1 ((\alpha \wedge \beta) \rightarrow \gamma) \rightarrow \gamma$. Any step conjoining α with β incurs a flag, whose transmission down the line blocks the application of $\rightarrow$+.[8]

Thus, our consequence relation is not a closure relation, as it fails cumulative transitivity. Nevertheless, it does a remarkable job. We will show that it is a conservative extension of classical consequence, closed under substitution, and outputs exactly the theses of the relevance logic R. Readers uninterested in the verifications leading up to this result may wish to skip to Theorem 11 itself.

Observation 3. $\vdash_1$ is compact and closed under substitution.

Proof. Compactness is immediate from the finite length of derivations. The rules are schematic with labels, flags and constraints. None of the latter is affected by uniform substitution of arbitrary formulae for elementary letters throughout a derivation. □

---

[8] There seems to be little prospect for recuperating cumulative transitivity by choosing different classical rules in Group II of our system, so long as they are flagged and the flags are used to block subsequent applications of $\rightarrow$+. Indeed, it would appear that, apart from degenerate cases, any introduction of projective constraint rules into a natural deduction system will create difficulties for cumulative transitivity in the induced consequence relation.

Observation 4. If β is derivable in the Anderson/Belnap system from undischarged assumptions A, then $A \vdash_1 \beta$.

Proof. The flagging regime does not obstruct any derivation that is carried out under the Anderson/Belnap rules alone. □

Observation 5. If β ∈ R then then $\emptyset \vdash_1 \beta$.

Proof. Suppose β ∈ R. Every thesis of R may be derived from the empty set of assumptions in the Anderson–Belnap natural deduction system. So by Observation 4, $\emptyset \vdash_1 \beta$. □

Observation 6. Let β and all formulae in A be from the classical language (that is, without →). If β is a classical consequence of A, then $A \vdash_1 \beta$.

Proof. Suppose β is a classical consequence of A. Then $\wedge A_0 \supset \beta$ is a classical tautology for some finite $A_0 \subseteq A$, so by Observation 5, $\emptyset \vdash_1 \wedge A_0 \supset \beta$, that is, there is a derivation of $\wedge A_0 \supset \beta$ in the projective constraint system from the empty set of assumptions. Add further lines for the elements of A_0 as assumptions, infer $\wedge A_0^{\#}$ with suitable label by unrestricted conjunction, then $\beta^{\#}$ with suitable label by an application of modus ponens for ⊃. This gives us a derivation of β from A, so $A \vdash_1 \beta$ as desired. □

We can also get the converses of Observation 5 and 6 (Observations 9, 10 below). For this we need a Lemma, and to state it succinctly we introduce some notation. Let X be any finite set of formulae, which we write as $\{\alpha_1,\ldots,\alpha_n\}$ with the understanding that $\alpha_i = \alpha_j$ only when $i = j$. We write $\downarrow(X, \beta)$ for the formula $\alpha_1 \to (\alpha_2 \to \ldots(\alpha_n \to \beta)\ldots)$, read as β when $n = 0$. As the relevance logic R allows permutation of embedded antecedents, the order of the formulae α_i does not matter, though it would for weaker logics such as E. We note that R contains $\downarrow(X, \beta) \to (\wedge X \supset \beta)$ as thesis.

Lemma 7. Consider any derivation of a formula β with label X in the projective constraint natural deduction system. Then (i) $\wedge X \supset \beta \in$ R and (ii) if β is not flagged then $\downarrow(X, \beta) \in$ R.

Proof. By induction on the derivation. For the basis, we need only note that if β is an assumption then it is not flagged and $\beta \to \beta \in$ R. Suppose that the Lemma holds for all derivations of length up to n. We need to show that it holds for any derivation of length $n+1$. *Case 1.* Suppose that β is not flagged. Then it is obtained by rules of group I alone and we use the well-known fact that in this situation $\downarrow(X, \beta) \in$ R for the Anderson/Belnap system, so that also $\wedge X \supset \beta \in$ R. *Case 2.* Suppose that β is flagged. Then β was not obtained by the rule →+; for that rule does not introduce a flag, nor does it transmit any flags since it is inapplicable if any of its inputs are flagged. For all other rules of the system it is easy to check using the induction hypothesis, classical tautologies, substitution and detachment wrt ⊃ (under which the thesis set of R is known to be closed), that $\wedge X \supset \beta \in$ R. □

Corollary 8. If $A \vdash_1 \beta$ then there is a finite subset $A_0 \subseteq A$ such that $\wedge A_0 \supset \beta \in \mathrm{R}$.

Proof. Suppose $A \vdash_1 \beta$. Then there is a derivation in the projective constraint natural deduction system with conclusion β all of whose undischarged assumptions are in A. Let X be the label of β in this derivation. By Lemma 7, $\wedge X \supset \beta \in \mathrm{R}$. Now, a straightforward induction tells us that in the projective constraint system (as in the Anderson–Belnap one), the label on the conclusion of a derivation is a subset of the set of all undischarged assumptions of that derivation. So $X \subseteq A$ and we may complete the proof by putting $A_0 = X$. □

Observation 9. $\emptyset \vdash_1 \beta$ iff $\beta \in \mathrm{R}$.

Proof. We already have right-to-left in Observation 5, so we need only the promised converse. But this is immediate by Corollary 8. □

Observation 10. Let β and all formulae in A be from the classical language. Then $A \vdash_1 \beta$ iff β is a classical consequence of A.

Proof. We already have right-to-left in Observation 6, so we need only the converse. Suppose $A \vdash_1 \beta$. Then by Corollary 8 there is a finite subset $A_0 \subseteq A$ such that $\wedge A_0 \supset \beta \in \mathrm{R}$. But the formula $\wedge A_0 \supset \beta$ is in the language of classical logic, so by Fact 1 it is a classical tautology, and so β is a classical consequence of $A \supseteq A_0$, and we are done. □

Putting together Observations 9, 10 and 3 we have immediately:

Theorem 11. The theses of the relevance logic R are exactly the consequences of the empty set under the consequence relation $\vdash_1$, which is a conservative extension of classical consequence, compact, closed under substitution, and generated by a natural deduction system suitably modifying that of Anderson and Belnap.

Thus, while our consequence relation fails unrestricted cumulative transitivity, it nevertheless works very nicely as a way of determining the theses of the relevance logic R from a classical starting point.

What is the relationship between $\vdash_1$ and the relation $\vdash_0$ defined in Sect. 2? They are generated differently: whereas $\vdash_0$ is defined directly from the Hilbertian axiomatization of R, $\vdash_1$ is obtained for adding rules to the Anderson–Belnap natural deduction system for R. They are not the same, since $\vdash_0$ satisfies cumulative transitivity while $\vdash_1$ does not. But there is a tight connection, as we now show.

Observation 12. The relation $\vdash_0$ is identical to the closure $\vdash_1^+$ of the relation $\vdash_1$ under cumulative transitivity.

Proof. First, we note that $\vdash_1 \subseteq \vdash_0$. For when $A \vdash_1 \beta$, by Corollary 8 there is a finite $A_0 \subseteq A$ such that $\wedge A_0 \supset \beta \in \mathrm{R}$, so $A \vdash_0 \beta$ by the definition of $\vdash_0$. Since $\vdash_1 \subseteq \vdash_0$, we have that $\vdash_1^+ \subseteq \vdash_0^+ = \vdash_0$ (by Theorem 2). For the converse inclusion, suppose $A \vdash_0 \beta$. Then by definition $\wedge A_0 \supset \beta \in \mathrm{R}$ for some finite $A_0 \subseteq A$, so $\emptyset \vdash_1 \wedge A_0 \supset \beta$ (by Observation 5), so by monotony (which, as we noted, holds for $\vdash_1$), $A \vdash_1 (\wedge A_0) \supset \beta$. Since $\vdash_1$ is closed under substitution (Observation 3) and extends classical consequence (Observation 6), we also have both $A \vdash_1 \wedge A_0$ and $\{\wedge A_0, (\wedge A_0) \supset \beta\} \vdash_1 \beta$. So by cumulative transitivity $A \vdash_1^+ \beta$ as required. □

6 Philosophical Perspectives

Theorems 2 and 11 make it clear that if one really wants to be a relevantist, then one may opt for the logic R whilst keeping classical logic intact—rejecting no tautology (as has long been known), impugning no tautological consequence (as shown here). The relevantist enterprise may, after all, be seen as similar to that of conventional modal logic where we add a non-truth-functional connective to a functionally complete set of classical ones and study the outcome; in relevance logic one likewise adds the arrow. This view may even make the relevantist project more attractive than it is at present, since migrating to a non-classical logic is less of a leap into the abyss if we don't have to give up classical logic into the bargain but can draw on it freely within an extended system.

On the other hand, as remarked earlier, the closure relation $\vdash_0$ figuring in Theorem 2 may not be one that all relevantists would be willing to call their own. Moreover, the natural deduction system that generates the consequence relation $\vdash_1$ in Theorem 11 is rather ungainly. Already, it inherits from the Anderson–Belnap system a very poor separation of the connectives: of the Anderson–Belnap rules in Group I of the system, only two are 'pure arrow' rules, six concern $\wedge$, $\vee$, $\neg$, and three are 'mixed rules' that combine $\rightarrow$ with $\neg$ or $\vee$. One might expect it to be easy to separate the connectives more elegantly but, unless one follows the path of Meyer (1974) and adds further non-classical connectives (see footnote 2), the task is tricky, as illustrated by an unsuccessful attempt of Prawitz.[9] Moreover, as remarked at the end of Sect. 4, the Group II rules introduce considerable redundancy into the system.

[9] Prawitz (1965) Chap. VII, Sect.2 presented a natural deduction system for relevance logic by modifying a straightforward system for classical logic (the modification uses global constraints on derivations rather than labels or flags). Unfortunately, however, the system does not output the system R or any of its well-known neighbours. As remarked by Dunn and Restall (2002) p. 26, for example, it outputs too little compared to R, since it does not yield the first-degree entailment of distribution of conjunction over disjunction. More seriously (because less easy to rectify) and apparently less well known (the author has not seen it in the literature), Prawitz' system also yields too much. On the one hand, the formula $p\rightarrow(\neg p\rightarrow q)$ is not a thesis of R or sister relevance logics. But on the other hand, in Prawitz' system, negation is defined from his primitive connectives $\bot,\rightarrow$ by treating $\neg\alpha$ as shorthand for $\alpha \rightarrow \bot$, so that $p\rightarrow(\neg p\rightarrow q)$ abbreviates $p\rightarrow ((p \rightarrow \bot) \rightarrow q)$. That formula can be derived from the empty set of premises using Prawitz' rules for the connectives occurring in it, as follows.

1. p_1	assumption
2. $p \rightarrow \bot_2$	assumption
3. $\bot_{1,2}$	from 1, 2 by $\rightarrow -$
4. $q_{1,2}$	from 3 by Prawitz' rule for $\bot$-elimination
5. $(p \rightarrow \bot) \rightarrow q_1$	$\rightarrow +$, unobstructed by Prawitz' global constraints
6. $p \rightarrow ((p \rightarrow \bot) \rightarrow q)_{\emptyset}$	$\rightarrow +$, unobstructed by Prawitz' global constraints

Whether the relevantist project itself is well-motivated is quite another matter. The author is inclined to suspect that the enterprise is engendered by confusion between two quite distinct problems: that of determining *what is implied* by a given set of assumptions, and that of deciding whether it *would be wiser* to carry out a particular inference in a given situation or to do something else instead.

The former may be investigated using purely mathematical tools. The latter also involves epistemic and pragmatic issues, and we need to take into account the purpose of the reasoning. Am I trying to *convince* my interlocutor (or myself) of the truth of the conclusion of the inference, or merely *exploring* the more salient consequences of the assumptions, or perhaps even constructing a *reductio ad absurdum* refutation of one among them? Each of these three contexts carries its own pragmatic features.

- In the first, it can happen that as our reasoning unfolds, it prompts second thoughts on the reliability of the assumptions on which it is based. Although we did not begin with the intention of carrying out a *reductio*, we may come to suspect that we are 'showing too much'; it could be wise to halt the inference train, back up, and do some revision or reformulation somewhere.
- In the second context, we may ask whether conclusion of the inference really worth knowing about, recording and communicating. Making an inference is like making a journey; we don't go just anywhere our feet can lead us, but travel in an economical or rewarding way to a priority destination.
- Even in the last of the three contexts, we may sometimes feel that it is unwise to keep slogging away in the direction we have been taking, when it does not seem to be getting us any closer to the desired contradiction; it might be a good idea to try proving the opposite.

There is thus a big difference between the logical question *Does it follow?* and the pragmatic one *Is it wise to continue carrying on with this line of inference?* Relevance logic is perhaps a quixotic attempt to tackle pragmatic questions with purely formal tools.

7 Other Logical Contexts

We end by looking briefly as some other contexts where the question of an alternative 'conservative extension' may arise. We first consider relevance logics other than R, and then intuitionistic logic.

How far do our results have analogues for the weaker system E of entailment? By running through the proof of Theorem 2, it is clear that it also holds when the relation $\vdash_0$ is redefined in terms of E rather than R. However, given the absence of the antecedent-permutation principle from E, it is questionable whether one can tweak our natural deduction system so as to output it. So the analogue of Theorem 11 remains an open question.

On the other hand, it is clear that everything in this paper carries over to any Hilbertian system that is stronger than R, so long as it is satisfies the analogue of

Fact 1 and is closed under substitution, adjunction, and detachment for both $\rightarrow$ and $\supset$. In particular, R-Mingle meets all those conditions.

Turning to intuitionistic logic, while it is ordinarily seen as a subsystem of classical logic Gödel (1933) observed that when the two logics are formulated with primitive connectives $\wedge$, $\vee$, $\neg$, $\rightarrow$, they agree on formulae using only the connectives $\wedge$, $\neg$, which of course are together functionally complete for classical logic. On this basis, Łukasiewicz (1952) proposed seeing intuitionistic logic as a conservative extension of classical logic.

His suggestion did not find much favour since it does not extend to intuitionistic consequence as it is usually defined, by putting $A \vdash \beta$ iff for some $\alpha_1,\ldots,\alpha_n \in A$ the formula $(\alpha_1 \wedge \ldots \wedge \alpha_n) \rightarrow \beta$ is a thesis of intuitionistic logic (analogously to the first of the eight candidates for a 'relevant consequence relation' considered in Sect. 2). The situation is reviewed with pointers to literature in Sect. 2.32 of Humberstone (2011).

It is natural to ask whether, in that context as well, we may redefine intuitionistic consequence in a manner parallel to that of $\vdash_0$ and prove for it a result like Theorem 2. The prospects do not seem good. Suppose we attempt to define a suitable closure relation $\vdash_*$ by putting $A \vdash_* \beta$ iff for some finite subset $\{\alpha_1,\ldots,\alpha_n\} \subseteq A$, the formula $(\alpha_1 \wedge \ldots \wedge \alpha_n) \supset \beta$ is a thesis of intuitionistic logic. We cannot understand the material conditional $\alpha \supset \beta$ here as $\neg\alpha \vee \beta$ as we did in the definition of $\vdash_0$, since our purpose is to prove an analogue of Theorem 2, and while intuitionistic logic agrees with classical logic over $\wedge, \neg$ it does not do so over $\vee, \neg$. So we need to understand $\alpha \supset \beta$ as abbreviating $\neg(\alpha \wedge \neg\beta)$.

But, as Lloyd Humberstone has remarked (personal communication), difficulties then arise for the limiting case $n = 0$. If one takes $(\alpha_1 \wedge \ldots \wedge \alpha_n) \supset \beta$, that is $\neg((\alpha_1 \wedge \ldots \wedge \alpha_n) \wedge \neg\beta)$, as reducing to $\neg\neg\beta$ in that case, then the analogue of step (5) in the proof of Theorem 2 fails: we are no longer able to show that $\vdash_*$ outputs intuitionistic logic. If we stipulate instead that it is in that case to be read as β, then the analogue of step (5) becomes immediate, but we are not out of the woods. The verification of step (6) is still in trouble, since intuitionistic logic is not closed under detachment for the material conditional understood as $\neg(\alpha \wedge \neg\beta)$, that is, we cannot in general go from theses α and $\neg(\alpha \wedge \neg\beta)$ to β. No matter which reading is followed, the proof hits a snag, and we conjecture that the intuitionistic analogue of Theorem 2 itself fails.

Acknowledgments The author is very much indebted to Lloyd Humberstone, whose comments on a badly crashed early text helped the author to start again, and whose remarks on later drafts were also valuable. Thanks also to Diderik Batens, Michael Dunn, Jim Hawthorne, João Marcos and Peter Verdée for helpful discussions, as well as two anonymous referees for their suggestions.

References

Anderson, A. R., & Belnap, N. D. (1975). *Entailment: The logic of relevance and necessity* (Vol. I). Princeton: Princeton University Press.

Anderson, A. R., Belnap, N. D., & Dunn, M. (1992). *Entailment: The logic of relevance and necessity* (Vol. II). Princeton: Princeton University Press.

Avron, A. (1992). Whither relevance logic? *Journal of Philosophical Logic*, *21*, 243–281.

Dunn, J. M. (1986). Relevance logic and entailment. In D. M. Gabbay & F. Guenthner (Eds.), *Handbook of philosophical logic*, (Vol. III, pp. 117–224). Dordrecht: Kluwer.

Dunn, J. M., & Restall, G. (2002). Relevance logic. In D. M. Gabbay & F. Guenthner (Eds.), *Handbook of philosophical logic*, (2nd ed., Vol. 6, pp. 1–128). Dordrecht: Kluwer.

Gödel, K. (1933). 'Zur intuitionistischen Arithmetik und Zahlentheorie' Ergebnisse eines mathematische Kolloquiums 4: 34–38. Translated as 'On intuitionistic arithmetic and number theory'. In S. Feferman et al. (Eds.), *Kurt Gödel: Collected works*, (Vol. I, Publications 1929–1936, pp. 286–295). Oxford: Oxford University Press, 1986.

Humberstone, L. (2011). *The connectives*. Cambridge: MIT Press.

Łukasiewicz, J. (1952). 'On the intuitionistic theory of deduction', Koninklijke Nederlandse Akademie van Wetenschappen, *Series A* 55: 202–212. Reprinted in pp. 325–40 of L. Borkowski (Ed.), *Jan Łukasiewicz: Selected Works*. North-Holland, Amsterdam 1970.

Mares, E. (2004). *Relevant logic*. Cambridge: Cambridge University Press.

Mares, E. (2012). Relevance logic. In E. N. Zalta (Ed.), *The Stanford Encyclopedia of Philosophy*.

Meyer, R. K. (1974). New axiomatics for relevance logics, I. *Journal of Philosophical Logic*, *3*, 53–86.

Meyer, R. K., & Dunn, J. M. (1969). E, R, and Gamma. *Journal of Symbolic Logic*, *34*, 460–474.

Pelletier, F. J., & Hazen, A. (2001). A history of natural deduction and elementary logic textbooks. In J. Woods & B. Brown (Eds.), *Logical consequence: Rival approaches* (pp. 105–138). Oxford: Hermes Science Publications.

Pelletier, F. J., & Hazen, A. (2012). Natural deduction. In D. Gabbay & J. Woods (Eds.), *Handbook of the history of logic* (Vol. 11) (Central Concepts)). Amsterdam: Elsevier North-Holland.

Prawitz, D. (1965). *Natural deduction: A proof-theoretical study*. Stockholm: Almqvist & Wiksell (Reprinted by Dover Publications, Mineola New York, 2006).

Prior, A. N. (1955). *Formal Logic*.Oxford: Oxford University Press.

Restall, G. (2000). *An introduction to substructural logics*. London and New York: Routledge.

Wolf, R. G. (1978). Are relevant logics deviant? *Philosophia*, *7*, 327–340.

Part VI
Responses

Reflections on the Contributions

David Makinson

Abstract The author reflects on some questions prompted by a reading of the contributions to the volume.

Keywords Logic of belief change · Uncertain reasoning · Normative systems

1 The Volume

The pages of this book provide much food for thought. They offer insights on what has been done on the logic of belief change, uncertain reasoning, and normative systems over the years, helping one appreciate how the pieces of the jigsaw fit together. At the same time, they contain serious research contributions of their own.

Logic is essentially a cooperative enterprise. There is not much that one can do without the stimulus and corrections, encouragement and inputs of others. Logic permits capital accumulation, progressing by construction, proof and comparison rather than polemic and contradiction. So this section is not one of 'replies to my critics' as in the old volumes of Schilpp's *Library of Living Philosophers*. Even to describe it as 'comments on the contributions' would be slightly off-centre. I will comment on few specific points in the chapters, but take the opportunity to share with readers, as well as with the authors themselves, some general reflections that came to one mind when thinking about the texts. I will do so in the order of the volume.

D. Makinson (✉)
Department of Philosophy, London School of Economics, Houghton St, London WC2A 2AE, UK
e-mail: david.makinson@gmail.com

S. O. Hansson (ed.), *David Makinson on Classical Methods for Non-Classical Problems*, Outstanding Contributions to Logic 3, DOI: 10.1007/978-94-007-7759-0_18,

2 The Logic of Belief Change

The logic of belief change by now forms a vast field, with many speciality areas. In this volume we have just five chapters on the subject. The first three deal with the specific topics of safe contraction, iterated belief revision, and ranking theory. Their authors, Hans Rott, Sven Ove Hansson, Pavlos Peppas, Wolfgang Spohn, may be described as 'insiders', for they have been leading investigators into these topics. The following two chapters are written, if I may say so, by 'neighbours'—distinguished logicians (Edwin Mares, Rohit Parikh) from related areas casting an independent but informed eye on what has been going on.

2.1 Hans Rott and Sven Ove Hansson "Safe Contraction Revisited"

The AGM trio elaborated three quite different-looking accounts of the formal logic of belief change. Of these, by far the best known, published in a paper with all three as authors and called the AGM account, is in terms of partial meet operations on maximally non-implying sets of formulae. Also fairly well known, especially in the computer science community, is the GM account using relations of 'epistemic entrenchment' between propositions. Much less familiar is the AM approach in terms of safe contraction.

Personally, I always found safe contraction particularly elegant with its use of a double minimality condition: an element a of a belief set K (not necessarily closed under consequence) is safe with respect to a proposition x that one wishes to eliminate from among the consequences of K, iff it is not a $\leq$-minimal element of any $\subseteq$-minimal subset of K that implies x. Here $\subseteq$ is just set-inclusion, while $\leq$ is a relation over K that is supposed to represent degrees of vulnerability (with lower elements read as being more vulnerable than higher). Simple properties of $\leq$ generate interesting behaviours for the set of x-safe elements and its closure under classical consequence, thereby validating associated principles of belief contraction.

It is something of a mystery why there have been enormous differences in the reception of the three approaches, even after one takes into consideration factors in the artificial intelligence community discussed by Raúl Carnota and Ricardo Rodríguez in their review 'AGM theory and artificial intelligence' (*Belief Revision Meets Philosophy of Science*, eds. E. J. Olsson and S. Enqvist, Dordrecht: Springer, 2011, 1–42). Perhaps it had something to do with their places of publication, but there also seem to have been some intrinsic features.

As early as 1988, Adam Grove transformed the internal order of the partial meet account. Instead of using the partial meet operations to define contraction and then obtain revision via the Levi identity, he showed how we may use them to define revision directly, passing to contraction via the Harper identity. As a result, the central items of attention changed. In the original version of AGM 1985 they had been maximal subsets of given belief set that fail to imply the proposition being removed; in the Grove version for revision they became maximal consistent sets of formulae of the language as a whole. These may be thought of as 'possible worlds', thus connecting belief revision with a tradition that had been initiated in modal logic in the 1960s, had continued into deontic logic and the logic of counterfactuals in the 1970s, and had given rise to some philosophically debated if rather obscure metaphysical readings.

Mathematically, these presentations can be mapped into each other, but their *gestalts* are quite different. Moreover, it is not immediately clear how one might replicate Grove's order-inversion in the realm of 'safe' operations, that is, first articulate some kind of safe revision and then obtain safe contraction via the Harper identity. This may have been one of the reasons why the 'safe' approach remained marginal. In any case, one result of its lower profile is that safe contraction has been less intensively studied than its partial meet and entrenchment counterparts, and the various results for it obtained in the literature have never been brought together for a clear overview of the state of play.

That gap has now been filled. In their quite extraordinary text, Hans and Sven Ove have marshalled in a very accessible way just about everything that is presently known about safe contraction and its variants such as kernel contraction. The results include several representation theorems and maps to and from the other two approaches. At the same time, the chapter provides solutions for some of the problems that had been left open, and identifies a surprising number of further ones that remain recalcitrant.

This paper will certainly be a central reference for those wishing to learn about safe contraction and its neighbours, and a starting point for those planning to take the research further.

2.2 *Pavlos Peppas "A Panorama of Iterated Revision"*

A panorama indeed! I am immensely grateful to Pavlos for having put together this overview of an area that I have never had the courage to enter as researcher, and for which I have always struggled to find clear guiding intuitions. The chapter has been very helpful to me, and I am sure that it will be so to many others. When a graduate student asks me for a good entry to iterated belief revision, I now know where to point!

One lesson that seems to emerge from Pavlos' review is that if we wish to find some order in the jumble of different proposals for logics of iterated belief change, it is not enough to classify the different logical principles that they validate or fail, or even the various semantic conditions that give rise to them in the model structures. In this area, the logical principles are just epiphenomena, and even their corresponding semantic conditions are little more than triggers. A serious inquiry should go deeper. It is to a considerable extent a matter of how much, and what kind of, structure we put into the very notion of a belief state, and likewise into that of an input element.

In the AGM account of one-shot contraction or revision, a belief state is no more than a set of propositional formulae closed under classical consequence, while the input for revision is a naked formula. There is little hope of extracting guidelines for iterated belief change from that meagre ontology; we need richer materials. The more structure one supplies, the more one can do with it, in particular to formulate, validate and vary logical principles. On the other hand, it would be unwise to supply more structure than is really needed, as that could distract from the essentials.

For iterated belief change, adding structure faces several decisions. Do we enrich the notion of belief state only, or also that of the input? Do we express the enrichment in purely qualitative terms (with auxiliary sets, relations, functions and so on) or in a quantitative mode in which operations such as addition, subtraction and multiplication make sense? And in the last case, do we remain discrete (as with the integers) or go continuous (with say the reals)?

For example, the 1997 Darwiche-Pearl constructions enriched the belief state in qualitative terms, while leaving inputs as bare formulae. The distance-based approach of Lehmann, Magidor and Schlechta 2001 also enriched the belief state alone, but quantitatively in a continuous manner. Spohn's 1988 models enhance both the belief state and the input in a discrete quantitative manner. Those of Nayak 1994 inject qualitative structure into the belief state and the input items so that the logic of belief change becomes essentially one of belief merging. And so on.

In many cases, the enriched models provide scope for graphic intuition. With them available, changes in belief can often be described in terms of geometric displacements. Diagrams, and indeed animated diagrams, begin to provide the clearest way to understand what is going on. This gives a particular flavour to the theory of iterated belief change as compared to most other areas of logic.

One topic that I find particularly fascinating, both in itself and in connection with belief change, is Parikh's 1989 notion of a finest splitting for a set of classical propositional formulae. This has little or nothing to do with relevant/paraconsistent logic as that is commonly understood, but is a matter of letter-sharing. I have myself participated in its development and application to one-shot belief change, in a joint paper with George Kourousias 'Parallel interpolation, splitting, and relevance in belief change' *Journal of Symbolic Logic* 2007, 72: 994–1002, continued in 'Propositional relevance through letter-sharing' *Journal of Applied Logic* 2009, 7: 377–387. But in the context of iterated belief change splitting presents a major challenge. In a paper of 2008, Peppas, Fotinopoulos and Seremetaki have shown that there is a quite radical conflict between Parikh's splitting-based principle of relevance-sensitivity and the Darwiche-Pearl postulates for iterated revision (DP1)–(DP4): the principle is inconsistent with *each* of the postulates! As Pavlos remarks, this really demands further clarification, and I hope that he, and readers of the present volume, will soon rise to the challenge.

2.3 *Wolfgang Spohn "AGM, Ranking Theory, and the Many Ways to Cope with Examples"*

Thanks to Wolfgang for this magnificent discussion of ways of coping with apparent counterexamples to the general postulates that are satisfied by both the AGM account of belief change and his own based on ordinal conditional functions.

Wolfgang's approach is just as old as the AGM one: it was outlined in his unpublished *Habilitationsschrift* of 1983, just after the Alchourrón/Makinson 'maxichoice' article of 1982 and before the AGM 'partial meet' paper of 1985. It saw print in 1988, the same year as Peter Gärdenfors' *Knowledge in Flux* took the AGM approach to book form. Actually, that volume already contains an extended discussion of ranking theory, for Peter had learned of it in 1985 when Wolfgang gave a talk while on a conference (plus honeymoon) trip to California. However, like the important discussion of the problem of revising probability functions, this topic did not attract as much attention among readers of *Knowledge in Flux* as did the central thread of the volume, namely the AGM constructions.

The chapter gives us three things. *First*: a concise explanation of what ordinal belief-ranking is, how one can define conditional belief from it, and how one may use such conditionality to define operations of belief change. *Second*: a brief examination of connections between belief change operations, so defined, and the corresponding operations in the AGM theory. *Third*, and in most detail: a discussion of some of the many 'counterexamples' to the AGM postulates—and thus also to Wolfgang's own theory, which also satisfies those postulates—that have appeared in the literature over the years. All three themes are developed in depth in Wolfgang's recent volume *The Laws of Belief, Ranking Theory and its Philosophical Applications* (Oxford University Press, 2012), and I would encourage readers aroused by the chapter here to continue exploring there.

As far as the counterexamples are concerned, I believe that Wolfgang and I see things in pretty much the same light, modulo some differences of attention and emphasis. Content-laden examples from ordinary discourse are graphic, thought-provoking and valuable; we are indebted to Sven Ove Hansson, Hans Rott, Wlodek Rabinowicz, Arthur Paul Pedersen and others for constructing so many intriguing ones for belief change. But they cannot be employed in quite the way as when one is working entirely *within* a logical or mathematical theory. There, a single counterexample suffices to sink a conjecture; for instance, the production of just one suitably behaved integer can suffice to knock down a statement about the behaviour of all of them. This is not so for the kinds of counterexample devised for the AGM postulates, for two main reasons.

In the first place, the example usually carries far more implicit content than is visible on the surface or can be expressed by a crude rendering of it in the language of classical propositional logic. What the examples show is that the theory is an idealization, which traces just some of the many strands entering into the fibre of our everyday utterances. We are perfectly familiar with this phenomenon from, say, the contrasts between the use of 'if...then...' statements in daily life and the idealized logic of truth-functional conditionals. Outside of logic, in economics or physics, extreme idealizations also abound. Articulation of a startling real-life counterexample does not serve as a cannon-shot sinking the theory, but as an alert signal, highlighting its perhaps underestimated degree of idealization, and encouraging reflection on when and how the abstract theory may still be useful.

In the second place, the theory may come with various parameters whose dials may be set at diverse positions; an example may well clash with the 'standard' setting of the dial, yet fit nicely with a variant one. Already, in the AGM constructions using partial meet operations to model contraction and revision, we may tweak a parameter to give us what I have called 'screened revision' (*Theoria* 1997, 63: 14-23) in which inputs are sometimes rejected. As Wolfgang remarks in his chapter, exactly the same may be done in ranking theory. In this fashion, the theory may be seen as providing a family of models, loosely centred on a standard one, that are available to represent a range of cognate notions of belief change.

Indeed, as Spohn suggests, ranking theory may be better than AGM at such dial-setting. This is because it has much more structure built into it—more parameters admitting fine tuning and so more dials to set. Whereas the partial meet operations in AGM only make use of set-intersection and a choice function (for satisfaction of the 'basic' postulates) accompanied by a relation (for the 'supplementary' postulates), ranking theory is from the beginning based on the positive integers or ordinal numbers, which admit operations of addition and subtraction (also multiplication and division which, however, are not used in the theory). Varying the way in which those two arithmetic operations are employed, or the values at which numerical parameters are fixed, provides a broad range of opportunities for adjusting the theory.

At this point, two comments on the notion of ordinality may be helpful. First comment: In his 1988 paper, Wolfgang took the ordinal numbers (less than whatever limit ordinal one chooses) as the range of what are now called ranking functions. The difficulty with this, as he has informed me in correspondence, is that if the limit ordinal is chosen to be greater than ω then addition is no longer commutative; this complicates and perhaps puts in jeopardy verifications needed for theorems about the system. At the same time, transfinite ordinals do not actually seem to be needed anywhere in the theory. For these reasons, he has come to prefer taking the range of the ranking functions to be the natural numbers or the reals. This preference is reflected in his 2012 book.

Second comment: In recent years, economists and political scientists working in agent/agenda choice theory have become interested in belief change, theory merging, and uncertain inference (aka nonmonotonic reasoning), because models of the latter use selection functions similar to those from their home territory, with similar opportunities for their representation with binary relations. There is, however, potential for terminological misunderstanding between the two communities, or their readers. When economists speak of an *ordinal* conception of preference (in contrast to a *cardinal* one), they are thinking of a notion for which no arithmetic operations are available: neither multiplication, division, addition, nor subtraction—only pairwise comparison. They are *not* thinking of the ordinal numbers in the sense of set theory, where such arithmetic operations are of course perfectly well defined and available for deployment, for example in Spohn's ranks. To speak paradoxically, it is really AGM theory that is restricted to ordinal comparison in the sense of the social scientists, whereas ranking theory uses what they would call cardinal comparison! So, in this terrain, readers should beware of the casual use of expressions such as 'we make ordinal comparisons only'; different cultures mean quite different things.

One word in the chapter particularly intrigues me. The ordinal ranking account of belief change is described as being *more general* than the AGM one. Already in 1988 Peter Gärdenfors had used the same expression. This is rather puzzling. A textbook account of comparative generality would presumably take one class of structures as being less general than another if all those in the former class are, or may be represented as, ones in the latter class but not conversely. In the context of belief change, Wolfgang's remark would then mean that all choice-and-preference structures may be represented as special cases of ordinal-number structures of ranking theory. That seems to me unlikely; if anything, it is the converse that may hold. So, I don't think that we can say that ranking theory is more general than partial meet theory in such a textbook sense.

However when comparing systems of logic, we also tend to use the notion of generality rather loosely to refer to how much the theory says and what it can do for us, and I imagine that this is what Wolfgang has in mind. For, as already mentioned, ranking theory with its built-in integer values is certainly more highly structured than its partial meet counterpart and, as he suggests, it may turn out that this additional structure provides more opportunities (or at least more natural ones) for resetting parameters to obtain interesting variants and extensions. In that sense it may indeed be 'more general'. Time will presumably tell; for the present I think that it is good to have both kinds of modelling in our toolkit.

2.4 Edwin Mares "Liars, Lotteries and Prefaces: Two Paraconsistent Theories of Belief Revision"

In its AGM formulation, the logic of belief contraction and revision makes use of a consequence operation that at least includes classical consequence and satisfies disjunction in the premises. The question naturally arises whether all the power of classical consequence is really needed, that is, whether one may adapt the AGM constructions to various sub-classical logics. Some work has been carried out with respect to logics of interest to computer science and artificial intelligence, see the overview of Márcio Moretto Ribeiro *Belief Revision in Non-Classical Logics* (London: Springer 2013). Edwin's chapter considers the same question but for relevance (aka paraconsistent) logic.

His first step is to define a paraconsistent consequence relation for the job. It is of set/set signature, and puts $A \Vdash X$, where A, X are sets of formulae, iff A, X are non-empty and there are finite sets $A_0 \subseteq A$, $X_0 \subseteq X$ with the formula $\wedge A_0 \rightarrow \vee X_0$ a thesis of a favoured relevance logic presented in the Hilbertian manner, say E or R. The second step is to redefine belief states in a manner that facilitates his task. They are taken as disjoint pairs (Γ, Δ) where Γ is an 'acceptance set' closed under $\Vdash$ and Δ is a 'reject set', dually constrained. With this apparatus, Edwin is able to define notions of expansion, contraction and revision so that they satisfy analogues of the 'basic' AGM postulates, although the status of the supplementary AGM postulates in this context is left unsettled.

Evidently, the enthusiasm that one has for sub-classical reconstruction projects of this kind will depend, to some degree, on one's attitude to the sub-classical logic itself. Those who feel that classical logic is just wrong and hope that relevance logic can do its job better, will see reconstruction as urgent. Those who regard classical logic as a satisfactory instrument in most contexts but not well adapted to managing inconsistency, may take the task as being less urgent but nevertheless important. Edwin appears to be working from this perspective. Finally, those who recognize the classical limitations for consistency management but feel that this should be dealt with by supplementation rather than excision may see the project as of technical more than foundational interest. I must confess that personally I am of the last persuasion.

As well as using relevance logic to reconstruct the formal logic of belief change, Edwin deploys it to tackle another question: can it throw light on the paradoxes of inconsistent belief, notably the lottery and preface paradoxes? For this task, he chooses a set/set consequence relation that is weaker than the one used in the reconstruction job. Whereas there he put $A \Vdash X$ iff there are finite sets $A_0 \subseteq A$, $X_0 \subseteq X$ with $\wedge A_0 \rightarrow \vee X_0$ a thesis of the chosen relevance logic, here he puts $A \Vdash X$ iff there are specific formulae $a \in A, x \in X$ with $a \rightarrow x$ a thesis. This has the result that the set of individual formulae that are consequences of a given set A is no longer closed under conjunction. Thus, using the relation in the context of a lottery, our grounds for believing of each individual ticket that it will not win the prize may no longer be grounds for believing that none of them will win.

I wonder about the indispensability of *relevance* for this approach. It seems to me that all the hard work is being done by the failure of conjunction on the right, irrespective of whether or not the logic engendering that failure is paraconsistent one. Paraconsistent behaviour seems to be needed only when, later in the paper, the basic construction is modified in an endeavour to obtain a partial reinstatement of conjunction by means of some rather complex maximality manoeuvres.

My own perspective on the two paradoxes has been set out in a paper 'Logical questions behind the lottery and preface paradoxes: lossy rules for uncertain inference' *Synthese* 2012, 186: 511–529. In the case of the preface, where serious appeal to probabilities is rather unrealistic, I am still inclined to draw the moral, already suggested in the original 1965 presentation, that indeed it is sometimes reasonable to believe all elements of an inconsistent set of propositions without believing their conjunction. Of course, this prompts the question *when* it is reasonable to do so, and some informal criteria are canvassed in the same 2012 paper. In the case of the lottery the same moral may be drawn, but the presence of probabilities also suggests another. When certain statements are taken to have probability (or conditional probability) just a little short of one, we should treat inferences to their conjunction as 'lossy' in the sense they lose a little of that probability on each application, but with a calculable limit that can be monitored. The idea of a lossy rule of inference is elaborated in the same *Synthese* paper with application and calculation of loss limits for conjunction, cumulative transitivity, cautious monotony, and unrestricted disjunction of premises. Further results on calculating loss are discussed in a paper with Jim Hawthorne, 'Lossy rules of inference: a brief review for philosophers', to appear in a Festschrift for Jean-Yves Béziau.

So our angles on the lottery and preface are quite different, as are those on the reconstruction of belief revision theory using relevance logic. But it would be unwise to be dogmatic or doctrinal on such matters, and I appreciate Edwin's insightful exploration of a radically different perspective.

2.5 Rohit Parikh "Epistemic Reasoning in Life and Literature"

If ever you are asked by someone at a dinner-party what you do for a living, don't answer that you are a logician. To do so is to invite a blank stare, followed by an earnest request to explain in a few sentences what logic is all about. It is not much better to say that you are a philosopher; you risk being dragged into an interminable account of your interlocutor's incoherent convictions and personal anxieties. It is far safer to say that you are a mathematician. Everyone thinks that they know what that is—after all, they had to do some at school—and would rather talk about something else, which is usually so much the better.

Rohit has given unsociable people like me a new strategy for coping with dinner parties. Tell your interlocutor straight out that you are a logician. After the blank stare and request for enlightenment, say that logic is just *much ado about nothing*, and charm with tales of iterated indexed epistemic gambits from Shakespeare's comedy.

The focus of the chapter is epistemic reasoning, but some of its examples are already well known for highlighting other, non-epistemic matters of logic. A Holmes deduction that Rohit quotes from *Silver Blaze* is often (and appropriately) flung before relevance logicians as an instance of perfectly acceptable disjunctive syllogism. These days, however, it may be more convincing to invite relevantist acquaintances to a game of Sudoku, where they will soon themselves be executing multiple disjunctive syllogisms with which they can be taunted.

On the other hand, Hamlet's stratagem to convince himself of his uncle's guilt, also recalled in the chapter, is an object lesson in logical error, illustrating the disasters that can ensue from illegitimately converting a conditional probability, or even a conditional certainty. If the uncle is guilty of murder, he may well react nervously to Hamlet's play, but such a reaction need not mean probability of guilt.

In Hamlet's defence it might be suggested along Bayesian lines that observation of Claudius' dismay will significantly increase the probability that Claudius is the murderer, provided that the probability of the former given the latter is close to one while its probability given the negation of the latter is close to zero. But these are big provisos! On the one hand a well-trained usurper may be immune to such naïve reactions, bringing the first conditional probability down quite low; on the other hand an innocent uncle may well see from his nephew's strange behaviour that he is implicitly being accused of a ghastly crime, bringing the second conditional probability up quite high. Even high-tech lie detectors have difficulty with these considerations. Famous literature contains bad reasoning as well as good.

3 Uncertain Reasoning

This section of the volume contains three papers on uncertain inference—two on probabilistic aspects and one purely qualitative. It begins with Jim Hawthorne's review of an important open problem about axiomatizing probabilistic inference; continues with Karl Schlechta's extended review of the analysis of qualitative uncertain inference using preference relations and selection functions; and ends with Hykel Hosni's discussion of how one might still make good use of probability in situations where the preconditions for its direct application are wanting.

3.1 James Hawthorne "New Horn Rules for Probabilistic Consequence: Is O^+ *Enough?"*

Jim's chapter concerns an unsolved problem in the analysis of probabilistic consequence—the problem of providing a reasonably transparent axiomatic basis for the class of finite-premise Horn rules that are sound for all probabilistic consequence relations. The issue is much more complicated than for the analogous qualitative one where soundness is respect to preferential consequence relations; in that context an elegant solution was obtained some time ago by Kraus, Lehmann and Magidor. I have dabbled in the problem without getting anywhere. Jeff Paris and his student Richard Simmonds have given a solution of sorts, with a recursively defined but extremely complex set of axioms, leaving open whether a more transparent axiomatic basis is possible.

I will not summarize Jim's concise review of the present state of play, nor can I add to it. But I would encourage any brave souls who are looking for a challenging logico-mathematical problem with both philosophical resonance and practical import, to put their minds to this one.

3.2 Karl Schlechta "Non-monotonic Logic: Preferential Versus Algebraic Semantics"

In logic, it is customary to distinguish between syntactic and semantic kinds of investigation, but sometimes I feel that it is more helpful to think of three levels rather than two.

On the purely *syntactic level* one studies expressions of a formal (or natural) language, irrespective of external structures that might be used to give it meaning. This is the level on which proof-theory is carried out, using primarily inductive arguments (sometimes of extraordinary intricacy) on syntactically defined objects such as the formulae of a language and derivations carried out within it. On the *model-theoretic level* one examines mathematical structures independently of their expression in any particular formal language; the work is thus carried out in a 'purely mathematical' way without mentioning logic. Finally, *semantic* investigations link formal languages with external structures. Its main results often take the form of soundness-and-completeness theorems. In the more customary way of speaking, the model-theoretic and semantic dimensions are bundled together, in opposition to the purely syntactic one.

In the particular case of so-called nonmonotonic logics—or, as I prefer to call them, qualitative logics for uncertain inference—the model-theoretic level has in turn its own two strata. One consists of *selection functions* on sets of items (called points, states, or possible worlds) which can be used to defined inference relations between propositions when we link with the syntactic level. The other layer consists of *preference relations* between those items. The two are also connected in a manner that is similar to, yet different from, that relating formal languages and model-theoretic structures. This time the connecting results are determination-and-representation theorems, to the effect that any preference relation satisfying certain conditions immediately determines a selection function with corresponding properties, while any selection function satisfying suitable conditions may be generated (perhaps with provisos or exceptions) out of an appropriate preference relation.

Karl is a grand master of this game. It is his conviction that a clear picture of what is really going on is best obtained by getting well below the syntax, working persistently with structures, and returning to the syntactic level on the basis of the results there established. He has followed this line of attack in many journal papers since the late 1980s, and has brought the main results together in the volume *Coherent Systems* (Elsevier 2004), and more recently in books co-authored with Dov Gabbay, *Logical Tools for Handling Change in Agent-based Systems* (Springer 2009) and *Conditionals and Modularity in General Logics* (Springer 2011). In those texts one can find the deepest analyses available of the roles of definability preservation (that is, the extent to which the selection function in a model structure preserves the definability-by-formulae of a set of models), and copies (that is, states that satisfy the same formulae but occupy different places in the model structure with respect to the preference relation or selection function). The former is an issue only in the infinite case, but the latter is already important in small finite model structures.

In his chapter Karl has provided a penetrating overview of this, covering what he and others have already done and highlighting the problems that remain open. I recommend it as a concise starting point for graduate students and mature logicians thinking of getting deeply into the qualitative logic of uncertain reasoning, before they turn to the details of the two books mentioned above.

3.3 Hykel Hosni "Towards a Bayesian Theory of Second-Order Uncertainty: Lessons from Non-standard Logics"

In the early twentieth century, when certain limitations of classical propositional connectives were noticed, two reactions were common. One was simply to ignore their existence or explain them away. Another was to propose very radical solutions, jettisoning classical logic in favour of some other construct, usually a sub-classical system.

With time, attitudes have changed, and more nuanced perspectives have developed. In some contexts, notably those of pure mathematics, classical connectives appear to be quite adequate. In some other contexts, shortcomings do manifest themselves, but one should not too readily abandon classical logic and fly to some rival system. Instead, one can consider the possibility of applying the classical framework in a more sophisticated manner, perhaps in association with additional ingredients.

In the cases of modalities, counterfactual conditionals, and preferential consequence relations for uncertain reasoning, the additional ingredients may be taken, at least to a first approximation, as points of reference (aka states or possible worlds) accompanied by a relation linking them or a selection function on collections of them. In the case of belief change, we may use similar additional ingredients or else (in my view more transparently, at least for non-iterated belief change) let subsets of the given belief set the work of the points. In the case of input/output logic, the additional ingredient is a production relation, which plays an even more central role in the analysis but is flanked by classical consequence in both the preparation of inputs and the packaging of outputs.

One cannot say that this way of looking at the limitations of classical logic is universal, but it does appear to be rather more widely accepted than it used to be. It is also the 'unequivocal message' of part 3 of Hykel's contribution to the present volume. Indeed, he goes further, suggesting that a similar situation may be observed in the realm of probability and its application in decision theory.

It has for long been remarked that in real life there are many situations in which the formal apparatus of probability is difficult or awkward to apply. For example, in a real-world problem there can be serious hesitations about how to define a satisfactory state space; quandaries in fixing a suitable initial probability function on that space in a non-arbitrary manner; and, for decision theory, issues about measuring the utilities of outcomes in a realistic manner. This, by the way, is all that Hykel means by the term 'second-order uncertainty' in the chapter title. As he makes clear towards the end of the chapter, he is not talking about second-order probabilities in the sense of 'probability of the truth of a given statement of probability'. He is referring to the kind of uncertainty that one may face when trying to specify a suitable state space and an appropriate probability function over it, for whatever practical problem is being tackled.

Rather than ignore the existence of this kind of problem, as once was common among Bayesian theorists, or abandon the notion of probability for a quite different kind of construct (whether quantitative like fuzzy logic, or qualitative as in preferential or default logic) it may be better, Hykel suggests, to distinguish those contexts in which we may directly apply the formalism of probability theory from those where application needs to be more sophisticated, and then try to work out the particular adaptations or complements that are needed for the latter.

This perspective has been less widely accepted for probability than in logic, perhaps because attitudes have tended to be rather more polarized and polemical. But proposals have been made for contexts where the state space is assumed to be clear while the specification of an initial probability function remains problematic. One strategy is to use the notion of maximum entropy to narrow down the range of candidate functions as much as possible. Another, in principle not excluded by the first, is to allow probability functions to take as values sub-intervals of [0, 1] rather than its elements. Of course, if we move in the latter direction, there is a good deal of reconstruction on the agenda, and some of its issues are discussed in the chapter; for example the question of how far Dutch book justifications of the Kolmogorov postulates for probability carry over to a modified version with interval values.

Here I would raise only one very general methodological question. In the chapter, Hykel remarks that questions of uncertainty matter to us primarily in so far as they enter into decisions regarding action, and draws the moral that theoretical analyses of uncertainty should always be conducted within the general context of some kind of decision theory. Personally, I am not so sure. Uncertainty and probability can already arise when we seek to establish plain facts, for example in matters of measurement, which may only indirectly, if ever, serve as a basis for practical action. And even when they are so used, it may still be preferable to analyse them in their own terms. After all, truth is also important for decision-making, but we are usually happy to study logic as a theory of truth-preservation independently of the theory of action. To take a rather more distant analogy, even if contracting a belief set rarely takes place except as part of a composite process of revision, it may be more transparent for analytical purposes to proceed in a Cartesian manner, separating the steps and studying the components before the composition.

4 Normative Systems

Why is this part of the collection called normative *systems*? Why not plain *norms* or just *deontic logic*? The reason is that norms typically occur as part of a more or less well-defined code or system, which one needs to keep in mind in any analysis of their logic. The analysis should not presuppose, as do the traditional 'possible worlds' semantics for deontic logic, that norms are capable of bearing truth-values. These two requirements were highlighted in my 1999 paper "On a fundamental problem of deontic logic", which also began the task of constructing a logic of normative systems that respects them both. This developed into the input/output logics developed in various papers with Leendert van der Torre from 2000 on. The contributions to the third part of the volume relate in one way or another to that development.

Audun Stolpe's chapter investigates how far certain concepts such as parallel interpolation and finest splitting, which were originally developed in the context of classical logic, may be carried over to the input/output framework. In contrast, Xavier Parent, Dov Gabbay and Leendert van der Torre examine how much of input/output logic remains when its background consequence relation is reduced to intuitionistic rather than classical strength. Jörg Hansen develops an approach to the logic of norms which, while influenced by ideas from input/output logic, is firmly based on his earlier work with imperatives. John Horty strikes out in a new direction with reasoning in the common law, where precedent plays a central role. Just as for the preceding chapters, I will hardly comment on specific details of the texts, but reflect on some general issues that they bring to mind.

4.1 Audun Stolpe "Abstract Interfaces of Input/Output Logic"

Audun's chapter continues his earlier work on adapting the notion of AGM contraction from classical belief sets to the codes that figure in input/output logic. His central question is the extent to which one may transpose to the input/output framework the notion of the finest splitting of a belief set, so as to carry out contractions not only on normative codes as they happen to be presented but also on their normal form versions with most finely split bases, so as to respect certain criteria of relevance in the contraction.

Back in classical logic, the notion of a splitting of a set of formulae was first formulated by Parikh in a paper of 1999, where he also proved the finest splitting theorem (that every set of formulae has a unique finest splitting) for the finite case. The result was extended to the infinite case by Kourousias and Makinson 2007, where the main tool of proof was a new version of the classical interpolation theorem baptised 'parallel interpolation'. So Audun sets himself a threefold task: (1) adapt to the input/output context the classical concepts of parallel interpolation and finest splitting; (2) prove, as far as possible that the classical results for them can be obtained in the input/output context; (3) use the finest splitting result to define and study a notion of 'relevant contraction' in that framework.

None of these tasks is easy, and while Audun obtains positive results for 'simple-minded' input/output logic, he also shows that parallel interpolation can fail for 'basic' input/output operations, and remains an open question for the stronger versions with reusability. However, on the positive side, since finest splittings are always available in the 'simple minded' case, operations of 'relevant contraction' may there be carried out.

This is an extremely rich investigation in difficult terrain. As well as extending what we know in the area, it articulates key problems still to be resolved. The chapter will be an essential reference point for those studying AGM-style belief change operations in the context of input/output logic, as well as for those exploring the repercussions of parallel interpolation and Parikh's finest splitting theorem outside their home territory of classical logic.

4.2 Xavier Parent, Leendert van der Torre and Dov Gabbay "An Intuitionistic Basis for Input/Output Logic"

A distinctive feature of input/output logic, distinguishing it from just about any other approach to logic in the literature, is that it does not see logic as an inference motor, but as a kind of secretarial assistant that services a production system, the latter being based on an arbitrary relation (set of ordered pairs) between propositions. The role of logic is to prepare the input propositions so that the production system can take them fully into account, unpack the outputs so that the user can appreciate their power and, for certain systems, feed output back into input.

It is natural to choose classical consequence to play the role of the secretarial assistant, and that is what Leendert and I did in our papers on the subject. But the four basic kinds of input/output are perfectly well-defined, if not necessarily well behaved, with an arbitrary closure relation playing the role of secretarial assistant in place of classical consequence. This raises the question of whether one might do input/output logic with a sub-classical background, and what it might look like. Xavier, Dov and Leendert investigate the question in the particular case of intuitionistic consequence. The enterprise thus has a family resemblance to that of Edwin Mares in an earlier chapter, except that he is reconstructing AGM belief change in relevant consequence, while Xavier and company are reconstructing input/output logic with intuitionistic consequence.

The results are mixed. Of the four main kinds of input/output logic, two ('simple-minded' and 'reusable') were originally defined without any appeal to maximality notions and, as the chapter shows, their representation theorems carry over (using disjunctively saturated sets in place of the maxisets that are evoked in the proof) in the context of intuitionistic consequence. But the other two kinds of input/output operation ('basic' and 'basic reusable') were already defined using maxisets, and for them the situation is more complex. In the 'basic' case, it is shown that if we adjust the definitions and rework the proofs sufficiently, we can get the corresponding representation theorem when intuitionistic logic is in play. But for the 'basic reusable' case, the completeness argument that was used in the classical context breaks down with little prospect for repair, due to its reliance on a curious inference pattern of 'ghost contraposition' that fails for intuitionistic logic.

I am grateful to Xavier, Dov and Leendert for examining so carefully the inner workings of the input/output representation proofs (whose original versions were in some cases given only in outline) revealing just what parts of classical logic are used, and how much survives or can be adapted if certain bits are left out.

Evidently, the same general question arises as in the case of Edwin's chapter: the results will be of particular interest for those who see intuitionistic (resp. relevantist) logic as a potential substitute for its classical counterpart, but less urgent for those who do not. I must admit belonging to the second category, and digress to explain why.

I see intuitionistic logic as motivated in two main ways, one philosophical and the other formal. The philosophical path sets out from the view that what passes for truth in mathematics should instead be seen as a matter of intuitively acceptable provability, and from that concludes that the usual truth-functional accounts of propositional connectives are without traction in mathematics (except, *post hoc*, for fragments of mathematics that can be shown intuitionistically to be decidable). Like many foundational positions, this is difficult to assess intelligibly, from any side, and I will not even attempt to do so here.

However, the formal motivation is much more amenable to analysis. It emerges from the combined effect of four methodological decisions on the part of the logician: (1) to define logics using inference patterns rather than via Hilbertian axiom systems, or semantically; (2) use only set/formula inferences (as contrasted with set/set ones) in the defining inference patterns; (3) within that range, allow only intelim rules (excluding less simple ones); (4) opt for the falsum (rather than negation) as a primitive connective alongside conjunction, disjunction and the conditional. If one makes the first three decisions together, one ends up with less than classical consequence; if at the same time one chooses the fourth option, the sub-classical system produced is exactly the intuitionistic one.

But while these four choices are natural, they are far from inevitable. In particular, the practical bent of the second sits uneasily with the purist aesthetics of the third, while the fourth seems to have no rationale other than its favoured outcome. As for the first choice, postulation with inference patterns is certainly a good way of presenting logic; but any disqualification of alternative presentations, in particular the usual semantic presentation of classical logic, would have to fall back on philosophical considerations.

Formal results relevant to these questions may be found in a paper with Lloyd Humberstone 'Intuitionistic logic and elementary rules'. *Mind* 120 (2011): 1035–1051, and in the chapter 'Intelim rules for classical connectives' of this volume.

4.3 Jörg Hansen "Reasoning About Permission and Obligation"

Jörg's contribution takes up a challenge issued in my 1999 paper "On a fundamental problem of deontic logic", where it was noted that there is a serious tension between the philosophy of norms and the formal work of deontic logicians. On a philosophical level, it is widely accepted that while declarative statements are capable of being true or false, norms are items of quite another kind: they may be applied, respected, or even assessed from the standpoint of other norms, but arguably they are not in themselves true or false. On the other hand, deontic logicians working in the popular 'possible worlds' tradition routinely assign truth-values to norms in possible worlds. While often conceding that while they do not actually bear truth-values, they assume that for the purposes of logic they may, for some mysterious reason, nevertheless be treated as if they did.

A fundamental problem of deontic logic, therefore, is to reconstruct it in accord with a positivistic philosophy of norms. My own work in that direction culminated in what Leendert van der Torre and I call 'input/output' logic. Jörg explores another path to reconstruction, which keeps more of the traditional syntactic apparatus of dyadic deontic logic. Taking just as seriously as do Leendert and I the maxim 'no logic of norms without a parameter for a code of which they form part', he considers possible rules for evaluating the status of conditional obligations $O(\beta/\alpha)$ and permissions $P(\beta/\alpha)$ with respect to a given code.

The exploration is systematic and technical, and I do not wish to comment on its details, advising the reader to turn directly to the chapter itself. But I would like to say a few words on three general issues in the logic of norms. One concerns the distinction between promulgating and noticing a norm in a system; another is about the problem of so-called free-choice permission; and the third concerns the implications of Ross' paradox for deontic inference. These remarks are not intended as criticisms of the chapter. Indeed Jörg does not discuss Ross' paradox at all (barely mentioning it in passing), and I imagine that he may well be in sympathy with what I wish to say about promulgation. However, our perspectives on free choice permission do seem to be rather different.

Evidently, there is a great difference between promulgating a permission in a normative code, and that of noticing that it is present in the code (whether explicitly so, or implicitly in the sense that the system is committed to it). The former is an act that *changes* the existing code by adding to it, whereas the latter makes no change, only recognizing that something is *already there*.

Now, when one adds a permission to a code, one runs the risk of creating conflict with obligations that are already present. So, if one wishes to avoid inconsistency in the enlarged corpus, one may have to carry out some kind of derogation (contraction, restriction or reinterpretation) of obligations within it. There are essentially two ways in which this may be conceptualized: as a two-step dance, first of contraction and then of addition, or as a single undivided step of revision. In either case, some of the older code must be discarded. Decisions about what is to be jettisoned should respect some logical regularity conditions in the spirit if not the letter of AGM belief change theory. They may also be assisted by a meta-code that supplies substantive rules of priority (later over earlier, higher authority over lower, specific content over general or vice versa, and so on). But none of these issues arise when merely checking that the code contains, or is committed to, a permission.

So much is well known: Bulygin and Alchourrón drew attention to it long ago, and much current work in the logic of normative systems endeavours to articulate it formally. But it has not been widely recognized that exactly the same distinction applies to obligation, with similar effects!

It is one thing to notice that a code is already committed to an obligation or imperative, quite another to introduce it into the code. The latter again involves a change in the system while the former does not. When we promulgate an obligation we are also liable to create contradiction (with already existing permissions and even prohibitions and obligations) modulo facts about what scenarios are physically realizable in the world, and in such cases we need to derogate if we are to avoid inconsistency. The upshot is that the logic of promulgating obligations can be just as complex as that for promulgating permissions, likewise being bound up with avoidance of inconsistency, whether by contraction, revision, or some other device.

This point is clearly recognized by Jörg in his chapter. But why is it usually overlooked? There are, I suspect, two factors obscuring it, one particular to Alchourrón and Bulygin, and the other an empirical phenomenon.

For Alchourrón and Bulygin, the situation was rendered less visible by a substantive rule of precedence that they proposed, reflecting their liberal philosophical outlook. This was that in case of conflict between permissions and obligations, one should give priority to the former (unless, presumably, other priority criteria, such as the level of authority of the issuing body, also come into play). If that rule is followed then, when we take a consistent code containing permissions and consider adding an obligation that conflicts with one of those permissions, avoidance of inconsistency will naturally leave us (under the same proviso) with the given code unchanged.

The other factor is practical, concerning the circumstances in which a permission or obligation is normally added to a code. When an authority promulgates a permission, it is seldom done 'in a void'. Usually, the existing code commits to certain obligations or prohibitions that the authority wishes to remove, and the permission is introduced expressly for that purpose. In contrast, when an obligation is introduced into a code, it is rarely done in order to remove an already promulgated permission or contrary prohibition. Usually it is to regulate behaviour that is believed not to have been regulated at all. Up to the time of introducing the obligation, such behaviour will have been permitted in the weak sense that it is not prohibited, but not in the strong sense of having been declared permissible. Thus it would seem that in practice, promulgating a permission *quite often* requires a contraction for consistency to be preserved, while promulgating an obligation does so *less often.* Nevertheless, from a logical point of view, the possible need for contraction is latent in both.

The second general issue on which I would like to comment is the problem of free-choice permission. It arises for statements such as "You may have fruit, dessert or coffee as part of the menu". In ordinary English, we would interpret this as expressing the same as the conjunction of the separate permissions "You may have a fruit" etc, leaving unspecified (or more commonly, by conversational implicature, excluding) any licence to take more than one of the items mentioned. So, why do we have an 'or' in the English?

I would suggest that the answer to this question has nothing to do with free choice, nor even with permission. For the question is much more general, and may be illustrated by examples in which neither of those features is involved.

Consider an advertisement for a wrist-watch, telling us that it is sufficiently waterproof to withstand a shower, or a swim, or a walk in a rainstorm. There is nothing here about choice (whether free or constrained); nor anything about permitting such activities. The advertisement simply claims that none of them will harm the watch. Thus any account of the role of 'or' in it that turns upon questions of freedom, choice, or permission will miss the point. The solution must be as general as the problem.

I would suggest that there is a general English idiom that may be called the *checklist conditional*. The wrist-watch advertisement in effect claims the following: for all x, if x is an act of showering, or of swimming, or of going out unprotected in the rain, then x will not harm the watch. Returning to the example from the restaurant, its 'or' is likewise an echo of multiple disjunction in a much longer conditional, namely: for all x, if x is a fruit or a dessert or a coffee, then consumption of x is covered by the menu. The undoubted presence of a permission to choose is no more than incidental.

Ordinary language is very flexible, and often one can express such examples with 'and' in place of 'or'. This should not be a surprise: after all, a conditional with disjunctive antecedent is classically equivalent to the conjunction of the respective conditionals with a common consequent. The disjunctions in the antecedent of the underlying checklist conditional, and the conjunctions in its equivalent formulation, compete with each other to survive as ‘or’ or ‘and’ in the abbreviated colloquial rendering. For more on this subject, see my paper “Stenius’ approach to disjunctive permission” *Theoria* 1984, 50: 136–145.

Finally, I would like to comment briefly on the supposed implications of Ross’ paradox for deontic inference. Suppose I say that the socks are in the top drawer; you infer that they are in one of the drawers, act on your conclusion and open the bottom drawer, finding the socks there. Should one conclude that your inference was faulty, or that the facts did not bear out my initial declaration, in other words, that my declaration was false? Evidently, the latter. Your inference was perfectly valid—though there may be hidden reasons why you chose to make it and act on its conclusion, which turned out to be true, rather than on my communication, which was false.

Now suppose I *instruct* you to put the socks in the top drawer. You infer from my instruction that they should be put into one of the drawers, and act on your conclusion by putting them in the bottom drawer. Should one conclude that your inference was faulty, or that the action did not satisfy my initial request? Strangely, many deontic logicians lean towards the former option, although the situation is just as clear as in the declarative example. Your inference was perfectly valid, but you have chosen to act on its conclusion rather than on my initial request. You have thus complied with the request that you inferred, but not with the initial one from which you inferred it. Your *logic* was fine, while your *action* was uncooperative. Inference from an item to a disjunction containing it as a disjunct remains as intact in deontic contexts as in declarative ones.

4.4 John Horty “Norm Change in the Common Law”

One of the original motivations for the AGM account of belief change, especially for Carlos Alchourrón, was to model ways in which a legal code or other normative system may change when one of its directives is abrogated or replaced by a contrary norm. Carlos was thinking in terms of statutory codes, which are built up with general rules, rather than systems of common law where the accumulation of decided cases plays an important role. This prompts the question: how might one go about modelling the latter?

This question has recently been receiving some attention, particularly in communities working on computation and the law, and John has been a participant in these investigations. A central concept in his formal analysis, with roots in legal theory and the philosophy of law, is that of a set of ‘legally relevant factors’, each with a plaintiff/defendant polarity, that may enter into the grounds for deciding a case.

Enter with what kind of mechanism? Two approaches have emerged. One, with a rather logical flavour, is to treat the presence and absence of such factors as predicate formulae and their negations, which may thus be conjoined to serve as antecedents of rules. The development of case law is seen as an evolution of those rules, under certain constraints. The other approach has a less syntactic and more model-theoretic flavour. It groups factors of common polarity into sets, which may be compared under a preference relation. The development of common law is seen as the evolution of that relation.

The purpose of John's contribution to this volume is to show that when appropriately defined, and under some basic conditions, the two approaches turn out to be equivalent in a natural sense. More specifically, any model of the logical kind, satisfying suitable conditions, determines an equivalent one of the structural kind.

Given John's results, two questions suggest themselves. One is whether there also exists a converse map, taking us from a given structural model to a logical one and, if so, how the two maps interact. The other is to ask what happens if we relax the assumption that there is a fixed stock of possible factors that may be appealed to throughout the development of a branch of common law, and allow that from time to time the judge in an individual case may declare a new factor to be legally relevant. Historically, this does seem to happen and have an influence on the life of the common law. But would it make any significant difference to the mathematical structure? Or would it have essentially the same effect as starting with a sufficiently large stock of factors, many of which happen to remain unused for a very long time? While in this vein, one might also ask what happens if in the model one allowed, more radically, that judges can declare some factors, previously taken as legally relevant, no longer to be so. Presumably, that could have quite important effects.

Appendix
David Makinson: Annotated List of Publications

Books and Monographs

Sets logic and Maths for Computing. Springer. Series: Undergraduate Topics in Computer Science. Second revised edition 2012.

A textbook on formal methods for first year university students, ostensibly those in computer science but in fact for those in any discipline requiring abstract reasoning. Chapters on sets, relations, functions, induction and recursion, combinatorics, probability, trees, propositional logic, quantificational logic, proof and consequence.

Bridges from Classical to Nonmonotonic Logic. London: College Publications. Series: Texts in Computing, 2005.

A textbook on methods of qualitative uncertain inference, directed to graduate students of philosophy, computer science, and mathematics.
Polish version *Od Logiki Klasycznej do Niemonotonicznej*, translated by Tomasz Jarmuzek, published 2008 by Wydawnictwo Naukowe Universytetu Mikolaja Kopernika, Torun.

Topics in Modern Logic. London: Methuen, 1973.

Directed to undergraduate students of philosophy with a limited background in mathematics but who wish to go beyond the elementary level.
Italian translation published by Boringhieri, Turin, 1979; Japanese translation published by Huritsu Bunka-Sha, Tokyo, 1980.

Aspectos de la Logica Modal. Universidad Nacional del Sur, Bahia Blanca, Argentina. Series: Notas de Logica Matematica, 1970.

A monograph directed to students of mathematics with a background in abstract algebra, who wish to learn about modal logics and their associated algebraic and relational structures.

S. O. Hansson (ed.), *David Makinson on Classical Methods for Non-Classical Problems*, Outstanding Contributions to Logic 3, DOI: 10.1007/978-94-007-7759-0,

Papers and Chapters

(73) "Intelim rules for classical connectives". In *David Makinson on Classical Methods for Non-Classical Problems*, ed. Sven Ove Hansson. Springer, New York, 2014.

Investigates introduction and elimination rules for truth-functional connectives, focusing on the general questions of the existence, for a given connective, of at least one such rule that it satisfies, and the uniqueness of a connective with respect to the set of all of them.

(72) "Relevance logic as a conservative extension of classical logic". In *David Makinson on Classical Methods for Non-Classical Problems*, ed. Sven Ove Hansson. Springer, New York, 2014.

Relevance logic is ordinarily seen as a subsystem of classical logic under the translation that replaces arrows by horseshoes. If, however, we consider the arrow as an additional connective alongside the classical connectives, it may be seen as a conservative extension of classical consequence.
An outline of the main results without proofs appeared as "Advice to the relevantist policeman" in *The Logica Yearbook 2012* ed. Vit Puncochar & Petr Svarny, College Publications, London, 2013.

(71) "On an Inferential Evaluation system for Classical Logic". *Logic Journal of IGPL* 2013; http://jigpal.oxfordjournals.org/ (November 7, 2013, doi:10.1093/jigpal/jzt038).

We seek a better understanding of why an evaluation system devised by Tor Sandqvist yields full classical logic, despite its inferential character, by providing and analysing a direct proof of the fact using a suitable maximality construction.

(70) "Logical questions behind the lottery and preface paradoxes: Lossy rules for uncertain inference". *Synthese* 186: 2012, 511–529.

We reflect on lessons that the lottery and preface paradoxes provide for the logic of uncertain inference. One concerns the use of probabilistically 'lossy' rules such as the conjunction of conclusions, cumulative transitivity, and disjunction of premises. Another concerns the criteria for when inconsistent belief sets may yet be rationally held.

(69) With Lloyd Humberstone, "Intuitionistic logic and elementary rules". *Mind* 120: 2011, 1035–1051.

The interplay of introduction and elimination rules for propositional connectives is often seen as suggesting a distinguished role for intuitionistic logic. We prove three formal results about intuitionistic propositional logic that bear on that perspective, and discuss their significance.

(68) "Conditional probability in the light of qualitative belief change" *Journal of Philosophical Logic* 40: 2011, 121–153.

Explores ways in which purely qualitative belief change in the AGM tradition can throw light on options in the treatment of conditional probability.
A preliminary version was published in *Probability, Uncertainty and Rationality*, eds H.Hosni & F. Montagna. Pisa; Edizioni della Scuola Normale Superiore, 2010.

(67) "Propositional relevance through letter-sharing" *J. Applied Logic* 7: 2009, 377–387.

Develops and extends ideas of Odinaldo Rodrigues and Rohit Parikh, which consider propositional relevance modulo the choice of a background belief set, putting the belief set into a canonical form called its finest splitting.

(66) "Levels of belief and nonmonotonic reasoning", pp 341–354 of *Degrees of Belief* ed. Franz Huber and Christoph Schmidt-Petri. Springer: Series Synthese Library vol 342, 2009.

Reviews connections between different kinds of nonmonotonic logic and the general idea of varying degrees of belief.

(65) With George Kourousias, "Parallel interpolation, splitting, and relevance in belief change". *Journal of Symbolic Logic* 72: 2007, 994–1002.

Devises a new 'parallel interpolation' theorem for classical propositional logic, strengthening standard interpolation, uses it to extend Parikh's 'finest splitting theorem' from the finite to the infinite case, and considers applications to the logic of belief change.

(64) With George Kourousias, "Respecting relevance in belief change". *Análisis Filosófico* 26: 2006, 53–61.

An explanation and discussion without proofs of results on parallel interpolation and splitting.

(63) With James Hawthorne, "The quantitative/qualitative watershed for rules of uncertain inference". *Studia Logica* 86: 2007, 249–299.

Charts the ways in which closure properties of consequence relations for uncertain inference take on different forms according to whether the relations are generated in a quantitative or a qualitative manner.

(62) "Completeness theorems, representation theorems: what's the difference?" In *Hommage à Wlodek: Philosophical Papers dedicated to Wlodek Rabinowicz*, ed. Rønnow-Rasmussen et al. 2007, http://www.fil.lu.se/hommageawlodek.

A discussion of connections and differences between completeness and representation theorems in logic. Illustrated by examples from classical and modal logic, the logic of friendliness and nonmonotonic reasoning.

(61) "Friendliness and sympathy in logic", pp 195–224 of *Logica Universalis*, 2nd edition, ed. J.-Y. Beziau. Basel: Birkhauser, 2007.

Defines and examines a notion of logical friendliness, a broadening of the familiar notion of classical consequence. Also reviews familiar notions and operations with which friendliness makes contact.
Also appeared under the title "Friendliness for logicians", pp 259–292 of *We Will Show Them! Essays in Honour of Dov Gabbay, vol 2*, ed S. Artemov et al. London: King's College Publications. 2005. A preliminary version appeared in the first edition of *Logica Universalis* (2005) under the title "Logical friendliness and sympathy".

(60) "How to go nonmonotonic", pp 175–278 In *Handbook of Philosophical Logic, Second Edition, volume 12*, ed. D. Gabbay and F. Guenthner. Amsterdam: Springer, 2005.

An overview of methods of analysing qualitative uncertain inference. Its material was elaborated into the 2005 textbook *Bridges from Classical to Nonmonotonic Logic*.

(59) "Natural deduction and logically pure derivations". *PhiNews*, April 2004, http://www.phinews.ruc.dk.

Discusses advantages and disadvantages of first-level, second-level, and split-level formalizations of rules for natural deduction.
A preliminary version appeared as "Logically pure derivations" in a Festschrift for Professor Norman Foo, November 2003 (http://www.cse.unsw.edu.au/~ksg/Norman).

(58) "Supraclassical inference without probability", pp 95–111 in P. Bourgine & J-P. Nadal eds, *Cognitive Economics: An Interdisciplinary Approach*. Springer Verlag, 2003.

An brief review of basic concepts of nonmonotonic inference, written for economists with some background in probability but little in logic.

(57) "Conditional statements and directives", pp 213–227 in P. Bourgine & J-P. Nadal eds, *Cognitive Economics: An Interdisciplinary Approach*. Springer Verlag, 2003.

An expository paper, describing some of the different kinds of conditional assertions and directives that appear in ordinary discourse, and ways in which logicians have sought to model them.

(56) With Leendert van der Torre, "Permission from an input/output perspective". *Journal of Philosophical Logic* 32: 2003, 391–416.

Shows how input/output operations may be used articulate clearly distinctions between negative and positive permission, as well as between dynamic and static positive permissions.

(55) "Ways of doing logic: what was different about AGM 1985?". *Journal of Logic and Computation* 13 (2003) 3–13.

Reflects from a very broad perspective on what, in 1985, was new or different about AGM belief revision as a way of doing logic.

(54) With Leendert van der Torre, "What is input/output logic?" pp 163–174 in *Foundations of the Formal Sciences II: Applications of Mathematical Logic in Philosophy and Linguistics*. Dordrecht: Kluwer, Trends in Logic Series, 2003.

Explains the *raison d'être* and basic ideas of input/output logic, sketching the central elements with pointers to other publications for detailed developments.

(53) "Bridges between classical and nonmonotonic logic". *Logic Journal of the IGPL* 11: 2003, 69–96. http://www3.oup.co.uk/igpl/Volume_11/Issue_01/

Sketch of the central ideas structuring the 2005 book of the same title.

(52) With Leendert van der Torre, "Constraints for input/output logics". *Journal of Philosophical Logic* 30: 2001, 155–185.

Studies ways of constraining input/output operations to avoid output that is inconsistent with a given item.

(51) With Leendert van der Torre, "Input/output logics". *Journal of Philosophical Logic* 29: 2000, 383–408.

Develops a general theory of propositional input/output operations. They are like consequence operations, but input propositions are not in general included among outputs. Attention is given to the special cases where outputs may be recycled as inputs, and where disjunctive inputs may be broken into their cases.

(50) "On a fundamental problem of deontic logic", pp 29–53 in *Norms, Logics and Information Systems. New Studies in Deontic Logic and Computer Science*, edited by Paul McNamara and Henry Prakken. Amsterdam: IOS Press, 1999.

A fundamental problem of deontic logic is to reconstruct it in accord with the philosophical position that norms direct rather than describe, and are neither true nor false. This text discusses the problem, and begins work towards a solution, which later culminated in the papers on input/output logic.

(49) "Screened revision", *Theoria* 63: 1997, 14–23.

Develops a concept of revision, akin in spirit to AGM partial meet revision, but in which the postulate of 'success' may fail.

(48) With Ramón Pino Pérez and Hassan Bezzazi "Beyond rational monotony: on some strong non-Horn conditions for nonmonotonic inference operations" *Journal of Logic and Computation* 7 (1997) 605–632.

Explores the effect of adding to the rules of preferential inference a number of non-Horn rules stronger than or incomparable with rational monotony, but still weaker than plain monotony, focussing on their representation in terms of special classes of preferential models.

(47) "On the force of some apparent counterexamples to recovery", in *Normative Systems in Legal and Moral Theory: Festschrift for Carlos Alchourrón and Eugenio Bulygin*, ed. E.Garzón Valdés et al. Berlin: Duncker & Humblot, 1997.

Examines the principal alleged counterexamples to the recovery postulate for operations of contraction on closed theories. Argues that the theories considered are implicitly accompanied by epistemic structure, and that recovery remains appropriate for closed theories considered without such additions.

(46) With Sven Ove Hansson "Applying normative rules with restraint", pp. 313–332 in *Logic and Scientific Methods*, M.L. Dalla Chiara et al eds. Dordrecht: Kluwer, 1997.

Investigates the logic of applying normative rules, and in particular those applications that are 'restrained', in the sense that they are carried through as fully as is compatible with avoidance of contradictions or other consequences specified as undesirable.
Also appeared as Uppsala Preprint in Philosophy, 1995 n°7, and as part of an informal *Festschrift* to Sten Linström.

(45) "Carlos Eduardo Alchourrón - a memorial note". *Rechtsphilosophie* 27: 1996, 125–131.

Obituary notice with overview of the work of Carlos Alchourrón, focussing on his contributions to the logic of norms and of normative systems.

(44) "In memoriam Carlos Eduardo Alchourrón". *Nordic Journal of Philosophical Logic* 1: 1996, 3–10.

Obituary notice with overview of the work of Carlos Alchourrón, focussing on his contributions to the logic of theory change.

(43) "Combinatorial versus decision-theoretic components of impossibility theorems". *Theory and Decision* 40: 1996, 181–190.

Separates the combinatorial part of Arrow's celebrated impossibility theorem from its decision-theoretic part. The separation leads to a particularly elegant proof of Arrow's result, via a new 'splitting theorem'.

(42) "General Patterns in Nonmonotonic Reasoning", pp. 35–110 in *Handbook of Logic in Artificial Intelligence and Logic Programming, vol.* 3, ed. Gabbay, Hogger and Robinson. Oxford University Press, 1994.

An extended review of the formal behaviour of nonmonotonic inference operations, as generated by the principal systems in the artificial intelligence literature.

(41) With Karl Schlechta, "Local and global metrics for the semantics of counterfactual conditionals". *Journal of Applied Non-Classical Logics* 4: 1994, 129–140.

Considers how far the family of world-indexed 'closeness' relations in a given model for counterfactual conditionals may be derived from a common source.
Also appeared in advance semi-published form as Technical Report RR37, September 1994, Lab. d'Informatique, URA CNRS 1787, Univ. Provence, Marseille, France.

(40) With Peter Gärdenfors, "Nonmonotonic inference based on expectations". *Artificial Intelligence* 65: 1994, 197–245.

Shows how nonmonotonic inference relations may be generated in natural ways from sets of, and relations between, expected propositions.

(39) "Five faces of minimality". Studia Logica 52, 1993: 339–379.

An extended discussion of similarities and residual differences, within the common semantic framework of minimalisation, between defeasible inference, belief revision, counterfactual conditionals, updating, and conditional obligation.

(38) With Jürgen Dix, "The relationship between KLM and MAK models for nonmonotonic inference operations". *Journal of Logic, Language, and Information* 1: 1992, 131–140.

Articulates the exact relationship between two variants of the notion of a preferential model for nonmonotonic inference.
Also appeared as Interner Bericht 16/91 of the Institut für Logik, Komplexität und Deduktionssysteme of the University of Karlsruhe, Germany, 1991.

(37) With Gerhard Brewka and Karl Schlechta, pp. 1–12 in "Cumulative inference relations for JTMS and logic programming", in Dix, Jantke and Schmitt eds, *Nonmonotonic and Inductive Logic*. Berlin: Springer Lecture Notes in Artificial Intelligence, n° 543, 1991.

Observes that JTMS and logic programming with negation (under the Gelfond-Lifschitz semantics) are non-cumulative, shows how to render them cumulative by an indexing device, and notes the residual drawback of floating conclusions.
A preliminary version appeared as "JTMS and logic programming" in Nerode and others eds, *Proceedings of the First International Workshop on Logic Programming and Nonmonotonic Reasoning*. Cambridge, Mass: MIT Press, 1991.

(36) With Karl Schlechta, "Floating conclusions and zombie paths: two deep difficulties in the directly skeptical approach to defeasible inheritance nets". *Artificial Intelligence* 48: 1991, 199–209.

Isolates and studies two recalcitrant phenomena in the theory of inference relations generated by defeasible inheritance nets.
A preliminary version appeared as an internal publication, under the title "On some difficulties in the theory of defeasible inheritance nets" in *Proceedings of the Tübingen Workshop on Semantic Networks and Nonmonotonic Reasoning* ed.

Michael Morreau (SNS Bericht 89–48: 1989, 58–68). An almost final version appeared as an internal publication in Brewka & Freitag eds, *Proceedings of the Workshop on Nonmonotonic Reasoning* (Arbeitspapiere der GMD n° 443, 1990, 201–216). Closely related material is also contained in the internal publication "On principles and problems of defeasible inheritance", Research report RR-92-59 of the DFKI, Saarbrücken, Germany.

(35) With Peter Gärdenfors, "Relations between the logic of theory change and nonmonotonic logic", pp. 185–205 in Fuhrmann & Morreau eds, *The Logic of Theory Change*. Berlin: Springer, 1991

Examines the link between nonmonotonic inference relations and theory revision operations, focusing on the correspondence between properties which each may satisfy.
Also appeared in two internal publications: *Proceedings of the Workshop on Nonmonotonic Reasoning* (Arbeitspapiere der GMD n° 443, 1990), and *Report of RP2: First Workshop on Defeasible Reasoning and Uncertainty Management Systems* (DRUMS), University of Toulouse, 1990.

(34) "The Gärdenfors impossibility theorem in nonmonotonic contexts". *Studia Logica* 49: 1990, 1–6.

Shows that the well-known impossibility theorem continues to be applicable in a range of contexts where the background logic is nonmonotonic.

(33) With Michael Freund and Daniel Lehmann, "Canonical extensions to the infinite case of finitary nonmonotonic inference relations", in Brewka & Freitag eds *Proceedings of the Workshop on Nonmonotonic Reasoning* (Arbeitspapiere der GMD n° 443, 1990, 133–138).

Examines the problem of finding good ways of extending a nonmonotonic inference relation defined on finite sets of propositions, to one defined also on infinite sets of propositions.

(32) "Logique modale: quelques jalons essentiels", pp. 75–97 in L. Iturrioz & A. Dussauchoy eds, *Modèles logiques et systèmes d'intelligence artificielle*. Paris: Hermes, 1990.

An expository review of central concepts and results of modal propositional logic for a readership of computer scientists.

(31) "General theory of cumulative inference", in M. Reinfrank et al eds *Nonmonotonic Reasoning*. Berlin: Springer, Lecture Notes on Artificial Intelligence n° 346, 1989, 1–17.

Studies the formal behaviour of inference operations which, whilst not monotonic, may nevertheless satisfy various other conditions, notably that of cumulativity.

(30) "Rights of peoples: point of view of a logician", pp. 69–92 in J. Crawford ed. *The Rights of Peoples*. Oxford University Press, 1988.

A review of the main logical questions around the notion of collective rights held by peoples, focussing on problems of indeterminacy and of inconsistency.
A condensed version published as "On attributing rights to all peoples: some logical questions", in *Law and Philosophy* **8**: 1989, 53–63. A preliminary version published in French translation under the title "Les droits des peuples: perspectives d'un logician" *Cahiers de la Fondation Lelio Basso* 7: 1986, 53–60.

(29) With Peter Gärdenfors, "Revisions of knowledge systems and epistemic entrenchment", pp. 83–95 in M.Vardi ed. *Proceedings of the Second Conference on Theoretical Aspects of Reasoning about Knowledge*. Los Altos: Morgan Kaufmann, 1988.

Relates the AGM postulates for contraction and revision functions to a relation of 'epistemic entrenchment' between propositions.

(28) "On the status of the postulate of recovery in the logic of theory change". *The Journal of Philosophical Logic* 16: 1987, 383–394.

Discusses the rather special behaviour of one of the postulates in the AGM account of theory change.

(27) "On the formal representation of rights relations: remarks on the work of Stig Kanger and Lars Lindahl". *The Journal of Philosophical Logic* 15: 1986, 403–425.

A discussion of work formalising Hohfeld's classic taxonomy of rights relations between two parties.

(26) With Carlos Alchourrón, "Maps between some different kinds of contraction function: the finite case". *Studia Logica* 45: 1986, 187–198.

Studies the relationship, in the finite case, between safe contractionand partial meet contraction.

(25) With Carlos Alchourrón, "On the logic of theory change: safe contraction". *Studia Logica* 44: 1985, 405–422.

Studies a way of generating contraction functions, in terms of preservation of elements that are in a certain sense safe.

(24) "How to give it up: a survey of some recent work on formal aspects of the logic of theory change". *Synthese* 62: 1985, 347–363 and 68: 1986, 185–186.

An overview of the work of Alchourrón, Gärdenfors and the author in the area of the logic of theory change.

(23) With Carlos Alchourrón and Peter Gärdenfors, "On the logic of theory change: partial meet contraction and revision functions". *The Journal of Symbolic Logic* 50: 1985, 510–530.

A study of formal aspects of the logic of theory change, defining partial meet contraction functions and establishing a representation theorem in terms of suitable postulates.
Russian translation in E. Lisanyuk ed. Normative Systems and Other Works in Legal Philosophy and Logic of Norms (Russian title) 2013.

(22) "Stenius' approach to disjunctive permission". *Theoria* 50: 1984, 136–145.

A critical review of Stenius' account of the logic of disjunctive permissions, leading to a closely related approach in terms of 'checklist conditionals'.

(21) "Individual actions are very seldom obligatory". *The Journal of Non-Classical Logic* 2: 1983, 7–13.

Discusses some striking differences between attributing deontic predicates to individual actions, on the one hand, and to generic actions on the other.

(20) With Carlos Alchourrón, "On the logic of theory change: contraction functions and their associated revision functions". *Theoria* 48: 1982, 14–37.

Defines and studies maxichoice contraction and revision operations.

(19) With Carlos Alchourrón, "Hierarchies of regulations and their logic", pp. 125–148 in Hilpinen ed. *New Studies in Deontic Logic*. Dordrecht: Reidel, 1981.

Investigates the resolution of contradictions and ambiguous derogations in a code, by means of the imposition of partial orderings. An exploration of initial ideas that later developed into the AGM account of 1985.
Russian translation 2013 in E. Lisanyuk ed. *Normative Systems and Other Works in Legal Philosophy and Logic of Norms* (Russian title) Saint Petersburg University Publishing.

(18) "Quantificational reefs in deontic waters", pp. 87–91 in Risto Hilpinen ed. *New Studies in Deontic Logic*. Dordrecht: Reidel, 1981, 87–91.

On the prevalence of implicit quantification in normative assertions of everyday language.

(17) "Non-equivalent formulae in one variable in a strong omnitemporal modal logic". *Zeitschrift für math. Logik und Grundl. der Math.* 27: 1981, 111–112.

A technical observation on a set of non-equivalent formulae in modal propositional logic. The last of the author's papers in that area.

(16) "A characterization of structural completeness of a structural consequence operation". *Reports on Mathematical Logic* 6: 1976, 99–102.

A lattice-theoretic characterisation of the notion mentioned in the title.

(15) With Krister Segerberg, "Post completeness and ultrafilters", *Zeitschrift für Math. Logik und Grundlagen der Mathematik* 20: 1974, 355–358.

A cardinality result in modal propositional logic.

(14) "Strengths and weaknesses of the Ajdukiewicz approach to grammar" (in Portuguese). *Discurso* (Brazil) 4: 1974, 155–165.

Discusses the powers and limitations of a simple mathematical model of grammatical structure.

(13) "A warning about the choice of primitive operators in modal logic". *The Journal of Philosophical Logic* 2: 1973, 193–196.

Draws attention to some unexpected consequences of using a zero-ary connective in modal propositional logic.

(12) "Some embedding theorems for modal logic". *Notre Dame Journal of Formal Logic* 12: 1971, 252–254.

Some observations about the upper end of the lattice of all modal propositional logics.

(11) With Luisa Iturrioz, "Sur les filtres premiers d'un treillis distributif et ses sous-treillis". *Comptes Rendus de l'Academie des Sciences de Paris* 270A: 1970, 575–577.

Solves a cardinality question about prime filters of infinite distributive lattices.

(10) "A generalization of the concept of a relational model for modal logic". *Theoria* 36: 1970, 331–335.

Generalises the concept of a relational model for modal logic, due to Kripke, obtaining a closer correspondence between relational and algebraic models.

(9) "On the number of ultrafilters of an infinite Boolean algebra". *Zeitschrift für Math. Logik und Grundlagen der Mathematik* 15: 1969, 121–122.

Solves a cardinality question about ultrafilters of infinite Boolean algebras.

(8) "A normal modal calculus between T and S4 without the finite model property". *The Journal of Symbolic Logic* 34: 1969, 35–38.

The first example of an intuitively meaningful propositional logic without the finite model property, and still the simplest one in the literature.

(7) "Remarks on the concept of distribution in traditional logic". *Nous* 3: 1969, 103–108.

Uses ideas of modern mathematical logic to clarify a controverted and rather obscure notion of traditional syllogistic.

(6) "On some completeness theorems in modal logic". *Zeitschrift für Math. Logik und Grundlagen der Mathematik* 12: 1966, 379–384.

Adapts the Lindenbaum/Henkin method of maximal consistent sets to establish the completeness of well-known modal propositional logics with respect to the relational models of Kripke.

(5) "How meaningful are modal operators?". *Australasian Journal of Philosophy* 44: 1966, 331–337.

A philosophical discussion of intuitive readings of the formalism of modal propositional logic.

(4) "There are infinitely many Diodorean modal functions". *The Journal of Symbolic Logic* 31: 1966, 406–408.

A cardinality result in the theory of modal logics.

(3) "An alternative characterization of first-degree entailment". *Logique et analyse* 8: 1965, 305–311, misprints corrected in 9: 1966, 394.

Axiomatizes the first-degree entailments of Anderson and Belnap, using the notion of a formula in which no variable occurs more than once.

(2) "The paradox of the preface". *Analysis* 25: 1965, 205–207.

By means of an example, shows the possibility of beliefs that are separately rational whilst together inconsistent.

(1) "Nidditch's definition of verifiability". *Mind* 74: 1965, 240–247.

Shows the failure of an attempt to define the notion of verifiability in purely deductive terms.

Miscellaneous

"A tale of five cities". In *David Makinson on Classical Methods for Non-Classical Problems*, ed. Sven Ove Hansson. Springer, New York, 2014.

An autobiographical text: The author looks back at his work in logic under the headings: early days in Sydney, graduate studies in Oxford, adventures in Beirut, midnight oil in Paris, and last lap in London.

"Reflections on the contributions". In *David Makinson on Classical Methods for Non-Classical Problems*, ed. Sven Ove Hansson. Springer, New York, 2014.

Reflects on some general issues of logic prompted by a reading of the contributions to the volume.

Foreword to the 2008 reprint of Peter Gärdenfors *Knowledge in Flux*. London: College Publications.

Recalls the gestation of Peter's book and discusses with hindsight some issues that arose at the time.

"Some properties of the lattice of all modal logics", in *Abstracts from the Fourth International Congress for Logic, Methodology and Philosophy of Science*. Bucharest, 1971.

Several hundred reviews for each of *Mathematical Reviews* and *Zentralblatt für Mathematik* from the 1960s to the present. Occasional reviews for other journals of logic and philosophy including in particular the *Journal of Symbolic Logic*.

MIX
Papier aus verantwortungsvollen Quellen
Paper from responsible sources
FSC® C105338

If you have any concerns about our products,
you can contact us on
ProductSafety@springernature.com

In case Publisher is established outside the EU,
the EU authorized representative is:
Springer Nature Customer Service Center GmbH
Europaplatz 3, 69115 Heidelberg, Germany

Printed by Libri Plureos GmbH
in Hamburg, Germany